CONCEPTS AND CALCULATORS
IN
CALCULUS
Second Edition

George Best

Stephen Carter

Douglas Crabtree

Venture Publishing
9 Bartlet Street, Suite 55
Andover, MA 01810

About the Cover

The photograph shows Phillips Academy's armillary sphere, the work of sculptor Paul Manship (1851-1966). The family group in the center symbolizes Time, Eternity and Humanity. Other elements in the sundial represent Earth, Water, and Fire. Carved on the hoops are symbols of the constellations of the zodiac; Twilight, with an owl; the moon, representing Night; and Dawn awakening to the cock's crow. Photo by Jennie Cline.

Acknowledgments

Many people have offered suggestions and encouragement for this project; they include: Sally Fischbeck, Rochester Institute of Technology; Harriet Scheuermann, National Cathedral School; Steve Olson, Hingham High School; and Phillips Academy colleagues Dave Penner, Don Barry, Doug Kuhlmann, Corbin Lang and Dick Lux. In addition we wish to thank Kathy Payne for her editorial assistance with the second edition. Thanks to Jennie Cline for her work on the cover design, and a special thanks to Fran Patterson for proofreading our manuscript with great care and her ongoing help in developing the text.

Preface

This new calculus book focuses on conceptual understanding and thinking skills. The graphing calculator is an integral part of the text and allows students to manipulate and compare graphical, numerical and algebraic representations of the ideas. The text is designed to include the best aspects of calculus reform along with the goals and well established methods of traditional calculus. It has been written to meet the needs of students enrolled in an AB advanced placement course as well as students who are enrolled in a non-advanced placement course.

The book features:

- All topics that are included in the new AP calculus course description.

- Useful programs for several topics.

- Labs that encourage students to explore patterns and then describe them.

- A wealth of applications.

The following resources are available, free of charge, to adopters of this text.

- Complete solutions manual that provides detailed solutions to all problems in the text.

- Test Items organized according to the main text that include both multiple-choice and opened-ended questions.

Table of Contents

Chapter 10   L'Hopital's Rule, Improper Integrals and Partial Fractions

Chapter 11   Taylor Polynomials and Series

Chapter 12   Parametric, Vector and Polar Coordinates

CHAPTER 1
FUNCTIONS AND GRAPHS

1.1   What is a Function?

The notion of a function is basic to the study of calculus. It is imperative that you have an understanding of how a function is defined, the notation used to describe functions and the operations on functions.

A function describes how one quantity depends on another. The amount of postage you pay for a letter depends on the weight of the letter. The population of the U.S. depends on what year it is. The amount of income tax you pay depends on the amount of your income. The area of a circle depends on its radius. This last relationship is a special function because it can be described by a formula. If $A(r)$ is the area of the circle and r is its radius, then $A(r) = \pi r^2$.

Informally, a mathematical function is a rule, or set of instructions, that specifies how the value of one quantity, the **input**, determines the value of a second quantity, the **output**. The set of all possible inputs is called the **domain** of the function and the set of corresponding outputs is called the **range** of the function.

Although most familiar functions are specified by formulas, like

$$f(x) = x^2 \text{ and } g(x) = \frac{x+1}{x-1} \text{ ,}$$

this is not the only possibility. Any description that pairs each input value with no more than one output value will suffice. Tables and graphs are common ways of defining functions.

Example 1:   *A Function Defined by a Table*

The correspondence between the national debt in billions of dollars and a given year is displayed in the table.

Year	Debt
1976	620.4
1979	826
1982	1142
1985	1823.1
1988	2602.3
1992	3502
1995	4705

If $D(x)$ is the debt in billions of dollars for a particular year x, then the function D is completely specified by the table. For example,

$$D(1976) = 620.4 \text{ and } D(1988) = 2602.3.$$

A scatter plot or graph of the data above can be created using the statistical features of the TI-83 and is shown below. To simplify the process the point for 1976 was plotted as (76, 620.4)

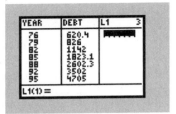 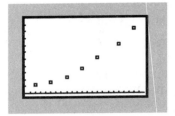

For the purposes of prediction, it would also be useful to have a mathematical formula that "fits" the given data. The TI-83 can fit a curve through the given data points. The viewing rectangle below shows the graph of a function D with formula

$$D(t) = 0.1498(1.1159)^t$$

superimposed on the data.

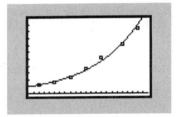

The constants 0.1498 and 1.1159 which determine the model function are called **parameters.** Finding the appropriate parameter values is a part of a process called **curve fitting**. Later, in Section 1.3, there will be an explanation of how to choose functions whose graphs "fit" observed data.

Graphs of Functions

Although the graph above was obtained from the table of data given, in some situations a graph is given and is used to define a function. The graph of a function is the collection of input-output pairs $(x, f(x))$ such that x is in the domain of f. Geometrically,

x = the directed distance from the y-axis.

$f(x)$ = the directed distance from the x-axis.

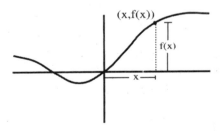

Defining a function graphically is a standard procedure in science. And frequently, as with a lie detector or seismograph, it is the only description of a particular function.

Example 2: *Analyzing Temperature Data*

Over a 24-hour period a recording device in a weather bureau produces the graph of the function *Temp* that relates temperature (in degrees Fahrenheit) to the time of day, as shown below.

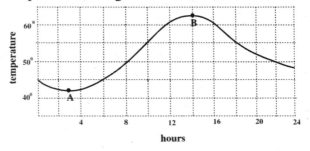

What can the graph tell us about the function *Temp*?

1) What is *Temp*(8)? Since (8, 50) is on the graph we know that *Temp*(8) = 50, that is, the temperature at 8 AM is 50°.

2) At what times of day was the temperature below 50°? One approach is to draw the horizontal line *Temp* = 50°. The graph shows that
$$Temp(t) < 50 \quad \text{whenever} \quad 0 < t < 8 \quad \text{and} \quad 22 < t < 24 .$$
So the temperature is below 50° between midnight and 8 AM and also between 10 PM and midnight.

3) When is the temperature rising? The graph rises from $t = 3$ to $t = 14$. The function *Temp* is said to be **increasing** for $3 < t < 14$. On the other hand, the function *Temp* is **decreasing** for $0 < t < 3$ and $14 < t < 24$.

4) The points A and B correspond to **local minimum** and **local maximum** values of the function Temp.

5) The graph of a function is **concave up** if it bends upward, and it is **concave down** if it bends downward. The *Temp* graph is concave up at point A and concave down at point B. Somewhere between A and B there is an **inflection point**, where the direction of the concavity changes.

Domain, Range and Zeros

Each time we consider a function we are going to be interested in its domain and range. The domain of a function is the set of all possible inputs. For example, the domain of the function $f(x) = x^2$ is all real numbers, whereas the domain of the function $f(x) = \frac{1}{x}$ is all real numbers except zero. Whenever the domain of a function is not specified, it is assumed to be the largest set of real number inputs for which the function produces real number outputs.

When functions are employed to model or explain real world phenomena, the physical situation often imposes limitations on the domain and range of the model function. For example, the domain in Example 2 is the set of hours in the day, that is, all real numbers from 0 to 24. The graph shows that the range consists approximately of numbers from 42 to 62.

When the graph of a function crosses the *x*-axis we say the function has a **zero** at that point. Thus, any input a is a **zero of a function** f if $f(a) = 0$. For example, the function $f(x) = x - 4$ has 4 as a zero because $f(4) = 0$.

The graph of a function provides a picture of its domain and range. The vertical projection onto the *x*-axis is the domain; the horizontal projection onto the *y*-axis is the range.

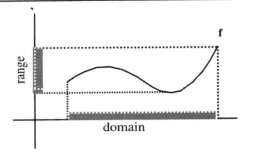

It is often difficult to determine the exact range of a function; however, a graph of the function will usually help.

For graphing on a calculator, it is necessary to have an idea of the range. A quick method for obtaining a rough guess is to display a table of values for the corresponding inputs on a given *x*-interval.

Example 3: *Finding the Domain, Range and Zeros of a Function*

Graph the function defined by $f(x) = (5x^3 - 6)\sqrt{4 - x^2}$.

Approximate the domain, range and zeros of this function.

Solution:

Start by entering $Y1 = (5x^3 - 6)\sqrt{4 - x^2}$. To find a viewing rectangle that contains the graph of f we note that the acceptable inputs are the real numbers x for which $4 - x^2 \geq 0$. Hence the domain of f is the interval [–2, 2], so we set Xmin = –2 and Xmax = 2 and use ZoomFit. The viewing rectangle determined by the calculator is shown below.

 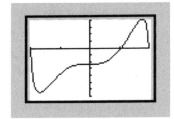

Thus, our estimate of the range is $-32.206 < y < 20.237$. How accurate are the Ymin and Ymax values as boundaries for the range?

To get a better approximation of the maximum range value of f we press [2nd] [CALC] and select item [4:maximum] from the Calc menu. At the prompt we set the Left bound of an interval that contains the maximum point and press [ENTER]. Next we set the Right bound and Guess. The result cursor is on the approximate maximum point and the coordinate values $x = 1.7800$ and $y = 20.2435$ are displayed. Using the [3:minimum] option from the Calc menu, the approximate minimum point is located at $x = -1.6797$ and $y = -32.2384$. We estimate the range to be all values $f(x)$ such that $-32.2384 \leq f(x) \leq 20.2435$.

There are zeros at $x = -2$ and $x = 2$. There also appears to be a zero between $x = 1$ and $x = 2$. Select [2:zero] from the Calc menu; set Left bound, Right bound and Guess as described for maximum. The TI-83 approximates a zero at $x = 1.0626586$.

1.1   Exercises

1.   For each functional situation described below, identify the two quantities that are related and decide which should be the input variable and which should be the output variable. Sketch a reasonable graph to show the relationship between the quantities and specify a domain and range.

   a) The amount of daylight each day of the year.

   b) The temperature of an ice-cold drink left in a warm room for a period of time.

   c) The height of a baseball after being hit into the air.

   d) The water level on the supports of a pier at an ocean beach on a calm day.

2.   Each function f shown graphically below has the interval $[-6, 6]$ as its domain. Find the range, zeros and $f(-2)$ for each function.

a)

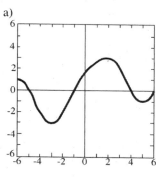

b)

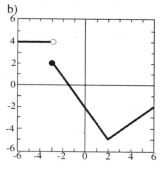

c)

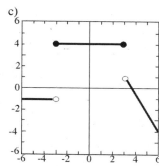

d)

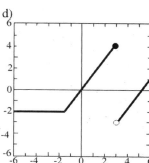

3.   Sketch the graph of a function that has the specified domain, range and zeros. There are many possible correct graphs.

   a) domain: $[-10, 10]$
      range: $[-5, 5]$
      zeros: $\{-1, 1\}$

   b) domain: $[-5, 5]$
      range: $[-1, 4]$
      zeros: $\{0\}$

4.   The function f shown graphically below has the interval $[-5, 5]$ as its domain.

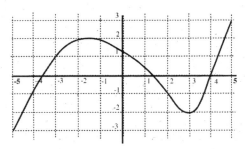

a) How many zeros does this function have?

b) Give approximate values for $f(-3)$ and $f(2)$.

c) Is the function increasing or decreasing near $x = -1$?

d) Is the graph concave up or concave down near $x = -2$?

e) List all intervals on which the function is increasing.

5.   The function f shown graphically below has the interval $[-5, 5]$ as its domain.

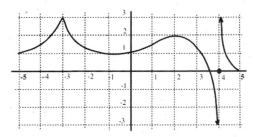

a) What is the range of f?

b) On which interval(s) is f increasing?

c) Approximate $f(1)$ and $f(4)$.

d) On which intervals is f concave up?

e) Is the function increasing or decreasing at $x = 0$?

In Exercises 6–13, use the TI-83 to enter and graph the given function. If necessary, use the Window menu or ZOOM [0:ZoomFit] to select an appropriate viewing rectangle. Use the graph to approximate

a) the domain;   b) the range;   c) the zeros of each function.

6. $f(x) = (x-1)^2 - 3$ 7. $f(x) = \sqrt{x+1} - 2$

8. $f(x) = 22 + 9x - x^2$ 9. $f(x) = 3x^4 + 4x^3$

10. $f(x) = (x-2)\sqrt{3-x}$ 11. $f(x) = \sqrt{(x-1)(x+2)}$

12. $f(x) = \sqrt{x^2 - x}$ 13. $f(x) = \dfrac{1}{x^2 - 9}$

14.   For each of the following functions use the TI-83 to estimate the largest and smallest values that f takes on in the given interval .

      a) $f(x) = x^3 - x$, $[-10, 10]$ b) $f(x) = x^{\sin x}$, $(0, 20]$ c) $f(x) = 2^x - x^2$, $[-10, 10]$

15.   Sketch a smooth, continuous curve which passes through the point $P(2, 3)$ and which satisfies each of the following conditions.

      a) Concave up to the left of P;

      b) Concave down to the right of P;

      c) Increasing for $x > 0$;

      d) Decreasing for $x < 0$.

16.   Sketch a smooth, continuous function that passes through the point $P(0, 3)$ and which satisfies each of the following conditions.

      a) Decreasing on $[-2, 3]$;

      b) Concave down to the left of P;

      c) Concave up to the right of P;

      d) Increasing on $[3, \infty)$.

17.   An open box is made by cutting squares of side x from the four corners of a sheet of cardboard that is 8.5 inches by 11 inches and then folding up the sides.

      a) Express the volume of the box as a function of x.

      b) Estimate the value of x that maximizes the volume of the box.

18.   A rectangular cookie sheet is to have a perimeter of 60 inches. Define a function whose input is the width of the cookie sheet and whose output is the corresponding area of the sheet. What is the domain of this function?

19.   Golf balls can be manufactured for sixty cents each. At a price of three dollars each, one thousand can be sold. For each ten cents the price is lowered, fifty more balls can be sold.

      a) Find a function that relates net profit and price.

      b) At what price is the maximum profit earned?

20.   An 8.5 in by 11 in piece of paper contains a picture with a uniform border. The distance from the edge of the paper to the picture is x inches on all sides.

a) Express the area of the picture as a function of x.

b) What are the domain and range of the function?

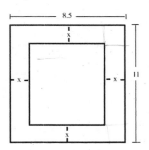

21.   A wire 6 meters long is cut into twelve pieces. The pieces are welded together to form the frame of a rectangular box with a square base.

a) Define a function that relates the height of the rectangular box to the length of one edge of the base.

b) Define a function that relates the total surface area of the box to the length of one edge of the base.

22.   Express the length of the chord in a circle of radius 8 inches as a function of the distance p from the center of the circle to the chord. What is the domain of this function?

23.   Build a table of values for each function f below with $x = 0, 0.5, 1, 1.5, ..., 5$. Use the table to estimate the x-coordinate of the maximum value of f on the interval $0 < x < 5$.

a) $f(x) = 2xe^{-x} - 3$ b) $f(x) = 2e^{-x} \sin x + 1$

24.   Define a function f that gives the distance from the point $P(3, 0)$ to a point on the unit circle. What are the domain and range of this function.

25.   Express the area A of an equilateral triangle as a function of the length s of one of its sides.

26.   Express the surface area of a cube as a function of its volume.

27.   Is the sum of two increasing functions increasing? Justify.

28.   Is the product of two increasing functions increasing? Justify.

1.2   Basic Functions and Transformations

The graphs of six basic functions are shown below. These functions are important as a set of tools for anyone who creates mathematical models to solve problems. You should be able to recognize these graphs.

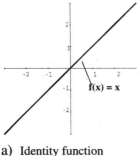

a) Identity function

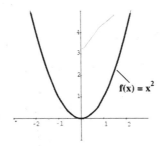

b) Squaring function

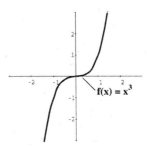

c) Cubing function

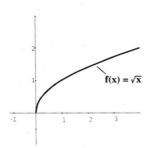

d) Square root function

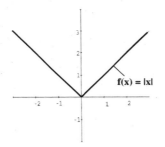

e) Absolute value function

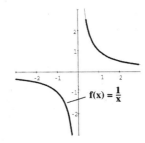

f) Reciprocal function

Graphing calculators and computers have so-called libraries of functions that can be accessed by pressing the appropriate keys. The libraries usually include the six basic functions defined above.

There are a number of operations that can be applied to the rule for a function. We shall now see that when the rule of a function is changed algebraically to produce a new function the graph of the new function can be obtained from the graph of the original function by a geometric transformation, such as a vertical or horizontal shift, or by stretching or shrinking.

Addition: Shifting

Given a function f a new function can be created by *adding* constants. For example, if $f(x) = x^2$, then shown below are graphs of new functions.

$$h(x) = x^2 + 2 \qquad\qquad \text{and} \qquad\qquad k(x) = (x+2)^2 .$$

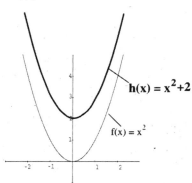
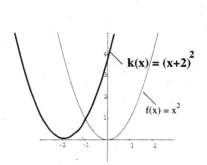

Adding a constant to the *output* of f results in a *vertical shift* of the f-graph.

Adding a constant to the *input* of f results in a *horizontal shift* of the f-graph.

Multiplication: Stretching or Compressing

If f is the periodic function $f(x) = \sin x$, then new functions may be created by *multiplying*. Shown below are graphs of the new functions.

$$h(x) = 2\sin x \qquad\qquad \text{and} \qquad\qquad k(x) = \sin(2x) .$$

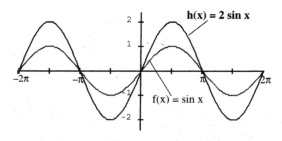
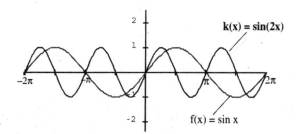

Multiplying the *output* of f by a constant results in a *vertical stretch/compression* of the f-graph.

Multiplying the *input* of f by a constant results in a *horizontal stretch/compression* of the f-graph.

Example 1:   *Adding Constants to the Input and Output Values of a Function*

Compare the graph of $f(x) = |x|$ with the graphs of $g(x) = |x| - 1$ and $h(x) = |x - 2|$.

Solution:

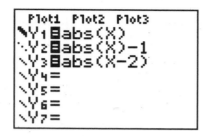 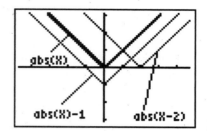

The graph of $g(x) = |x| - 1$ is a translation of $f(x) = |x|$ by a downward shift of 1 unit.
The graph of $h(x) = |x - 2|$ is a translation of $f(x) = |x|$ by a horizontal shift to the right of 2 units.

Example 2:   *Multiplying the Output of a Function by a Constant*

Compare the graph of $f(x) = \cos x$ with the graphs of $g(x) = 3\cos x$ and $h(x) = \frac{1}{3}\cos x$.

Solution:

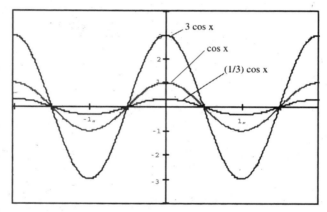

The graph of $g(x) = 3\cos x$ results from the vertical stretching of the f-graph by a factor of 3.

The graph of $h(x) = \frac{1}{3}\cos x$ results from a vertical compression of the f-graph by a factor of 3.

1.2 **Exercises**

In Exercises 1–6, use the basic functions and transformations to guess a rule of each function from its graph. Verify your answer with your calculator.

1.   2.   3.

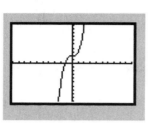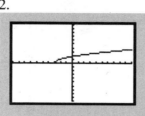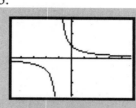

4.   5.   6.

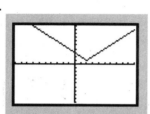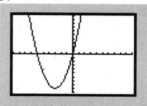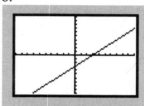

In Exercises 7–12, graph each function by determining the basic function and then using transformations. Confirm your answer by graphing the function on the TI-83.

7. $y = |x + 3| - 3$ 8. $y = \sqrt{x - 1} + 2$

9. $y = 3x^2 - 4$ 10. $y = \dfrac{1}{x - 3}$

11. $y = 2\sqrt{x + 2} + 1$ 12. $y = .5|x| + 2$

In Exercises 13–19, the order in which transformations are to be applied to the graph of a given function graph is specified. Give an equation for the transformed function.

13. $y = |x|$; vertical stretch by 3, shift up 4

14. $y = x^2$; shift left 2, shift down 1

15. $y = \dfrac{1}{x}$; shift right 1, shift up 3

16. $y = x^3$; vertical stretch by 2, shift down 4

17. $y = \sqrt{x}$; shift left 2, shift up 3

18. $y = x^2$; horizontal compression by a factor of 2, shift up 1

19. $y = \sqrt{x}$; vertical stretch by 2, shift down 3

20. a) Plot each pair of functions in the same viewing rectangle.

   i) $f(x) = \dfrac{1}{x}$ and $g(x) = -\dfrac{1}{x}$ ii) $f(x) = \sqrt{x}$ and $g(x) = -\sqrt{x}$

   b) Use the results in part a) to describe the effect that multiplying the *outputs* of a function by -1 has on the graph of the original function.

21. a) Plot each pair of functions in the same viewing rectangle.

   i) $F(x) = x^3$ and $G(x) = (-x)^3$ ii) $F(x) = \sqrt{x}$ and $G(x) = \sqrt{-x}$

   b) Use the results in part a) to describe the effect that multiplying the *inputs* of a function by -1 has on the graph of the original function.

22. The function f is defined by $f(x) = x^2 - x$. Graph each of the following functions noting its relationship to the original function f.

   a) $y = f(|x|)$ b) $y = |f(x)|$ c) $y = \dfrac{1}{f(x)}$

23. The function g is defined by $g(x) = \dfrac{1}{x}$. Use the results of the previous exercise to predict the appearance of each of the graphs of the following functions, then check by graphing each function.

   a) $y = g(|x|)$ b) $y = |g(x)|$ c) $y = \dfrac{1}{g(x)}$

24. The graph of a function f with domain [0, 3] and range [1, 2] is shown in the figure. Sketch a graph of the following functions and specify the domain and range.

   a) $y = f(x) + 2$ b) $y = f(x) - 1$

   c) $y = 2f(x)$ d) $y = -f(x)$

   e) $y = f(x + 2)$ f) $y = f(x - 1)$

   g) $y = 1 + f(x + 3)$ h) $y = 3f(x - 2) + 1$

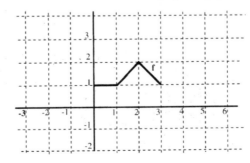

25.   The graph of a function g with domain $[-2, 4]$ is shown in the figure below. Sketch a graph of the following functions and specify the domain and range.

     a) $y = g(x) - 1$ b) $y = g(x - 1)$

     c) $y = -g(x)$ d) $y = g(x + 2)$

     e) $y = -g(x) + 2$ f) $y = 2g(x)$

     g) $y = 0.5g(x)$ h) $y = g(x + 1) - 2$

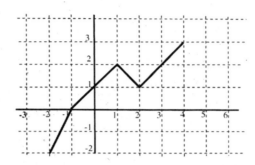

26.   The order in which transformations are applied to the graph of the given function is specified. Give an equation for the transformed function in each case.

     a) $f(x) = x^2 + x$;   shift horizontally 6 units to the right, stretch vertically from the x-axis by a factor of 2.

     b) $g(x) = x^3 + 2x^{-1}$;   reflect in the y-axis, shift down 3.

     c) $h(x) = \sqrt{x^3 + 5}$;   shift horizontally 2 units to the left, then shift vertically upward 3 units.

1.3   Linear Functions and Mathematical Modeling

Lines and linear functions play a surprisingly important role in calculus. A linear function f is one whose rule can be written in the form

$$f(x) = mx + b$$

where m and b are constants. The graph of a linear function is a line with slope m and y-intercept $(0, b)$.

The slope of a line represents the number of units the line rises or falls vertically for each unit of horizontal change (rise/run). For instance, consider any two points (x_1, y_1) and (x_2, y_2) on the non-vertical line shown below.

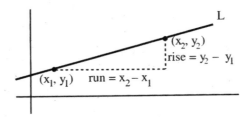

The slope m, or the rate of change of y with respect to x, of the line L is defined to be the ratio

$$m = \frac{\text{rise}}{\text{run}} = \frac{\Delta y}{\Delta x} = \frac{y_2 - y_1}{x_2 - x_1} .$$

Using the slope definition it can be shown that:

     a) a line has the same slope everywhere;

     b) parallel lines have equal slopes and the slopes of perpendicular lines are negative reciprocals;

     c) horizontal lines have zero slope; vertical lines have undefined slopes.

Point-Slope Form

Suppose the value y_0 of a linear function f is known at some initial input x_0 so that $y_0 = f(x_0)$ and the slope m is also known. Then the graph of f is a straight line with slope m passing through the point (x_0, y_0). A **point-slope formula** for f is

$$y = m(x - x_0) + y_0.$$

If (x, y) is any point on the line, then for the given point (x_0, y_0) we have

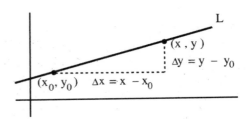

$$\frac{y - y_0}{x - x_0} = m$$

$$y - y_0 = m(x - x_0)$$

$$y = m(x - x_0) + y_0$$

Example 1:   *Using the Point Slope Equation*

Find an equation in point-slope form for the secant line that intersects the graph

of $y = \dfrac{1}{x}$ at points where $x = \dfrac{1}{2}$ and $x = 3$.

Solution:

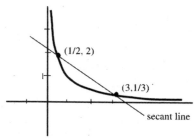

The graph shows that the secant line passes through the points $(\frac{1}{2}, 2)$ and $(3, \frac{1}{3})$. The slope is

$$m = \frac{2 - \dfrac{1}{3}}{\dfrac{1}{2} - 3}.$$

Multiplying the numerator and denominator by 6, we have

$$m = \frac{12 - 2}{3 - 18} = \frac{10}{-15} = \frac{-2}{3}.$$

Using $(x_0, y_0) = (\frac{1}{2}, 2)$ as the initial point, the secant line has point-slope equation:

$$y = m(x - x_0) + y_0 = -\frac{2}{3}(x - \frac{1}{2}) + 2.$$

Many functions, especially those arising out of experimental data in the lab, are defined by tables. To recognize that a function f given by a table is linear, look for differences in output values that are constant for equal differences in input.

Example 2:   *Using a Linear Function*

The weight of a person on the earth and on the moon is given in the table below.

Wt. on earth (lbs)	120	132	144	156	168
Wt. on moon (lbs)	20	22	24	26	28

Plot the data and find a function that predicts the weight of a person on the moon, given his or her weight on the earth.

Solution:

From the table data we see that for each 12 pound increase in a person's weight on earth (input) there is a corresponding 2 pound weight increase on the moon (output). This constant rate of increase, $\dfrac{\text{rise}}{\text{run}} = \dfrac{2}{12} = \dfrac{1}{6}$, is a clear indication that a linear function M fits the data and, using the point $(120, 20)$, we have the point-slope formula

$$M = \frac{1}{6}(E - 120) + 20.$$

Mathematical Models

Functions are indispensable to scientists to describe the relationships between measurements. Suppose that a physicist stretches a spring many times and each time measures the extension of the spring as well as the corresponding stretching force. The collected data, recorded in the table below, can be quickly recognized as a set of ordered pairs. If the numbers in the left column are taken as inputs and those in the right column as corresponding outputs, the table defines a set that we call a **data function**.

E (extension) cm	F (force) gm
0.0	0.0
2.0	3.2
4.0	6.7
6.0	10.0
8.0	13.4
10.0	17.2
12.0	20.4
14.0	23.1
16.0	27.0

Its graph, called a scatter plot, looks like this:

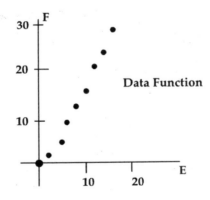

The physicist might now ask herself what force would extend the spring 9 cm, or how many centimeters would the spring be stretched by a force of 22 gm? Encouraged by the linear appearance of the points on the graph of the data function, she constructs a new graph by drawing a line segment that lies as close as possible to all of the data points.

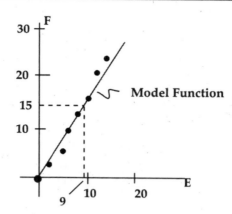

This segment is part of the graph of a linear function that we shall call a **model function**. Just as the data function records *fact*, the model function conveys *fiction*, namely, the physicist's guess as to how stretching forces are *in general* related to extensions. On the basis of her model function she would predict that a 9 cm extension of the spring would require approximately a 15 gm force and a 22 gm force would cause an extension of about 13 cm. The ability to predict is the ultimate test for a model function.

Often the relationship between measured quantities is complex and unclear; but an approximate linear pattern may emerge, especially when we see the results of plotting the data on a coordinate plane.

Global warming is currently a major environmental concern. Carbon dioxide (CO_2) traps heat better than other gases in the atmosphere, so more CO_2 makes the atmosphere like a greenhouse. Increased industrial production of CO_2 and decreased conversion of CO_2 by plants during the photosynthesis process are therefore considered to be causes of a general deleterious trend around the world. This is the so-called greenhouse effect.

The graph below shows average monthly CO_2 concentrations at the South Pole plotted at six month intervals between 1958 and 1968. Units on the vertical axis are parts per million (PPM). For example, 320 PPM means that, in every million molecules of air, there are 320 molecules of CO_2. The data oscillates but there is an increasing trend that appears to be linear. It seems natural to model the data with a linear function.

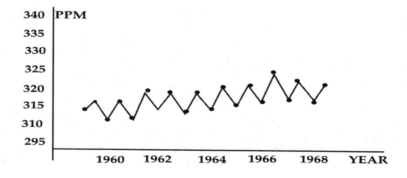

One way to make a linear approximation is to lay a ruler on the graph and visually balance the wiggles above and below the edge of the ruler. This will give a reasonable approximation; however, there are statistical techniques for finding from the data points the equation of the best fitting line (with various meanings of "best"). The most popular of these techniques, called **least squares regression**, produces a best-fitting line that minimizes errors in prediction. This line is referred to as the **least squares line** or the **regression line**. All statistical computer software and some calculators will produce the regression line for given data. For example, inputting the carbon dioxide data into a TI-83 calculator produces the slope $m = 0.6783$ and the y-intercept -1013. The corresponding regression line,

$$y = 0.6783x - 1013,$$

is plotted in the following diagram along with the data points.

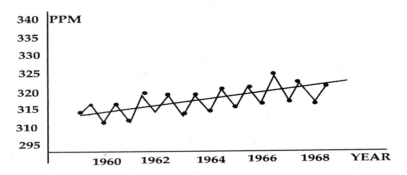

The linear function $y = 0.6783x - 1013$, or, equivalently, its straight line graph, is our mathematical model of the carbon dioxide data. The model approximates the data reasonably well and gives us a tool for predicting future concentrations of CO_2. For example, to predict the 1973 CO_2 average, let $x = 1973$ and calculate as follows:

$$y = 0.6783(1973) - 1013 \approx 325.$$

Example 3: *Fitting a Linear Model to Data*

A data function relating heights and weights is displayed in the following table.

height (in)	60	64	68	70	73
weight (lbs)	132	135	155	183	192

Enter the data into a TI-83 and graph the scatter plot and the corresponding regression line.

Solution:

Pressing ⃞ STAT ⃞ accesses the statistical menu.

To clear any existing lists, first press [5:SetUpEditor] (to copy SetUpEditor to the Home screen) and then press ⃞ ENTER ⃞ to remove old list names; then press ⃞ STAT ⃞ [4:ClrList] (to copy ClrList to the Home screen) and then press ⃞ 2nd ⃞ [L1] ⃞ , ⃞ ⃞ 2nd ⃞ [L2] ⃞ ENTER ⃞.

To enter data pairs from the height and weight chart above, press ⃞ STAT ⃞ [1:Edit] to display the STAT list editor.

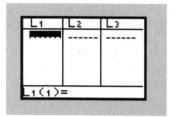

Press 60. As you type, the value is displayed on the bottom line. Press ENTER. The value is shown as the first element of L1 and the cursor moves to the second element in the same list. Press 64 ENTER 68 ENTER 70 ENTER 73 ENTER to enter the remaining heights into list L1. Press ▶ to move the cursor to the first element of list L2. Press 132 ENTER 135 ENTER 155 ENTER 183 ENTER 192 ENTER to enter the weights into list L2. To quit entering data press 2nd [QUIT].

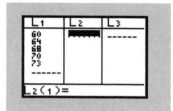 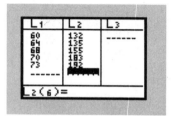

Before a scatter plot can be drawn it is necessary to define the plot. Press 2nd [STAT PLOT] [1:Plot 1] to display the Plot 1 screen.

Press ENTER to turn Plot 1 On. Leave Type as the first option: scatter plot, Xlist as L1, Ylist as L2 and Mark as a □. Press ZOOM 9 (to select ZoomStat). ZoomStat examines the data and adjusts the viewing rectangle to include all points.

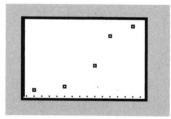 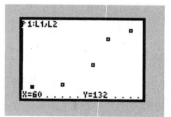

Press TRACE ▶ to see the data values. The P1 in the upper left corner indicates the cursor is tracing on points from Plot 1.

To compute the slope and y-intercept of the regression line, press STAT ▶ [4:LinReg(ax+b)]. The instruction is copied to the Home screen.

Press 2nd [L1] , 2nd [L2] , VARS ▶ [1:Function] [1:Y1]. Adding the argument Y1 causes the equation of the regression line to be copied to Y1.

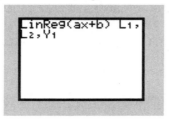

Press ENTER and the following results appear.

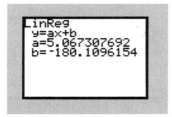

The slope of the regression line is $a = 5.07$ and its y-intercept is $b = -180.11$.

Now press GRAPH to view the graph of the regression equation.

 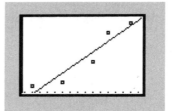

In developing a mathematical model of the physical world, we select what we consider to be essential characteristics of the real situation and represent them in an idealized form. The same situation may give rise to different models, depending upon what characteristics we consider important. Then we study the model to gain insight into the original phenomena and their relationships in order to predict future behavior. To model different phenomena successfully we need to have a large tool kit of functions and mathematical techniques at our disposal. We will discuss some of these functions and techniques later in the chapter.

Technology Tip: *Turning Off Stat Plots*
It is important to use 2nd [STAT PLOT] [4:PlotsOff] and ENTER to turn off all activated Stat Plots when you are done using the statistical features of the TI-83. Otherwise, the points from the Stat Plots will interfere with graphing and tracing other functions.

1.3   Exercises

1.   Find the slope and intercepts of the line whose equation is $2y + 3x - 12 = 0$.

2.   Find an equation of the line that passes through the points $(6, 1)$ and $(-3, 2)$.

3.   Find an equation of the line that has intercepts $(0, 4)$ and $(-3, 0)$.

4.   Find an equation of the line through the point $(2, 1)$ that is

     a) perpendicular to the line $y = 3x - 5$;

     b) parallel to the line $y = 2x + 3$.

5.   Do the points in the table below lie on a line? If so, find an equation of the line.

x	4.2	4.3	4.4	4.5	4.6
y	7.82	8.03	8.24	8.45	8.66

6.   The table below shows the sum of the measures of the interior angles of a polygon as a function of the number of sides.

Number of sides	3	4	5	6	7
Sum of angle measures	180°	360°	540°	720°	900°

     a) Explain how the data in the table tells us that the function is linear.

     b) What is the domain of the function?

     c) Write a formula for the function that relates the sum of measures of the interior angles to the number of sides.

7.   Let f be the function defined by $f(x) = x^2$. Using f, define the function m by $m(x) = $ slope of the line from $(0, 0)$ to $(x, f(x))$.

     a) Find $m(-1)$, $m(1)$, $m(2)$.

     b) Write a formula for $m(x)$ in terms of x.

     c) Draw a graph of m for $-5 \le x \le 5$.

8.   A function f starts out at $x = 0$ with a value of 7.9 and goes up at a constant rate of 0.3 units on the y-axis for each unit on the x-axis. Find a formula for f.

9.   Water freezes at 32 degrees Fahrenheit and at 0 degrees Celsius. Water boils at 212 degrees Fahrenheit and at 100 degrees Celsius. Find the linear function that takes Celsius degrees as input and produces the corresponding Fahrenheit temperature as output.

10.   A truck costs $12,000. After 5 years of heavy use, it is predicted to be worth $4,000. Give a reasonable estimate of its value 3 years after the purchase.

11.   Consider the following tables of data.

i)

x	y
.9	1.81
1	2
1.1	2.21
1.3	2.69
1.4	2.96
1.5	3.25

ii)

x	y
1.01	1.505
1.03	1.515
1.05	1.525
1.07	1.535
1.09	1.545
1.11	1.555

iii)

x	y
.75	.866
.85	.921
.95	.974
1.05	1.025
1.15	1.072
1.25	1.118

a) Which of the tables above represents a linear relationship? Justify your answer.

b) Create a model function for the table with the linear data.

c) Use the function of part b) to find y when $x = 2.65$.

12.   Match each of the functions $f(x) = 0.6x$, $g(x) = x^{0.6}$, and $h(x) = 0.6^x$ with the appropriate table of data below.

i)

x	y
.2	.903
.7	.699
1.1	.570
1.3	.515
1.8	.399

ii)

x	y
.5	.3
1.0	.6
1.5	.9
2	1.2
2.5	1.5

iii)

x	y
.31	.495
.61	.743
.91	.945
1.11	1.065
1.31	1.176

13.   Given the table below.

t	1.1	1.5	1.7	2.1
d	3.4	5.9	8.5	10.3

a) Create a scatter plot for the data.

b) Approximate $d(1.3)$ and $d(1.8)$.

c) Create a linear model function for the above data.

d) Use the function created in part c) to evaluate $d(1.3)$ and $d(1.8)$. Do the values agree with your approximation in part b) ?

14.   A bicycle store opened and sold 11 bicycles in the first month. After a few months, the store owner calculates that bicycle sales have been growing at a rate of 3 bicycles per month since opening.

a) Assuming that the rate of growth of sales continues, write a function that relates the number of bicycles sold in a month to the number of months since the store opened.

b) Use your function in part a) to predict the sales per month one year after the store opened.

c) How many bicycles have been sold in the first year of operation?

15.   You need a Lear jet for 1 day. Knowing that Swissair rents a Lear jet with pilot for $2,000 a day and $1.75 per mile, while Air France rents a Lear jet for $1,500 a day and $2.00 per mile, find the following:

a) For each company, write a formula giving cost as a function of distance travelled.

b) Sketch graphs of both functions, labeling intercepts and point of intersection.

c) If cost were the only issue, when would you rent from Air France?

16.   The table below gives information about the odometer readings and the model year of various Honda cars.

year	mileage
1994	8,000
1992	20,000
1985	88,000
1991	29,000
1979	153,000
1993	21,000
1989	92,000

   a) Make a scatter plot of the data with your calculator.

   b) Use your calculator to find a regression line.

   c) What would you expect the mileage of a 1988 car to be?

   d) If you wanted to buy one of the cars, which car is the best deal? Why?

   e) What is the real world meaning of the slope of the regression line?

   f) Describe the meaning of the intercepts.

17.   The following data were obtained in a study of the relationship between number of years that students studied German and the scores they received on a proficiency test.

x (years)	2	3	4	5	6
y (score)	57	78	72	84	91

   a) Enter the data into the TI-83 and graph the scatterplot and the corresponding regression line.

   b) Use the regression equation to predict the proficiency score for a student who has studied German for 7 years.

18.   The following data were obtained in a study of the relationship between rainfall and yield of wheat.

Rainfall (in.)	12.9	7.2	8.8	10.3	13.1
Wheat (bushels)	62.5	28.7	27	41.6	54.4

   a) Enter the data into the TI-83 and graph the scatterplot and the corresponding regression line.

   b) Use the regression equation to predict the wheat yield if the rainfall is 11 inches.

   c) What is the slope of the regression line? What is the real world meaning of this number?

1.4   Exponential Functions

Many of the problems we face today relate to populations and their changes over time. For example, the world population reached 3 billion in 1960, 4 billion in 1975, 5 billion in 1986 and will reach 6 billion in 1997. Surprisingly, it will have taken from the beginning of time until 1960 to reach 3 billion and only 37 years more to increase another 3 billion.

Listed in the table below is the United States population data for years 1980 to 1986. Column one shows the year and column two gives an estimate of the population in millions for the given year.

Year	Pop (millions)
1980	226
1981	229.16
1982	232.37
1983	235.63
1984	238.92
1985	242.27
1986	245.66

US Population 1980-1986

To measure how the population is growing, we can compute the increase in population from one year to the next and display it, as in column three below.

Year	Pop	Increase
1980	226	–
1981	229.16	3.16
1982	232.37	3.21
1983	235.63	3.36
1984	238.92	3.29
1985	242.27	3.35
1986	245.66	3.39

If the population had been growing linearly, the rate of increase would be a constant and all numbers in column three would be the same. But populations grow faster as they get larger because there are more people to contribute to the population growth.

If each year's population is divided by the previous year's population, the ratio is interesting and is displayed in column three below.

Year	Pop	Ratio
1980	226	–
1981	229.16	1.014
1982	232.37	1.014
1983	235.63	1.014
1984	238.92	1.014
1985	242.27	1.014
1986	245.66	1.014

From the table, it appears that there is a constant growth factor of 1.014, or 1.4% per year. Thus, if t is the number of years since 1980 and the initial population P is 226 million, we have

$$P(0) = 226 = 226(1.014)^0 \qquad \text{when } t = 0$$

$$P(1) = 229.16 = 226(1.014)^1 \qquad \text{when } t = 1$$

$$P(2) = 232.37 = 226(1.014)^2 \qquad \text{when } t = 2$$

and so t years after 1980, the population is given by

$$P(t) = 226(1.014)^t.$$

This is an example of an **exponential function** with base 1.014. The base represents the factor by which the population grows each year.

To develop a better sense of how population grows over time, consider the US population data for years 1800 to 1990. List L1 below contains the year and list L2 is the official census figure in millions for the given year.

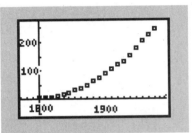

A scatter plot of the population data is shown below.

US Population Data from 1800 to 1990 in Millions

Note that the 190 years of population data does not appear to be linear. In fact, the data strongly suggests a curve that is growing faster as the population grows.

We now consider an example in which a quantity is *decreasing*. Nearly every year the price of a new car goes up. However, as time passes, once a car has been purchased its value begins decreasing. In fact, for most cars the decline in value, or depreciation, is about 20% per year. In other words, the value of a one-year old car is 80% of its purchase price. With a car that sells for $12,000, this rate of depreciation gives the following table of values.

t (age)	V(t) (car value)
0	12000
1	9600
2	7680
3	6144
4	4915.20
5	3932.16

In this example the growth factor is 80% or .80, a number less than one. Thus, if V_0 is the initial value of the car and $V(t)$ is the value of the car after t years we have

$$V(0) = V_0 = 12,000$$

$$V(1) = (0.8)V_0$$

$$V(2) = (0.8)(0.8)V_0 = (0.8)^2 V_0$$

$$V(3) = (0.8)(0.8)^2 V_0 = (0.8)^3 V_0$$

and so after t years

$$V(t) = V_0(0.8)^t = 12000(0.8)^t.$$

The viewing rectangle below shows a graph of the function $V(t) = 12000(0.8)^t$ superimposed on the data in our table.

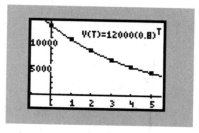

Each of the examples above involves quantities related to each other by a rule of the form $A(t) = A_0 b^t$ where A_0 is the initial quantity (when $t = 0$) and b is the factor by which A changes when t is increased by 1. A function of this form is called an **exponential function** because the input variable t is used as an exponent in determining the value of the output variable.

Exponential Functions

Any function of the form $A(x) = A_0 b^x$, where $A_0 > 0$, $b > 0$, and $b \neq 1$ is called an **exponential function with base b**. Its domain is the set of all real numbers; its range is the set of positive numbers. Graphs of increasing and decreasing exponential functions are sketched below.

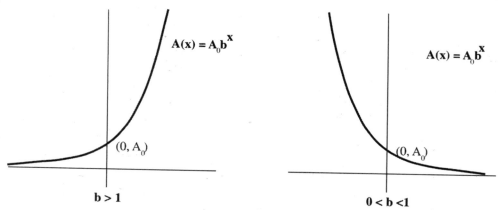

All exponential functions have graphs similar to these two graphs. The base b, which is as important for an exponential function as the slope is for a linear function, determines whether the exponential function is increasing or decreasing. If $b > 1$, $A(x) = A_0 b^x$ is an increasing function. These functions are useful for modeling growth behavior and are called **growth functions.** If $0 < b < 1$, $A(x) = A_0 b^x$ is a decreasing function. These functions are useful for modeling decay (negative growth) and are called **decay functions**. Notice that the x-axis is a horizontal asymptote for both increasing and decreasing exponential functions.

To recognize that a function f given by a table is exponential, look for ratios of outputs that are constant for equally spaced inputs.

Example 1: *Fitting an Exponential Model to Data*

An exponential function f was used to generate the values in the table below.

x	0	1	2	3
$f(x)$	4.50	3.60	2.88	2.30

Plot the data and find a possible formula for the exponential function.

Solution:

The inputs are equally spaced and the constant ratio of successive outputs confirms that the data is exponential.

$$\frac{3.60}{4.50} = 0.8, \qquad \frac{2.88}{3.60} = 0.8, \qquad \frac{2.30}{2.88} = 0.8.$$

Since f is an exponential function, $f(x) = A_0 b^x$ for some constants A_0 and b, we use a pair of data points to find A_0 and b.

Given $f(0) = 4.5$, we have $A_0 b^0 = 4.5$. Thus $A_0 = 4.5$.

Given $A_0 = 4.5$ and $f(1) = 3.60$, we have $4.5 b^1 = 3.60$. Thus we get the expected base of $b = 0.8$.

Therefore, a formula for f is $f(x) = 4.5(0.8)^x$.

In addition to population growth and car depreciation, exponential functions can be used to model and explain compound interest, radioactive decay and a host of other phenomena. The following example is about the growth of money in a bank account. In general, if P dollars are invested at an annual interest rate r and compounded annually, then at the end of t years the accumulated amount A is given by the formula $A(t) = P(1 + r)^t$.

Example 2: *Modeling Bank Interest*

If Columbus had deposited \$1.00 five hundred years ago in a bank that paid 5% interest compounded annually, how much money would have accumulated by now?

Solution:

The accumulated balance in the account is given by $A(t) = P(1 + r)^t$

where $P = \$1.00$, $t = 500$ and $r = 0.05$.

Thus

$$A = 1(1.05)^{500} = 3.932 \cdot 10^{10} \qquad \text{(approximately 39 billion dollars).}$$

Radioactive decay is another process that can be represented by an exponential function. When a radioactive substance decays, its mass decreases as it emits radioactive particles. The rate of decay diminishes as the mass grows smaller so that it takes progressively longer for the substance to lose a given mass. When a substance decays exponentially, the time required for the substance to lose half its current mass is fixed. This time is called the **half life**, and is a characteristic of each particular radioactive isotope.

Example 3: *Modeling Radioactive Decay*

Carbon-14, used by archaeologists for dating skeletal remains, is radioactive and decays exponentially. The percentage y of carbon-14 present after x years is given by the equation:

$$y = 10^{2-0.00005235x}.$$

Graph this equation and determine the half life of carbon-14.

Solution:

Graph the exponential function and the constant function $y = 50$ (percent). Using the intersection option from the Calc menu, we find that $y = 50\%$ when $x = 5750$, so the half life of carbon-14 is 5750 years.

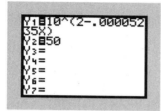
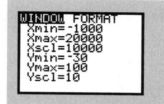
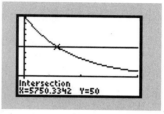

Why does this exponential function defined by $y = 10^{2-0.00005235x}$ decrease? Simplifying the formula for y reveals that $0 < b < 1$ for this exponential function.

$$10^{2-0.00005235x} = 10^2 \cdot 10^{-0.00005235x} = 100 \cdot \left(\frac{1}{10}\right)^{0.00005235x} = 100 \cdot (0.99879^x)$$

$$[\, b = \left(\frac{1}{10}\right)^{0.00005235} = .999879 \,] \; .$$

As shown in the last example, when studying exponential functions it is frequently necessary to apply the laws of exponents. Listed below are the essential rules for operating with exponents.

> For all positive real numbers a and b and all real numbers x and y:
>
> $$b^x b^y = b^{x+y}$$
>
> $$\frac{b^x}{b^y} = b^{x-y}$$
>
> $$\left(b^x\right)^y = b^{xy}$$
>
> $$(ab)^x = a^x b^x$$
>
> $$b^0 = 1$$
>
> $$b^{-x} = \frac{1}{b^x}$$

Some exponent laws become interesting function properties for exponential functions. If $f(x) = b^x$ and k is any constant, then

$$f(x+k) = b^{x+k} = b^x b^k = f(x) \cdot f(k) \text{, and}$$

$$f(x \cdot k) = b^{x \cdot k} = \left(b^x\right)^k = (f(x))^k.$$

1.4   Exercises

1.   Given below is a list of six exponential functions, each with a rule of the form $y = b^x$.

$$y_1 = 2^x, \quad y_2 = 0.5^x, \quad y_3 = 1.2^x, \quad y_4 = 0.2^x, \quad y_5 = 0.8^x, y_6 = 3.5^x$$

To help answer the following questions, display a graph and label values for each function.

a) Which of the functions are increasing? Decreasing?

b) Among those increasing, which has the greatest rate of increase? Which has the least?

c) Among those decreasing, which has the smallest rate of decrease?

d) Is there any input for which two of the six functions have the same output?

e) For any exponential function $f(x) = b^x$, describe how the base b affects the shape of the curve.

2.   Given below is a list of six exponential functions, each with a rule of the form $y = A_0 b^x$.

$$y_1 = 0.6^x, \quad y_2 = 4(0.6)^x, \quad y_3 = 8(0.6)^x, \quad y_4 = 2.5^x, \quad y_5 = 4(2.5)^x, y_6 = 8(2.5)^x$$

a) Which of the functions are increasing? Decreasing?

b) Among those increasing, which has the greatest rate of increase? Which has the least?

c) Among those decreasing, which has the smallest rate of decrease?

d) Is there any input for which two of the six functions have the same output?

e) For any exponential function $f(x) = A_0 b^x$, describe the effect the value of b has on the graph of the function.

3.   Describe how the graph of each of the following functions results from transformation on the graph of $y = 2^x$.

a) $y = 2^{x-1}$ b) $y = \left(\dfrac{1}{2}\right)^x$ c) $y = 2\left(2^{x+1}\right)$

d) $y = -\left(2^{-x}\right)$ e) $y = 2^{x-1} - 1$ f) $y = 2^{x/2}$

4.   Use your calculator to graph each of the following functions. Use the graph to determine i) the domain, ii) the range and iii) the zeros of each function.

a) $y = 2^{x-3} - 1$ b) $y = 2\left(3^{x+2}\right) + 3$

5.   Find a rule for the function that results from each of the following transformations on the graph of $y = 3^x$.

a) Shift up 3 and then reflect about the x-axis.

b) Reflect about the y-axis.

c) Reflect about the y-axis, then reflect about the x-axis.

d) Shift right 2, then shift up 3.

6. Find a function whose graph is shown in each of the following. Use exponential functions, perhaps shifted, and confirm your answer with your calculator.

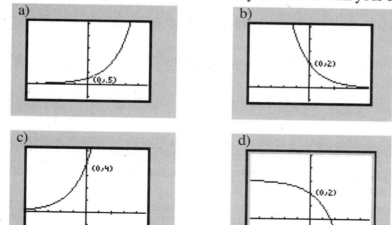

a)   b)

c)   d)

7. Each of the tables of data below can be represented by an exponential function of the form $f(x) = A_0 b^x$. Find an appropriate formula for each table of data.

a)

x	0	1	2	3
y	3	6	12	24

b)

x	-2	0	2	4
y	50	2	.08	.0032

c)

x	1	2	3	4
y	4.5	13.5	40.5	121.5

d)

x	1	3	5	7
y	1.8	1.458	1.181	.9566

8. Determine a model exponential function for each graph shown below.

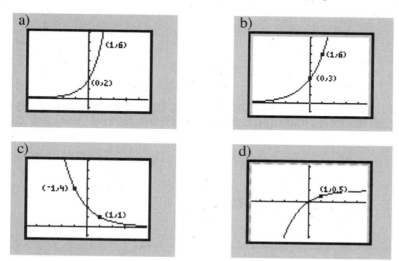

a)   b)

c)   d)

9. Create an exponential function such that $f(0) = 120$ and the function output grows 15% for each unit increase in input.

10. Create an exponential function for which $f(0) = 12,000$ and for which the output decreases by 18% for each unit increase in input.

11. The new Phillips sports car sells for $14,000 new. Because it is so well built it only depreciates at 12% per year.

   a) If the Phillips depreciates linearly at 12% of the original value each year, how much will it be worth in 5 years?

   b) If the Phillips depreciates exponentially at 12% each year, how much will it be worth in 5 years?

12. The population of mosquitoes around Rabbit Pond grows exponentially in the early summer. The population was measured to be 1000 mosquitoes per 100 sq feet on May 23. The mosquito population doubles every 6 days during this time of the year around Rabbit Pond.

   a) Write an exponential function that relates the mosquito population per 100 sq feet to the time, t days, after May 23.

   b) Determine the mosquito population on June 4.

   c) How long does it take for the mosquito population to reach 5000 per 100 sq feet?

13. George has $7,000 to invest in a money market fund. He decides that his best choice is to invest in a fund that pays 5.75% interest compounded annually.

   a) Write a function that relates the amount of money George has in his account to the number of years it has been invested.

   b) How much money will George have after 3 years?

   c) How long will it take George to double his money?

14. Steve is trying to decide how to invest his $300 in savings. One investment option grows linearly at 35% of the original investment per year while a second option grows exponentially at 7% per year.

   a) If Steve pursues the first option, write a function to describe the amount of money he will have in his account after t years.

   b) If Steve pursues the second option, write a function to describe the amount of money he will have in his account after t years.

   c) Compare the investment strategies.

In Exercises 15–18, the data in the tables is either linear or exponential. For each table, determine whether the data is linear or exponential, briefly justify your answer, and find a model function that best fits the table.

15.

x	0	.1	.2	.3	.4	.5
y	2.14	2.18	2.22	2.26	2.30	2.34

16.

x	0	.1	.2	.3	.4	.5
y	2.50	3.05	3.72	4.54	5.54	6.76

17.

x	0	.2	.4	.6	.8
y	1.70	1.47	1.23	1.00	.76

18.

x	0	.2	.4	.6	.8
y	4.57	4.43	4.30	4.17	4.05

19.   A student invests $2,000 in a certificate of deposit which compounds its interest once a year. If the student has $3,310 after 8 years, what was the advertised annual interest rate paid by the certificate of deposit?

20.   Water lilies have started growing on Rabbit Pond. At the moment, they cover about 3% of the pond. The lilies will cover 12% of the pond in 14 days.

   a) If the area covered by the lilies grows exponentially, create a model function to describe the amount of area of the pond covered by lilies as a function of time ($t = 0$ now).

   b) After how many days will the pond be 50% covered?

   c) How long does it take to cover the second 50% of the pond?

21.   The following table has information on the tuition costs at two universities in the 1990's. One university has been raising tuition linearly while the other has raised tuition exponentially. Let $t = 0$ correspond to the year 1990.

   a) Determine a function that best describes the cost of tuition at Mesa U.

   b) Determine a function that best describes the cost of tuition at Prairie U.

   c) What is the projected tuition at Mesa U in the year 1999 ?

   d) Which university will cost more to attend in the year 2010 ?

Year	Tuition at Mesa U	Tuition at Prairie U
1990	$15780	$15720
1991	$16610	$16430
1992	$17440	$17170
1993	$18270	$17940
1994	$19100	$18750
1995	$19930	$19600

1.5   Compound Interest and the Number e

One of the most familiar examples of exponential growth is that of an investment earning compound interest. Suppose P dollars (the principal) is invested at an interest rate of $r\%$ per time period. If the interest is added to the principal at the end of each time period, then at the end of one time period you would have P dollars, plus the interest rP, for a balance of $P + rP = P(1 + r)$ dollars. At the end of the second time period you have $P(1 + r)$ dollars, plus interest $r[P(1 + r)]$, for a total of

$$P(1 + r) + r[P(1 + r)] = P(1 + r)(1 + r) = P(1 + r)^2.$$

Your balance continues to grow by a factor of $(1 + r)$ each time period. In general, if P dollars is invested at an interest rate of $r\%$ per time period, then the balance after t time periods is given by the exponential function

$$A(t) = P(1 + r)^t.$$

This formula is the basis of virtually all financial calculations, whether they apply to bank accounts, loans, mortgages, or annuities. Banks often state an annual interest rate, but compound more than once a year. Instead of adding all the annual interest at the end of each year, a bank that compounds quarterly, for example, would apply one-fourth the annual rate four times a year or every three months. Suppose the compounding is done n times per year. For each time period the bank uses the annual interest rate divided by n, that is, $\frac{r}{n}$. Since in t years there are nt time periods, a principal P after t years will yield an amount :

$$A(t) = P(1 + \frac{r}{n})^{nt}$$

Example 1: *Finding an Account Balance if the Interest is Compounded Quarterly*

If \$5000 is invested at 8% annual interest, compounded quarterly, what is the balance after 3 years?

Solution:

Interest is compounded 4 times per year, so $n = 4$ and the interest rate per period is

$r = \frac{.08}{4} = .02$. Substituting in the compound interest formula $A(t) = P(1 + \frac{r}{n})^{nt}$

we have

$$A(3) = 5000(1 + .02)^{4 \cdot 3} = \$6341.21.$$

As a general rule, the more often interest is compounded the larger the account balance. But there is a limit.

To explore the effect that n, the number of times a year that interest is compounded, has on the yield from a given principal, we consider a special case. Let $r = 1$ (annual interest rate of 100%), $P = \$1$ and $t = 1$ year. Then the compound interest formula becomes

$$A(1) = (1 + \frac{1}{n})^n.$$

Using a calculator, the value of A is computed for increasing values of n. The results are displayed in the table.

n	1	2	4	12	365	1000	20000
$(1+1/n)^n$	2	2.25	2.4414	2.6130	2.7146	2.7169	2.7182

It appears that the quantity $(1+\frac{1}{n})^n$ has a limiting value as n grows. The last entry in the table is the special number e approximated to four decimal places. In mathematical terms we have

$$\lim_{n\to\infty} (1+\frac{1}{n})^n = e \approx 2.71828.$$

This is just one example of how the number e arises naturally in a real world situation. Later we will see that e is especially useful in dealing with logarithms and exponentials in a calculus course.

Like π, e is an irrational number. Its decimal expansion begins

$$e = 2.718281828459...$$

Continuous Compounding

Suppose that P dollars is invested at 8% annual interest compounded n times per year for t years. Then at the end of t years, P dollars will have grown to

$$A(t) = P(1+\frac{.08}{n})^{nt}$$

which can be rewritten as

$$A(t) = P\left[\left(1+\frac{.08}{n}\right)^{n/.08}\right]^{.08t}$$

To see the role that e plays as n gets larger, let $k = \frac{n}{.08}$. Then $\frac{.08}{n} = \frac{1}{k}$.

Thus, the compound interest equation in terms of k is

$$A(t) = P\left[\left(1+\frac{1}{k}\right)^{k}\right]^{.08t}.$$

As the number of compoundings, n, increases without bound, the way that k depends on n $(k = \frac{n}{.08})$ means that k also increases without bound. And since $\lim_{k\to\infty} (1+\frac{1}{k})^k = e$, it follows that if interest is compounded frequently, future values of $P\left[(1+\frac{1}{k})^k\right]^{.08t}$ can be approximated by $Pe^{.08t}$. Thus,

$$\lim_{n\to\infty} P\left(1+\frac{.08}{n}\right)^{nt} = \lim_{k\to\infty} P\left[(1+\frac{1}{k})^k\right]^{.08t} = Pe^{.08t}.$$

This approximation using e is called **continuous compounding**. The term continuous compounding means that interest earned on the account is added to the account at every instant of time, which is more frequently than once every hour, or once every minute, or even once every second. In general, if interest on an initial balance of P dollars is compounded continuously, at an annual rate of r, the balance, $A(t)$, t years later is:

$$A(t) = Pe^{rt}$$

The letter e was chosen in honor of the Swiss mathematician Leonhard Euler (1707–1783), who investigated the limit $\lim\limits_{n \to \infty} \left(1 + \dfrac{1}{n}\right)^n$ and explained a number of applications in which the limit played a useful role. At first glance, there may seem to be very little about the number e that is "natural"; however, later in the text it will become clear that e is an extremely useful exponential base.

Example 2:   *Comparing Continuous and Monthly Compounding*

     If $1000 is invested at an annual interest rate of 6%, find the balance 5 years later if it is compounded

     a) monthly;   b) continuously.

Solution:

     a) For monthly compounding, we have $n = 12$. Thus in 5 years at 6% the balance is

$$A(t) = P\left(1 + \frac{r}{n}\right)^{nt}$$

$$A(5) = 1000\left(1 + \frac{.06}{12}\right)^{12 \cdot 5}$$

$$= 1348.85$$

     b) Compounding continuously, the balance is

$$A(t) = Pe^{rt} \quad \text{so that} \quad A(5) = 1000e^{.06 \cdot 5} = 1349.86.$$

Note that continuous compounding yields $1349.86 – $1348.85 = $1.01 more than monthly compounding.

     The number e and exponential functions with base e, $f(x) = A_0 e^x$, play a central role in mathematics. For example, the standard normal probability curve, which is introduced later in the text, is the graph of an exponential function with base e.

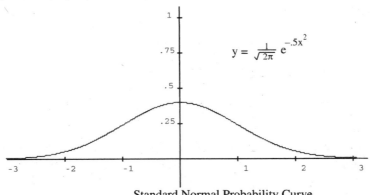

$$y = \frac{1}{\sqrt{2\pi}}\, e^{-.5x^2}$$

Standard Normal Probability Curve

1.5   Exercises

Convert the following functions from the form $f(x) = A_0 e^{rx}$ to the form $f(x) = A_0 b^x$.
(Hint: $e^r = b$.) Indicate which functions are increasing which are decreasing.

1. $f(x) = 13e^{0.2x}$

2. $g(x) = 20.5e^{1.15x}$

3. $P(t) = \dfrac{e^{-0.67t}}{2}$

4. $k(t) = \dfrac{2e^{1.8t}}{5}$

5. $h(x) = 1705e^{-2x}$

6. $A(t) = \left(e^t\right)^{-0.5}$

Convert the following functions from the form $f(x) = A_0 b^x$ to the form $f(x) = A_0 e^{rx}$.
(Hint: Graphically solve the equation $e^r = b$ for r .)

7. $A(x) = A_0 \left(2^x\right)$

8. $f(x) = 1153\left(\dfrac{1}{2}\right)^x$

9. $P(x) = 16.5(1.2)^x$

10. $g(t) = \dfrac{3\left(2^{2t}\right)}{2}$

11. $k(x) = \dfrac{275}{3^x}$

12. $h(x) = 5\left(\dfrac{2}{3}\right)^{-x}$

13.   Suppose \$100 is invested at 7% annual interest. Write a model function for this situation if the interest is compounded

   a) annually;

   b) quarterly;

   c) monthly;

   d) daily;

   e) continuously.

14.   Sarah deposits \$400 in her bank account. The account pays interest at an annual rate of 6.75%.

   a) How much money will she have in the account after 5 years if the interest is compounded quarterly?

   b) If the interest is compounded continuously, write a formula relating the amount of money in the account to the time since Sarah deposited the money.

   c) How long will it take Sarah to double her money, if the bank compounds her interest continuously?

15.   A virus grows continuously by 11% per day. If the virus population is 200 at $t = 0$, write a model function of the form $P(t) = P_0 e^{rt}$ to describe the population of the virus as a function of time. How long will it take for the virus population to exceed 10,000?

16.   The population of a small country is 1.5 million and is growing continuously at $2\frac{1}{2}\%$
       per year.
       a) Write a function to model the population as a function of time.
       b) What will be the country's population in 7 years?
       c) How long will it take for the population to reach 3 million? 4.5 million? 6 million?

17.   What is the smallest positive integer n so that $\left(1+\dfrac{1}{n}\right)^{n}$ approximates e to the nearest
       hundredth?

18.   Compound interest works in favor of credit card companies. Most companies take the
       advertised interest rate, say 18%, and then charge you $\dfrac{18\%}{12}$ per month. Suppose you
       have a $500 balance on your credit card and make no payments over a 12 month period.
       What do you owe?

19.   If P dollars are borrowed for N months compounded monthly at an annual interest rate of
       r, then the monthly payment m is found by the formula

$$m = \frac{P\left(\dfrac{r}{12}\right)}{1-\left(1+\dfrac{r}{12}\right)^{-N}}.$$

       Use the formula to determine the monthly car payment for a new car costing
       $12,487 with a down payment of $2487. The car is financed for 4 years at 12%.

1.6   Inverse Functions

Sometimes it is necessary to reverse a function's process. That is, starting with a function output value, we would like to recover the original input value.

Suppose a car is being driven at a constant rate of 55 mph. The table below describes a function that pairs time and distance.

t hr (input)	0	0.5	1	1.5	2	2.5	3	3.5	4
d miles (output)	0	27.5	55	82.5	110	132.5	165	192.5	220

In this case, t represents the number of hours traveled at 55 mph and d represents the distance traveled in that amount of time.

The function is currently defined so that t is the input variable and d is the output; that is $d = f(t)$ where f is a rule for pairing each time with a corresponding distance. The function f, represented by the table, is set up to answer the question, "How far can one travel in t hours driving at a constant rate of 55 mph?"

An equally valid question is one which asks, "If I need to go d miles, how long will it take me?" Assuming a constant speed of 55 mph, the same table could be used. However, our inputs are now distances and our outputs are the times required to cover the distances. Hence, we have a new function g of the form $t = g(d)$ defined by the new table.

d miles (input)	0	27.5	55	82.5	110	132.5	165	192.5	220
t hr (output)	0	0.5	1	1.5	2	2.5	3	3.5	4

When two functions are related so that for every ordered pair (a, b) in one function the ordered pair (b, a) is always in the other function, and vice versa, the two functions are known as **inverses** of each other. **The inverse of a function f is denoted as f^{-1}.** In our example, $g(d) = f^{-1}(d)$ so that we can write that $d = f(t)$ and $t = f^{-1}(d)$.

In general, the inverse of a function is another function, denoted by f^{-1}, that "undoes" the effect of the original function.

Fahrenheit and Celsius temperatures are related by the formula

$$F = f(C) = \frac{9}{5}C + 32.$$

That is, the function f defined by $f(C) = \frac{9}{5}C + 32$ converts Celsius degrees into Fahrenheit degrees. For example, $f(20°) = 68°$ or $20°C = 68°F$.

We often want to reverse the process and convert from Fahrenheit to Celsius temperature. In the language of functions, this means finding the inverse function f such that $C = f^{-1}(F)$. For a linear function such as f, this is easy. First, solve the formula for C, giving

$$C = \frac{5}{9}(F - 32)$$

and then define the function

$$f^{-1}(F) = \frac{5}{9}(F - 32) .$$

The function f^{-1} converts Fahrenheit degrees to Celsius degrees and so is the "undoing" or inverse function for f.

Not every function has an inverse. For example, consider the squaring function f where $f(x) = x^2$. To see why f does not have an inverse, try to determine a value for $f^{-1}(4)$. Unhappily, there are two such values, 2 and –2, such that $f(2) = 4$ and $f(-2) = 4$. Furthermore, every number except 0 in the range of this function is an output for two distinct inputs.

Since a function must take a single output for every input there is no inverse function f^{-1}. In contrast, the cubing function has an inverse because each output corresponds to only one input. Graphs of the squaring and cubing function are shown below.

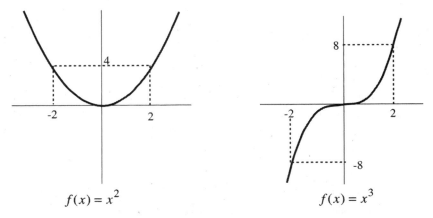

$$f(x) = x^2 \qquad\qquad f(x) = x^3$$

Notice that the cubing function is increasing everywhere and never assigns the same output to two different inputs. A function f is called **one-to-one** if different inputs always result in different outputs ($x_1 \neq x_2$ implies $f(x_1) \neq f(x_2)$).

Now, an important fact: **Every one-to-one function has an inverse.**

A graph is helpful in determining whether or not a given function is one-to-one. If every horizontal line intersects the graph of a function f at most once, then f is one-to-one. So if any horizontal line intersects the graph of f more than once, the function f is not one-to-one.

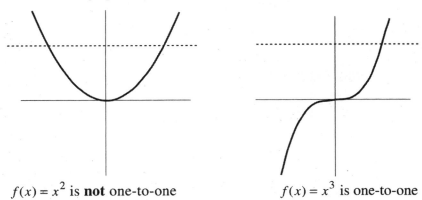

$$f(x) = x^2 \text{ is \textbf{not} one-to-one} \qquad\qquad f(x) = x^3 \text{ is one-to-one}$$

This graphical test for a one-to-one function is called the **horizontal line test**. Any function that is strictly increasing or strictly decreasing on its domain passes this test, and so has an inverse.

Example 1:   *Finding the Inverse of a Function*

A one-to-one function f is defined by $f(x) = \sqrt[3]{2x - 3}$. What is its inverse?

Solution:

Here f is described by the instruction, take an input, double it, subtract 3 and take the cube root. In order to reverse or undo this process, we must first cube an input, add 3 and then divide by 2. This suggests that the inverse of f is the function

$$g(x) = \frac{x^3 + 3}{2} .$$

Since f^{-1} reverses the input-output process, there is a simple but important relationship between the graph of a one-to-one function f and its inverse f^{-1}.

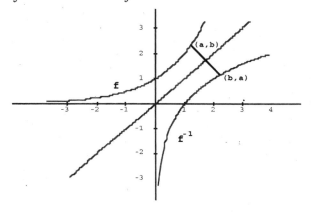

For every point (a, b) that belongs to the graph of f, the point (b, a) belongs to the graph of f^{-1}. But (a, b) and (b, a) are symmetric with respect to the line $y = x$; that is, the line $y = x$ is the perpendicular bisector of the segment joining these two points. We call (b, a) the reflection of (a, b) in the line $y = x$.

Hence, the graph of f^{-1} is obtained by reflecting the graph of f in the line $y = x$.

Because an inverse function is obtained from a one-to-one function by interchanging coordinates, it follows that the domain of a one-to-one function f is the range of f^{-1}, and the range of f is the domain of f^{-1}.

1.6   Exercises

A function is $y = f(x)$ is defined by each of the tables. List the inverse of f, f^{-1}, and state the domain and range of f^{-1}.

1.

x	−1	1	3	5	7
y	0	2	10	26	50

2.

x	1.1	1.01	1	0.99	0.9
y	2.2	2.02	2	1.98	1.8

3.

x	0	$\pi/6$	$\pi/4$	$\pi/3$	$\pi/2$
y	0	1/2	$\sqrt{2}/2$	$\sqrt{3}/2$	1

4.

x	−2	−1	0	1	2
y	1/4	1/2	1	2	4

For each of the following, sketch the given function and its inverse on the same coordinate system. (Hint: Use the fact that if $(a, b) \in f$, then $(b, a) \in f^{-1}$.) State the domain and range of f and f^{-1}.

5.

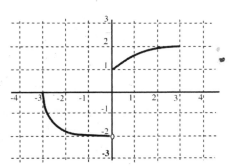

6.

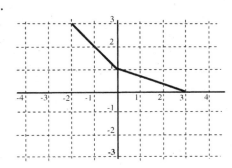

7.

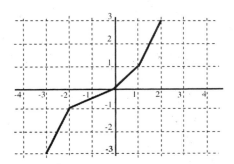

8.

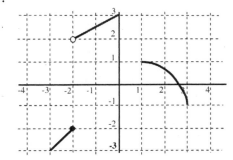

9.   Use a calculator and the horizontal line test to determine whether or not the given function f is one-to-one.

a) $f(x) = x^3 + x$

b) $f(x) = x^3 - 4x^2 + x - 10$

c) $f(x) = 0.1x^3 - 0.1x^2 - 0.005x + 1$

d) $f(x) = x^5 + 2x^4 - x^2 + 4x - 5$

10.   The function $f(x) = (x-1)^2$ is not one-to-one. Define a new function h by $h(x) = (x-1)^2$ but only on the restricted domain $(-\infty, 1]$. Show that the function $g(x) = 1 - \sqrt{x}$, with domain $[0, \infty)$ is the inverse of h.

11.   Let the function k be defined on the domain $[1, \infty)$ by $k(x) = (x-1)^2$. Show that k is one-to-one and determine a rule for its inverse $k^{-1}(x)$.

Determine which of the following functions has an inverse. If the function has an inverse, sketch the graph of the inverse. If the function does not have an inverse, explain how to limit the domain of the function so that it will have an inverse.

12.

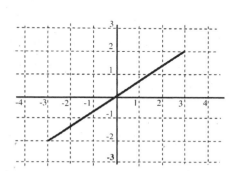

13.

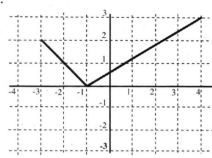

14.

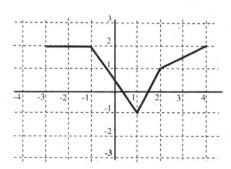

15.

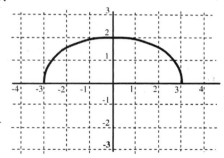

In working with pairs of functions which are inverses of each other, it is often useful to think of one function, f, performing a certain series of operations on a given input, while the inverse function, f^{-1}, undoes whatever the function f does. Use this concept to determine the inverses of the following functions. Specify the domain and range of both the function and its inverse.

16. $f(x) = 7x$ 17. $F(x) = \sqrt[3]{x}$

18. $g(x) = 2x - 1$ 19. $H(x) = \sqrt{x-1}$

20. $K(x) = \dfrac{x+2}{3}$ 21. $g(x) = x^2 - 2 \quad (x \geq 0)$

22. $D(t) = 55t$ 23. $J(x) = \dfrac{1}{x}$

Determine which of the following functions have inverses. If the function has an inverse, sketch the function and its inverse on the same coordinate plane. If the function does not have an inverse, briefly explain why.

24. $f(x) = x^3 - 2$

25. $g(x) = e^x$

26. $h(x) = |x - 3|$

27. $g(x) = \begin{cases} 2x - 1, & x < 0 \\ x^2 - 1, & x \geq 0 \end{cases}$

28.   If f is an increasing (or decreasing) function, then f is one-to-one. Why?

29.   Show by example that not every one-to-one function is an increasing function or a decreasing function.

30.   If f is a one-to-one function and $f(k) = k$ for some real number k, then the point (k, k) will be on the graph of f and f^{-1}. Show by example that a function f and its inverse f^{-1} may also intersect at points not on the line $y = x$.

31.   Let $f(x) = 1 - x^3$.

a) Find $f^{-1}(x)$.

b) How many solutions are there to the equation $f(x) = f^{-1}(x)$?

32.   Give an example of a function, other than $f(x) = x$, that is its own inverse.

1.7   Logarithms

Before the advent of calculators and computers, certain arithmetic computations such as $(1.37)^{13}$ and $\sqrt[7]{3.09}$, were difficult to perform. The computations could be easily performed (approximately) using logarithms, which were developed in the 16th century by John Napier, or by using a slide rule, that is based on logarithms. The use of logarithms as a computing technique has all but disappeared, but logarithm functions continue to be useful in modeling real-world phenomena.

Logarithmic Functions

If $b > 0$ and $b \neq 1$, then the exponential function $f(x) = b^x$ is either increasing or decreasing. Thus, f is one-to-one and has an inverse function f^{-1} whose graph is the reflection of the graph of f in the line $y = x$.

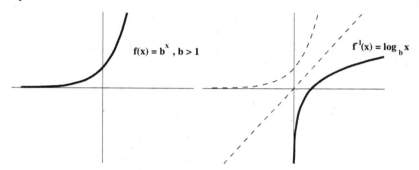

The inverse of the exponential function $y = b^x$ is known as the **logarithm function to the base b**. The value $f^{-1}(x)$ of this function is denoted by $\log_b x$ and called **the logarithm to the base b of the number x**. Since the domain of $f(x) = b^x$ is the set of all real numbers and the range is the set of all positive numbers, we can interchange these to obtain the domain and range of $f^{-1}(x)$.

$$\text{Domain of } \log_b x \ = \ (0, \infty) = \text{ Range of } b^x$$
$$\text{Range of } \log_b x \ = R \ = \text{ Domain of } b^x$$

Since the exponential and logarithmic functions are inverses of each other, the number $\log_b x$ is the exponent that b must be raised to in order to produce x. Thus, for example,

$$\log_4 16 \ = 2 \ \text{ since } \ 4^2 = 16$$
$$\log_{10} 1000 \ = 3 \ \text{ since } \ 10^3 = 1000$$
$$\log_{10} \frac{1}{100} \ = -2 \ \text{ since } \ 10^{-2} = \frac{1}{10^2} = \frac{1}{100}$$

In general, if $b > 0$ and $x > 0$ then

$$y = \log_b x \ \text{ if and only if } \ x = b^y \ .$$

Most calculators have two logarithm functions immediately available, $\log_{10}$ (known as log) and $\log_e$ (known as ln).

 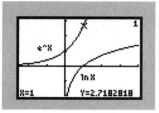 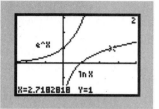

Tables of values and a graph of $y = \ln x$ are shown below.

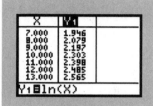 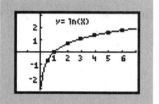

Notice that

$$\ln(0.5) + \ln(2) = \ln(1),$$

$$\ln(2) + \ln(3) = \ln(6),$$

and

$$2\ln(2) = \ln(4),$$

$$3\ln(2) = \ln(8),$$

$$2\ln(3) = \ln(9).$$

These are examples of some properties of logarithm functions which can be shown to be related to some familiar properties of exponents.

Properties of Logs

If $f(x) = b^x$ is any exponential function we know that

$$b^{x+y} = b^x b^y, \quad b^{x-y} = \frac{b^x}{b^y}, \quad \text{and} \quad b^{xy} = \left(b^x\right)^y.$$

These properties of exponents lead to a corresponding set of properties for logarithms.

If x and y are positive numbers, then

1) $\log_b(xy) = \log_b x + \log_b y$

2) $\log_b\left(\dfrac{x}{y}\right) = \log_b x - \log_b y$

3) $\log_b x^r = r\log_b x,$ where r is any real number.

Proof: We prove property 1 and leave properties 2 and 3 for the exercises.

If $c = \log_b x$ and $d = \log_b y$,

then $b^c = x$ and $b^d = y$.

Thus, $x \cdot y = b^c \cdot b^d = b^{c+d}$

and it follows that

$$\log_b(x \cdot y) = c + d = \log_b x + \log_b y.$$

The steps in this proof used only properties of logs and exponents from this section. You should go through the proof and determine which property was used at each step.

These log properties simplify computation involving logs.

Example 1: *Using Logs to Simplify Computation*
Solve for x.

a) $100 \cdot (1.03)^x = 375$

b) $\ln x + \ln x^2 = 5$

Solution:

a) Take the natural logarithm of both sides.

$$\ln\left[100 \cdot (1.03)^x\right] = \ln 375$$

$$\ln 100 + \ln(1.03)^x = \ln 375$$

$$\ln 100 + x \ln(1.03) = \ln 375$$

$$x = \frac{\ln 375 - \ln 100}{\ln(1.03)} \approx 44.716$$

b)

$$\ln x + \ln x^2 = 5$$
$$\ln x + 2\ln x = 5$$
$$3\ln x = 5$$
$$\ln x = \frac{5}{3}$$
$$x = e^{5/3} \approx 5.294$$

Changing Base for Log Functions

It is sometimes necessary, or convenient, to convert logarithms in one base to logarithms in another base. For example, to change $f(x) = \log_2 x$ to a natural logarithm function let

$$y = \log_2 x,$$

and then rewrite this equation as

$$2^y = x.$$

Now take the natural logarithm of both sides

$$\ln\left(2^y\right) = \ln x$$

$$y \ln 2 = \ln x$$

$$y = \frac{\ln x}{\ln 2}$$

Therefore,

$$f(x) = \log_2 x = \frac{\ln x}{\ln 2}.$$

In general, the **Change of Base** formula is

$$\log_b x = \frac{\log_a x}{\log_a b}$$

This formula can be employed to compute logarithms with any base and to obtain a graph of any logarithmic function.

Example 2:   *Using the Change of Base Formula*

Use a calculator and the Change of Base formula to approximate $\log_5 7$.

Solution:

Since most calculators have keys for natural and common logarithms, we can use ln or log. We obtain the same answer either way

$$\log_5 7 = \frac{\ln 7}{\ln 5} = \frac{\log 7}{\log 5} \approx 1.2091.$$

1.7   Exercises

1.   Use the fact that $y = \log_b x$ if and only if $b^y = x$ to evaluate the following.

     a) $\log_2 8$ b) $\log_3 81$

     c) $\log_{1/2} 16$ d) $\log_e e^7$

2.   Sketch a graph of each of the following functions. How does each function graph relate to the graph of $y = \ln x$?

     a) $y = \ln(x - 1)$ b) $y = \ln x + 2$

     c) $y = 2 \ln x$ d) $y = -\ln(x + 1)$

3.   Use the properties of logarithms to express the following as an algebraic expression involving $\log x$, $\log y$ and/or $\log z$.

     a) $\log \dfrac{x^y}{z}$ b) $\log \dfrac{x^2 y^3}{z}$

     c) $\log x^4 y \sqrt{z}$ d) $\log \dfrac{z}{\sqrt{xy}}$

4.   Express each of the following as the logarithm of a single expression.

     a) $2 \log x + 4 \log y - \log 13$ b) $\log(x + 1) - 2 \log x - \log y$

     c) $\log 7 + 5 \log y - \dfrac{\log x}{2}$ d) $2 \log_6 2 + \log_6 3 + \log_6 18$

Use logarithms to solve the following:

5. $5^x = 7$ 6. $2^x + 5 = 17$

7. $7.01 = 3^{2x} - 2.4$ 8. $(8.1)\left(7.4^x\right) = (21.8)\left(3.7^x\right)$

9. $5320 e^x = 14756$ 10. $5 e^{x+1} = 27$

11. $3 \cdot 2^{2x+1} = 24$ 12. $e^{x^2 + 2x} = 2$

13.   A certain radioactive isotope has a half–life of 16 days. If there are initially 240 grams of the isotope, in how many days will there be only 20 grams left?

14.   A student has \$1430 in summer earnings to invest in a savings account.

     a) If the student places the money in a bank account that accrues 6.5% annual interest compounded annually, how long does it take for the money in the account to grow to \$2,000 ?

     b) If the student places the money in a bank account that accrues 6.75% annual interest compounded continuously, how long does it take for the amount of money in the account to double?

15. Patty buys a car for $14,300. If the vehicle depreciates at 23% per year, in how many years will the car be worth only 10% of its original value?

16. The population of Arkansas is currently estimated to be 2,500,000 and growing continuously at 3% per year.

   a) Write a model function that relates the population of Arkansas to the number of years from now.

   b) How many years will it take for the population of Arkansas to reach 3,000,000 ?

17. A new state-of-the art computer is currently worth $4,300. It is known that computers depreciate at 25% per year.

   a) Write a model function that relates the value of the computer to the number of years from now.

   b) How many years from now will the computer be worth only $100 ?

18. It has been estimated that at a certain school, the overall grade point average for the school is currently 4.3 and increasing at 2% per year. How many years will it take for the overall grade point average to reach 5.0 ?

The calculator is only able to graph directly $y = \log_{10} x$ and $y = \ln x$. Use the change of base formula to enable you to graph the following on your calculator.

19. $y = \log_2 x$ 20. $y = \log_3 x$

21. $y = 2\log_{1/2} x$ 22. $y = 2\log_7 x + 1$

Use your calculator and the change of base formula where necessary to graph the following pairs of equations. What do you notice about these pairs?

23. $y = 2^x$ and $y = \log_2 x$ 24. $y = 3^x$ and $y = \log_3 x$

25. $y = e^x$ and $y = \ln x$ 26. $y = 3 \cdot 2^x$ and $y = \log_2 x - \log_2 3$

27. Starting with $\log_b x = c$ and $\log_b y = d$, and x and y both positive, prove that

$$\log_b\left(\frac{x}{y}\right) = \log_b x - \log_b y .$$

28. Prove that if $x > 0$ and r is any real number, $\log_b\left(x^r\right) = r\log_b x$.

29. Solve for x:

   a) $\log_3(x - 4) \leq 2$

   b) $\log_2(3x - 2) - \log_2(x + 1) = 3$

   c) $\log_2(7 - x) - \log_2(5 - x) = 3$

   d) $\left(\log_2 x\right)^2 - 3\log_2 x - 4 = 0$

1.8   Combining Functions; Polynomial and Rational Functions

Just as numbers can be combined arithmetically to produce new numbers, functions can be added, subtracted, multiplied, and divided (except when the denominator is 0) to produce new functions. If f and g are functions, then for every x that belongs to the domain of both f and g we have:

$$(f+g)(x) = f(x) + g(x)$$

$$(f-g)(x) = f(x) - g(x)$$

$$(f \cdot g)(x) = f(x) \cdot g(x)$$

$$\left(\frac{f}{g}\right)(x) = \frac{f(x)}{g(x)} \quad \text{provided } g(x) \neq 0.$$

Example 1:   *Evaluating Functions*

Functions f and g have the graphs given below.

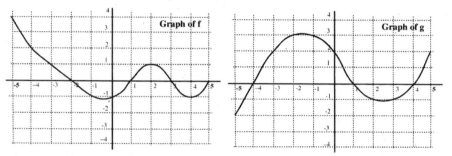

Using the graphs, evaluate the following:

a) $(f+g)(3)$ b) $(f-g)(2)$ c) $(f \cdot g)(-3)$ d) $\left(\dfrac{f}{g}\right)(0)$

Solution:

a) $(f+g)(3) = f(3) + g(3) = 0 + (-1) = -1$

b) $(f-g)(2) = f(2) - g(2) = 1 - (-1) = 2$

c) $(f \cdot g)(-3) = f(-3) \cdot g(-3) = 1 \cdot 2 = 2$

d) $\left(\dfrac{f}{g}\right)(0) = \dfrac{f(0)}{g(0)} = \dfrac{-1}{2}$

Recall that when a function is given by a formula and there is no mention of the domain, it is understood that the domain consists of all real numbers where the formula makes sense and yields a real number.

Example 2:   *Writing Formulas for Functions*

The functions f and g are defined by $f(x) = \sqrt{x+1}$ and $g(x) = \dfrac{x-1}{x}$. Find a formula for the values of $(f+g)(x)$ and $\left(\dfrac{f}{g}\right)(x)$, and specify the domain of each function.

Solution:

$$(f+g)(x) = f(x) + g(x) = \sqrt{x+1} + \frac{x-1}{x} \quad \text{and} \quad \left(\frac{f}{g}\right)(x) = \frac{f(x)}{g(x)} = \frac{\sqrt{x+1}}{\frac{x-1}{x}} = \frac{x\sqrt{x+1}}{x-1}$$

For $(f+g)(x)$, x must be greater than or equal to -1 (so that $\sqrt{x+1}$ is defined), and x must not be 0 (so that $\frac{x-1}{x}$ is defined). Hence, the domain of $f+g$ is $[-1, \infty) - \{0\}$.

For $\left(\frac{f}{g}\right)(x)$, we must again have $x \ge -1$ and $x \ne 0$. However, we must now also be sure x is not 1, for then the denominator function would have the value 0. Thus the domain of $\frac{f}{g}$ is $[-1, \infty) - \{0, 1\}$.

From the graphs of $Y1 = f(x)$ and $Y2 = g(x)$ we have visual support that the x-values common to both of their domains is $[-1, \infty) - \{0\}$.

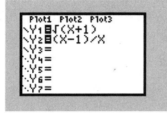
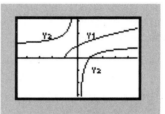

f and g using ZDecimal

From the following calculator table, we confirm that 0 is not in the domain of $f+g$ or $\frac{f}{g}$.

The x value of 1 is also excluded from the domain of $\frac{f}{g}$ so that the denominator is not zero.

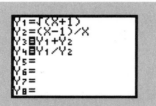
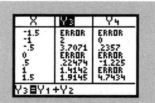

To see the domain from the graphs, trace along $f+g$ and $\frac{f}{g}$ in the viewing rectangle $[-4.7, 4.7] \times [-3.1, 6]$. Undefined y-values are blank.

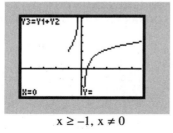
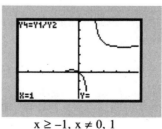

x ≥ -1, x ≠ 0   x ≥ -1, x ≠ 0, 1

Polynomial Functions

A function f is a polynomial function if $f(x)$ has the form

$$f(x) = a_n x^n + a_{n-1} x^{n-1} + \ldots + a_0,$$

where n is a nonnegative integer and the coefficients a_n, a_{n-1}, ... , a_0 are real numbers.

If $a_n \neq 0$, then f has **degree n**. Thus $p(x) = x^3 + 5x^2 + \dfrac{1}{2}$ and $q(x) = -x + 5x^{17}$ are

polynomials, but $r(x) = 3 - \dfrac{1}{x}$ and $s(x) = \sqrt{x}$ are not.

The domain of a polynomial function is the set of all real numbers, whereas the range varies and depends on the characteristics of the specific polynomial.

A number r is a **zero** of a polynomial p if $p(r) = 0$. Graphically speaking, real roots, x-intercepts and zeros are all the same thing.

Given the polynomial $y = x^5 - x = x(x+1)(x-1)(x^2+1)$, the following graph shows the connection among real roots, factors and intercepts.

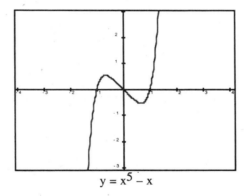

$$y = x^5 - x$$

Notice that for each linear factor there corresponds one zero; however, the quadratic factor $x^2 + 1$ is always positive, so it contributes no real zeros. A polynomial of degree n can have at most n zeros. In this example we see that a polynomial of degree n may have fewer than n real zeros.

Zeros and factors of a polynomial have an important connection: $(x - r)$ is a factor of polynomial $p(x)$ if and only if the number r is a zero of p.

Polynomial functions are extremely useful; for example, they can be employed to approximate non-polynomial functions. For instance, it can be shown that if x is close to 0, then $\sin x$ can be approximated closely by the polynomial function

$$p(x) = x - \frac{x^3}{6} \ .$$

In fact, values in most mathematical tables for non-polynomial functions like $\sin x$, e^x and $\ln x$ are computed using polynomial approximation methods.

Graphing Polynomial Functions

Polynomials of various degrees have graphs with characteristic shapes: lines, parabolas, cubic curves, etc.. In general, as the degree of a polynomial function increases it has more zeros and its graph exhibits more maximum and minimum points (turning points). Sometimes graphing higher degree polynomials can be made easier if they can be factored; however, in general, they are best left to a calculator or computer.

Example 3: *Approximating the Turning Point of a Function Graph*

Graph the function defined by $f(x) = x^3 - x^2 - 3x + 7$.
Approximate the coordinates of the turning point in the first quadrant.

Solution:

Press $\boxed{Y=}$. Enter the polynomial function f in Y1 and graph it. Press $\boxed{2nd}$ [CALC] to display the Calc menu and select [3:minimum]. Choose a Left and Right Bound and a Guess. Press $\boxed{ENTER}$ and the coordinates (1.3874243, 3.583498) of the turning point are displayed.

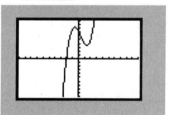 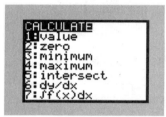 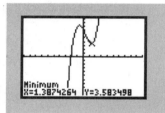

Rational Functions

If two polynomial functions are added or multiplied, the result is a polynomial function. On the other hand, the quotient of two polynomial functions is not a polynomial function. The quotient of two polynomial functions is called a **rational function.** For example, $f(x) = \dfrac{x-1}{x^2 + 3}$ is a rational function. In general, f is a rational function if it is expressible in the form

$$f(x) = \frac{p(x)}{q(x)}$$

where p and q are polynomial functions. The domain of a rational function consists of all real numbers for which the denominator $q(x)$ differs from zero. If in the rational function $f(x) = \dfrac{p(x)}{q(x)}$ the numerator and denominator polynomials $p(x)$ and $q(x)$ have no common factors, then the zeros of f are the same as the zeros of the numerator polynomial $p(x)$.

What makes rational functions different from polynomial functions is that rational functions often exhibit **asymptotic behavior.** A vertical asymptote is a vertical line $x = k$ such that if an argument x is chosen close to k the corresponding function value is very far from 0. A horizontal asymptote is a horizontal line $y = b$ with the property that if the argument x is very far from 0, then the corresponding value $f(x)$ is close to b.

The reciprocal function $f(x) = \dfrac{1}{x}$ is the simplest rational function; its graph has both a horizontal asymptote $(y = 0)$ and a vertical asymptote $(x = 0)$.

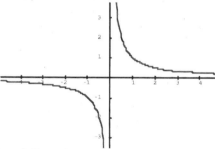

Example 4: *Graphing a Rational Function*

Graph the rational function f defined by $f(x) = \dfrac{x^2 - x}{x^2 - 1} = \dfrac{x(x-1)}{(x-1)(x+1)}$. State the domain and range of f. Determine what value $f(x)$ approaches asymptotically as $x \to \infty$ and as $x \to -\infty$.

Solution:

Press $\boxed{\text{Y=}}$. Enter $f(x)$ in Y1 and graph f in the ZDecimal viewing rectangle.

   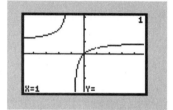

Since $x^2 - 1$ equals 0 when $x = 1$ *or* $x = -1$, we know that the domain of f consists of all real numbers except 1 and -1. In this example, the graph confirms this information about the domain of f. The range of f is all real numbers except 1 and $\dfrac{1}{2}$.

To investigate graphically the behavior of the values f as x gets larger, press $\boxed{\text{TRACE}}$. Hold down the $\boxed{\blacktriangleright}$ key and watch the cursor move along the curve to the right. Hold down the key until the cursor reaches the point with x-coordinate 20.00. Note that the y-coordinate is .95238095 and is approaching the value 1.

To investigate numerically the behavior of f as x gets larger, set up and display a table of function values for Y1 $= f(x)$.

Note in the table that as x gets larger, Y1 $= f(x)$ is approaching 1. Graphically and numerically investigating the left branch of the curve and noting that $f(-10000) = 1.00010001$ and $f(10000) = .99990001$, we conclude that the line $y = 1$ is a horizontal asymptote.

Even and Odd Functions

Some functions possess special properties that make it easy to sketch their graphs. Symmetries with respect to the vertical axis and the origin are examples of such special properties.

Consider the functions f and g defined by $f(x) = x^2$ and $f(x) = x^3$.

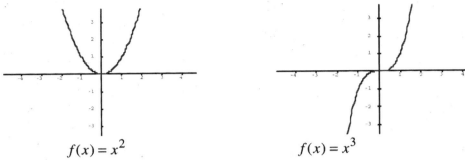

$$f(x) = x^2 \qquad\qquad\qquad f(x) = x^3$$

Comparing the graphs of f and g, we observe that the graph of the parabola $f(x) = x^2$ can be made to coincide with itself by reflecting the curve about the vertical axis; the graph is said to be symmetric with respect to the y-axis. On the other hand, the graph of $f(x) = x^3$ can be made to coincide with itself by rotating the graph $180°$ about the origin, and is therefore said to be symmetric with respect to the origin. The degree of f is **even** whereas the degree of g is **odd**.

In fact, every function of the form $f(x) = x^{2n}$ (where n is a positive integer) demonstrates the same symmetry as the specific example $f(x) = x^2$. In addition, every function of the form $g(x) = x^{2n-1}$ has the same symmetry as the particular function $f(x) = x^3$. This prompts the following definition:

> If f is a function whose domain contains $-x$ whenever it contains x, then
>
> a) f is **even** if $f(-x) = f(x)$ for all x in the domain of f;
>
> b) f is **odd** if $f(-x) = -f(x)$ for all x in the domain of f.

Example 5: *Showing that a Function is Odd*

Demonstrate that $f(x) = x^3 - 2x$ is an odd function.

Solution:

Graph $y = f(x)$ and $y = f(-x)$ in the ZDecimal viewing rectangle.

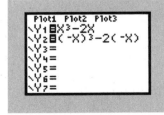 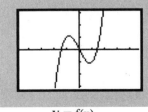 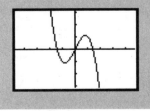

$$y = f(x) \qquad\qquad\qquad y = f(-x)$$

Since the graph of $y = f(x)$ can be made to coincide with itself by rotating the graph $180°$ about the origin, the function f is odd. We also observe that the graph of $y = f(-x)$ is the reflection of the graph of $y = f(x)$ about the *x*-axis. Thus, $f(-x) = -f(x)$ and, by definition, *f* is odd.

The Table feature shows numerical support of the fact that $f(-x) = -f(x)$ since $-Y_1(X) = Y_1(-X)$. For example, $-f(3) = f(-3) = 21$.

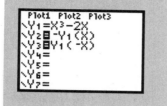

In this example you can also demonstrate algebraically that *f* is odd:

$$f(-x) = (-x)^3 - 2(-x)$$
$$= -x^3 + 2x$$
$$= -(x^3 - 2x)$$
$$= -f(x).$$

1.8   Exercises

1.   A table of values for functions f and g is given below.

x	$f(x)$
-1	3
0	4
1	-2
2	6
3	2
4	-1

x	$g(x)$
-1	3
0	1
1	-7
2	0
3	-1
4	2

Using the tables, find

a) $(f+g)(4)$

b) $(g-f)(-1)$

c) $(f \cdot g)(1)$

d) $(\frac{f}{g})(0)$

e) all x such that $(f-g)(x) = 3$

2.   The functions f and g are defined by $f(x) = \dfrac{1-2x}{x}$ and $g(x) = \dfrac{x}{x-2}$. Find

a) the zeros and domain of f;

b) the zeros and domain of g;

c) the domain of $f \cdot g$;

d) a simplified formula for $f \cdot g$.

3.   Graph each of the following functions with your calculator. Determine from the graph whether the function is odd, even, or neither. Support answers numerically using tables of values and algebraically using the definitions.

a) $f(x) = 2x^2 - 1$

b) $f(x) = \dfrac{2x}{x^2 - 1}$

c) $f(x) = x - x^3$

d) $f(x) = x^4 - x^2 + 1$

e) $f(x) = \dfrac{x^2}{x^3 - 1}$

f) $f(x) = \dfrac{x}{x^3 + 1}$

4.   Graph each of the following rational functions. Use the graph and table of values to determine the (approximate) domain and range of the function and to investigate the behavior of the value $f(x)$ as $x \to \infty$ and as $x \to -\infty$.

a) $f(x) = \dfrac{x}{9 - 7x + x^2}$

b) $f(x) = \dfrac{2x^2 + 3}{x^2 - 5}$

5.   Find a polynomial of degree 5 with $x = -1$, $x = 2$ and $x = 3$ as its only real roots. Sketch its graph.

6.   Given the cubic polynomial functions

$$\text{i) } f(x) = x^3 + x + 1$$

$$\text{ii) } f(x) = x^3 - 2x^2$$

$$\text{iii) } f(x) = x^3 + 2x^2 - x - 2$$

a) Using the TI-83, plot graphs of each function.

b) Do cubic functions always have turning points?

c) Using [2:zero] from the Calc menu, approximate (correct to two decimal places) the zeros of each function.

d) It is possible for a cubic function to have exactly two zeros. Try to create a rule for such a cubic function.

e) Graphs of cubic functions always have a **point of inflection** where the graph stops bending one way and starts bending the other way. If

$$f(x) = ax^3 + bx^2 + cx + d,$$

then the x-coordinate of the point of inflection is $x = \dfrac{-b}{3a}$. Calculate this x-value for the functions f, g, and h above. When each graph is displayed, select [4:Vertical] from the Draw Draw menu and draw a vertical line at these values of x. What symmetry do you see?

7.   Solve each of the equations below graphically. In each case, define function Y_1 to be the expression on the left of the equation, define function Y_2 to be the expression on the right, and define function Y_3 to be $Y_1 - Y_2$. Graph only the function Y_3 and find its zeros.

a) $|x - 1| = x^2 - 4x + 3$ b) $x^3 - 4x + 1 = x^4 - 1$ c) $\left|x^2 - 1\right| = x^3 - 3x - 2$

8.   Graph the function defined by $f(x) = \dfrac{1}{x^2 - x - 6}$ in the ZDecimal viewing rectangle.

a) Use the tracing function to determine the values of x for which $f(x)$ is not defined.

b) Use the result in part a) to factor $x^2 - x - 6$.

c) Locate any asymptotes of f.

9.   a) Graph $Y_1 = -2x^2 + 7x + 4$ and $Y_2 = -Y_1$.

b) How do the zeros of Y_1 compare with the zeros of Y_2?

c) In general, how are the graphs of f and $-f$ related?

10. Find possible polynomial functions whose graphs are sketched in a $[-4.7, 4.7] \times [-3.1, 3.1]$ viewing rectangle.

a) 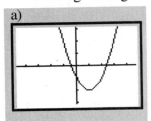   b) 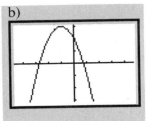   c) 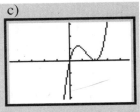   d)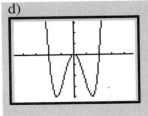

11. Determine a rational function that has zeros at -2 and 3, vertical asymptotes at $x = 2$ and $x = -1$, and a horizontal asymptote at $y = 1$. Check the function with your calculator.

12. Find a rational function with vertical asymptotes at $x = -1$ and $x = 2$ and with the line $y = 3$ as a horizontal asymptote. Check your function with your calculator.

13. Compare and contrast the behavior of the graphs of the rational functions

$$f(x) = \frac{x^2 - 1}{x + 1} \quad \text{and} \quad g(x) = \frac{x^2 - 2}{x + 1}.$$

14. a) Use a calculator to plot graphs of the functions

$$f_n(x) = x^n \quad \text{for} \quad n = 1, 2, 3, 4, 5, 6$$

on the interval $[-1.2, 1.2]$. Describe the behavior of $f_n(x)$ as n increases for values of x near $-1, 0$ and 1.

b) Do the same for $g_n(x) = x^{1/n}$.

15. Any three points on the graph of a quadratic function $y = ax^2 + bx + c$ will uniquely determine the coefficients a, b and c. Use the STAT menu to find a, b and c so that the graph of $y = ax^2 + bx + c$ passes through the points $(1, 8), (-1, 2)$ and $(3, 6)$. Proceed as follows:

i) Before entering points in the lists L1 and L2, clear the lists by employing the [4:ClrList] command from the STAT menu. Execute ClrList L1, L2 on the Home screen.

ii) Select [1:Edit] from the STAT menu and enter inputs in list L1 and outputs in list L2.

L1	L2	L3
1	8	------
-1	2	
3	6	
------	------	

L1(1)=1

iii) To calculate the quadratic regression equation, key in the sequence ⟦STAT⟧ ⟦▶⟧ [5:QuadReg] L1 ⟦,⟧ L2 ⟦,⟧ ⟦VARS⟧ ⟦▶⟧ [1:Function] [1:Y1] ⟦ENTER⟧. The regression equation is stored in Y1.

16.   Any four points on the graph of a cubic function $y = ax^3 + bx^2 + cx + d$ will uniquely
        determine the coefficients a, b, c and d. Use the STAT menu and follow the steps in the
        previous exercise to find the cubic function that passes through the points $(-3, 0)$, $(-2, 0)$,
        $(3, 3)$ and $(6, 0)$.

17.   a) For all x near 0, the polynomial function $p(x) = x - \dfrac{x^3}{6}$ is a good approximation of
        the sine function. How well does $p(x)$ approximate $\sin x$ on the interval $[0, 1]$? On
        the interval $[0, 2]$? For what values of x can we be sure that $p(x)$ differs from $\sin x$
        by less than 0.01? Use graphical and numerical solutions.

        b) Show graphically that the function $p(x) = \dfrac{5 + 15x - 5x^2 + x^3}{16}$ approximates
        $f(x) = \sqrt{x}$ closely near $x = 1$. Also use the Table feature to investigate.

18.   The following pairs of polynomial functions have "reversed coefficients".

        i) $f(x) = x^2 + 2x - 3$ $g(x) = -3x^2 + 2x + 1$

        ii) $f(x) = 6x^2 - 5x - 4$ $g(x) = -4x^2 - 5x + 6$

        iii) $f(x) = x^3 - 2x^2 + x - 1$ $g(x) = -x^3 + x^2 - 2x + 1$

        a) Find the zeros of f and g for each pair above. What is the relationship between the
        zeros of f and g?

        b) Does the relationship, for example, hold for quadratic functions in general? Prove
        your answer.

19.   Consider the functions $g(x) = x^2$ and $g(x) = 2^x$.

        a) How many positive solutions are there to the equation $f(x) = g(x)$?

        b) In the viewing rectangle $0 \le x \le 3$, $0 \le y \le 10$, which function is growing faster?

        c) In the viewing rectangle $0 \le x \le 8$, $0 \le y \le 100$, which function is growing faster?

        d) What are the positive solution(s) to $f(x) = g(x)$?

20.   Consider the functions $f(x) = x^4$ and $g(x) = 3^x$.

        a) On the interval $0 \le x \le 5$, which function grows faster?

        b) On the interval $0 \le x \le 10$, which function grows faster?

        c) Find the positive solution(s) to $f(x) = g(x)$ to the nearest hundredth.

        d) When $x = 3$, which function has the larger value?

        e) When $x = 10$, which function has the larger value?

21.   Prove that if f and g are both even functions, then their sum $f + g$ is also even.

22.   Prove that the sum of two odd functions is odd.

23.   Prove that the product of two even functions is even.

24.   What can you say about the product of two odd functions?

1.9   Composition of Functions

There is an operation on functions, called **composition**, that has no counterpart in the algebra of numbers; however, it is by far the most important function operation. Consider the function defined by

$$h(x) = (x - 4)^2 .$$

To evaluate $h(x)$ for a given input x, we would first compute $x - 4$ and then square the result. If we consider the functions $g(x) = x - 4$ and $f(x) = x^2$, then

$$h(x) = (x - 4)^2 = [g(x)]^2 = f[g(x)].$$

Note that h has the same effect as g and f evaluated successively. We can think of h as being "composed" of two functions g and f where we use the output of one function as the input for the other.

If oil is spreading from a leaking offshore well, the area of the approximately circular oil slick will grow with time. Suppose a function g, whose graph is given in the first figure, gives the radius as a function of time, $r = g(t)$. Time is measured in hours and the radius is measured in miles. Suppose a second function f, whose graph is given in the second figure, gives the area of the circular oil slick as a function of its radius, $a = f(r)$. The area is measured in square miles.

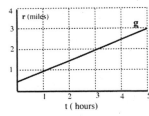
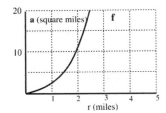

The graphs can be used to find the area of the oil slick after 3 hours. Using the graph of g we find the radius is 2 miles when $t = 3$ hr. And, using the graph of f, a radius of 2 miles is paired with approximately 12 sq miles. Thus, given two functions, a solution is found by using the output from one function as the input in the other function. This is an example of function composition.

Function composition is defined as follows.

If f and g are functions, then their **composite** $f \circ g$ is the function with

$$[f \circ g](x) = f[g(x)]$$

for each x in the domain of g such that $g(x)$ is in the domain of f.

The function g is called the **inner** function, and its input is x;
f is the **outer** function and its input is $g(x)$.

We can use a "machine diagram" to illustrate function composition.

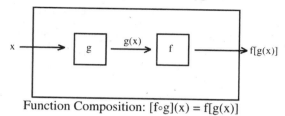

Function Composition: $[f \circ g](x) = f[g(x)]$

Example 1: *Finding Composites of Functions*

If $f(x) = 2x + 1$ and $g(x) = \dfrac{1}{x}$, find $[f \circ g](x)$ and $[g \circ f](x)$.
Describe the domain of each composite function.

Solution:

$$[f \circ g](x) = f[g(x)] = f\left(\frac{1}{x}\right) = \frac{2}{x} + 1.$$

The domain of $f \circ g$ is the set of all real numbers except 0.

$$[g \circ f](x) = g[f(x)] = g(2x + 1) = \frac{1}{2x + 1}.$$

The domain of $g \circ f$ is the set of all real numbers except $-\dfrac{1}{2}$ since the domain cannot contain any number that makes $f(x) = 0$.

Below are the graphs of $y = f[g(x)]$ and $y = g[f(x)]$ in the ZDecimal viewing rectangle.

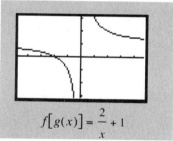

$$f[g(x)] = \frac{2}{x} + 1$$

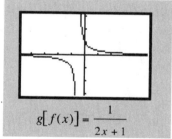

$$g[f(x)] = \frac{1}{2x + 1}$$

Note that $f \circ g$ and $g \circ f$ are different functions. Thus the order in which functions are composed makes a difference.

To check the accuracy of the algebraically determined formula for $f[g(x)]$, use the TI-83's functional notation capabilities and the Table feature.

Plot1	Plot2	Plot3
\Y₁	= 2X+1	
\Y₂	= 1/X	
\Y₃	= Y₁(Y₂(X))	
\Y₄	= 2/X+1	
\Y₅	=	
\Y₆	=	
\Y₇	=	

X	Y₃	Y₄
-3	.33333	.33333
-2	0	0
-1	-1	-1
0	ERROR	ERROR
1	3	3
2	2	2
3	1.6667	1.6667

Y₄=2/X+1

     To analyze a function with a complicated formula it may be helpful to think of it as the composite of simpler functions. Determining the simpler functions is made easy if you consider the order in which computations must be performed in the evaluation of the composite function.

For example, let h be the function defined by $h(x) = \sqrt{\sin(2x)}$. Suppose we were to evaluate this function at $x = \pi$. First we would have to double π, then use 2π as input for sine and, finally, take the square root to get our final output $h(x) = \sqrt{\sin(2x)}$. Note that if three functions k, f and g are defined such that

$$g(x) = 2x$$
$$f(x) = \sin x$$
$$k(x) = \sqrt{x}$$

then

$$h(x) = \big[k \circ f \circ g\big](x) = k\big[f(g(x))\big] = \sqrt{\sin(2x)}$$

Function Inverses Revisited   (See page 39)

The **identity function** $I(x) = x$ or $y = x$, the function that assigns each number to itself, has an important connection to function inverses.

      The word inverse is always used in reference to some operation and some identity element for that operation. For example, the additive inverse of any real number x is $-x$, since $x + (-x) = 0$ and 0 is the identity element for real number addition. Similarly, the multiplicative inverse of a nonzero real number x is its reciprocal $\dfrac{1}{x}$, since $x \cdot \dfrac{1}{x} = 1$ and 1 is the identity element for real number multiplication.

      The inverse function f^{-1} is an inverse in the same sense. The operation is function composition, and the identity is the identity function $I(x) = x$. For example, if $f(x) = 2x - 5$ and $f^{-1}(x) = \dfrac{x + 5}{2}$, then

$$\big[f \circ f^{-1}\big](x) = f\big[f^{-1}(x)\big] = 2\left(\frac{x + 5}{2}\right) - 5 = x.$$

In general, for a one-to-one function f and its inverse f^{-1}

> 1) $\big[f^{-1} \circ f\big](x) = x$ for all x in the domain of f
>
> 2) $\big[f \circ f^{-1}\big](x) = x$ for all x in domain of f^{-1} .

Thus, for suitable x,

$$\big[f^{-1} \circ f\big](x) = \big[f \circ f^{-1}\big](x) = I(x) \text{ where } I \text{ is the identity function } I(x) = x.$$

Example 2:   *Showing that Functions are Inverses of each Other*

Demonstrate that $f(x) = \ln(3x + 1)$ and $g(x) = \dfrac{e^x - 1}{3}$ are inverses of each other by showing that their composite is the identity function.

Solution:

$$[f \circ g](x) = f[g(x)] = \ln\left[3\left(\frac{e^x - 1}{3}\right) + 1\right] = \ln(e^x) = x$$

and

$$[g \circ f](x) = g[f(x)] = \frac{e^{\ln(3x+1)} - 1}{3} = \frac{(3x + 1) - 1}{3} = x, \quad \text{for all } x > -\frac{1}{3}$$

Example 3:   *Finding the Inverse of a Function*

Use the fact that $f\!\left(f^{-1}(x)\right) = x$ to find $f^{-1}(x)$ if $f(x) = \dfrac{2}{3x + 1}$.

Solution:

$$f\!\left(f^{-1}(x)\right) = x$$

$$\frac{2}{3f^{-1}(x) + 1} = x$$

$$2 = 3x \cdot f^{-1}(x) + x$$

$$2 - x = 3x \cdot f^{-1}(x)$$

$$f^{-1}(x) = \frac{2 - x}{3x}.$$

Caution: The function f^{-1} must not be confused with the function $\dfrac{1}{f}$. For example, if $f(x) = x^3$, then

$$f^{-1}(x) = x^{1/3}, \quad \text{whereas} \quad \left(\frac{1}{f}\right)(x) = \frac{1}{x^3}.$$

In particular,

$$f^{-1}(8) = 8^{1/3} = 2, \quad \text{whereas} \quad \left(\frac{1}{f}\right)(8) = \frac{1}{8^3} = \frac{1}{512}.$$

1.9   Exercises

1.   Each function below has the interval $[-5, 5]$ as its domain.

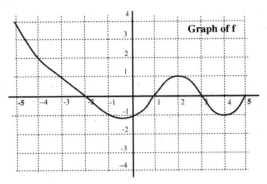 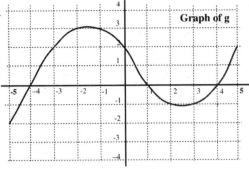

Find

a) $[f \circ g](-2)$ b) $[f \circ g](2)$ c) $[g \circ f](-1)$

d) $[f \circ f](5)$ e) $[g \circ g](-2)$ f) all inputs x for which $g[g(x)] = -1$.

2.   Given tables for functions f and g.

x	$f(x)$
-1	2
0	4
1	3
2	0
3	1
4	-1

x	$g(x)$
-1	3
0	4
1	2
2	6
3	2
4	-1

Find

a) $[f \circ g](3)$ b) $[g \circ f](2)$ c) $[f \circ f](4)$

d) $[g \circ g](4)$ e) all inputs x such that $[f \circ g](x) = 2$

3.   Tables of values for functions f and g are given below.

x	$f(x)$
-2	0
0	1
2	-1
4	2

x	$g(x)$
-1	-2
0	2
1	3
2	1

Find

a) $[f \circ f](-2)$ b) $[f \circ g](0)$ c) $[g \circ f](1)$

d) the domain of $f \circ g$ e) the domain of $g \circ f$

4.   Let $f(x) = 2x - 3$, $g(x) = e^x$ and $h(x) = \ln x$. Find a formula for each function and specify its domain.

a) $f \circ f$ b) $f \circ g$ c) $g \circ h$ d) $h \circ g$ e) $h \circ h$

5.   Let $f(x) = x^3$, $g(x) = 5x + 1$, and $h(x) = 2^x$. Find a formula for each function and specify its domain.

   a) $f(g(x))$ b) $h(f(x))$ c) $h(g(x))$ d) $g(h(x))$ e) $g(f(x))$ f) $f(g(h(x)))$

6.   In each of the following, write formulas for $f(x)$ and $g(x)$ so that $h(x) = f[g(x)]$.

   a) $h(x) = (x + 4)^3$ b) $h(x) = e^{x-1}$ c) $h(x) = \ln(2x + 5)$

   d) $h(x) = \dfrac{1}{(2x - 1)^2}$ e) $h(x) = \sqrt{x + 3} - \sqrt[3]{x + 3}$ f) $h(x) = e^{\ln(2x+3)}$

7.   For the composition $[f \circ g](x) = f(g(x))$, g is known as the inner function and f is the outer function. For each of the problems below, specify inner and outer functions.

   a) $y = (7x - 3)^3$ b) $y = \ln(x^2 + 4)$ c) $y = 2^{3x-5}$

   d) $y = e^{\ln x}$ e) $y = 7e^x + 13$ f) $y = (\ln x)^2 + 1$

8.   For the composition $[f \circ g](x) = f(g(x))$, g is the inner function and f is the outer function. For each of the compositions below, identify inner and outer functions.

   a) $y = (x^2 - 4)^3$ b) $y = e^{x^2+2}$ c) $y = \ln(x - 4)$

   d) $y = 2 \cdot 3^x - 4$ e) $y = \dfrac{1}{2x - 1}$ f) $y = \dfrac{2}{(x - 7)^2}$

9.   Verify that $[f \circ g](x) = I(x)$ and $[g \circ f](x) = I(x)$, where I is the identity function whose rule is $I(x) = x$.

   a) $f(x) = 2x - 3$, $g(x) = \dfrac{x + 3}{2}$ b) $f(x) = \sqrt[3]{x} + 2$, $g(x) = (x - 2)^3$

   c) $f(x) = \ln(2x + 1)$, $g(x) = \dfrac{e^x - 1}{2}$ d) $f(x) = \dfrac{1}{x}$, $g(x) = \dfrac{1}{x}$

   e) $f(x) = x^5 - 3$, $g(x) = \sqrt[5]{x + 3}$

10.   a) If $[f \circ g](x) = \ln(x^2 + 1)$ and $f(x) = \ln x$, what is $g(x)$?

   b) If $[f \circ g \circ h](x) = e^{x^2+1} - 7$, and $f(x) = x - 7$, determine $g(x)$ and $h(x)$.

11.   Specify how the graphs of the functions below are each related to the graph of $f(x) = e^x$.

   a) $y = e^x + 1$ b) $y = -e^{x-2}$ c) $y = 2e^x - 3$ d) $y = -e^{-x}$

12.   If $f(x) = x - 2$, $g(x) = x + 3$, $h(x) = \ln x$, create compositions of f, g, and h to shift the graph of h as specified below.

       a) Shift left 3   b) Shift down 2

       c) Reflect through x-axis then shift up 3   d) Shift right 2 and up 3

13.   Use the fact that $\left(f \circ f^{-1}\right)(x) = f\left(f^{-1}(x)\right) = x$ to find f^{-1} in each case below.

       a) $f(x) = 5x - 2$ b) $f(x) = e^{2x-1}$ c) $f(x) = \ln\left(x^3 + 4\right)$ d) $f(x) = \dfrac{1}{x - 1}$

14.   In each case below, sketch f. Then sketch f^{-1} using the fact that f^{-1} is the reflection of f over the line $y = x$. Guess a rule for f^{-1} in each case.

       a) $f(x) = e^x + 1$ b) $f(x) = 2x + 4$ c) $f(x) = (x - 2)^3$ d) $\ln x - 1$

15.   Let $f(x) = x^2 - 4$, $g(x) = x - 2$, and $h(x) = |x|$. Sketch a graph of each of the following functions.

       a) $y = \left[f \circ g\right](x)$ b) $y = \left[g \circ f\right](x)$ c) $y = \left[h \circ g\right](x)$

       d) $y = \left[h \circ f \circ g\right](x)$ e) $y = \left[h \circ g \circ f\right](x)$

16.   a) Is the composition of two linear functions a linear function? Justify.

       b) Is the composition of two odd functions an odd function? Justify.

       c) Is the composition of two polynomial functions a polynomial function? Justify.

       d) Is the composition of two quadratic functions a quadratic function? Justify.

1.10   Trigonometric Functions

Many natural phenomena repeat in regular patterns or periods. For example, there is the recurring pattern in the rising and setting of the sun, the changing of seasons, the circulation of blood through the heart, and the rise and fall of tides. The functions that are employed to describe or model such phenomena also have a periodic or repeating pattern.

A non-constant function f is periodic if there is a positive number p, called **a period of** f, such that

$$f(x + p) = f(x)$$

for all x in the domain of f. It follows that a function with period p is completely defined if we know its values for all arguments in one interval of length p. The most important examples of periodic functions are the trigonometric functions.

The graphs of the two basic trigonometric functions – the sine and cosine – are sketched below.

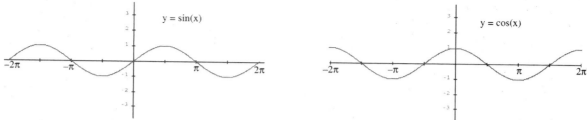

The domain of each function is the set of all real numbers and the range of each is the interval $[-1, 1]$. Note that the period of sine is 2π. The graph of cosine is identical to the graph of sine, shifted left by $\frac{\pi}{2}$ radians. Thus the period of cosine is also 2π. Observe that the graph of sine is symmetric about the origin. It is an odd function: $\sin(-x) = -\sin x$. On the other hand, the graph of cosine is symmetric about the y-axis; it is an even function: $\cos(-x) = \cos x$.

Unit Circle and the Graphs of Sin and Cos

The two basic circular functions, **sine** and **cosine**, are defined using the **unit circle**; a circle of radius 1 centered at the origin with equation $x^2 + y^2 = 1$. For any real number t, let $P(t)$ be the point on the unit circle obtained by moving a distance $|t|$ from the point $A(1, 0)$, measured along the circle in a counterclockwise direction if $t > 0$ and in a clockwise direction if $t < 0$. If $P(t)$ has coordinates (x, y), then the cosine of t and the sine of t (abbreviated by cos t and sin t) are defined by

$$\cos t = x \quad \text{and} \quad \sin t = y.$$

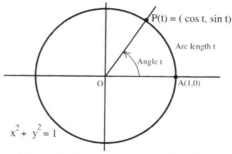

The definition of cos t and sin t for $t > 0$.

In the figure above, the **radian measure** of angle AOP is the real number t. According to this definition, any real number can be interpreted as an angle measure, and its cosine and sine can be determined. Since the circumference of the unit circle is 2π, the relationship between radian and degree measure is $2\pi = 360^\circ$, and the conversion formulas between degrees and radians are:

$$1^\circ = \frac{\pi}{180} \text{ radians} \qquad \text{and} \qquad 1 \text{ radian} = \frac{180^\circ}{\pi}.$$

Notice that if the point $P(t)$ lies in the first quadrant, a right triangle is formed, one of whose acute angles is t radians.

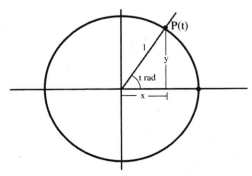

Now it is easy to see that the unit circle definition agrees with the right triangle definition of sine and cosine. For example,

$$\cos t = \frac{\text{adjacent}}{\text{hypotenuse}} = \frac{x}{1} \quad \text{and} \quad \sin t = \frac{\text{opposite}}{\text{hypotenuse}} = \frac{y}{1}.$$

The relationship between the circular functions and the unit circle is easily visualized through the use of parametric equations and a graphing calculator. Enter the parametric equation form for the unit circle, $X1T = \cos T$ and $Y1T = \sin T$ as one function and $X2T = T$ and $Y2T = \cos T$ for the second function. Then set the following window parameters: Tmin = 0, Tmax = 2π, Tstep = $\pi/12$, Xmin = -2, Xmax = 2π, Xscl = $\frac{\pi}{2}$, Ymin = -2.43, Ymax = 2.43 and Yscl = 1 and graph.

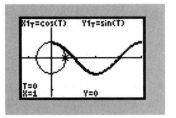

By using the TRACE feature and the ▶ and ◀ keys we can move the cursor around the unit circle. The values of the sine and cosine of the angle in radian measure are displayed at the bottom of the screen. To show the corresponding point on the cos curve, select curve 2 by pressing ▼. To return to the corresponding point on the unit circle, press ▲.

Trigonometric functions can be evaluated in one of two ways: decimal approximations with a calculator or, in certain special cases, exact values using formulas from geometry.

The table below lists exact values of sine and cosine for some frequently used inputs. We assume that you have encountered these special inputs before and that you have learned, or can quickly figure out, how they are calculated.

t	0	$\dfrac{\pi}{6}$	$\dfrac{\pi}{4}$	$\dfrac{\pi}{3}$	$\dfrac{\pi}{2}$	$\dfrac{2\pi}{3}$	$\dfrac{3\pi}{4}$	$\dfrac{5\pi}{6}$	π	$\dfrac{3\pi}{2}$	2π
$\cos t$	1	$\dfrac{\sqrt{3}}{2}$	$\dfrac{\sqrt{2}}{2}$	$\dfrac{1}{2}$	0	$-\dfrac{1}{2}$	$-\dfrac{\sqrt{2}}{2}$	$-\dfrac{\sqrt{3}}{2}$	-1	0	1
$\sin t$	0	$\dfrac{1}{2}$	$\dfrac{\sqrt{2}}{2}$	$\dfrac{\sqrt{3}}{2}$	1	$\dfrac{\sqrt{3}}{2}$	$\dfrac{\sqrt{2}}{2}$	$\dfrac{1}{2}$	0	-1	0

The remaining trigonometric functions – **tangent**, **cotangent**, **cosecant**, and **secant** –are defined in terms of sine and cosine.

$$\tan x = \frac{\sin x}{\cos x} \qquad\qquad \cot x = \frac{\cos x}{\sin x}$$

$$\sec x = \frac{1}{\cos x} \qquad\qquad \csc x = \frac{1}{\sin x}$$

Note that there are domain restrictions implied by these definitions. For example,

$$\text{domain } \tan = \left\{ x \mid \cos x \neq 0 \right\} = R - \left\{ \frac{\pi}{2} + k\pi; \ \ k \ \text{ is any integer} \right\}$$

Identities

There are many equations, called trigonometric identities, that describe connections between the various trigonometric functions. The most commonly used ones are summarized here.

Pythagorean Identities
$$\sin^2(x) + \cos^2(x) = 1 \qquad\qquad \tan^2(x) + 1 = \sec^2(x)$$

Double-angle Identities
$$\sin(2x) = 2\sin x \cos x \qquad\qquad \cos(2x) = \cos^2(x) - \sin^2(x)$$

Half-angle Identities
$$\sin^2\left(\frac{x}{2}\right) = \frac{1 - \cos x}{2} \qquad\qquad \cos^2\left(\frac{x}{2}\right) = \frac{1 + \cos x}{2}$$

Sum and Difference Identities
$$\sin(x + y) = \sin(x)\cos(y) + \cos(x)\sin(y)$$

$$\sin(x - y) = \sin(x)\cos(y) - \cos(x)\sin(y)$$

$$\cos(x + y) = \cos(x)\cos(y) - \sin(x)\sin(y)$$

$$\cos(x - y) = \cos(x)\cos(y) + \sin(x)\sin(y)$$

Many periodic phenomena can be modeled using the sine and cosine functions; however, most periodic phenomena do not have period 2π and oscillate between 1 and −1. Consequently, it is necessary to incorporate horizontal and vertical stretches, compressions and shifts into our model functions.

Example 1:

Describe how the function $f(x) = 2\sin\left(2x - \dfrac{\pi}{3}\right)$ can be obtained from the graph of $y = \sin x$ by applying transformations.

Solution:

Notice that $f(x)$ can be rewritten as $f(x) = 2\sin\left(2x - \dfrac{\pi}{3}\right) = 2\sin\left[2\left(x - \dfrac{\pi}{6}\right)\right]$.

Apply the following transformations to the graph of $y = \sin x$ to obtain the graph of f.

1) A vertical stretch by a factor of 2 to obtain $y = 2\sin x$.

2) A horizontal compression by a factor of 2 to obtain $y = 2\sin(2x)$.

3) A horizontal shift of $\dfrac{\pi}{6}$ units to the right to obtain $f(x) = 2\sin\left[2\left(x - \dfrac{\pi}{6}\right)\right]$.

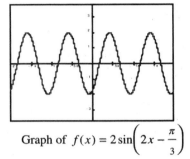

Graph of $f(x) = 2\sin\left(2x - \dfrac{\pi}{3}\right)$

The TI-83 command SinReg in the STAT CALC menu may be employed to fit a function to periodic data.

Example 2:

The average monthly temperature for the city of Atlanta, Georgia is given below.

Jan	Feb	Mar	Apr	May	Jun	Jul	Aug	Sep	Oct	Nov	Dec
44.7	46.1	51.4	60.2	69.1	76.6	78.9	78.2	73.1	62.4	51.2	44.8

Number the months 0, 1, ..., 11 and plot the data. Use sine regression to obtain a periodic function that fits these data.

Solution:

The TI-83 command SinReg (item C:SinReg in the STAT CALC menu)may be used to fit the model function $y = a\sin(bx + c) + d$ to the periodic data above.

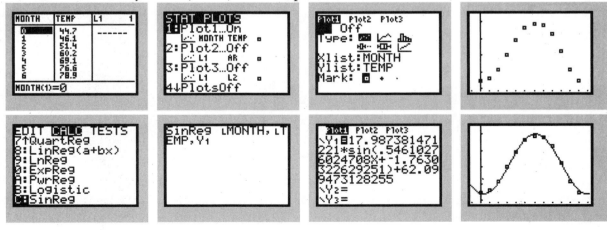

1.10   Exercises

1.   a) Graph the following functions in the viewing rectangle $-\pi \leq x \leq \pi$ and $-4 \leq y \leq 4$.

   i) $y = \sin x$ ii) $y = 2\sin x$ iii) $y = 3\sin x$ iv) $y = \dfrac{1}{2}\sin x$

   b) How does the graph of $y = A\sin x$ $(A > 0)$ change as A is increased? decreased?
      Note: $|A|$ is called the **amplitude** of the graph.

2.   Graph each of the following functions in the viewing window $-\pi \leq x \leq \pi$ and $-2 \leq y \leq 2$.

   i) $y = \sin x$ ii) $y = \sin(2x)$ iii) $y = \sin(3x)$ iv) $y = \sin(\dfrac{1}{2}x)$

   a) How many complete periods occur in the viewing rectangle for each function?

   b) How does the graph of $y = \sin(Bx)$ compare with the graph of $y = \sin x$ if B is a
      positive integer? What about $y = \sin(\dfrac{x}{B})$?

   c) What is the period of $y = \sin(Bx)$?

3.   Graph the following functions using the $\boxed{\text{ZOOM}}$ [7: Trig] setting.

   i) $y = \sin x$ ii) $y = \sin(x + \dfrac{\pi}{4})$ iii) $y = \sin(x + \dfrac{\pi}{3})$

   a) Describe what happens to the graph of $f(x) = \sin(x + C)$ as C is increased $(C > 0)$.

   b) Describe what happens when $C < 0$ and C is decreased.

4.   Predict the domain, range and period of the following functions. Sketch a graph on the
      calculator to check your answer.

   a) $y = 2\sin\left(x - \dfrac{\pi}{4}\right)$ b) $y = -\cos(2x - \pi)$

   c) $y = -2\sin(\pi x + \pi)$ d) $y = 6\cos\left(3x + \dfrac{\pi}{2}\right)$

5.   Is applying a vertical stretch factor of 2 to the graph of $y = \cos x$ followed by a vertical
      shift up 3 units the same as applying a vertical shift up 3 units to the graph of $y = \cos x$
      followed by a vertical stretch factor of 2? Explain briefly.

6.   Find a possible formula for each graph.

a) 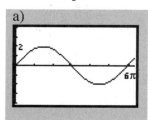 b)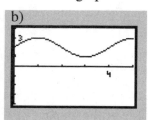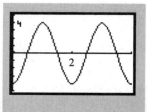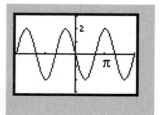

7.   Find the number of solutions for the following equations.

     a) $\sin x = 0.1x$ b) $\sin x = 1.11 \cdot \log x$

8.   Create a display for visualizing the relationship between points on the unit circle and the tan function similar to that shown for the cos function in the text.

9.   Given the function $f(x) = x \sin x$.

     a) Determine the zeros of f in the interval $[0, 2\pi]$.

     b) For what arguments a does $f(a) = a$?

10.   A ball is bouncing up and down. Its height (feet) above the ground at time x (sec) is given by

$$y = |3\cos(4\pi x)|$$

How many times does the ball hit the ground from $x = 0$ to $x = 4$?

11.   Sketch appropriate graphs to estimate solutions of the following.

     a) $\cos x = \sin x$ b) $\cos x \le \sin x$ c) $\cos x + \sin x = 1$

     d) $\cos x \cdot \sin x > 0$ e) $\sin x = \dfrac{x}{2}$ f) $\cos x = \dfrac{x}{2}$

12.   On January 1, 1992, high tide in Seattle was at midnight. The water level at high tide was 9.9 feet and, at time t hours, the height h of the water was given by

$$h(t) = 5 + 4.9\cos\left(\frac{\pi t}{6}\right).$$

Approximate the time periods in the next 24 hours when the height of the water was above 6 feet.

13.   The number y of daylight hours on the xth day of the year in Boston is given by

$$y = 11.7 + 2.3\sin\left(\frac{2\pi(x-81)}{365}\right).$$

a) Graph this function, using the window settings $1 \le x \le 365$ and $5 \le y \le 15$.

b) Approximate the dates when there is a maximum and a minimum number of hours of daylight.

14.   The populations of some kinds of predatory animals vary periodically. The wolf population in a national park over a 11 year period is listed below.

Year	0	1	2	3	4	5	6	7	8	9	10
Pop	129	186	288	379	407	356	256	162	127	171	266

Plot the data and use the SinReg feature on your TI-83 to find a function to fit the data.

15.   Use an isosceles right triangle with hypotenuse 1 to verify the values given in text for the sine and cosine of $\frac{\pi}{4}$.

16.   In a right triangle with hypotenuse 2 and acute angles with measures of 30° and 60°, why are the sides opposite these angles 1 and $\sqrt{3}$ respectively? Use the triangle to verify the values given in the text for sine and cosine of $\frac{\pi}{6}$ and $\frac{\pi}{3}$.

17.   The order in which transformations are applied to the graph of the function f is specified. Give an equation for the transformed function.

a) $f(x) = \sin x$;   Vertical stretch by a factor of 3; a reflection through the x-axis; a horizontal compression by a factor of 3; a horizontal shift right $\frac{\pi}{3}$ units.

b) $f(x) = \cos x$;   A reflection in the x-axis; a horizontal stretch by a factor of 2: a horizontal shift left $\frac{\pi}{4}$ units.

Chapter 1   Supplementary Problems

1. True/False

_____a) $\log 45 - 2\log 3 = \log 5$

_____b) The function f defined by $f(x) = -2\sin\left(\dfrac{1}{3}x\right)$ has amplitude 2 and period 6π.

_____c) $f(x) = x^2 - 2x$, then $f(a+1) = a^2 + 3$.

_____d) If the function g is odd and $g(-2) = 5$ then $g(2) = -5$.

_____e) $e^{\ln(1/2)} - e^{3\ln 2} = \dfrac{7}{2}$

2.   Given $f(x) = 2x - 3$ and $g(x) = \log_2 x$, evaluate

    a) $f[g(4)] =$ b) $g[f(2)] =$

    c) $f\left[g^{-1}(0)\right] =$ d) $g\left[f^{-1}(1)\right] =$

3.   Find a possible formula for each of the following graphs.

a) 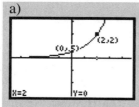   b) 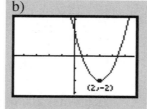   c) 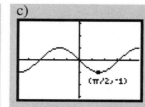   d)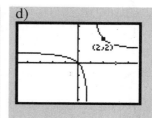

4.   The order in which the transformations are applied to the graph of the given function is specified. Give an equation for the transformed function in each case.

    a) $f(x) = |x|$; vertical stretch 3 and shift up 4.

    b) $g(x) = \ln x$; vertical shrink by $\dfrac{1}{2}$, shift left 2 and shift down 1.

    c) $h(x) = e^x$; reflect through the x-axis, shift right 1 and shift down 1.

5.   The function f is defined by $f(x) = \dfrac{x-1}{2x^2 - 8}$.

    a) Determine the x- and y-intercepts, if any.

    b) Write equations for any horizontal and vertical asymptotes.

    c) Sketch a graph of f.

6. Given $f(x) = x^3 - 6x^2 + 9x$ and $g(x) = 4$.

    a) Find the coordinates of the points common to the graphs of f and g.

    b) Find all the zeros of f.

    c) If the domain of f is limited to the closed interval $[0, 2]$, what is the range of f? Show your reasoning.

7. Let f be the function given by $f(x) = \dfrac{3x}{\sqrt{x^2 - x + 1}}$.

    a) Find the domain of f. Justify your answer.

    b) In the viewing rectangle provided below, sketch a graph of f.

    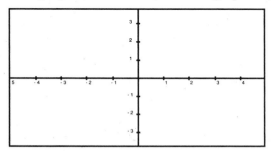

    c) Write an equation for each horizontal asymptote of the graph of f.

    d) Find the range of f.

8. A population is growing continuously at 3.75% per year. Given that the population in 1980 was 11 million, find

    a) a formula that gives the population as a function of t in years after 1980;

    b) the size of the population in the year 2000;

    c) the approximate doubling time for this population.

9. An experiment yields the following data

t (hours)	0	1	2	3	4	5
P (number)	8	8.37	8.75	9.16	9.58	10.02

    a) Is the data linear or exponential? Why?

    b) Based on your conclusion in part a), determine a formula for the function that defines a relationship between P and t.

    c) Plot the data points and function to confirm your findings.

    d) Use your formula in part b) to determine approximately when P will equal 9, to the nearest tenth of an hour.

10.   a) Plot graphs of $y = x^4$ and $y = 3^x$. Determine the points where they intersect correct to three decimal places.

      b) For what values of x is $3^x > x^4$?

11.   One of the following tables of data is linear and one is exponential. Say which is which and give an equation that best fits each table.

      a)

x	0	.50	1.00	1.50	2.00
y	3.12	2.62	2.20	1.85	1.55

      b)

x	0	.50	1.00	1.50	2.00
y	2.71	3.94	5.17	6.40	7.63

12.   Decide whether the following functions have inverses. If a function has an inverse,

      i) calculate $f^{-1}(4)$ and $f^{-1}(2)$, and ii) plot f and its inverse on the same coordinate plane.

      a) $f(x) = \dfrac{x^4}{20} - 5x$.   b) $f(x) = x^5 + 2x + 1$

13.   Determine functions f and g such that $h(x) = f(g(x))$.

      a) $h(x) = \sin^2 x$ b) $h(x) = \sin x^2$

      c) $h(x) = (x+1)^3$ d) $h(x) = e^{\cos x}$

14.   Each function below has the interval $[-5, 5]$ as its domain.

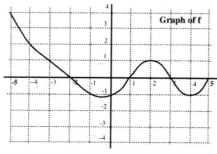 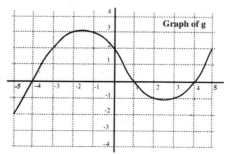

Graph of f   Graph of g

      a) Specify the range of f and g.

      b) For what inputs x does $f(x) = g(5)$?

      c) List all intervals (approximately) on which the function f is increasing.

      d) Evaluate, if possible: i) $f(-3) \cdot g(0)$ ii) $2f(2) - g(-2)$ iii) $\dfrac{f(3)}{g(-2)}$

                                       iv) $\dfrac{f(-3)}{g(-4)}$ v) $\dfrac{f(3)}{g(-3)}$

      e) Evaluate, if possible: i) $f[g(4)]$ ii) $g[f(4)]$ iii) $f[f(-4)]$ iv) $g[g(2)]$

      f) Sketch a graph of $1 - g(x)$.

15.   i) Graph each function by determining the basic function and then using transformations. Label two points on each graph.

ii) Specify the domain and range of each function.

a) $y = -2\sqrt{x+2}$ b) $y = \dfrac{1}{x-2} + 1$

c) $y = 1 + 2\ln x$ d) $y = -2\sin(x - \pi)$

16.   Draw the graph of a function f that satisfies the following four conditions:

i) domain of $f = [-2, 4]$

ii) range of $f = [-3, 3]$

iii) $f(-1) = f(2)$

iv) $f(\tfrac{1}{2}) = 0$

17.   Let the function f be defined by the formula $f\,f(x) = x^2 - \dfrac{3}{x}$ and let the function g be defined by the following table.

x	-3	-2	-1	0	1	2	3	4
$g(x)$	-4	-9	-1	2	0	8	3	-2

a) Evaluate, if possible: i) $f(1) + g(1)$

ii) $f(3) \cdot g(3)$

iii) $\dfrac{f(0)}{g(0)}$

iv) $f[g(0)]$

b) Specify the domain of the functions $f + g$ and $\dfrac{f}{g}$.

18.   a) If $g(x) = 2x^2 + 3x$, determine and simplify:

i) $g(7.13)$ ii) $g(2+h) - g(2)$

b) Solve for x to the nearest hundredth: $2^x = 2^{x-1}$.

19.   If $f(x) = \dfrac{3x}{5+x}$ find

a) a rule for $f^{-1}(x)$;

b) all inputs x such that $f(x) = f^{-1}(x)$.

20. Sketch the graph of a function that is continuous on the domain $0 \le x \le 10$ and has all of the following properties:

    i) the range of f is the interval [0, 5]

    ii) $f(0) = 1$

    iii) the graph of f is concave up on the interval (0, 4)

    iv) the graph of f is concave down on the interval (4, 10)

21. Given $f(x) = \sin x \sqrt{25 - x^2}$ and $g(x) = -\frac{1}{2}x^3 + x^2 + 4x - 0.3$.

    a) Find coordinates of the points common to the graphs of f and g.

    b) Find all zeros of f.

    c) If the domain of f is limited to the closed interval [0, 2], what is the range of f?

22. Give an example of a rational function that has vertical asymptotes at $x = -1$ and $x = 2$ and that has the line $y = 3$ as a horizontal asymptote.

23. Let $f(x) = \ln(3 + \sin x)$.

    a) Evaluate $f\left(\frac{\pi}{6}\right)$ and $f\left(\frac{\pi}{2}\right)$ using a calculator.

    b) Find the domain and range of f.
    c) Is f even? Odd? Neither? Justify your answer.
    d) Is f periodic? If so, what is its period? If not, explain why not.

24. You have $500.00 invested in a bank account earning 5% interest compounded annually.

    a) Write an equation for the money M in your account at the end of t years.

    b) How long will it take the balance in the account to triple?

    c) Suppose the interest compounds monthly instead; that is, you earn $\frac{5\%}{12}$ each month. What interest will you earn for 1 year?

25. For each of the following functions, determine the

    i) period   ii) amplitude   iii) viewing rectangle   [Xmin,Xmax] x [Ymin,Ymax].

    a) $y = -2\sin\left(\dfrac{x}{3}\right)$ b) $y = \dfrac{1}{2}\sin(\pi x)$ c) $y = -3\sin(2x - \pi)$

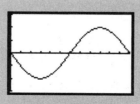   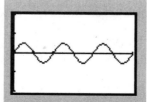   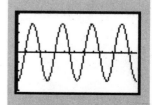

26. Sketch the graph of an even rational function with all of the following properties:

    i) zeros at $x = 1$ and $x = -1$;

    ii) vertical asymptotes $x = 3$ and $x = -3$;

    iii) a horizontal asymptote $y = 1$;

    iv) $f(0) = \dfrac{1}{9}$.

27. Use a graph to approximate the domain, range and zeros of the function g defined by $g(x) = \sqrt{(x - 12)(x + 2)}$.

28. Match each of the following functions with the appropriate graph.

    i) $f(x) = x^3 - 4x$ ii) $f(x) = 3x^4 - 5x^3 + 2x$

    iii) $f(x) = \dfrac{1}{3}x^3 + x^2$ iv) $f(x) = x^4 + x - 2$

    v) $f(x) = 2(x - 1)(x - 2)^2$ vi) $f(x) = 2(x + 1)^2(x + 2)$

a)
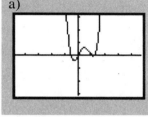

b)
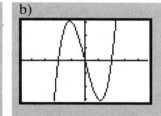

c)
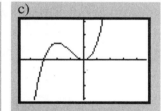

d)
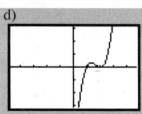

e)
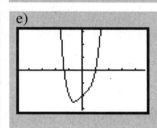

f)
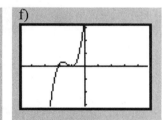

29. Values of the functions f and g for several arguments x are given in the table to the right.

Let $h(x) = \dfrac{1}{x^2 + 1}$. Use the information about f and g given in the table and the definition of h to evaluate the following.

x	$f(x)$	$g(x)$
1	10	5
2	9	6
3	8	7
4	7	8
5	6	7
6	5	6
7	4	5

a) $[f + g](1)$ b) $\left[\dfrac{f}{g}\right](4)$ c) $[f \circ g](2)$ d) $h[f(7)]$ e) $[f \circ g \circ h](0)$

30.   The consumer price index compares the costs of goods and services over various years. The same goods and services that cost $42 in 1938 cost $100 in 1967.

    a) Assume a linear relationship between time and cost.

        i) Write an equation for the cost of goods and services as a function of the number of years since 1938.

        ii) Use your linear function to estimate the cost of goods and services in the year 2000.

    b) Assume an exponential relationship between time and cost.

        i) Write an equation for the cost of goods and services as a function of time since 1938.

        ii) Use your exponential function to estimate the cost of goods and services in the year 2000.

31.   Let f be the function given by $f(x) = 5xe^{-x} - 1$.

    a) Find the domain of f.

    b) In the viewing rectangle provided below, sketch a graph of f.

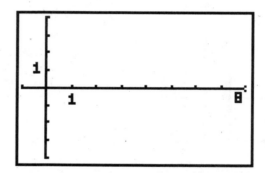

    c) Write an equation for each horizontal asymptote of the graph of f.

    d) Find the range of f.

CHAPTER 2
The Derivative Function

The rate at which one quantity changes with respect to another is a central concept in calculus. The purpose of this chapter is to study the connection between a function and its rate of change. The calculus tool that makes this possible is called the derivative function; the derivative f' measures the rate of change of a given function f.

2.1   Average and Instantaneous Velocity

Suppose you drive a car from one city to another. The cities are 90 miles apart and it takes 2 hours. You say that your average velocity is 45 mph because by the familiar formula

$$\text{rate} = \frac{\text{distance}}{\text{time}},$$

we have

$$\text{average velocity} = \frac{\text{distance traveled}}{\text{time elapsed}} = \frac{90}{2} = 45.$$

A possible graph of the distanced traveled by a car over a 2 hour time interval is shown below.

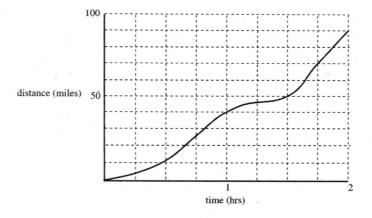

If f is the function that pairs distance d and time t, $d = f(t)$, we have $f(0) = 0$, $f(1) = 40$ and $f(2) = 90$. Notice that the graph starts out flat, representing that the car began the trip from a stopped position. Then the graph steepens gradually corresponding to the increase in speed of the car. Subsequent variations in the graph occur at times when the car is speeding up or slowing down. Flat places in the graph correspond to the car being stopped.

We can visualize average velocity on this graph. For example, consider the second hour of the trip $1 \le t \le 2$,

$$\text{average velocity} = \frac{\text{distance traveled}}{\text{time elapsed}} = \frac{f(2) - f(1)}{2 - 1} = \frac{90 - 40}{1} = 50.$$

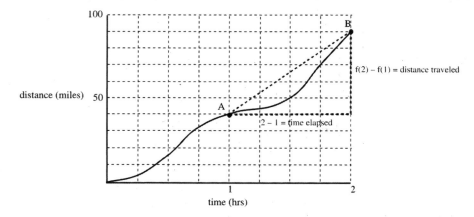

In the figure above, point A has coordinates $(1, 40)$ and point B has coordinates $(2, 90)$. Now $f(2) - f(1)$ is the change in height over the interval or the distance traveled and is marked vertically (rise). The number $2 - 1 = 1$ is the time elapsed and is marked horizontally (run). Thus,

$$\text{average velocity} \; = \; \frac{\text{distance traveled}}{\text{time elapsed}} \; = \; \frac{\text{rise}}{\text{run}} \; = \; \text{slope of the line } AB.$$

In general,

> The average velocity over any interval is the slope of the line joining the endpoints of the interval. The line is called **a secant line.**

For a moving car there is an exact or instantaneous velocity at each instant of time that corresponds to the reading of the car's speedometer. So far, we have no method for finding instantaneous velocity. However, the instantaneous velocity can be approximated by computing the average velocity for some small intervals of time.

Suppose we want to calculate the instantaneous velocity of our car at time $t = 1$ and suppose we are given the following data showing the position of the car over small intervals on either side of 1.

time (hr)	0.90	1.00	1.10
distance (miles)	38.76	40.00	41.16

Computing average velocities we have

$$\text{Average velocity on } [0.90, 1.00] \; = \frac{40.00 - 38.76}{0.10} \; = 12.40,$$

$$\text{Average velocity on } [1.00, 1.10] = \frac{41.16 - 40.00}{0.10} \; = 11.60.$$

Geometrically we can see that these two average velocities are the slopes of two secant lines, one backward and one forward, drawn from $t = 1$.

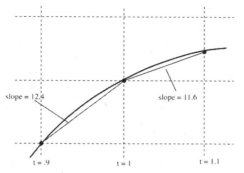

Since the curve is concave down, the first estimate is too big and the second is too small. We conclude that the instantaneous velocity at $t = 1$ must lie between 11.60 mph and 12.40 mph. To get a better estimate of the velocity at $t = 1$ we should take smaller and smaller intervals on either side of $t = 1$ until the forward and backward average velocities are approximately the same.

In general, if we want the instantaneous velocity at $t = a$, we examine the average velocities over smaller and smaller intervals near $t = a$. For example, consider the following distance-time graphs and the time interval from $t = a$ to $t = a + h$ where h is the length of the time interval. In the first figure below, points A and B mark the endpoints of the interval.

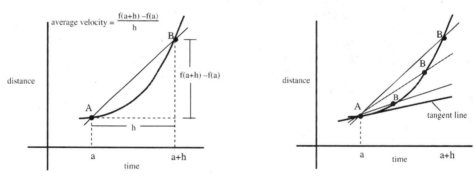

Over the interval $a \le t \le a + h$

$$\text{average velocity} \; = \; \frac{\text{distance traveled}}{\text{time elapsed}} \; = \; \frac{f(a + h) - f(a)}{h} \, .$$

This ratio is the slope of the secant line AB. As the time intervals get shorter (i.e., as h approaches zero) the secant line AB may change position (as shown in the figure to the right). If the slope of the secant line approaches a limiting value as the intervals decrease in size, that is, as h gets smaller, then the secant line becomes a **tangent line to the curve** at point A. This limiting value will be the slope of the tangent line and also the instantaneous velocity at $t = a$. The slope of the tangent line is also called the **slope of the graph** at $t = a$. Symbolically,

$$\text{instantaneous velocity (at } t = a) \; = \; \lim_{h \to 0} \frac{f(a + h) - f(a)}{h} \, .$$

If the instantaneous velocity of the car at any point is the slope of the distance-time curve, you can see how the car's velocity varies by drawing tangents to the curve.

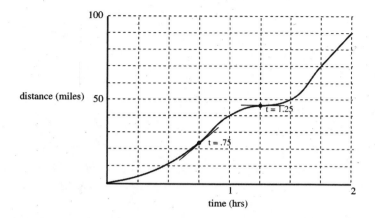

For example, at $t = 0.75$ the distance-time graph has a large positive slope indicating that the car is traveling fast. At $t = 1.25$, the slope of curve is approximately zero meaning that the car is momentarily stopped.

In the discussion above, the term "velocity" rather than "speed" was used. In a careful study of motion, a distinction is made between speed and velocity. The word **velocity** is commonly used when direction is taken into account. **Speed**, on the other hand, is used when you are simply concerned with how fast an object is moving and not bothering with whether it is moving forward or backwards. Thus a car traveling forward at a rate of 10 mph has a velocity of +10 mph; when backing up at 10 mph, it has a velocity of −10 mph. In both cases the speed is 10 mph. Speed is the magnitude (i.e., absolute value) of the velocity and so is always positive or zero.

2.1   Exercises

1.   Suppose a car is driven at a constant rate of 30 mph. Sketch a graph of the distance the car has traveled as a function of time.

2.   Suppose a car is driven at an increasing speed. Sketch a possible graph of the distance the car has traveled as a function of time.

3.   A commuter drives to work and encounters delays at two red lights. Sketch a graph of the distance the car has traveled as a function of time.

4.   Suppose a train from Lancaster to Los Angeles leaves at 4:30 PM and takes 2 hours to reach Los Angeles, waits one hour at the station and then returns, arriving back in Lancaster at 9:30 PM. If the distance from Lancaster to Los Angeles is 80 miles, draw a graph representing the distance of the train from Lancaster as a function of time.

5.   The graph below shows the distance traveled by a car over the first two hours of a trip.

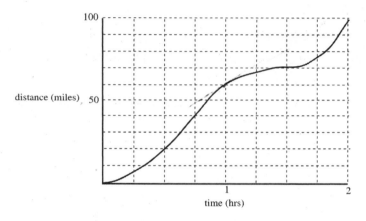

a) Find the average velocity of the car for the entire trip.

b) Find the average velocity of the car in the second hour of the trip.

c) Is the car slowing down or speeding up during the first half-hour of the trip?

d) Estimate the instantaneous velocity of the car at $t = 1$ hr and $t = 1.5$ hr.

e) Estimate when the instantaneous velocity of the car was the same as the average velocity for the entire trip.

6.   The distance traveled by a car is tabulated below.

t (sec)	0	0.2	0.4	0.6	0.8	1.0
d (ft)	0	0.7	1.8	3.6	6.7	9.4

a) Find the average velocity over the interval $0 \le t \le 0.4$.

b) Find the average velocity over the interval $0.2 \le t \le 0.4$.

c) Find the average velocity over the interval $0.4 \le t \le 0.6$.

d) Estimate the velocity at $t = 0.4$.

7.   The graph below shows the distance traveled by a car over time.

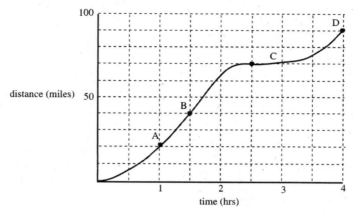

a) At which labeled point is the car traveling the fastest?

b) At which labeled point is the car stopped?

c) Find the average velocity between times $t = 1$ and $t = 4$.

d) Estimate the car's speed at point A and point D.

8.   The graph below shows the distance traveled by a car over time.

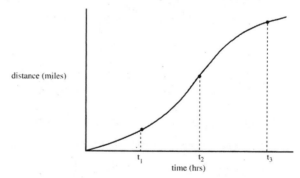

a) When is the car moving fastest?

b) Is the car speeding up or slowing down in the interval $[t_1, t_2]$?

c) Is the car speeding up or slowing down in the interval $[t_2, t_3]$?

9.   A ball is tossed in the air and its height h (in feet) above the ground t seconds after it is thrown is given by

$$h(t) = -16t^2 + 96t + 6.$$

a) What is the average velocity of the ball in the first second?

b) Approximate the velocity of the ball at $t = 1$ second.

c) At what time does the ball reach its maximum height?

10.   The distance a person can walk in time t is given by $d = \sqrt{t}$ where d is measured in miles and t is measured in hours.

   a) Find the average velocity from $t = 1$ to $t = 4$ hours.

   b) Estimate the person's velocity at $t = 4$ hours.

11.   A ball is dropped from a tall building and d is the number of feet it falls in t seconds. The following table shows some of the values of t and d.

t (sec)	0	1	2	3	4	5
d (ft)	0	15.07	57.90	127.03	201.76	333.00

   a) Plot the six data points. Label the axes and indicate the units that are being used.

   b) The slope of any secant line on the graph is a measure of velocity. Why?

   c) Make three estimates of the velocity of the falling ball at $t = 2$ seconds using the distance fallen between these times:

      i) from $t = 1$ to $t = 2$;   ii) from $t = 2$ to $t = 3$;   iii) from $t = 1$ to $t = 3$.

12.   A rock is dropped from a height of 576 ft and falls toward earth in a straight line. In t seconds the rock drops a distance of $d = 16t^2$ feet.

   a) How many seconds after release does the rock hit the ground?

   b) What is the average velocity of the rock during the time it is falling?

   c) Estimate the instantaneous velocity of the rock when it hits the ground.

13.   Mary is a swimmer who trains in a 50 meter pool. Let f be a function that pairs her distance from one end of the pool with time t (in seconds), $d = f(t)$. Below are some values for a recent swim.

t (sec)	0	3.0	8.4	14.6	20.8	27.6	31.9	38.1	45.8	53.9	60
d (meters)	0	10	20	30	40	50	40	30	20	10	0

   a) Find Mary's average speed and average velocity over the whole swim.

   b) Estimate Mary's velocity at $t = 14.6$ by taking the average of two average velocities.

14.   Suppose a particle is moving along a straight line and $s = f(t)$ represents the distance of the particle from its starting point as a function of time, t. Sketch a possible graph of f if the average velocity ot the particle between t = 3 and t = 7 is the same as the instantaneous velocity at t = 6.

2.2   Average and Instantaneous Rate of Change; the Derivative of a Function at a Point

The preceding discussion of velocity illustrates a more general idea. In general, the average rate of change of a function over an interval is the amount of change divided by the length of the interval. Let f be any function, not just distance traveled as a function of time, with inputs x and outputs y so that $y = f(x)$. If h is the change in input from a to $a + h$, $h \neq 0$, then the change in x is given by

$$h = (a + h) - a = \Delta x,$$

and the corresponding change in y is

$$f(a + h) - f(a) = \Delta y.$$

The ratio

$$\frac{\Delta y}{\Delta x} = \frac{f(a + h) - f(a)}{h}$$

is called the **difference quotient**. It gives the **average rate of change of y with respect to x** over the interval $[a, a + h]$.

$$\text{average rate of change of } y \text{ with respect to } x \text{ over the interval from } a \text{ to } a + h = \frac{f(a + h) - f(a)}{h}$$

Graphically, the difference quotient is the slope of the secant line connecting points $A(a, f(a))$ and $B(a + h, f(a + h))$.

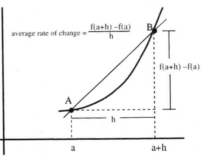

Example 1: *Finding Average Rate of Change*

Let the function f be defined by $f(x) = x^2$.

a) Find the average rate of change of f over the interval $1 \leq x \leq 3$.

b) Find an equation of the corresponding secant line.

c) Plot a graph of f and the secant line.

Solution:

a) $\Delta x = 3 - 1 = 2$ and $\Delta y = f(3) - f(1) = 8$;

average rate of change of f over $[1,3]$ $= \dfrac{\Delta y}{\Delta x} = \dfrac{8}{2} = 4.$

b) The secant line is the line passing through the two points $(1, f(1))$ and $(3, f(3))$. Its slope is the average rate of change found in part a). Thus the point-slope equation of this secant line is

$$y = 4(x-1) + f(1) = 4x - 4 + 1 = 4x - 3.$$

c) The graphs of f and the secant line are displayed in the following viewing rectangle.

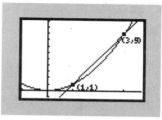

Graph of $f(x) = x^2$ and $y = 4x - 3$ in
$[-1, 3.7]$ x $[-1, 11]$ Viewing Rectangle

The difference quotient gives the average rate of change over an interval. If the average rate of change has a limiting value as the interval decreases in size, then this limit is called the **instantaneous rate of change** of outputs with respect to inputs. A difference quotient calculated over a small interval usually gives a good approximation for an instantaneous rate of change.

Example 2: *Estimating the Instantaneous Rate of Change*

Let $f(x) = 2^x$. Estimate the instantaneous rate of change of f at $x = 1$ by computing the average rate of change for intervals of length $h = $ 0.1, 0.01, 0.001, 0.0001.

Solution:

Calculating the average rate of change over intervals to the right of $x = 1$ we obtain the following table.

Interval size, h	average rate of change $= \dfrac{f(1+h) - f(1)}{h}$
0.1	$\dfrac{f(1.1) - f(1)}{0.1} = 1.4355$
0.01	$\dfrac{f(1.01) - f(1)}{0.01} = 1.3911$
0.001	$\dfrac{f(1.001) - f(1)}{0.001} = 1.3868$
0.0001	$\dfrac{f(1.0001) - f(1)}{0.0001} = 1.3863$

Notice that the values of these difference quotients get more and more alike. Rounding off to three decimal places suggests that the limiting value is 1.386; therefore, we guess that the instantaneous rate of change at $x = 1$ is 1.386.

As we did with velocity, the instantaneous rate of change of a function at a point is defined as the limiting value of the average rate of change. A traditional notation for this limiting value is

$$\begin{array}{c} \text{Instantaneous rate of change} \\ \text{of } f \text{ at } x = a \end{array} = \lim_{h \to 0} \frac{f(a+h) - f(a)}{h}$$

which is read "the limit as h approaches zero of the difference quotient" .

The Derivative of a Function at a Point

The limit used to define the instantaneous rate of change of a function is also used to define one of the key ideas in calculus – the derivative. For a function f and a particular input a, if the difference quotient

$$\frac{f(a+h) - f(a)}{h}$$

approaches a single limiting value as h approaches zero, that number is called **the derivative of** f **at** $x = a$ or $f'(a)$ (read "f prime of a"). Thus, when the limit exists, the definition of $f'(a)$ is

---

The **Derivative of a function** f **at** $x = a$, denoted by $f'(a)$, is

$$f'(a) = \lim_{h \to 0} \frac{f(a+h) - f(a)}{h}.$$

The derivative of a function f at an input $x = a$, $f'(a)$, is the limit of the difference quotients; the derivative represents the instantaneous rate of change of outputs relative to a unit change in input. The statement $f'(3) = 4$ means, for instance, that when $x = 3$, outputs from f increase 4 times as fast as inputs to f.

Computing the derivative of a function at a point requires finding the limiting value of difference quotients.

Example 3: *Finding the Derivative of a Function at a Point*

Find the value of the derivative of $f(x) = x^2$ at the point where $x = 1$.

Solution:

Calculating the difference quotient over intervals to the right and left of $x = 1$ we obtain the following tables.

h	$\dfrac{f(1+h) - f(1)}{h}$
0.1	$\dfrac{f(1.1) - f(1)}{0.1} = 2.1$
0.01	$\dfrac{f(1.01) - f(1)}{0.01} = 2.01$
0.001	$\dfrac{f(1.001) - f(1)}{0.001} = 2.001$
0.0001	$\dfrac{f(1.0001) - f(1)}{0.0001} = 2.0001$

h	$\dfrac{f(1+h) - f(1)}{h}$
−0.1	$\dfrac{f(0.9) - f(1)}{-0.1} = 1.9$
−0.01	$\dfrac{f(0.99) - f(1)}{-0.01} = 1.99$
−0.001	$\dfrac{f(0.999) - f(1)}{-0.001} = 1.999$
−0.0001	$\dfrac{f(0.9999) - f(1)}{-0.0001} = 1.9999$

The limiting value of the difference quotients appears to be 2, so we have $f'(1) = 2$.

We can confirm this derivative algebraically. For any $h \neq 0$, the difference quotient for $f(x) = x^2$ at $x = 1$, is

$$\frac{f(1+h) - f(1)}{h} = \frac{(1+h)^2 - 1^2}{h}$$

$$= \frac{1 + 2h + h^2 - 1}{h}$$

$$= 2 + h$$

Now as h approaches zero, the value of $2 + h$ approaches 2, so we have $f'(1) = 2$. Thus, at $x = 1$ outputs of f increase 2 times as fast as inputs to f.

Tangent Lines

Slope is the key idea that links graphs and derivatives. Given a function f and an input a, if f is defined on the interval from a to $a + h$, we know that the slope of the secant line joining the endpoints $(a, f(a))$ and $(a + h, f(a + h))$ is

Slope of secant line = $\dfrac{f(a + h) - f(a)}{h}$.

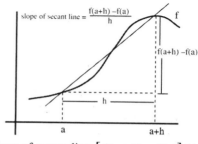

slope of secant line $[f(a + h) - f(a)] / h$

As h gets smaller the secant line may approach a limiting position which we call the tangent line. So we expect the limiting value of the slopes of the secant lines to equal the slope of the tangent line. That is:

the limiting value of slopes of secant lines

$$= \text{ slope of the tangent line}$$

$$= \lim_{h \to 0} \frac{f(a + h) - f(a)}{h}.$$

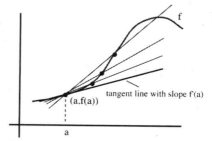

secant lines approaching the tangent line at $x = a$

This limit is the derivative of f at $x = a$.

> The tangent line to the graph of f at $x = a$ is the line
> passing through the point $(a, f(a))$ with slope $f'(a)$.

The slope of the tangent line to the graph of f at the point $(a, f(a))$ is also called **the slope of the graph of f at $x = a$**.

Example 4: *Using the Derivative to Find the Slope at a Point*

Let $f(x) = 2x^2$.

a) Find the slope of the tangent line to the graph of f at (1, 2).

b) Find an equation of the tangent line.

c) Plot a graph of f and the tangent line.

Solution:

a) To find the slope of the tangent line at the point (1, 2) we estimate the derivative $f'(1)$ by examining difference quotients as follows:

$$\frac{f(1+h) - f(1)}{h} = \frac{2(1+h)^2 - 2 \cdot 1^2}{h}$$

$$= \frac{2 + 4h + 2h^2 - 2}{h}$$

$$= 4 + 2h$$

Now as h gets smaller, the limiting value of $4 + 2h$ is 4, and $f'(1) = 4$.

b) The tangent line passes through the point (1, 2) with slope $f'(1) = 4$. Thus, the point-slope form of the tangent line is

$$y = 4(x - 1) + f(1) = 4x - 4 + 2 = 4x - 2.$$

c) A graph of f and the tangent line are displayed in the following viewing rectangle.

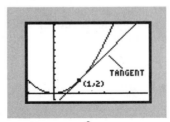

Graph of $f(x) = 2x^2$ and tangent $y = 4x - 2$
in [−1, 3.7] x [−1, 11] viewing rectangle

In summary, the derivative of a function at a point $(a, f(a))$ is its rate of change at $x = a$ and may be interpreted as

• the slope of the tangent line to the graph of f at $x = a$;

• the slope of the graph of f at $x = a$.

Zooming to Estimate Slope

Drawing a straight line on a curved graph is one way to estimate the graph's slope at a point. Another strategy is to zoom in on the point in question until the graph itself looks like a straight line.

Consider taking the graph of a function near a point and enlarging the region by zooming in. For most functions, the higher the magnification the more the curve will appear to be a straight line. If a curve resembles a straight line when you zoom in at a point, then it is called **locally linear** at that point. For example, the graph of $f(x) = |x^2 - 1|$ is locally linear at all points except $(-1, 0)$ and $(1, 0)$.

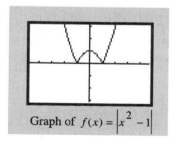

Graph of $f(x) = |x^2 - 1|$

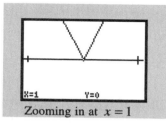

Zooming in at $x = 1$

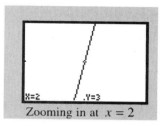

Zooming in at $x = 2$

At $x = 1$, we find that the graph does not look like a straight line. Moreover, no matter how much we zoom in, there is always a sharp corner in the graph at $x = 1$. This means that $f'(1)$ is undefined.

If a function f is locally linear at the point $(a, f(a))$, then the slope of the magnified line is an estimate the slope of the tangent line. Thus to estimate the derivative value $f'(a)$ graphically, zoom in on the function graph until the graph appears straight. The slope of this line will be an approximation of $f'(a)$.

Example 5: *Zooming to Approximate the Slope of a Tangent Line*

Approximate graphically the slope of the tangent line to the graph of $f(x) = \sin x$ at the point where $x = 0$.

Solution:

Graphing f and zooming in near the point $(0, 0)$ we find that the graph soon looks like a straight line. By using the Trace feature we can obtain the coordinates of two points on this line and thereby calculate the slope of the line to be approximately 1. Thus we conclude that $f'(0) = 1$.

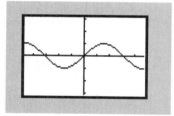
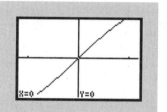
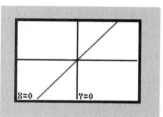

Zooming in on the graph of $\sin x$ near $x = 0$.

The following TI-83 program may be used to approximate the slope of a line segment. After graphing a function and zooming until the graph resembles a straight line, execute the program SLOPE. The program works interactively so that the user only needs to place the cursor on a point and press $\boxed{\text{ENTER}}$. The Trace command followed by a blank gets the screen coordinates from the graphics screen and stores the coordinates in x and y. The next program commands store the x-coordinate and corresponding function value in the variables A and B.

Moving the cursor to another point on the line segment and pressing $\boxed{\text{ENTER}}$ updates x and y. The new x-coordinate and corresponding function value are saved in C and D respectively; the slope (rise/run) is computed, stored in S and then displayed.

```
PROGRAM:SLOPE
:Trace
:X→A
:Y₁→B
:Trace
:X→C
:Y₁→D
:(D-B)/(C-A)→S
:Disp"SLOPE=",S
```

Example 6:   *Using Zooming and the Program SLOPE to Approximate the Slope of a Tangent Line*

Let $f(x) = e^x$. Use zooming and the program SLOPE to approximate $f'(0)$, the slope of the tangent line to the graph of $f(x) = e^x$ at the point (0, 1).

Solution:

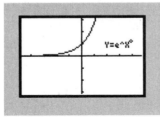

Graph of $y = e^x$

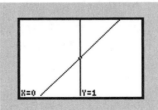

Zoom at the point (0,1) until the graph resembles a straight line.

Execute the SLOPE program

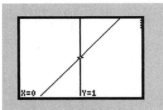

Move the cursor to one endpoint of a line segment and press $\boxed{\text{ENTER}}$.

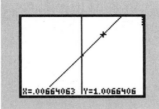

Move the cursor to the other endpoint and press $\boxed{\text{ENTER}}$.

The slope approximation is displayed.

2.2 Exercises

1.   Let $f(x) = 3x - x^2$.

    a) Find the average rate of change of f from $x = 1$ to $x = 3$.

    b) Find an equation of the corresponding secant line.

    c) Plot graphs of f and the secant line.

2.   The graph below shows the squirrel population in a certain wilderness area. The population increases as the squirrels reproduce, but then decreases sharply as predators move into the area.

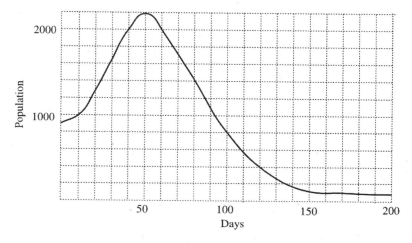

    a) What is the average growth rate from day 10 to day 50? What does this number mean?

    b) During what approximate time period, beginning at day 0, is the average growth rate of the squirrel population positive?

    c) During what approximate time period, beginning at day 0, is the average growth rate of the squirrel population 0?

    d) Approximate the instantaneous rate of change of the squirrel population on day 30 and on day 100.

3.   Given the following data for a function f.

x	3	3.5	4	4.5	5	5.5	6
$f(x)$	10	8	7	4	2	0	−1

    a) Find the average rate of change of f from $x = 4$ to $x = 6$.

    b) Estimate the instantaneous rate of change of f at $x = 4$.

    c) Find, approximately, an equation of the tangent line at $x = 4$.

    d) Use the tangent line to estimate $f(4.5)$.

4. Make a table of values, rounded to two decimal places, for $f(x) = \sqrt{x}$ with $x = 1, 1.5, 2, 2.5, 3$. Use the table to

    a) find the average rate of change from $x = 1$ to $x = 3$;

    b) estimate the instantaneous rate of change of f at $x = 2$.

In Exercises 5–9, calculate the average rate of change $\dfrac{f(a+h)-f(a)}{h}$ for the function f at the given value of a with $h = 0.001$ and also with $h = -0.001$. On the basis of your calculations, make a guess at the instantaneous rate of change at the given value of a.

5. $f(x) = 2x^2 + x$ at $a = 1$.   6. $f(x) = \sqrt{x}$ at $a = 9$.

7. $f(x) = 2^x$ at $a = 2$.   8. $f(x) = x \ln x$ at $a = 2$.

9. $f(x) = \sin\left(x^2\right)$ at $a = 1$.

10. Zoom in on the graph of the given function at the specified point until the graph appears straight. Use the slope of the resulting line to estimate the derivative at the point.

    a) $f(x) = x^3 + 2x$, $P(1,3)$ b) $f(x) = \arctan x$, $P(1, \frac{\pi}{4})$

    c) $f(x) = e^{2x}$, $P(0,1)$ d) $f(x) = x \cos x$, $P(\frac{\pi}{3}, \frac{\pi}{6})$

    e) $f(x) = \ln x$, $P(1,0)$ f) $f(x) = \dfrac{1}{2x+1}$, $P(1, \frac{1}{3})$

11. Graph the function f. By looking at the graph and zooming in on points determine at which points f has a derivative and at which points it does not.

    a) $f(x) = \left| x^3 - 3x^2 \right|$ b) $f(x) = \sqrt{x^2 + .01} + |x - 1|$

12. Find an equation of the tangent line to the graph of $f(x) = x^2 - 2x$ at $x = 1$. Sketch a graph of the function and its tangent line.

13. Suppose that f is a function for which $f'(x)$ exists for all x. Use the values in the table below to estimate $f'(2.99)$, $f'(3)$, and $f'(3.01)$. Explain how you obtained your estimates.

x	2.9	2.99	2.999	3.0	3.001	3.01	3.1
$f(x)$	15.34	23.97	24.88	25	25.10	26.05	36.18

14. The graph of the function f is given in the figure. Arrange the following quantities in ascending order:

    $0,\ 1,\ f'(1),\ f'(2),\ f'(3)$.

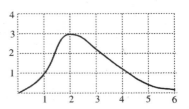

15. Below is the graph of the function f. Draw lines on the graph with the slopes given in parts a), b) and c). Label the lines.

   a) $\dfrac{f(2)}{2}$

   b) $\dfrac{f(4) - f(2)}{4 - 2}$

   c) $\lim\limits_{h \to 0} \dfrac{f(5+h) - f(5)}{h}$

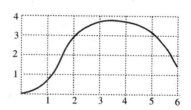

16. Let $f(x) = \ln x$.

   a) Find the average rate of change of f between $x = 0.99$ and $x = 1$. Use this result to estimate $f'(1)$. Explain why the estimate is too large.

   b) Find the average rate of change of f between $x = 1$ and $x = 1.01$. Use this result to estimate $f'(1)$. Explain why the estimate is too small.

17. Explain why the function f defined by $f(x) = |x|$ has a derivative at $x = 2$ and $x = -2$ but does not have a derivative at $x = 0$.

18. Let $f(x) = \text{int}(x)$.

   a) Explain why $g'(2.5) = 0$.

   b) Explain why $g'(2)$ does not exist.

19. A function f is defined on the interval $[0, 5]$ and its graph is shown below. At which inputs $x = a$ does $f'(a)$ not exist? Explain.

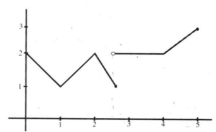

20. Suppose the average rate of change of a function f over the interval from $x = 3$ to $x = 3 + h$ is given in each of the following. Determine $f'(3)$.

   a) $h^2 + 9h + 25$

   b) $\dfrac{3h + 8}{3(h + 3)}$

   c) $3e^h - \cos(2h)$

2.3   The Derivative Function

In the previous section we concentrated on calculating the derivative of a given function f at a fixed point. Now we will discover that, in general, the derivative takes on different values at different points and is itself a function.

Recall that the derivative of a function at a point is the slope of the tangent line at the point.

Example 1:   *Estimating Slopes of Tangent Lines*

Let $f(x) = x^2$. Estimate the slopes of the tangent lines to the graph of f at the points where $x = -2, -1, 0, 1, 2$.

Solution:

Here is the graph of f with tangent segments drawn at each point. Using a ruler and the grid squares on the graph one can estimate the slope of each line segment.

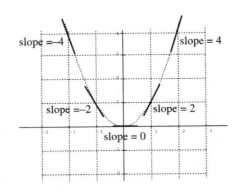

The work of Example 1 may be summarized in a table.

X	y	tangent slope
-2	4	-4
-1	1	-2
0	0	0
1	1	2
2	4	4

We see that the first and third columns determine a new function. This new function, that is derived from the original function f and predicts the slope of a tangent line to the f-graph, is denoted by f' and called the **derivative of** f. From the third column of the table above we see that in this case the outputs for the derivative function are simply the doubling of the inputs. That is

$$f'(x) = 2x.$$

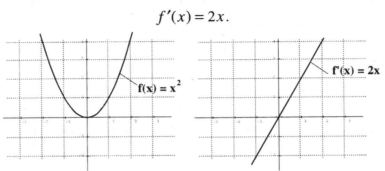

Because the derivative function f' gives the slopes of tangent lines to the graph of f we can predict, for example, that at the point $(3, 9)$ the slope of the tangent line is 6.

Derivative of a Function Defined Graphically

Suppose a function f is defined graphically. To see the connection between a function and its derivative f', it is convenient to sketch the derivative function and compare it with the graph of f

Example 2:   *Graphing f' from f*

The graph of f is shown below.

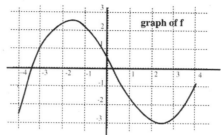

a) Estimate the derivative of f at $x = -4, -2, 0, 2, 4$.

b) Sketch the graph of the derivative, f', of the function f.

Solution:

a) Approximate derivative values are obtained by placing a straight edge so that it represents the tangent at a given point and then using the grid squares to estimate the slope of the straight edge.

x	-4	-2	0	2	4
$f'(x)$	3.5	0.6	-1.5	$-.5$	3

b) In addition to plotting the data points above it is helpful to identify the key features of the derivative graph from the original function f.

- *Locate all points where the tangent line is horizontal.* For instance, at $x = -1.5$ and $x = 2.5$ the slope of the f-graph is zero, and so f' must be zero.

- *Identify the intervals where f is increasing or decreasing.* We can see that the f-graph is increasing from $x = -4$ to $x = -1.5$ and from $x = 2.5$ to $x = 4$. Thus, tangent lines at points in those intervals will have positive slopes and the f'-graph must be drawn above the x-axis. Between $x = -1.5$ and $x = 2.5$ the f-graph is decreasing, so the f'-graph must be drawn below the x-axis.

- *On the interval where f is decreasing, identify the point where the graph of f is the steepest.* At approximately $x = 0.5$ the f-graph is decreasing most steeply. Thus the f'-graph should have a minimum value at $x = 0.5$.

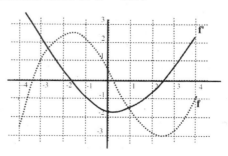

The Derivative of a Function Defined Numerically

If a function is defined by a table instead of a graph, we can approximate the derivative by using the difference quotient.

Example 3: *Graphing the Derivative From Data*

The unemployment rate $u(t)$ varies with time. The following table gives the percentage of unemployed in the U.S. labor force from 1986 to 1994. Estimate the values of $u'(t)$, the rate of change of $u(t)$ with respect to time.

t	1986	1987	1988	1989	1990	1991	1992	1993	1994
$u(t)$	7.0	6.2	5.5	5.3	5.6	6.8	7.5	6.9	6.1

Solution:

Using the data in the table, we can estimate the derivative by employing the difference quotient.

We know that $u'(a) \approx \dfrac{u(a+h) - u(a)}{h}$. So, to estimate $u'(1990)$ for example, we calculate the average rate of change between 1991 and 1990 with $h = 1$.

$$\frac{u(1991) - u(1990)}{1} = \frac{6.8 - 5.6}{1} = 1.2$$

The average rate of change between 1990 and 1989 with $h = 1$ is

$$\frac{u(1990) - u(1989)}{1} = \frac{5.6 - 5.3}{1} = 0.3$$

Taking the average of these forward and backward rates of change we get

$$u'(1990) \approx \frac{1}{2}(1.2 + 0.3) = 0.75.$$

Making similar calculations for the other years we have following table of approximate values for the derivative.

t	$u'(t)$
1986	−0.80
1987	−0.75
1988	−0.45
1989	0.05
1990	0.75
1991	0.95
1992	0.05
1993	−0.70
1994	−0.80

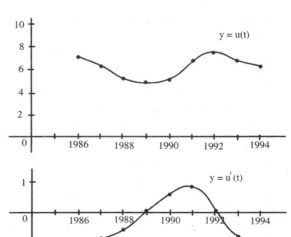

The Derivative of a Function Given by Formula

If a function is defined graphically or numerically, the derivative, or instantaneous rate of change, can be found only approximately. However, if we are given an exact formula for a function, the definition can be employed to find an exact formula for the derivative. The value of the derivative, f', at each x in the domain of f is defined as the limiting value of the average rate of change as h approaches zero, provided the limiting value exists.

$$f'(x) = \lim_{h \to 0} \frac{f(x+h) - f(x)}{h}.$$

This definition can be used to find the derivative of a specified function.

Example 4: *Using the Definition*

Find a formula for the derivative of $f(x) = x^3$.

Solution:

The average rate of change or difference quotient for any input x is

$$\frac{f(x+h) - f(x)}{h} = \frac{(x+h)^3 - x^3}{h}$$

$$= \frac{(x^3 + 3x^2 h + 3xh^2 + h^3) - x^3}{h}.$$

$$= 3x^2 + 3xh + h^2$$

Thus $f'(x) = \lim_{h \to 0} \left(3x^2 + 3xh + h^2\right) = 3x^2.$

Alternative Notations for the Derivative

Several 18th century mathematicians had a hand in inventing symbols for the derivative function. Each had his own notation for the derivative so today we have:

1) $f'(a)$, the symbol invented by Lagrange (1736-1813);

2) $\dfrac{dy}{dx}$, the symbol invented by Leibniz (1646-1716);

3) $D(f)$, the symbol invented by L. Arbograst (1759-1803).

For example, if $f(x) = x^2$, we have

1) $f'(x) = 2x$;

2) $\dfrac{dy}{dx} = 2x$;

3) $D\left(x^2\right) = 2x.$

The f'-notation makes it easy to distinguish between a derivative function, f', and the specific value of the derivative at a particular point, $f'(a)$. On the other hand, Leibniz's notation is more suggestive. It reminds us that the derivative is a limit of ratios of the form

$$\frac{\text{Difference in } y\text{-values}}{\text{Difference in } x\text{-values}} = \frac{\Delta y}{\Delta x}$$

Example 5:   *Using the Definition of the Derivative*

Use the limit definition of the derivative to verify that $\dfrac{dy}{dx} = -\dfrac{1}{x^2}$ if $y = \dfrac{1}{x}$.

Solution:

The average rate of change or difference quotient for any input x ($x \neq 0$ and $x + h \neq 0$) is

$$\frac{f(x+h) - f(x)}{h} = \frac{\dfrac{1}{x+h} - \dfrac{1}{x}}{h}$$

$$= \frac{\dfrac{1}{x+h} - \dfrac{1}{x}}{h} \cdot \frac{x(x+h)}{x(x+h)}$$

$$= \frac{x - (x+h)}{h \cdot x(x+h)}$$

$$= \frac{-h}{h \cdot x(x+h)}$$

$$= \frac{-1}{x(x+h)}$$

Thus, $\dfrac{dy}{dx} = \lim_{h \to 0} \dfrac{-1}{x(x+h)} = -\dfrac{1}{x^2}$.

Alternative Definition of the Derivative

The limit definition of the derivative of a function f at an input $x = a$, $f'(a) = \lim\limits_{h \to 0} \dfrac{f(a+h) - f(a)}{h}$, can be written in the following equivalent form.

$$f'(a) = \lim_{x \to a} \frac{f(x) - f(a)}{x - a}$$

Example 6:   *Applying the Alternative Definition of the Derivative*

Differentiate $f(x) = x^2$ .using the alternate definition.

Solution:

$$f'(a) = \lim_{x \to a} \frac{f(x) - f(a)}{x - a} = \frac{x^2 - a^2}{x - a} = \frac{(x-a)(x+a)}{x-a}$$

$$= \lim_{x \to a} (x + a) = 2a$$

2.3   Exercises

In Exercises 1–8, sketch a graph of the derivative function for each of the given functions.

1.

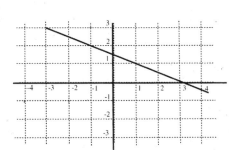

2.

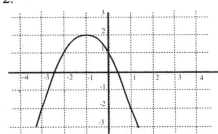

3.

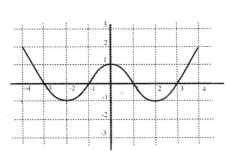

4.

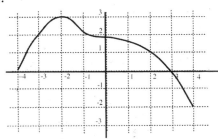

5.

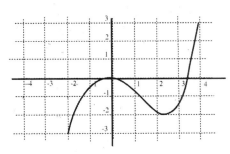

6.

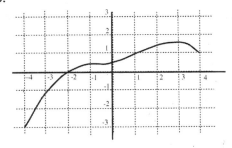

7.

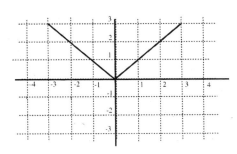

8.

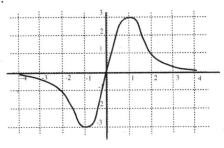

9.   The graph of f is shown below

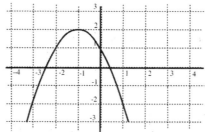

a) Fill in the missing entries in the table below.

x	−3	−2	−1	0	1
$f'(x)$	4			−2	

b) Draw a scatterplot of the data in part a). Is the data linear? quadratic? cubic?

c) Find a model function $y = f'(x)$ that best fits the data and plot its graph. Use the model function to predict the slope of a tangent line to the graph of f at $x = -1.5$.

10.   The graph of f is shown below.

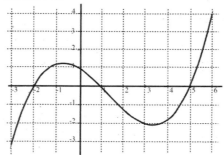

a) Fill in the missing entries in the table below.

x	−3	−2	−1	0	1	2	3	4	5	6
$f'(x)$	4.4			−.7				.9		

b) Draw a scatterplot of the data in part a). Is the data linear? quadratic? cubic?

c) Find a model function $y = f'(x)$ that best fits the data. Use the model function to predict the slope of a tangent line to the graph of f at $x = 4$.

11.   Given the following data for a function f.

x	1.1	1.3	1.5	1.7	1.9	2.1
$f(x)$	12	15	21	23	24	25

a) Estimate $f'(1.7)$.

b) Write an equation for the tangent line to the graph of f at $x = 1.7$.

c) Use your answer in b) to predict the value of f at $x = 1.8$.

12.   Sketch the graph of a function f that is consistent with the following data.

x	-2	−1	1	2
$f(x)$	1	−1	−2	−2
$f'(x)$	−3	0	−1	1

13.   Sketch a graph of $f(x) = x^3 - 4x^2 + x - 1$ and $g(x) = x^3 - 4x^2 + x + 3$ in the same viewing rectangle. How are the graphs related? Does $f'(x) = g'(x)$ for every x ?

In Exercises 14–18, sketch the graph of f and use it to sketch the graph of f'.

14. $f(x) = x - x^2$ 15. $f(x) = \sin x$

16. $f(x) = 3x - 1$ 17. $f(x) = e^x$

18. $f(x) = \dfrac{1}{x}$

In Exercises 19–22, use the limit definition of the derivative to find a formula for $\dfrac{dy}{dx}$.

19. $y = 3x + 1$ 20. $y = 2x^2$

21. $y = x^2 - x$ 22. $y = \sqrt{x}$ [Hint: You need to rationalize the numerator (i.e., multiply the numerator and denominator of the difference quotient by the quantity ($\sqrt{x+h} + \sqrt{x}$).]

23.   Suppose $f(x) = x^3$ and $g(x) = f(x) - 3$.

      a) Explain how the graphs of f and g are related.

      b) How is the graph of g' related to the graph of f' ?

      c) If $f'(2) = 12$, what is $g'(2)$?

24.   Suppose that $f(x) = x^2$ and $g(x) = f(x - 3)$.

      a) Explain how the graphs of f and g are related.

      b) How is the graph of g' related to the graph of f' ?

      c) If $f'(1) = 2$, what is $g'(4)$?

25.   Suppose that $f(x) = x^2 + x$ and $g(x) = 3f(x)$.

      a) Explain how the graphs of f and g are related.

      b) How is the graph of g' related to the graph of f' ?

      c) $f'(1) = 3$, what is $g'(1)$?

26.   Let $f(x) = x^3$.

   a) Use zooming to estimate $f'(0)$, $f'(1)$, $f'(2)$ and $f'(3)$.

   b) Use your results in part a) to sketch a graph of f' over the interval $[0, 3]$.

   c) Use your results in parts a) and b) to guess a formula for $f'(x)$. Compare your
      result with Example 4, page 103.

27.   Let $f(x) = \ln x$.

   a) Use zooming to estimate $f'\left(\frac{1}{2}\right)$, $f'(1)$, $f'\left(\frac{3}{2}\right)$, $f'(2)$ and $f'\left(\frac{5}{2}\right)$.

   b) Use your results in part a) to sketch of graph of f' over the interval $[\frac{1}{2}, \frac{5}{2}]$.

   c) Use your results in parts a) and b) to guess a formula for $f'(x)$.

28.   Let $f(x) = \sin x$.

   a) Use zooming to estimate $f'(0), f'\left(\frac{\pi}{4}\right), f'\left(\frac{\pi}{2}\right), f'\left(\frac{3\pi}{4}\right), f'(\pi), f'\left(\frac{3\pi}{2}\right)$ and $f'(2\pi)$.

   b) Use your results in part a) to sketch of graph of f' over the interval $[0, 2\pi]$.

   c) Use your results in parts a) and b) to guess a formula for $f'(x)$.

29.   The interest rate on U.S. Treasury bills is a function of time. The table below gives the
       midyear values of this function $y = f(t)$ from 1985 to 1992.

t	1985	1986	1987	1988	1989	1990	1991	1992
$f(t)$	7.49	5.97	5.83	6.67	8.11	7.51	5.41	3.46

   a) What is the meaning of $f'(t)$? What are the units?

   b) Create a table of values for $f'(t)$.

   c) Sketch a graph of f and f'.

30.   In Example 5 the limit definition was used to show that $f'(x) = -\dfrac{1}{x^2}$ for $f(x) = \dfrac{1}{x}$.

   a) Use the limit definition of the derivative to find $f'(x)$ for $f(x) = \dfrac{1}{x^2}$.

   b) Guess the derivative of $f(x) = \dfrac{1}{x^3}$. Check your conjecture for $f'(x)$ by analyzing the
      graphs of $y = f(x)$ and $y = f'(x)$

   c) Write a general formula for the derivative of $f(x) = \dfrac{1}{x^n}$, where n is a positive integer.

31.   Use the alternative definition of the derivative to show that

   a) if $f(x) = \frac{1}{x}$, then $f'(2) = -\frac{1}{4}$ b) if $f'(x) = \sqrt{x}$, then $f'(3) = \dfrac{1}{2\sqrt{3}}$.

2.4   Calculating the Derivative Numerically

We have used the difference quotient for small h to approximate the derivative $f'(a)$. Geometrically, this means that the slope of the secant line passing through the two points $(a, f(a))$ and $(a+h, f(a+h))$ on the graph of f approximates the slope of the tangent line at the point $(a, f(a))$.

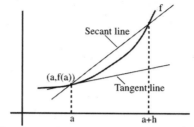

The slope of the secant line approximates
the slope of the tangent line for small h.

Another numerical technique for estimating the value of the derivative $f'(a)$ is the **symmetric difference quotient:**

$$f'(a) \approx \frac{f(a+h) - f(a-h)}{2h}$$

for small values of h. This is just the average of a right and left difference quotient

$$\frac{f(a) - f(a-h)}{h} \quad \text{and} \quad \frac{f(a+h) - f(a)}{h}.$$

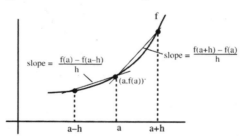

The TI-83 does not compute derivatives exactly, but does produce a numerical approximation using the built-in **nDeriv** function (option 8 under the Math menu). **nDeriv** approximates the derivative by employing the symmetric difference quotient. The standard format of the **nDeriv** function is

$$\textbf{nDeriv}(\textit{function, variable, value}).$$

nDeriv returns an approximate derivative of *function* with respect to the specified *variable*, given the *value* at which to calculate the derivative. An optional input for **nDeriv** is h, the h that is used in the symmetric difference approximation of $f'(x)$. If h is not specified, .001 is used. This alternate format is

$$\textbf{nDeriv}(\textit{function, variable, value, h})$$

Example 1: *Using the Symmetric Difference Quotient*

Use the symmetric difference quotient to numerically approximate the slope of the tangent line to the graph of the function $y = x^3$ at the point where $x = 2$.

Solution:

From the Home screen enter the expression **nDeriv**(X^3, X, 2) and press $\boxed{\text{ENTER}}$. The number 12.000001 is displayed as an approximation of the tangent slope at $x = 2$. Thus we have obtained a very good approximation to the exact slope value $f'(2) = 12$.

The last example suggests that **nDeriv** is very accurate. This is usually the case; however, for some functions **nDeriv** will produce wrong results. For example, the derivative of the function $f(x) = |x|$ at $x = 0$ does not exist since the graph of f has a sharp corner at $x = 0$. However, **nDeriv**(abs (X), X, 0) produces the value 0 because the symmetric difference estimate is

$$f'(0) \approx \frac{f(.001) - f(-.001)}{.002} = \frac{.001 - .001}{.002} = 0.$$

Moreover, any nonzero value of h will give the same estimate of 0 using the symmetric difference quotient because the absolute value function is even.

Technology Tip: *Drawing a Tangent Line*

While a graph is displayed, you can draw the tangent line at a specified point using the cursor. For example, if $Y1 = X^3 - 2X$ is the only selected function, choose **Tangent** (from the Draw Draw menu – option 5). Use $\boxed{\blacktriangleright}$ and $\boxed{\blacktriangleleft}$ to move the cursor to the point on the graph at which you want the tangent line drawn and press $\boxed{\text{ENTER}}$.

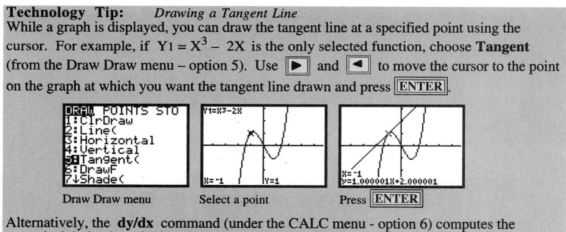

Draw Draw menu   Select a point   Press $\boxed{\text{ENTER}}$

Alternatively, the **dy/dx** command (under the CALC menu - option 6) computes the numerical derivative of a function at a point.

Example 2: *Using a Logistic Curve to Model a Population*

A lake is stocked with 500 fish and their population increases according to the logistic curve

$$f(x) = \frac{10000}{1 + 19e^{(-x/5)}}$$

where $f(x)$ is the current fish population and x is measured in months. Use **nDeriv** to approximate the rate at which the fish population is changing at the end of 10 months.

Solution:

Enter the function in the Y= menu as Y1. From the Home screen enter the expression **nDeriv**(Y1, X, 10) and press ⟦ENTER⟧. The number 403.2040142 is displayed as an approximation of $f'(10)$, the rate at which the fish population is changing at the end of 10 months. Thus at the end of ten months, the fish population is increasing at the rate of about 403 fish per month.

Graphs of Derivative Functions

Graphing a function and its derivative in the same viewing rectangle is a useful way to develop an understanding of the significance of a derivative. For any function defined by a formula, **nDeriv** will quickly produce the graph of the numerical derivative.

Example 3: *Using **nDeriv** to graph a Derivative*

Graph $f(x) = x^3 - 2x$ and its numerical derivative in the ZDecimal viewing rectangle.

Solution:

Clear the Y= menu and change this screen to the following.

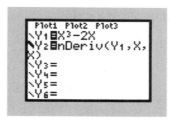

Press ⟦ZOOM⟧[4:ZDecimal] to see this function and its derivative.

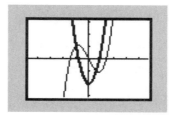

2.4   Exercises

In Exercises 1–6, use **nDeriv** to find an equation for the tangent line to the curve at the given point. Then graph the curve and the tangent line in the same viewing rectangle.

1. $f(x) = x\sqrt{4 - x^2}, \quad x = 0$

2. $f(x) = x^3 - 2x^2, \quad x = 2$

3. $f(x) = x \ln x, \quad x = e$

4. $f(x) = \dfrac{x^2 - 4}{x^2 + 1}, \quad x = 2$

5. $f(x) = \sin(x^2), \quad x = 1$

6. $f(x) = 2^{\tan x}, \quad x = \dfrac{\pi}{4}$

7.   Given the following data about a function f.

x	0	.1	.2	.3	.4
$f(x)$	.10	.19	.27	.34	.38

     a) Use the symmetric difference quotient to estimate $f'(0.1)$ and $f'(0.3)$.

     b) Find, approximately, an equation of the tangent line at $x = 0.1$.

8.   Graph $f(x) = \sqrt[3]{x - 2}$ and its numerical derivative in the ZDecimal viewing rectangle.

     a) Where does the $f'(x)$ fail to exist? Why?

     b) For what values of x is $f'(x) > 0$? Is the f-graph increasing or decreasing for all x?

     c) Explain the connection, if any, among the answers to the questions in part b).

9. $f(x) = \begin{cases} x, & \text{if } x < 0 \\ -x^2, & \text{if } x \geq 0 \end{cases}$

     Graph f and its numerical derivative in the ZDecimal viewing rectangle. What is $f'(-1)$, $f'(0)$ and $f'(1)$? (Hint: Y1 = X (X < 0) + $^-$X² (X ≥ 0) , using the TEST menu.)

10.   Graph $f(x) = \sin x$ and its numerical derivative in the ZDecimal viewing rectangle. Make a conjecture about the formula for the function determined by the numerical derivative **nDeriv**(sin (X), X, X).

11.   Verify that the symmetric difference quotient produces the exact derivative value for a quadratic function $f(x) = ax^2 + bx + c$. [Hint: Show that $\dfrac{f(x+h) - f(x-h)}{2h} = 2ax + b$ provided $h \neq 0$]

12.   Let $f(x) = 2^x$.

     a) Find the average rate of change of f between $x = 0$ and $x = 0.01$.

     b) Find the average rate of change of f between $x = -0.01$ and $x = 0$.

     c) Using the results of parts a) and b), estimate $f'(0)$.

13.   The function f is defined by $f(x) = (x-1)^{2/3}$.

     a) Graph f and its numerical derivative in the ZDecimal viewing rectangle. Where does $f'(x)$ fail to exist?

     b) Show by zooming that $f'(1)$ does not exist.

     c) Evaluate on the Home screen: nDeriv((X − 1)^(2/3), X, 1). Explain.

14.   Let $f(x) = x \cdot |x|$.

     a) For what values of x is the function f differentiable?

     b) Find a formula for $f'(x)$.

15.   Let $f(x) = \begin{cases} 0, & \text{if } x < 0 \\ x^2, & \text{if } x \geq 0 \end{cases}$

     a) Explore graphically and numerically if $f'(0)$ exists.

     b) Find a formula for $f'(x)$.

16.   If the function f is defined by $f(x) = x^2 - 2x$, find $\displaystyle\lim_{h \to 0} \frac{f(3+h) - f(3-h)}{2h}$.

17.   Let functions f and g be defined by $f(x) = x^{1/3}$ and $g(x) = x^{2/3}$.

     a) Explain why neither function is differentiable at $x = 0$.

     b) Use **nDeriv** to estimate the $f'(0)$ and $g'(0)$.

     c) By analyzing the definition of the symmetric quotient, explain why **nDeriv** returns incorrect values that are so different from each other for these two functions.

2.5   Critical Numbers;   Relative Maximum and Minimum Points

The important geometric connection between the function f and its derivative f' is that:

$f'(a)$ is the slope of the tangent line to the graph of f at $x = a$.

This tangent slope interpretation of f' has several useful geometric implications. For example, the values of the derivative enable us to detect when a function is increasing and when it is decreasing.

If $f'(x) > 0$ on an interval (a, b) then f is increasing on that interval.

If $f'(x) < 0$ on an interval (a, b) then f is decreasing on that interval.

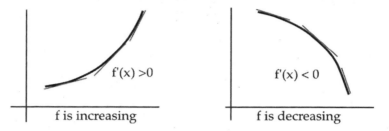

f'(x) >0   f'(x) < 0

f is increasing   f is decreasing

A number c in the interior of the domain of a function f is called a **critical number** if either

i) $f'(c) = 0$ or   ii) $f'(c)$ does not exist.

The point $(c, f(c))$ is called a **critical point** of the graph of f. The only inputs, $x = c$, at which the derivative of a function can change sign are where $f'(c)$ is either 0 or undefined; i.e., at the critical numbers.

A **local maximum** (or **relative maximum**) for a function f occurs at a point $x = a$ if $f(a)$ is the largest value of f in some interval centered at $x = a$. A **local minimum** (or **relative minimum**) for a function occurs at a point $x = a$ if $f(a)$ is the smallest value in some interval centered at $x = a$. Collectively, local maximum and minimum values are called local **extreme values**, and local maximum and minimum points are called **local extreme points**.

Graphs of f and its derivative f' are shown below with three interesting points labeled on each.

To the left of point A the f-graph is increasing and the derivative f' is positive valued. Between points A and B the f-graph is decreasing and f' is negative valued. To the right of B the f-graph is again increasing with a positive derivative. Points A and B, where the f-graph is horizontal, are critical points. Point A is a local maximum point; point B is a local minimum point. Point C on the f-graph, corresponding to the local minimum point on the f'-graph, is an example of a so-called **point of inflection**.

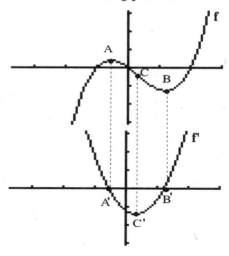

Example 1: *Using the Calculator to Find a Local Minimum*

Use **nDeriv** to find the local minimum of $f(x) = x^3 - 4x^2 + 3x - 5$.

Solution:

Enter and graph $Y1 = X^3 - 4X^2 + 3X - 5$ and $Y2 = \mathbf{nDeriv}(Y1, X, X)$ in the ZStandard viewing rectangle.

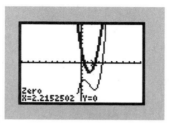

Notice that the zeros of the derivative correspond to the x-coordinates of the local maximum and minimum points on the graph of f. Using the [2:zero] option under the CALC menu, we see that the local minimum value is $y = -7.1126$ at $x = 2.2152502$.

Example 2: *Finding Local Extreme Values*

Suppose the derivative of a polynomial function $y = f(x)$ is $\dfrac{dy}{dx} = 2x - x^2$. Determine the values of x, if any, where f has local extreme values.

Solution:

A critical number is not necessarily a local maximum or local minimum. On the other hand, local extreme values only occur at critical points. Thus, we begin our search for local extreme values by examining critical numbers. Because $f'(x) = x(2 - x)$, the only critical numbers for f are 0 and 2. To the left of a local maximum a function is increasing ($f'(x) > 0$), while to the right the function is decreasing ($f'(x) < 0$). Similarly, a function is decreasing ($f'(x) < 0$) to the left of a local minimum and increasing ($f'(x) > 0$) to the right. For each critical number we evaluate the derivative a little to left and a little to the right and conclude that f has a local minimum at $x = 0$ since f is decreasing ($f'(x) < 0$) for $x < 0$ and increasing ($f'(x) > 0$) for $0 < x < 2$. Similarly, f has a local maximum at $x = 2$ since f is increasing ($f'(x) > 0$) for $0 < x < 2$ and decreasing ($f'(x) < 0$) for $x > 2$.

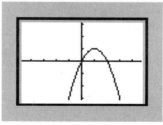

$$f'(x) = 2x - x^2$$

2.5   Exercises

In Exercises 1–6, sketch the graphs of f and f' by entering f in Y1 and **nDeriv**(Y1, X, X) in Y2. List the critical numbers for each function and note at which critical numbers the derivative changes sign from positive to negative and at which critical numbers the derivative changes from negative to positive. Then list the points where the function has a local maximum or local minimum value.

1. $f(x) = x^4 - 2x^2$

2. $f(x) = (1 + \cos x)\sin x, \quad [-2\pi,\ 2\pi]$

3. $f(x) = \sin x - \cos x, \quad [-2\pi,\ 2\pi]$

4. $f(x) = 2x - 3x^{2/3}$

5. $f(x) = \dfrac{2}{3}x^3 - x^2$

6. $f(x) = \dfrac{x}{1 + x^2}$

7.   a) On the basis of your answers to Exercises 1–6, describe how the first derivative of a function could be used to determine whether a local maximum or a local minimum occurs at a critical number.

    b) If c is a critical number, is it true that $(c, f(c))$ is either a local maximum or minimum point?

8.   The graph of a function f is shown below with several points labeled.

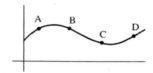

    a) At which labeled points is f' positive? Explain.

    b) Between which pairs of labeled points does f have a critical point?

    c) Between which pairs of labeled points is f' increasing?

    d) Between which pairs of labeled points does f' achieve its minimum value?

9.   a) Sketch the graph of a function f for which $f(x) > 0$ and $f'(x) > 0$ for all x.

    b) Sketch the graph of a function g for which $g(x) > 0$ and $g'(x) < 0$ for all x.

    c) Give formulas for functions with these properties.

10.   The numerical derivative **nDeriv** frequently calculates misleading values at critical numbers where the derivative is undefined. For example, let $f(x) = \sqrt[3]{x - 2}$ and $g(x) = \sqrt[3]{(x - 2)^2}$. At $x = 2$ both $f'(2)$ and $g'(2)$ do not exist. (Can you verify that?) However, **nDeriv** gives numerical values of $f'(2) = 100$ and $g'(2) = 0$. Show the calculations the TI-83 does to arrive at these incorrect answers.

11.   The graph of the function f shown below has horizontal tangents at $x = -1$ and $x = 2$.

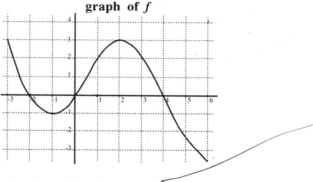

graph of f

a) Where does f have critical numbers?

b) On what intervals is f' negative?

c) On what intervals is f' increasing?

d) Where does f' achieve its maximum value? Estimate this value of f'.

e) Sketch a graph of f'.

12.   The graph of the function g shown below has horizontal tangents at $x = -2$, 0, and 2.

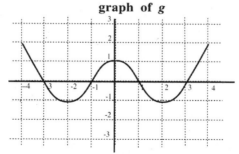

graph of g

a) Where does g have critical numbers?

b) On what intervals is g' negative? Positive?

c) On what intervals is g' increasing? Decreasing?

d) Sketch a graph of g'.

13.   The slope of a curve at any point (x, y) is given by $\dfrac{dy}{dx} = (x - 2)^2 (x - 3)$. Determine whether the following statements are *true* or *false*.

a) The curve has a horizontal tangent at the point where $x = 2$.

b) The curve has a local minimum at the point where $x = 2$.

c) The curve has a local maximum at the point where $x = 3$.

d) The curve is increasing at $x = 2$.

14.   The graph of f', the derivative of f, is shown below for $-3 \le x \le 6$. The graph of f'
       has horizontal tangents at $x = -2, 0$ and 4.

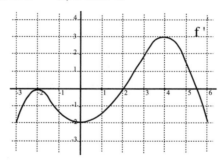

a) On which intervals is f increasing? Decreasing?

b) Where does f have critical numbers?

c) Where does f have a local maximum? A local minimum?

d) Suppose $f(0) = 0$. Sketch a possible graph of f.

15.   The graph of g', the derivative of a function g is shown below for $-4 \le x \le 4$. The graph
       of g' has horizontal tangents at $x = -3, 0$ and 3.

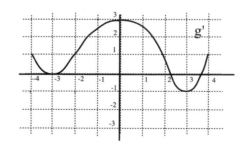

a) On which intervals is g increasing? Decreasing?

b) Where does g have critical points?

c) Where does g have a local maximum? A local minimum?

d) Suppose $g(0) = 1$. Sketch a possible graph of g.

16.   Suppose the derivative of a function $y = f(x)$ is $\dfrac{dy}{dx} = x + 3$.

a) On what intervals, if any, is f increasing?

b) At which x, if any, does f have a local minimum?

c) Suppose $f(-2) = 1$. Sketch a possible graph of f.

17.   Let $f(x) = e^{\cos x}$. How many zeros does f' have in the interval $[0, 2\pi]$?
       Where are they located?

In Exercises 18–21, the derivative of a function f is given. Use the graph of the derivative to answer the following.

   a) On which intervals is f increasing?

   b) At what values of x, if any, does f have a local maximum? A local minimum?

   c) On which intervals is f' decreasing?

   d) Sketch a possible graph of f.

18. $f'(x) = (x-2)(x+4)$ 19. $f'(x) = \dfrac{x+1}{x-2}$

20. $f'(x) = (x+2)^2(x-1)$ 21. $f'(x) = \sin(2x)$

22.   In the left-hand column below are graphs of several functions. In the right-hand column are graphs of derivative functions. Match each function with its derivative.

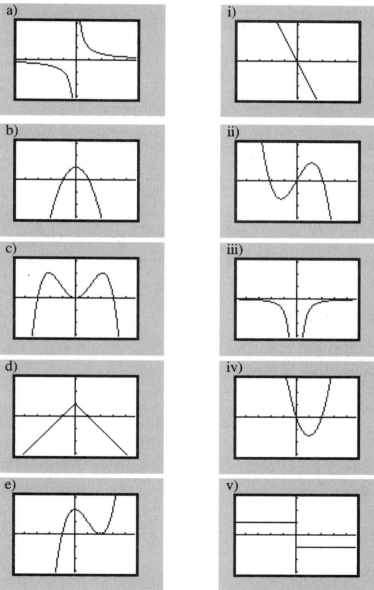

23. A function f is defined on the interval $[-4, 4]$ and some inputs and corresponding outputs for f are shown in the table below.

x	-4	-3	-2	-1	0	1	2	3	4
$f(x)$	-51	-17	-1	3	1	-1	3	19	53

Based on this data, and assuming that the derivative of f exists at each input in the domain of f, answer the following questions.

a) On what intervals is f' positive? Explain.

b) Between what inputs does f have a critical number? Explain.

24. A function g is defined on the interval $[-2, 2]$ and some inputs and corresponding outputs for g are shown in the table below.

x	-2	-1.5	-1	$-.5$	0	.5	1	1.5	2
$g(x)$	5.0	1.4	0	0.1	1	1.9	2	.6	-3

Based on this data, and assuming that the derivative of g exists at each input in the domain of g, answer the following questions.

a) On what intervals is g' positive? Explain.

b) Between what inputs does g have a critical number? Explain.

25. A function f is defined and differentiable on the interval $[-2, 2]$ and some inputs and corresponding outputs for the derivative f' are shown in the table below.

x	-2	-1.5	-1	$-.5$	0	.5	1	1.5	2
$f'(x)$	8.0	2.8	-1	-3.3	-4	-3.3	-1.0	2.8	8.0

Based on this data, answer the following questions.

a) On which intervals does f appear to be increasing? Explain.

b) Between what inputs does f seem to have a local minimum?

c) Sketch a possible graph of f.

26. A function g is defined and differentiable on the interval $[-4, 4]$ and some inputs and corresponding outputs for the derivative g' are shown in the table below.

x	-4	-3	-2	-1	0	1	2	3	4
$g'(x)$	$-.3$	1.9	-1.3	$-.8$	2	$-.8$	-1.3	1.9	$-.3$

Based on this data answer the following questions.

a) On which intervals does g appear to be increasing? Explain.

b) Between what inputs does g seem to have a local minimum?

c) Sketch a possible graph of g.

2.6   Inflection Points and the Second Derivative

If f is a function, then f' is the function that assigns the number $f'(x)$ to each input x at which f is differentiable. Since the derivative, f', is also a function, we can calculate its derivative and define $f''(a)$ by the formula

$$f''(a) = \lim_{h \to 0} \frac{f'(a+h) - f'(a)}{h} = (f')'(a).$$

We call $f''(a)$ **the second derivative of f at a**. In the Leibniz notation the second derivative is written as $\dfrac{d}{dx}\left(\dfrac{dy}{dx}\right) = \dfrac{d^2y}{dx^2}$.

Recall that the derivative of a function tells you whether a function is increasing or decreasing. Since f'' is the derivative of f',

1) if $f''(x) > 0$ on an interval (a, b), then f' is increasing on that interval;

2) if $f''(x) < 0$ on an interval (a, b), then f' is decreasing on that interval.

When the tangent slopes are increasing the graph of f is said to be **concave up**. When the tangent slopes are decreasing the graph of f is called **concave down**. In other words, the sign of the second derivative determines whether the graph of f is concave up or concave down. Concavity is discussed in greater detail in Section 5.2.

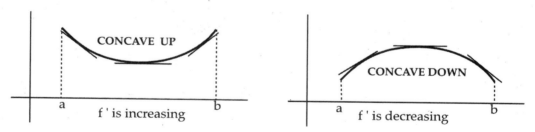

A point on the graph of a function where the curve changes concavity is called an **inflection point**.

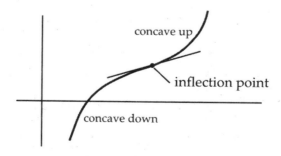

Since a point of inflection occurs where the concavity of a graph changes, it must be true that the sign of f'' changes at such a point. Thus to locate possible points of inflection, we need first to find where the second derivative is zero or undefined, since these are the places where a function can change sign. Then we must check to see if $f''(x)$ has opposite signs on either side of such a place.

Warning: Not every point where $f''(x) = 0$ (or f'' is undefined) is an inflection point. For example, $f(x) = x^4$ has $f''(x) = 12x^2$ so that $f''(x) = 0$; however, $f''(x) > 0$ for $x < 0$ and for $x > 0$, so there is no change in concavity at $x = 0$.

The following TI-83 program may be used to locate an inflection point on a curve. After graphing the function execute the program INFLECPT. The current graph is displayed along with a prompt to graphically select the Lower bound – the left endpoint of an interval that contains an inflection point. Move the cursor to the point you want as a lower bound and press ENTER . Set the Upper bound of the interval that contains the inflection point in a similar way. Press ENTER and the coordinates of the inflection point are displayed.

```
PROGRAM: INFLECPT
:ClrDraw:FnOn 1
:StorePic Pic1
:Text(42,2,"LOWER BOUND")
:Trace:X→A
:Text(42,2,"UPPER BOUND")
:Trace:X→B:(A+B)/2→C
:FnOff :ClrDraw
:RecallPic Pic1
:solve(nDeriv(nDeriv(Y1,X,X),X,X)-0,X,C,{A,B})
:Text(40,2,"INFLECT PT")
:Text(48,2,"X=",Ans)
:Text(48,50,"Y=",Y1(Ans))
```

Example 1:   *Using the Program INFLECPT to locate an Inflection Point*

Let $f(x) = x^3 - 2x^2$. Use the program INFLECPT to locate the inflection point on the graph of f.

Solution:

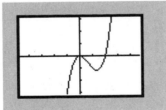

Graph of $f(x) = x^3 - 2x^2$

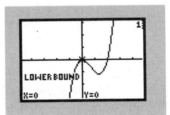

Execute program INFLECPT, select a lower bound and press ENTER

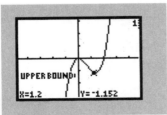

Select an upper bound and press ENTER

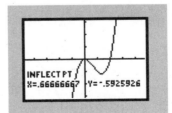

The Second Derivative as a Rate of Change

The first derivative gives information about immediate change, whereas the second derivative gives information about long-term trends. Thus, in long range planning, it is often the second derivative that is important.

The exponential model for population growth is flawed since it predicts faster and faster growth. In most cases, the limited amount of space and resources will eventually force a slowing of the growth rate. In such cases a more realistic model for population growth is the so called **logistic curve** whose graph is shown below.

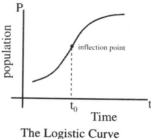

The Logistic Curve

The graph suggests that for $t < t_0$, the population P is growing at an increasing rate, thus $\dfrac{dP}{dt} > 0$

and $\dfrac{d^2 P}{dt^2} > 0$. By contrast, if $t > t_0$, the population continues to grow but at a decreasing rate.

Thus, $\dfrac{dP}{dt} > 0$, but $\dfrac{d^2 P}{dt^2} < 0$. The curve changes concavity at the inflection point $t = t_0$,

suggesting that $\dfrac{d^2 P}{dt^2} = 0$.

Example 2: *A Virus Infection*

A cold virus has invaded a small school community. The table below shows the total number of people who have become infected by a given day .

t (day)	1	2	3	4	5	6	7
N (number infected)	4	11	30	56	76	85	88

a) Estimate the value of $\dfrac{dN}{dt}$ for each day.

b) What can be said about the second derivative during the spread of the cold virus?

Solution:

a) Approximating the derivative with average rates of change we have on day 1, for example,

$$\frac{dN}{dt} \approx \frac{11-4}{2-1} = 7.$$

Approximate derivative values are tabulated below.

t (day)	1	2	3	4	5	6	7
$dN/dt \approx$ avg rate of change	7	19	26	20	9	3	-

b) The total number of people infected is increasing each day and, up to day 3 the rate of infection is also increasing. However, after the third day, the rate of infection is decreasing. This means that $\dfrac{d^2N}{dt^2} > 0$ prior to day 3 and $\dfrac{d^2N}{dt^2} < 0$ after day 3.

Acceleration

While the first derivative can be interpreted as the velocity of a moving object, the second derivative can be interpreted as acceleration.

The velocity of a moving object is not necessarily constant. For example, as a car approaches a red light, its velocity is usually decreasing. A change in velocity is called **acceleration.**

If $s = f(t)$ denotes the position of an object at time t then the velocity, v, is the rate of change of position with respect to time:

$$v = \frac{ds}{dt} .$$

The acceleration, a, of the moving object describes how fast the velocity is changing with time:

$$a = \frac{dv}{dt} .$$

Since velocity is a derivative, we have

$$a = \frac{dv}{dt} = \frac{d^2s}{dt^2} .$$

The acceleration of an object will be positive if the velocity is increasing, zero if the velocity is constant, or negative if the velocity is decreasing. If velocity is measured in feet per second, the acceleration is given in feet per second, per second; this is usually written ft/sec^2.

Example 3:   *Motion Along a Straight Line*

An object is moving along a straight line and its distance to the right of the point where it started is graphed in the following figure.

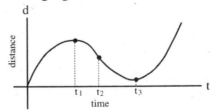

a) When is the object moving to the right and when is it moving to the left?

b) When does the object have positive acceleration? Negative acceleration?

Solution:

a) The object is moving to the right whenever the distance d is increasing. From the graph this appears to be for $0 < t < t_1$ and for $t > t_3$. For $t_1 < t < t_3$, the distance d is decreasing, so the object is moving to the left.

b) The object has positive acceleration whenever the curve is concave up, which appears to be for $t > t_2$. The particle has negative acceleration when the curve is concave down, for $t < t_2$.

2.6   Exercises

1.   The graph of f appears below.

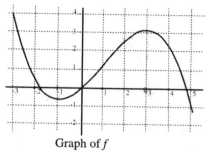

Graph of f

    a) For which x, if any, is $f''(x)$ positive? negative? zero?

    b) Rank the four numbers $f'(0)$, $f(2)$ and $f'(2)$ and $f''(2)$ in increasing order.

2.   Below are the graphs of the second derivatives of functions f and g. For each, locate points of inflection and intervals where the original function is concave up or concave down.

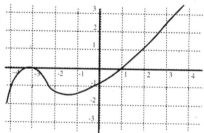   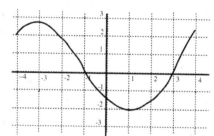

     Graph of second derivative of f Graph of the second derivative of g

3.   Suppose a function f is defined so that its derivative is

$$f'(x) = (\ln x)^2 - 2(\sin x)^2 \quad \text{for } 0 < x \le 6.$$

    Use the TI-83 to sketch a graph of f' and its derivative f''.

    a) On which intervals is f increasing?

    b) On which intervals is f concave up?

    c) Given that $f(0.1) = 0$, sketch a possible graph of f .

4.   Given $f(x) = e^{x/2} - \ln(x^3 + 1)$ for $x > -1$, find the x-coordinates of all local maximum points, local minimum points and inflection points to two decimal places.

5.　The graph of the second derivative of g is shown below. Use the graph to answer the following questions about g and g' on the interval $[-4, 4]$.

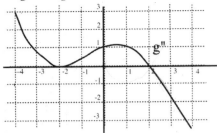

a) Where is the graph of g concave down?

b) Where does g have points of inflection?

c) Suppose that $g'(0) = 0$. Is g increasing or decreasing at $x = 2$? Justify your answer briefly.

6.　The graph of the derivative of the function f is shown below. Use the graph to answer questions about f on the interval $[-1, 8]$.

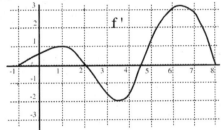

a) Suppose that $f(1) = 5$. Find an equation of the tangent line to the graph of f at the point $(1, 5)$.

b) On what intervals is f both increasing and concave down?

c) Where does f have a local minimum?

d) Assume that $f(0) = 0$. Sketch a possible graph of f.

7.　The derivative of a function f is given. Use a graph of the derivative to answer the following questions.

　　i)　On what intervals is f increasing? decreasing?

　　ii)　On what intervals is the graph of f concave up? concave down?

　　iii)　Where does f have points of inflection?

a) $f'(x) = (x-1)^2(x-2)$; $-4 \le x \le 4$

b) $f'(x) = x \cos x$; $-2\pi \le x \le 2\pi$

c) $f'(x) = \sqrt{x^3 + 2x^2} - 1$; $-2 \le x \le 2$

8.   Sketch the graph of $f(x) = x^{1/3}$.

       a) Locate any inflection points.

       b) Does f'' exist at each inflection point?

9.   Assume the function f is defined on the closed interval $[0, 8]$ so that $f(2) = 1$, $f(4) = 3$ and $f(6) = 6$. Using the following information, sketch a graph of f.

	$0 \le x < 2$	2	$2 < x < 4$	4	$4 < x < 6$	6	$6 < x \le 8$
$f'(x)$	−	0	+	+	+	0	−
$f''(x)$	+	+	+	0	−	−	−

10.   In the left-hand column below are graphs of several functions. In the right-hand column are graphs of the second derivative functions. Match each function with its second derivative.

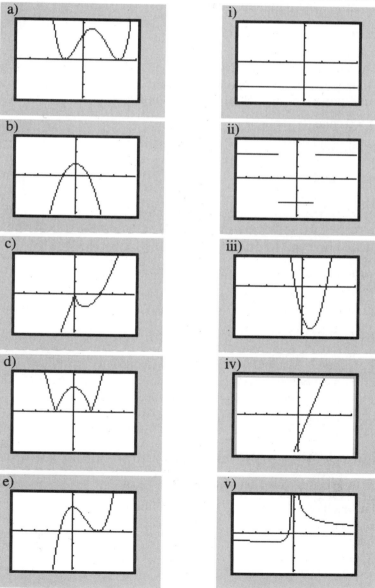

11.   Draw a possible graph of a function f that has all of the following properties:

   i) for $x < 0$, $f'(x) > 0$ and $f''(x) > 0$;

   ii) $f(0) = 3$, $f'(0) > 0$, $f''(0) = 0$;

   iii) for $0 < x < 2$, $f'(x) > 0$ and $f''(x) < 0$;

   iv) $f'(2) = 0$;

   v) for $x > 2$, $f'(x) < 0$ and $f''(x) < 0$.

12.   The function f is defined on the interval $[0, 2\pi]$ by $f(x) = \sin^2 x + 2 \sin x$.
       Use your calculator to locate

   a) the maximum and minimum values of $f(x)$ on $[0, 2\pi]$;

   b) the inflection points on the graph of f.

13.   Several students at Dartmouth College become ill with a cold virus. Let $N = f(t)$ represent
       the total number of students who have become infected with the virus at the end of t days.
       Give an interpretation for each of the following conditions.

   a) $f'(t) > 0$ and $f''(t) < 0$.

   b) $f'(t) < 0$ and $f''(t) < 0$.

   c) $f'(t) > 0$ and $f''(t) > 0$.

14.   Let $y = f(t)$ be the number of people who have contracted a contagious disease at time t.
       Describe each of the following situations by assigning numerical signs (positive, negative
       or zero) to the first and second derivatives of f.

   a) The number of ill people is increasing, but at a slower rate than in the previous time
      period.

   b) The number of ill people is decreasing at a faster rate than in previous time periods.

   c) The number of ill people is increasing, but at a constant rate.

15.   A patrol car travels east and west on the Massachusetts Turnpike. The distance function
       $d = f(t)$, plotted below, gives the car's distance, in miles, west of Worcester at time t, in
       hours.

   a) Estimate the car's velocity at $t = 1$. What
      direction is the car traveling?

   b) Estimate the car's velocity at $t = 2$.

   c) At what time is the patrol car moving fastest east?

   d) Translate each of the following statements about
      the distance function f into a sentence about the
      patrol car.

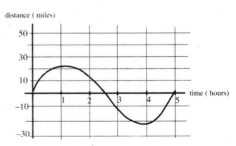

   i) $f(2) = 12$ ii) $f'(2) = -30$ iii) $f''(2.5) = 0$

16. A car starts from rest and its distance traveled is recorded every 2 seconds.

t (seconds)	0	2	4	6	8	10	12
s (feet)	0	10	38	92	180	240	290

a) Estimate the speed at 4 seconds.

b) Estimate the coordinates of the inflection point of the graph of the position function. What is its significance?

17. The figure shows the graphs of f, f', and f''. Identify each curve and explain your choices.

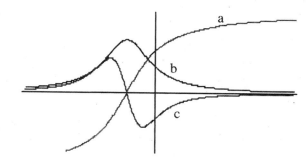

18. The figure below shows the graph of three functions. One is the position function of a car, one is the velocity of the car, one is its acceleration. Identify each curve.

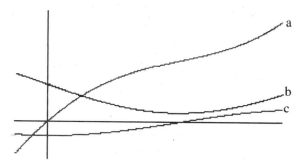

19. Water is flowing into a spherical tank at a constant rate. Let $V(t)$ represent the volume of water in the tank and $H(t)$ represent the height of water in the tank at time t.

a) What is the meaning of $V'(t)$ and $H'(t)$? Are these derivatives positive, negative or zero?

b) Are the values of $V''(t)$ positive, negative or zero?

c) Is the value of $H''(t)$ positive, negative or zero when the tank is

     i) one quarter full?   ii) half-full?   iii) three-quarters full?

2.7   The Limit of a Function

Having seen how limits arise when we want to define the tangent line to a curve, we now take a closer look at limits in general and the methods for computing them. Then we will use limits to define another important concept: **continuity** of a function at a point of its domain.

Limits are used to describe the behavior of a function's values, or outputs, near a number $x = a$ where the function may not be defined.

> The function f has a limit L at $x = a$, written
>
> $$\lim_{x \to a} f(x) = L$$
>
> if we can make outputs $f(x)$ as close to L as we wish by taking inputs x sufficiently close to a but not equal to a.

Example 1:   *Finding the Limit of a Function*

Use the TI-83 to examine the behavior of the function $f(x) = \dfrac{x^2 - 9}{x - 3}$ near $x = 3$.

a) Give the domain of f.

b) Graphically and numerically examine f at points near $x = 3$ and use the results to estimate $\displaystyle\lim_{x \to 3} f(x)$.

Solution:

a) The domain of f is all real numbers except 3.

b) Graphing f with a viewing rectangle of [–14.1, 14.1] by [–9.3, 9.3] gives the following picture.

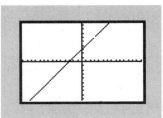

The graph appears to be a straight line, and it looks like the function has values near 6 when x is near 3, although the point (3, 6) is missing.

Numerically we can examine values of $f(x)$ for x close to 3. First press $\boxed{\text{2nd}}$ [TblSet] and set TblStart = 2.97 and ΔTbl = .01. Then press $\boxed{\text{2nd}}$ [TABLE] to see the following table.

Notice that we have taken values of x near 3 from below ($x < 3$) and from above ($x > 3$). Both the graphical and numerical evidence suggest that the limit is 6.

Observe, finally, that the graph of f looks like the line $y = x + 3$. A little algebra shows why. If $x \neq 3$, we have

$$f(x) = \frac{x^2 - 9}{x - 3} = \frac{(x+3)(x-3)}{x-3} = x + 3.$$

Thus for $x \neq 3$, the graph of f is identical to the graph of the straight line $g(x) = x + 3$. Since 3 isn't in the domain of f, we have a hole; however, 3 is in the domain of g, with $g(3) = 6$. Since f is the identical to g except for whatever value g has at $x = 3$,

$$\lim_{x \to 3} f(x) = \lim_{x \to 3} g(x) = 6.$$

Limits are important in calculus. One reason is that functions with "holes" in their domain, like $f(x) = \dfrac{x^2 - 9}{x - 3}$ in the example above, arise when the notion of tangent lines to curves is discussed.

X

Example 2: *Finding the Limit of Secant Slopes*

The figure below shows the secant line from $x = 3$ to $x = a$ on the graph of $y = x^2$.

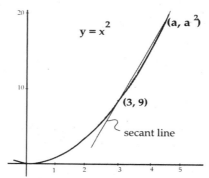

Let $m(a) = $ the slope of this secant line joining $(3, 9)$ and (a, a^2).

a) What is the domain of the function m?

b) Find $\lim_{a \to 3} m(a)$.

Solution:

The slope of a line is determined by two different points. Thus, the function m is defined by

$$m(a) = \frac{\text{rise}}{\text{run}} = \frac{a^2 - 9}{a - 3} .$$

a) The domain of the function m is all real numbers except $a = 3$.

b) From the previous example, $\lim_{a \to 3} (\text{secant slope}) = \lim_{a \to 3} m(a) = 6$.

Notice that as the point (a, a^2) is chosen closer to the fixed point $(3, 9)$, the secant line approaches the tangent line to the graph at $x = 3$. Thus, $\lim_{a \to 3} m(a) = 6$ means that the slope of the tangent line at $x = 3$ is 6 and $f'(3) = 6$.

These examples illustrate an important point about limits. A limit is used to describe the behavior of a function near a point but not at the point. The function need not even be defined at the point. If it is defined there, the value of the function at the point does not affect the limit.

The graphs of functions f, g and h are shown in the figure below. Note that $g(a) \neq L$ and $h(a)$ is not defined. However, in each case, regardless of what happens at $x = a$, the limit as x approaches a is L.

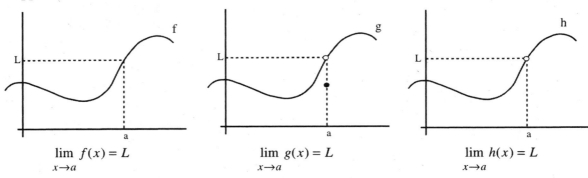

$$\lim_{x \to a} f(x) = L \qquad \lim_{x \to a} g(x) = L \qquad \lim_{x \to a} h(x) = L$$

The slope of the secant line connecting the point $(0, 0)$ and a nearby point $(x, \sin x)$ on the graph $y = \sin x$ is determined by the quotient $\dfrac{\sin x - 0}{x - 0} = \dfrac{\sin x}{x}$. Therefore, the slope of the tangent line to the graph of $y = \sin x$ at the point $(0, 0)$ is related to $\lim\limits_{x \to 0} \dfrac{\sin x}{x}$.

Example 3: *A Limit Involving a Trigonometric Function*

Graphically and numerically examine $y = \dfrac{\sin x}{x}$ at points near $x = 0$ and use the results to estimate $\lim\limits_{x \to 0} f(x)$.

Solution:

Graphing f in the **ZOOM** [4:ZDecimal] viewing rectangle gives the following picture.

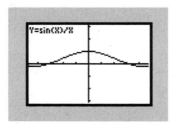

This graph indicates a value of 1 at $x = 0$; however, we know that 0 is not in the domain of f. If we zoom in on the point of the curve near $x = 0$, it more and more resembles the horizontal line $y = 1$; graphically we guess

$$\lim_{x \to 0} \frac{\sin x}{x} = 1.$$

Numerically we can examine values of $f(x)$ for inputs close to 0.

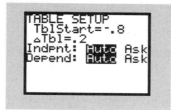
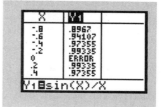

From both directions, $x > 0$ and $x < 0$, the values Y1 $= f(x)$ are close to 1. Both the graphical and numerical evidence suggest that the limit is 1.

This limit means that the line tangent to the graph of $y = \sin x$ at the point where $x = 0$ (i.e., at $(0, 0)$) has a slope of 1. Hence the tangent line is $y = x$.

Since a TI-83 generated graph is only a collection of dots, a graph can be misleading. For example, holes between adjacent points may go undetected. In the example above, the hole in the graph of $\dfrac{\sin x}{x}$ at $x = 0$ is not evident because the y-axis fills it in. If we turn **Axes Off** in the FORMAT menu, and then plot $y = \dfrac{\sin x}{x}$, this graph clearly shows that $\dfrac{\sin x}{x}$ is not defined at $x = 0$.

In general, the TI-83 may be employed to numerically or graphically investigate limits; however, the precision limitations of the calculator do not permit evaluating a function at points arbitrarily close to a given number.

One-sided Limits

It is sometimes useful to consider the existence of limits when inputs are restricted to one side or the other of a particular point.

Consider the function f defined by $f(x) = \dfrac{|x|}{x}$.

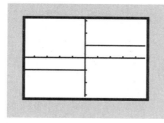
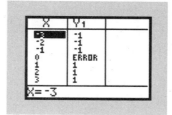
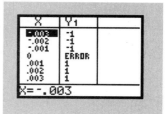

Notice that if x is close to, but greater than 0, then $\dfrac{|x|}{x}$ is equal to 1. If x is close to, but less than 0, then $\dfrac{|x|}{x}$ is equal to -1. We represent this condition by the statements:

$$\lim_{x \to 0^+} \frac{|x|}{x} = 1 \quad \text{and} \quad \lim_{x \to 0^-} \frac{|x|}{x} = -1$$

and refer to the **right-hand limit** and **left-hand limit** of the function $f(x) = \dfrac{|x|}{x}$ at $x = 0$.

Right-hand limit: $\lim\limits_{x \to a^+} f(x) = L$ means that $f(x)$ can be made as close to L as we wish by taking x sufficiently close to a but greater than a.

Left-hand limit: $\lim\limits_{x \to a^-} f(x) = L$ means that $f(x)$ can be made as close to L as we wish by taking x sufficiently close to a but less than a.

Whenever the limit of a function exists and equals L, then the right- and left-hand limits will exist and equal L. Notice that in Example 3 on page 132, we have

$$\lim_{x \to 0^+} \frac{\sin x}{x} = \lim_{x \to 0^-} \frac{\sin x}{x} = \lim_{x \to 0} \frac{\sin x}{x} = 1$$

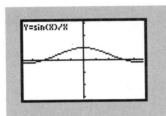
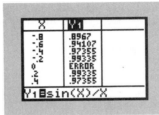
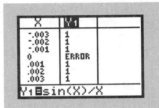

On the other hand, if the left- and right-hand limits exist, then the two-sided limit exists only if the one-sided limits are equal.

$$\lim_{x \to 0^+} \frac{|x|}{x} \ne \lim_{x \to 0^-} \frac{|x|}{x}; \text{ therefore, } \lim_{x \to 0} \frac{|x|}{x} \text{ does not exist.}$$

Rules for Calculating Limits

The following rules make it easy to calculate limits (and one-sided limits) of combinations of functions.

If $\lim\limits_{x \to a} f(x) = L$, $\lim\limits_{x \to a} g(x) = M$, and k is a constant, then

1. **Sum Rule:** $\lim\limits_{x \to a} \left[f(x) + g(x) \right] = L + M.$

2. **Difference Rule:** $\lim\limits_{x \to a} \left[f(x) - g(x) \right] = L - M$

3. **Product Rule:** $\lim\limits_{x \to a} \left[f(x) \cdot g(x) \right] = \lim\limits_{x \to a} f(x) \cdot \lim\limits_{x \to a} g(x) = L \cdot M.$

4. **Constant Multiple Rule:** $\lim\limits_{x \to a} k \cdot f(x) = k \cdot L.$

5. **Quotient Rule:** $\lim\limits_{x \to a} \dfrac{f(x)}{g(x)} = \dfrac{\lim\limits_{x \to a} f(x)}{\lim\limits_{x \to a} g(x)} = \dfrac{L}{M}$, if $M \ne 0$.

In words, the Sum rule says that the limit of a sum of functions is the sum of their limits. Similarly, the Quotient Rule says that the limit of the quotient of two functions is the quotient of their limits, provided that the limit of the denominator is not zero. Try to state the other parts in words.

Example 4: *Using Graphs to Approximate Function Limits*

The graphs of functions f and g are given below. Use the graphs to evaluate each limit, if it exists. If it does not exist explain why.

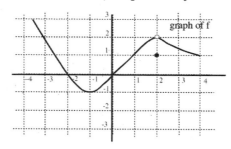 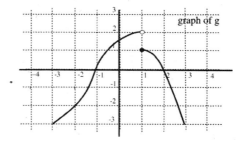

a) $\lim_{x \to 2} [f(x) + 3g(x)]$

b) $\lim_{x \to 1} [f(x) \cdot g(x)]$

c) $\lim_{x \to -1} \dfrac{f(x)}{g(x)}$

Solution:

a) From the graphs of f and g we see that $\lim_{x \to 2} f(x) = 2$ and $\lim_{x \to 2} g(x) = 0$.

   Therefore, we have

$$\lim_{x \to 2} [f(x) + 3g(x)] = \lim_{x \to 2} f(x) + 3 \lim_{x \to 2} g(x) = 2 + 3 \cdot 0 = 2.$$

b) We see that $\lim_{x \to 1} f(x) = 1$, but $\lim_{x \to 1} g(x)$ does not exist because the left- and right-limits are different:

$$\lim_{x \to 1^-} g(x) = 2 \quad \text{and} \quad \lim_{x \to 1^+} g(x) = 1.$$

   So the Product Rule can not be used. In this case, it can be shown that the $\lim_{x \to 1} [f(x) \cdot g(x)]$ does not exist.

c) The graphs show that

$$\lim_{x \to -1} f(x) = -1 \quad \text{and} \quad \lim_{x \to -1} g(x) = 0$$

   Because the limit of the denominator is 0, the Quotient Rule can not be used. The given limit does not exist.

The condition in Quotient Rule that $\lim_{x \to a} g(x) \neq 0$ is essential. This does not necessarily mean that the quotient $\dfrac{f(x)}{g(x)}$ has no limit if $\lim_{x \to a} g(x) = 0$, but it does mean that the Quotient Rule may not be used to find it.

Two special limits: i) $\lim\limits_{x \to a} c = c$, (where c is any constant), and

ii) $\lim\limits_{x \to a} x = a$,

along with the five limit rules make it possible to calculate the limit of any polynomial function and most rational functions.

Example 5:   *The Limit of a Rational Function*

Find: $\lim\limits_{x \to 2} \dfrac{x^2}{3x + 4}$.

Solution:

$$\lim_{x \to 2} \frac{x^2}{3x + 4} = \frac{\lim\limits_{x \to 2} x^2}{\lim\limits_{x \to 2} (3x + 4)} \qquad \text{Rule 5}$$

$$= \frac{\lim\limits_{x \to 2} x \cdot \lim\limits_{x \to 2} x}{3 \lim\limits_{x \to 2} x + \lim\limits_{x \to 2} 4} \qquad \text{Rule 1, 3, and 4}$$

$$= \frac{2 \cdot 2}{3 \cdot 2 + 4} \qquad \text{i) and ii) above}$$

$$= \frac{2}{5}$$

Note that if the function f is defined as $f(x) = \dfrac{x^2}{3x + 4}$, then $f(2) = \dfrac{2}{5}$. In other words, the limit in Example 5 could have been found by evaluating the function f at $x = 2$. This is true in general for polynomial functions and most rational functions.

> If f is a polynomial or rational function and a is in the domain of f, then
>
> $$\lim_{x \to a} f(x) = f(a).$$

Functions whose limit at an input $x = a$ is found by evaluating the function at a are called *continuous at a* and will be studied in the next section.

Not all limits can be evaluated by direct substitution as the following example illustrates.

Example 6:   *Factoring to Obtain the Limit of a Rational Function*

Find: $\lim\limits_{x \to 3} \dfrac{x^2 - x - 6}{x - 3}$.

Solution:

Although we are taking the limit of a rational function, we are unable to apply the quotient rule for limits because the limit of the denominator is 0. However, by factoring the numerator and canceling, we obtain

$$\lim_{x \to 3} \frac{x^2 - x - 6}{x - 3} = \lim_{x \to 3} \frac{(x - 3)(x + 2)}{x - 3} = \lim_{x \to 3} (x + 2) = 5.$$

2.7   Exercises

In Exercises 1–10, use the TI-83 to find numerical or graphical evidence that the given function does or does not have a limit. If the limit does not exist, decide whether the left-hand and right-hand limits exist.

1. $\displaystyle\lim_{x\to0}\frac{\sqrt{x+1}-1}{x}$

2. $\displaystyle\lim_{x\to0}\frac{\sin(2+x)-\sin 2}{x}$

3. $\displaystyle\lim_{x\to0}\frac{e^{1+x}-e^{1}}{x}$

4. $\displaystyle\lim_{x\to2}\frac{x^{2}-4}{x^{2}-5x+6}$

5. $\displaystyle\lim_{x\to0}\frac{\cos x-1}{x}$

6. $\displaystyle\lim_{x\to0}\frac{x^{2}+x}{\sqrt{4x^{2}+x^{3}}}$

7. $\displaystyle\lim_{x\to0} x\sin\left(\frac{1}{x}\right)$

8. $\displaystyle\lim_{x\to0}\frac{\sin x-x+x^{3}/6}{x^{5}}$

9. $\displaystyle\lim_{x\to0}(1+x)^{1/x}$

10. $\displaystyle\lim_{x\to2}\sqrt{x-2}$

11.   Each of the following limits is $f'(a)$ for some function f and some number a. Identify f and a and use your calculator to estimate the value of the limit.

a) $\displaystyle\lim_{h\to0}\frac{2^{(3+h)}-8}{h}$

b) $\displaystyle\lim_{h\to0}\frac{2\sin(3+h)-2\sin 3}{h}$

c) $\displaystyle\lim_{h\to0}\frac{\sqrt{1+h}-1}{h}$

d) $\displaystyle\lim_{h\to0}\frac{\tan\left(\frac{\pi}{4}+h\right)-\tan\left(\frac{\pi}{4}\right)}{h}$

e) $\displaystyle\lim_{h\to0}\frac{(2+h)^{2}-4}{h}$

f) $\displaystyle\lim_{h\to0}\frac{\ln(1+h)-\ln(1)}{h}$

12.   Let f be the function whose graph is shown below. Use the graph to evaluate each limit, if it exists. If it does not exist explain why.

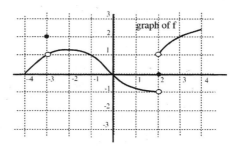

a) $\lim_{x \to -3^-} f(x)$

b) $\lim_{x \to -3^+} f(x)$

c) $\lim_{x \to -3} f(x)$

d) $\lim_{x \to 2^-} f(x)$

e) $\lim_{x \to 2^+} f(x)$

f) $\lim_{x \to 2} f(x)$

13.   Let g be the function whose graph is shown below. Use the graph to evaluate each quantity if it exists. If it does not exist explain why.

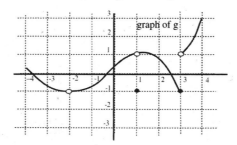

a) $\lim_{x \to -2} g(x)$

b) $\lim_{x \to 1} g(x)$

c) $g(1)$

d) $\lim_{x \to 3^-} g(x)$

e) $\lim_{x \to 3^+} g(x)$

f) $\lim_{x \to 3} g(x)$

In Exercises 14–27, use the limit rules to find the indicated limits when they exist. Confirm your answers graphically.

14. $\lim_{x \to 2} (2x^2 + x - 1)$

15. $\lim_{x \to 0} \dfrac{x}{x^2 + x + 1}$

16. $\lim_{x \to 2} \dfrac{x^2 + 4}{x - 1}$

17. $\lim_{x \to 0} \dfrac{x^3 + x}{x}$ [Hint: Factor and cancel]

18. $\lim_{x \to 0} x \cos(2x)$

19. $\lim_{h \to 0} \dfrac{(3 + h)^2 - 3^2}{h}$ [Hint: Expand and cancel]

20. $\lim\limits_{x \to 0} e^{-x}(x^2 - 5)$

21. $\lim\limits_{x \to e} \dfrac{\ln x}{x}$

22. $\lim\limits_{x \to 4} \dfrac{x - \sqrt{x}}{x + \sqrt{x}}$

23. $\lim\limits_{x \to k} \dfrac{x^3 - kx^2}{x^2 - k^2}$ [$k \neq 0$]

24. $\lim\limits_{x \to 0} \dfrac{x^2 - 3x + 2}{x^2 - x - 2}$

25. $\lim\limits_{x \to 1} \dfrac{1 - \frac{1}{x}}{x - 1}$

26. $\lim\limits_{x \to 0} \dfrac{e^x - e^{-x}}{e^x + e^{-x}}$

27. $\lim\limits_{x \to 0} e^{-x} \cos(3x)$

28.   Solve each of the following for the constant h.

a) $\lim\limits_{x \to 4} (3x^2 + h) = 6$

b) $\lim\limits_{x \to 4} \dfrac{x + 2h}{x + h} = 3$

c) $\lim\limits_{x \to 2} \left[3(x + h)^2 - 2(x + h)\right] = 0$

d) $\lim\limits_{x \to -1} (hx^2 + h) = 2$

29.   Assume $\lim\limits_{x \to 0} \sin x = 0$ and $\lim\limits_{x \to 0} \cos x = 1$. Find each of the following limits.

a) $\lim\limits_{x \to 0} \sin^2 x$

b) $\lim\limits_{x \to 0} \sin(2x)$

c) $\lim\limits_{x \to 0} \tan x$

30.   Let f be the piecewise function defined by

$$f(x) = \begin{cases} x^2 & , \ x < 1 \\ |x - 5| & , \ 1 < x < 5 \\ \dfrac{1}{x - 5} & , \ x > 5 \end{cases}$$

Enter f as $Y_1 = (X < 1)X^2 + (X > 1)(X < 5)\text{abs}(X - 5) + (X > 5)(1/(X - 5))$. Determine

a) domain of f

b) range of f

c) $\lim\limits_{x \to 5^-} f(x)$

d) $\lim\limits_{x \to 5^+} f(x)$

e) $\lim\limits_{x \to 5} f(x)$

f) $\lim\limits_{x \to 0} f(x)$

31.   Show by means of an example that $\lim\limits_{x \to a} \left[f(x) \cdot g(x)\right]$ may exist even though neither $\lim\limits_{x \to a} f(x)$ nor $\lim\limits_{x \to a} g(x)$ exists.

2.8   **Continuity**

In the preceding section we discovered that the limit of a function as x approaches a can often be found by simply calculating the value of the function at a . This is not always true; however, when $\lim\limits_{x \to a} f(x) = f(a)$ we say the function f is continuous at the input a. More explicitly, there are three requirements for a function to be continuous at $x = a$.

A function f is **continuous at** $x = a$ if

1) $\lim\limits_{x \to a} f(x)$ exists,

2) f is defined at a, and

3) $\lim\limits_{x \to a} f(x) = f(a)$.

If a function f is not continuous at a, we say f is **discontinuous at** a, or f has a **discontinuity at** a.

Continuity is a point property of a function. And a function may be continuous at some of its inputs and discontinuous at other locations. However, if a function f is continuous at each point in an interval I, we say that f is continuous on the interval I. Geometrically we consider a function continuous if its graph contains no holes, jumps or gaps.

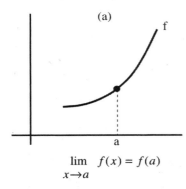

(a)

$$\lim\limits_{x \to a} f(x) = f(a)$$

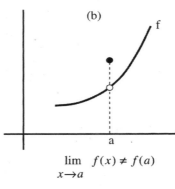

(b)

$$\lim\limits_{x \to a} f(x) \neq f(a)$$

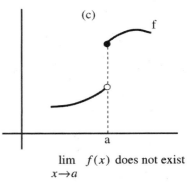

(c)

$$\lim\limits_{x \to a} f(x) \text{ does not exist}$$

Consider the figure above. In (a), f is continuous at $x = a$. In (b), f is discontinuous at $x = a$ because $\lim\limits_{x \to a} f(x) \neq f(a)$. In (c), f is discontinuous at $x = a$ because $\lim\limits_{x \to a} f(x)$ does not exist. The kind of discontinuity illustrated in part (b) is called **removable** because the discontinuity can be "removed" by redefining f at $x = a$.

Example 1:   *Continuity of a Function at a Point*

Show that if the function f is defined by $f(x) = \begin{cases} \dfrac{x^2 - 4}{x - 2} & , \ x \neq 2 \\ 4 & , \ x = 2 \end{cases}$ then f is continuous at

$x = 2$.

Solution:

For $x \neq 2$, $f(x) = \dfrac{(x-2)(x+2)}{x-2} = x+2$. Thus, $\lim\limits_{x \to 2} f(x) = 4$. Also, $f(2) = 4$, so that $\lim\limits_{x \to 2} f(x) = f(2)$ and f is continuous at $x = 2$.

Example 2: *Investigate Continuity Graphically*

Investigate the continuity of the functions $f(x) = \dfrac{x(x-1)}{x-1}$ and $g(x) = \dfrac{x+1}{x-1}$ on the interval $[-4, 4]$.

Solution:

Enter and graph the functions in the ZDecimal viewing rectangle. Both functions are defined and continuous everywhere except at $x = 1$, where they are undefined and therefore discontinuous. However, the function f has a removable discontinuity at $x = 1$.

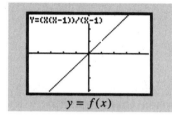
$y = f(x)$

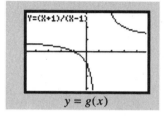
$y = g(x)$

Since the screen resolution of the TI-83 is low, discontinuities from breaks in the graph of the function may not show up. Thus to determine conclusively whether a function is continuous we must use our knowledge of its domain and limits.

Example 3: *A Discontinuous Function*

Investigate the continuity of the function $f(x) = \begin{cases} 3-x, & \text{if } x \geq 1 \\ x^2 + 2x, & \text{if } x < 1 \end{cases}$.

Solution:

Enter and graph the function in the ZDecimal viewing rectangle. The function f is defined everywhere but the left-hand and right-hand limits as x approaches $x = 1$ are not equal. Therefore the limit as x approaches 1 does not exist and the function is not continuous at $x = 1$. The jump in the graph of f at $x = 1$ is more easily seen when using Dot Mode, instead of Connected Mode.

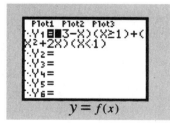
$y = f(x)$

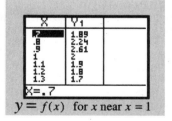
$y = f(x)$ for x near $x = 1$

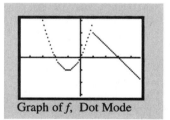
Graph of f, Dot Mode

Most of the elementary functions are continuous at all points where they are defined. For example, polynomial functions and rational functions (quotients of polynomial functions) are continuous at every number in their domains. Exponential functions are continuous everywhere, as are the sine and cosine. And from the limit rules it follows that if functions f and g are continuous at $x = a$ and c is a constant, then so are the functions $f + g$, $f \cdot g$, cf, and f / g if $g(a) \neq 0$.

Example 4: *The Limit of a Rational Function*

Find $\lim\limits_{x \to 2} \dfrac{x^3 - x + 1}{5 - x}$.

Solution:

The function $f(x) = \dfrac{x^3 - x + 1}{5 - x}$ is rational and is therefore continuous on its domain, which consists of all real numbers $x \neq 5$. Thus

$$\lim_{x \to 2} \frac{x^3 - x + 1}{5 - x} = \lim_{x \to 2} f(x) = f(2)$$

$$= \frac{(2)^3 - 2 + 1}{5 - 2} = \frac{7}{3}.$$

Continuity on an Interval

This section includes the statements of two important and geometrically obvious theorems. The hypothesis of each theorem includes an assumption that a function f is continuous on a closed interval.

The concept of a one-sided limit makes it possible to extend the idea of continuity to closed intervals.

> **A function f is continuous on the closed interval $[a, b]$** if it is continuous at every number in the open interval (a, b) and
>
> $$\lim_{x \to a^+} f(x) = f(a) \quad \text{and} \quad \lim_{x \to b^-} f(x) = f(b).$$
>
> In this case, we say that the function f is **right continuous** at $x = a$ and **left continuous** at $x = b$.

For example, the function g defined by $g(x) = \sqrt{4 - x^2}$ is continuous on the closed interval $[-2, 2]$, since it is continuous on the open interval $(-2, 2)$ and

$$\lim_{x \to -2^+} g(x) = g(-2) \quad \text{and} \quad \lim_{x \to 2^-} g(x) = g(2).$$

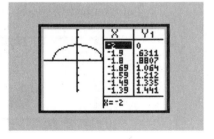

We tend to think of continuous functions as having graphs which are uninterrupted. Thus, if a function is continuous on an interval $[a, b]$ and k is a number between $f(a)$ and $f(b)$, then we would expect that there is an input c in the interval $[a, b]$ for which $f(c) = k$.

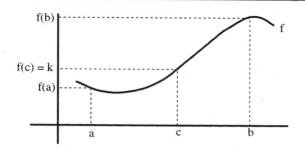

> **The Intermediate Value Theorem**: If the function f is continuous on the closed interval $[a, b]$, with $f(a) \neq f(b)$, and k is a number between $f(a)$ and $f(b)$, then there exists at least one number c in (a, b) for which $f(c) = k$.

The Intermediate Value Theorem is useful in locating zeros of continuous functions. For example, consider the polynomial function $f(x) = x^3 - 3x + 1$.

Since polynomials are continuous at all inputs x, and $f(0) = 1$ and $f(1) = -1$, the Intermediate Value Theorem guarantees the existence of at least one c in the interval $[0, 1]$ such that $f(c) = 0$, as show in the figure.

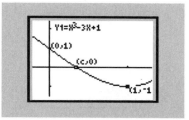

The **bisection method** for approximating the zeros of a continuous function also uses the Intermediate Value Theorem and is similar to the method just used in locating the zero of the function $f(x) = x^3 - 3x + 1$ on the interval $[0, 1]$. If $f(a)$ and $f(b)$ have different signs, then either $f(a)$ and $f\left(\dfrac{a+b}{2}\right)$ have different signs or $f\left(\dfrac{a+b}{2}\right)$ and $f(b)$ have different signs, where of course $\dfrac{a+b}{2}$ is the midpoint of the interval $[a, b]$.

From the signs of $f(a)$, $f\left(\dfrac{a+b}{2}\right)$ and $f(b)$ it can be determined which half of the original interval contains the zero. By repeatedly bisecting intervals, an approximation of the function zero can be found.

A second property which holds for a function that is continuous on a closed interval is the existence of a largest and a smallest function value. That is, if f is continuous on $[a, b]$, there exist numbers c and d in the interval $[a, b]$ such that $f(c) \leq f(x) \leq f(d)$ for all x in $[a, b]$.

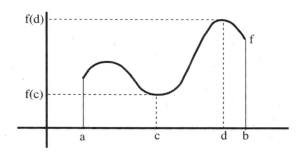

The Extreme Value Theorem : If the function f is continuous on the interval $[a, b]$, then there exist numbers c and d in $[a, b]$ such that for all x in $[a, b]$,

$$f(c) \le f(x) \quad \text{and} \quad f(d) \ge f(x).$$

Often the key step in solving important problems in mathematics is finding the maximum or minimum value of a function. Notice, however, that the Extreme Value Theorem above merely asserts that minimum and maximum values exist; it does not tell us how to find them. In Chapter 5 techniques for calculating maximum or minimum function values will be developed.

Differentiable Functions are Continuous

It is important to recognize that a function can be continuous at $x = a$ without being differentiable at that point. For example, if $f(x) = |x|$, then $f'(0)$ does not exist but f is continuous at $x = 0$ (and everywhere). However, a function that is differentiable at $x = a$ is automatically continuous at that point because the difference quotients $\dfrac{f(a+h) - f(a)}{h}$ could not possibly have a limiting value unless $\lim\limits_{h \to 0} f(a+h) = f(a)$.

Theorem : If the function f has a derivative at $x = a$, then f is continuous at $x = a$.

For many functions the easy way to demonstrate continuity at $x = a$ is to show that the function is differentiable at a.

Example 5: *Graphically Investigating the Continuity and Differentiability of a Function*

State whether the following functions are continuous, differentiable, both, or neither at $x = a$.

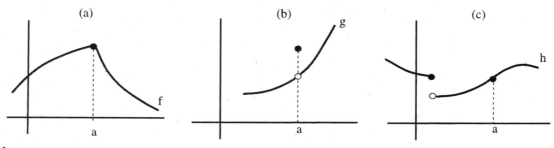

Solution:

a) The function f is continuous but not differentiable. At the cusp, the slopes of approximating secant lines approach different numbers from the left and right.

b) The function g is neither continuous nor differentiable.

c) The function h is continuous and differentiable at $x = a$. The discontinuity elsewhere does not affect the behavior of h at $x = a$.

2.8   Exercises

1.   Let g be the function whose graph is shown below.

    a) State the numbers at which g is discontinuous and explain why.

    b) State the intervals on which g is continuous.

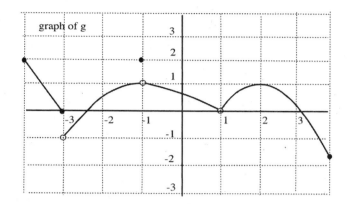

2.   Let f be the function whose graph is shown below. Use the graph to answer the following questions.

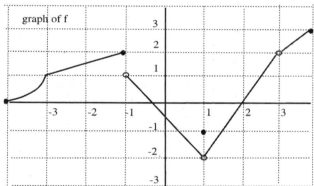

    a) Is f continuous on the interval $[-4,4]$? Justify your answer briefly.

    b) $\displaystyle\lim_{x \to -1^+} f(x) =$

    c) $\displaystyle\lim_{x \to -1^-} f(x) =$

    d) On which intervals is $f'(x) < 0$?

    e) $\displaystyle\lim_{x \to 0} f(x) =$

    f) Is f continuous at $x = 1$? Explain briefly.

3.   Sketch a graph of a function that is continuous everywhere except at $x = 2$ and is continuous from the right at 2.

In Exercises 4–9, find the numbers, if any, at which the given function is discontinuous.

4. $f(x) = \dfrac{1}{x-1}$ 5. $f(x) = \dfrac{x^2 - 1}{x+1}$

6. $f(x) = |x+2|$ 7. $f(x) = \begin{cases} 1-x, & \text{if } x \le 2 \\ x^2 - 2x, & \text{if } x > 2 \end{cases}$

8. $f(x) = e^x \sin x$ 9. $f(x) = \ln|x|$

In Exercises 10–14, find a number b that makes the given function continuous at $x = 1$.

10. $f(x) = \begin{cases} bx^2 - 1 & \text{, if } x < 1 \\ x & \text{, if } x \ge 1 \end{cases}$

11. $f(x) = \begin{cases} -2x + b & \text{, if } x < 1 \\ \ln x & \text{, if } x \ge 1 \end{cases}$

12. $f(x) = \begin{cases} 2x^2 - x - b & \text{, if } x < 1 \\ 2e^{x-1} & \text{, if } x \ge 1 \end{cases}$

13. $f(x) = \begin{cases} (x+b)^2 & \text{, if } x < 1 \\ (x - 2b)^2 & \text{, if } x \ge 1 \end{cases}$

14. $f(x) = \begin{cases} \dfrac{b}{x+1} & \text{, if } x < 1 \\ \cos(x-1) & \text{, if } x \ge 1 \end{cases}$

15.   Show that if the function g is defined by $g(x) = \begin{cases} \dfrac{x^2 - 9}{5} & \text{, if } x \ne 3 \\ 5 & \text{, if } x = 3 \end{cases}$

then g is not continuous at $x = 3$.

16.   Let f be the function whose graph is shown to the right. Use the graph to answer the following questions.

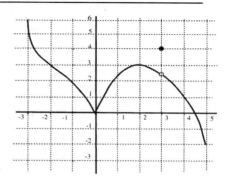

a) Is f continuous on the interval $[-3, 5]$? Justify your answer briefly.

b) Does f have any critical numbers? Explain briefly.

c) On which intervals is $f'(x) > 0$?

d) On which intervals is f' increasing?

e) Identify any inflection points.

f) $\lim\limits_{x \to 0} f(x) =$

g) Is f continuous at $x = 0$? Explain briefly.

h) $\lim\limits_{x \to 2} f'(x) =$

i) $\lim\limits_{h \to 0^+} \dfrac{f(0 + h) - f(0)}{h}$

j) $\lim\limits_{h \to 0^-} \dfrac{f(0 + h) - f(0)}{h}$

k) $\lim\limits_{x \to 0} f'(x) =$

17.   Sketch the derivative for the function in Exercise 16.

18.   Let f be the function whose graph is shown below. Use the graph to answer the following questions.

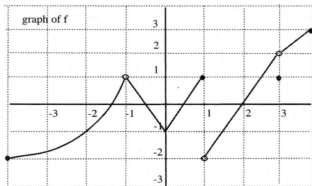

graph of f

a) Is f continuous on the interval $[-4,4]$? Justify briefly.

b) Does f have any critical numbers? Explain briefly.

c) On which intervals is $f'(x) > 0$?

d) $\lim\limits_{x \to 3} f(x) =$

e) Is f continuous at $x = -1$? Explain briefly.

f) $\lim\limits_{x \to 2} f'(x) =$

g) $\lim\limits_{h \to 0^+} \dfrac{f(0 + h) - f(0)}{h} =$

h) $\lim\limits_{h \to 0^-} \dfrac{f(0 + h) - f(0)}{h} =$

19.   Sketch the derivative for the function in Exercise 18.

20.   Let f be the function whose graph is shown to
        the right. Use the graph to answer the following
        questions.

        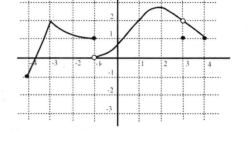

        a) For what values of x is f discontinuous?

        b) On which intervals is $f'(x) > 0$?

        c) Show that f is not continuous at $x = 3$.

        d) On which intervals is $f''(x) > 0$?

        e) Find $\lim\limits_{x \to -1^+} f(x)$.

21.   Let g be the function whose graph is shown to
        the right. Use the graph to answer the following
        questions.

        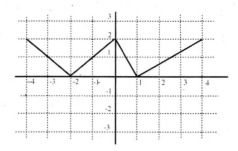

        a) Find $g'(-1)$.

        b) For what values of x is $g'(x) > 0$?

        c) Show that g is continuous at $x = 0$.

        d) Find $\lim\limits_{h \to 0} \dfrac{g(2+h) - g(2)}{h}$.

        e) Sketch a graph of g'.

        f) If possible, determine the extreme values of
             g on $[-4, 4]$.

22.   Functions f and g are defined on the interval $[-3, 3]$ and their graphs are shown below.

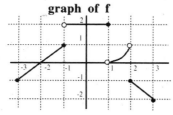

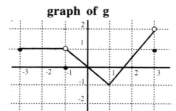

        Neither f nor g is continuous on the entire interval $[-3, 3]$, so neither satisfies the
        hypotheses of the Intermediate Value Theorem or the Extreme Value Theorem. However,
        it is possible that either theorem's conclusion might hold. Explain your answers briefly.

        a) Does the conclusion of the Intermediate Value Theorem hold for f on the interval $[-3, 3]$?

        b) Does the conclusion of the Intermediate Value Theorem hold for g on the interval $[-3, 3]$?

        c) Does the conclusion of the Extreme Value Theorem hold for f on the interval $[-3, 3]$?

        d) Does the conclusion of the Extreme Value Theorem hold for g on the interval $[-3, 3]$?

Chapter 2   Supplementary Problems

1.   A particle moves along a line so that its distance in meters to the right of the starting
     point after t seconds is $s(t) = t^2 - 5t$.

     a) How far does the particle move in the first two seconds?

     b) What is the average velocity over the first two seconds?

     c) What is the instantaneous velocity at $t = 1.5$?

     d) When is the particle stopped?

     e) When is the particle moving to the left?

     f) What total distance does the particle travel during the first 5 seconds?

2.   At the right is a graph of $y = P(t)$,
     which shows the growth over time in
     the population of an endangered
     species.

     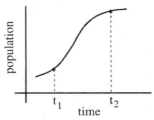

     a) Compare the signs of $P'(t_1)$ and $P''(t_1)$ with the signs of $P'(t_2)$ and $P''(t_2)$.

     b) What conclusion might you expect scientists who are studying the endangered species
        to make at times t_1 and t_2?

3.   A ball is dropped from a bridge. In t seconds the ball drops to a height of

     $$d(t) = 45 - 4.9t^2$$

     meters above the ground.

     a) How high is the ball when it is released?

     b) When is the ball exactly 6 meters off the ground?

     c) Find when the ball hits the ground.

     d) Find the average velocity of the ball between $t = 1$ and $t = 2$.

     e) Estimate the instantaneous velocity of the ball when it is 10 meters off the
        ground.

4.   The pressure P of a gas depends upon its volume. The relationship is given by Boyle's
     law, $P = \dfrac{C}{V}$, where C is a constant. Suppose that $C = 2000$.

     a) Find the average rate of change of P with respect to V as V increases from 100 cubic
        inches to 125 cubic inches.

     b) Find the rate of change of P with respect to V at the instant when $V = 100$ cubic inches.

5.   Sketch a graph of a function f that is consistent with the following data.

x	-2	-1	1	3	4
$f(x)$	-2	0	-1	1	2
$f'(x)$	2	0	0	0	3

6.   Find an equation of the tangent line to the graph of the given function at $x = 1$.

(Hint: Use **nDeriv** to approximate $\dfrac{dy}{dx}$.)

a) $y = x \cdot 2^x$ b) $y = \dfrac{x^2 - 2x}{x+1}$

7.   Determine whether the following statements are True or False.

a) There exists a tangent line to the graph of $y = |x-1|$ at $x = 0$.

b) A tangent line may intersect a graph in more than one point.

c) There exists a tangent line to the graph of $f(x) = \sqrt[3]{x}$ at $x = 0$.

d) The slope of the tangent line to the graph of f at $x = a$ is the instantaneous rate of change of f at $x = a$.

e) The graph of f is concave up at $x = a$ if $f'(a) > 0$.

f) If $f(2) = 3$, $f'(2) = 1$ and $f''(2) = 0$, then the graph of f has an inflection point at $(2, 3)$.

8.   Find the y-intercept of the line tangent to the graph of $f(x) = \dfrac{2}{x}$ at the point $(1, 2)$.

9.   A pair of functions is graphed in each of the following viewing rectangles. Label f and f' in those exercises where one function is the derivative of the other.

a) 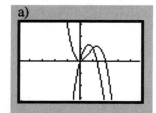   b) 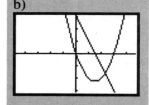   c) 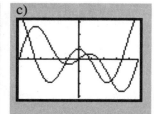   d)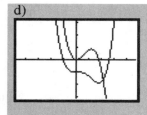

10.   Make a table of values, rounded off to two decimal places for $f(x) = x \sin x$ with $x = 2, 2.5, 3, 3.5, 4$. Use the table to

a) Find the average rate of change of f from $x = 2$ to $x = 3$.

b) Estimate the instantaneous rate of change of f at $x = 3$.

11.   Let $f(x) = \sqrt{x} + \sin x + 10$, $0 \le x \le 10$.

a) Find an equation of the line tangent to the graph of f at $x = 2$.

b) Determine the critical numbers of f to the nearest hundredth.

c) On what intervals is f decreasing?

12.   Suppose the population P of seals on an island at year $t = 0$ (the present) is 50, and the population is growing yearly at a rate of 7% .

    a) Find a formula for the population function $P(t)$ that gives the number of seals t years from now.

    b) How many seals per year are being born at $t = 5$ years?

13.   a) Given $f(x) = \dfrac{1}{x-1}$, use the alternate definition of the derivative $\left[f'(a) = \lim\limits_{x \to a} \dfrac{f(x)-f(a)}{x-a} \right]$ to find $f'(3)$.

    b) On what intervals is $f(x) = x^4 - 4x^3$ both decreasing and concave down?

14.   The graph of f shown to the right has horizontal tangents at $x = -3, -1, 1$ and 3.5.

    a) Where does f have critical numbers?

    b) On what intervals is f' negative?

    c) On what intervals is f' increasing?

    d) Sketch a graph of f' .

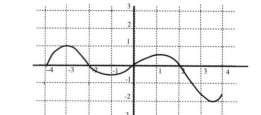

15.   The population, P, of Canada in millions can be approximated by the function

$$P(t) = 22.14(1.015)^t$$

where t is the number of years since the start of 1990. According to this model, how fast is the population growing at the start of 1990 and at the start of 1995?

16.   Use **nDeriv** to find an equation for the tangent line to the graph of the function at the given point. Then graph the function and the tangent line in the same viewing rectangle.

    a) $y = x \cos x$ at $x = 2$;   b) $y = \dfrac{x^2 - 4}{x^2 + 1}$ at $x = 2$.

17.   The graph of g', the derivative of a function g, is shown in the figure for $-4 \le x \le 4$. The graph of g' has horizontal tangents at x = $-3, -1.5$, and 1.5.

    a) On which intervals is g increasing?

    b) Where does g have critical points?

    c) Where does g have a local maximum?

    d) Suppose $g(0) = 0$. Sketch a possible graph of g.

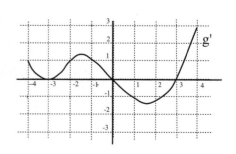

18.   The graph of the derivative of f is shown in
       the figure.

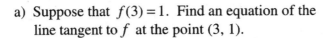

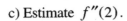

   a) Suppose that $f(3) = 1$. Find an equation of the
      line tangent to f at the point (3, 1).

   b) Where does f have a local minimum? Explain
      briefly.

   c) Estimate $f''(2)$.

   d) Where does f have an inflection point? Explain briefly.

   e) Where does f achieve its maximum on the interval [1, 4]?

19.   The graph of f is shown below.

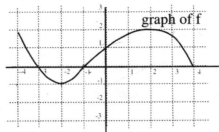

   a) For what values of x, if any, is $f''(x)$ negative? Positive? Zero?

   b) Rank the four numbers $f'(0)$, $f(2)$, $f'(2)$ and $f''(2)$ in increasing order.

20.   The graph of the second derivative of a function f is shown below.

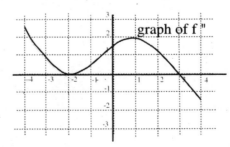

   a) Where is the graph of f concave up?

   b) Where does f have points of inflection?

   c) Suppose $f'(0) = 0$. Is f increasing or decreasing at $x = 2$? Justify your answer.

   d) Suppose $f'(1) = 0$. Is f increasing or decreasing at $x = -1$? Justify your answer.

21.   The graph of f is shown below. Using the graph, evaluate each of the limits or explain why the limit does not exist.

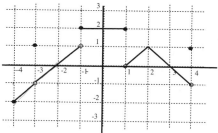

a) $\lim\limits_{x \to 2} f(x) =$

b) $\lim\limits_{x \to -3} f(x) =$

c) $\lim\limits_{h \to 0^+} \dfrac{f(2+h) - f(2)}{h} =$

d) $\lim\limits_{h \to 0^-} \dfrac{f(2+h) - f(2)}{h} =$

e) $\lim\limits_{x \to 0} f'(x) =$

f) $\lim\limits_{x \to 3} \dfrac{f(x) - f(3)}{x - 3} =$

22.   a) The function f is defined by $f(x) = \begin{cases} x \sin\left(\dfrac{1}{x}\right) + 1, & \text{if } x \neq 0 \\ 1, & \text{if } x = 0 \end{cases}$.

Is f locally linear at $x = 0$? Explain your answer.

b) The function g is defined by $g(x) = \begin{cases} x^2 \sin\left(\dfrac{1}{x}\right) + 1, & \text{if } x \neq 0 \\ 1, & \text{if } x = 0 \end{cases}$

Is g locally linear at $x = 0$? Explain your answer.

23.   Gina takes a trip from Andover to Hartford, a distance of 80 miles. Due to road conditions, she drives the first 10 miles at a constant speed of 20 mph. For the next 30 miles, she maintains a constant speed of 60 mph and then stops at McDonalds for a 15 minute snack. She then drives the last 40 miles at a constant speed of 40 mph.

a) Draw a graph that shows her distance from Andover as a function of time.

b) What is her average speed for the entire trip?

c) Draw a graph that shows her velocity as a function of time.

24.   The position $p(t)$ (in meters) of an object at time t (in seconds) that is moving along a line is given by

$$p(t) = 3t^2 + t.$$

a) Find the change in position from $t = 1$ to $t = 3$.

b) Find the average velocity of the object between $t = 1$ and $t = 3$.

c) Explain how you would estimate the instantaneous velocity at $t = 1$.

d) Use your strategy in part c) to estimate the instantaneous velocity at $t = 1$.

25.   Sketch a graph of the derivative function of each given function.

a)   i)

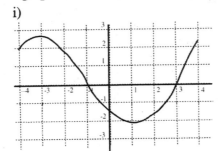

ii)

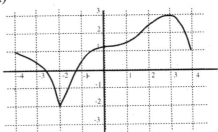

b)   Sketch the graph of a function that is consistent with the following data.

x	$f(x)$	$f'(x)$
-3	2	-1
-1	0	0
1	-1	-1
3	-2	0

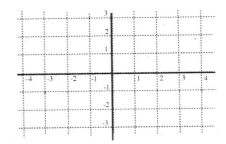

26.   a) Answer the following questions about the function f, whose graph is given in the figure.

i)   Find $f(3)$.

ii) Find $f'(3)$.

iii) What is the average rate of change of f over the interval $[0, 3]$?

b) Sketch on the graph a line whose slope is equal to:

i) $\dfrac{f(3) - f(2.5)}{3 - 2.5}$

ii) $\dfrac{f(3)}{3}$

iii) $\lim\limits_{h \to 0} \dfrac{f(1+h) - f(1)}{h}$

27.   Let a function f and its derivative f' be defined by

$$f(x) = \begin{cases} 1 + 3bx + 2x^2, & \text{if } x \le 1 \\ mx + b, & \text{if } x > 1 \end{cases} \quad \text{and} \quad f'(x) = \begin{cases} 3bx + 4x, & \text{if } x \le 1 \\ m, & \text{if } x > 1 \end{cases}$$

Determine values for m and b so that f is both continuous and differentiable at $x = 1$.

28.   The graphs of functions f and g are given below.

graph of f

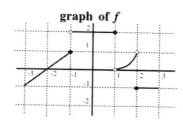

graph of g

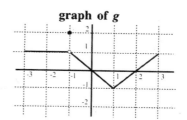

Determine whether the following limits exist. If they do, then find the limit.

a) $\lim\limits_{x \to -1} [f(x) + g(x)]$

b) $\lim\limits_{x \to -1} [f(x) \cdot g(x)]$

c) $\lim\limits_{x \to -1} \dfrac{f(x)}{g(x)}$

d) $\lim\limits_{x \to 0} \dfrac{f(x)}{g(x)}$

e) $\lim\limits_{x \to 1^+} f(x)$

f) $\lim\limits_{x \to 1^-} f(x)$

g) True or False: i) g is continuous at $x = 1$ ii) g has a derivative at $x = 1$

29.   Use the limit rules to find the indicated limits when they exist.

a) $\lim\limits_{x \to 1} \left(3x^3 - x^2 + 5x \right)$

b) $\lim\limits_{x \to 0} \dfrac{x - 2}{x^2 + x + 3}$

c) $\lim\limits_{x \to -2} \dfrac{x^2 - 1}{x + 5}$

d) $\lim\limits_{x \to 0} (x - 2)^2 \cos(2x)$

30.   Use a calculator as an aid in finding the following limits.

a) $\lim\limits_{x \to 0} \dfrac{\sqrt{x + 9} - 3}{x}$

b) $\lim\limits_{x \to 0} \dfrac{3 \sin x}{2x}$

c) $\lim\limits_{x \to 5} \left| x^2 - 25 \right|$

d) $\lim\limits_{x \to -1} \dfrac{x^2 - 5x - 6}{x + 1}$

31.   Use the Intermediate Value Theorem to show that there is a zero for the given
        function in the specified interval.

a) $f(x) = x^3 - 3x + 1$, $[0, 1]$

b) $g(x) = \ln x - e^{-x}$ $[1, 2]$

32.   The function f is defined on the interval [–4, 4] and its graph is shown below. Use the
        graph of f to answer the following questions.

graph of f

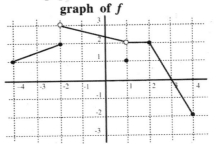

a) On what intervals is f continuous?

b) Evaluate $\lim\limits_{x \to 1} f(x)$.

c) Find i) $f'(1.5)$ ii) $f'(-3)$

d) For what values of x does $f'(x)$ not exist? Why?

e) For what values of x is $f'(x) < 0$?

f) Evaluate $\lim\limits_{h \to 0} \dfrac{f(3+h) - f(3)}{h}$.

g) Sketch a graph of f'.

33.   The function g is defined on the interval [–4, 4] and its graph is shown below. Use the
        graph of g to answer the following questions.

graph of g

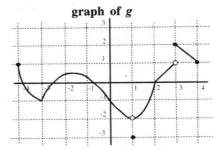

a) For what values of x is g discontinuous?

b) For what values of x is $g'(x) = 0$?

c) Evaluate $\lim\limits_{x \to 1} g(x)$.

d) For what values of x does $g'(x)$ not exist? Why?

e) For what values of x is the derivative g' increasing?

f) Evaluate $\lim\limits_{h \to 0} \dfrac{g(3.5+h) - g(3.5)}{h}$.

CHAPTER 3
The Definite Integral

Calculus is largely about the study of the connection between a function and its derivative. In Chapter 2, we began by calculating the velocity from the distance traveled. And that led to the notion of a derivative, or rate of change, of a function. Now we consider the reverse problem: given the velocity can we determine the distance traveled? This will lead us to the notion of a definite integral, which is the limiting value of a special kind of sum.

3.1 Calculating Distance Traveled

It is easy to tell how far a car has traveled by reading its odometer. The problem is more complicated for an airplane. Most airplanes have instruments that can measure their velocity at any time. Thus, during any time interval in which the velocity of the plane is constant, the distance traveled is given by the formula

$$\text{distance} = \text{velocity} \cdot \text{elapsed time} .$$

Suppose that a jet takes off, becomes airborne at a velocity of 180 mph and climbs to its cruising altitude. The following table gives the velocity every minute for the first 5 minutes, a time during which its speed increases smoothly.

Time(minutes)	0	1	2	3	4	5
Velocity(mph)	180	240	300	360	420	480

How far has the jet flown? Since the velocity is changing every moment the distance cannot be calculated exactly; however, an estimate can be computed. Our strategy is to divide the time interval up into one minute subintervals and assume that the velocity is constant on each. Then calculate the distance traveled on each subinterval and add them up. For example, if we use the velocity at the beginning of each time interval, then during the first minute of flight the jet goes at least

$$180\,\frac{\text{miles}}{\text{hour}} \cdot 1 \text{ minute} = 180\,\frac{\text{miles}}{\text{hour}} \cdot \frac{1}{60} \text{ hour} = 3 \text{ miles.}$$

In the second minute it goes at least

$$240\,\frac{\text{miles}}{\text{hour}} \cdot 1 \text{ minute} = 240\,\frac{\text{miles}}{\text{hour}} \cdot \frac{1}{60} \text{ hour} = 4 \text{ miles.}$$

In the third minute it goes at least 5 miles, in the fourth minute at least 6 miles, and at least 7 miles in the last minute. Thus, in the first five minutes of flight the jet goes at least

$$3 + 4 + 5 + 6 + 7 = 25 \text{ miles.}$$

Using the slowest speed in each subinterval makes 25 miles a lower estimate of the total distance traveled in the first 5 minutes.

Our next step is to make a similar calculation, this time using the highest velocity obtained in each subinterval. Thus, we take 240 mph as the constant velocity in the first subinterval and calculate that the jet goes at most

$$240\,\frac{\text{miles}}{\text{hour}} \cdot 1 \text{ minute} = 240\,\frac{\text{miles}}{\text{hour}} \cdot \frac{1}{60} \text{ hour} = 4 \text{ miles.}$$

Using velocities of 300 mph in the second minute, 360 mph in the third, 420 mph in the fourth, 480 mph in the fifth, and the fact that each applies to a 1 minute interval, the estimate of the distance traveled is the sum

$$4 + 5 + 6 + 7 + 8 = 30 \text{ miles.}$$

Since we used the largest velocity in each subinterval, 30 miles is an upper estimate of the distance traveled in the first 5 minutes. Thus, the true distance traveled by the jet is between 25 and 30 miles.

$$25 \text{ miles} \leq \text{Total Distance Traveled} \leq 30 \text{ miles.}$$

Notice that there is a difference of 5 miles between the lower and upper estimates. To improve the estimate of the distance traveled by the jet in the first 5 minutes of flight, we record the velocity more frequently, say every half minute.

Time(minutes)	0	.5	1	1.5	2	2.5	3	3.5	4	4.5	5
Velocity(mph)	180	210	240	270	300	330	360	390	420	450	480

To calculate a new lower estimate of the distance traveled we use the velocity at the beginning of each half-minute subinterval as the constant velocity. The 10 velocities we get are 180, 210, ... , 450, and since each applies to a half minute interval (or $\frac{1}{120}$ hours), the estimate of the distance traveled is

$$\begin{aligned}
\text{Lower Estimate} = {} & 180 \cdot \frac{1}{120} + 210 \cdot \frac{1}{120} + 240 \cdot \frac{1}{120} + 270 \cdot \frac{1}{120} + 300 \cdot \frac{1}{120} \\
& + 330 \cdot \frac{1}{120} + 360 \cdot \frac{1}{120} + 390 \cdot \frac{1}{120} + 420 \cdot \frac{1}{120} + 450 \cdot \frac{1}{120} \\
= {} & 26.25 \text{ miles.}
\end{aligned}$$

Notice this estimate is larger than the previous lower estimate of 25 miles.

Taking the largest (right endpoint) velocity as the constant velocity in each subinterval gives us a new upper estimate of the distance traveled.

$$\begin{aligned}
\text{Upper Estimate} = {} & 210 \cdot \frac{1}{120} + 240 \cdot \frac{1}{120} + 270 \cdot \frac{1}{120} + 300 \cdot \frac{1}{120} + 330 \cdot \frac{1}{120} \\
& + 360 \cdot \frac{1}{120} + 390 \cdot \frac{1}{120} + 420 \cdot \frac{1}{120} + 450 \cdot \frac{1}{120} + 480 \cdot \frac{1}{120} \\
= {} & 28.75 \text{ miles.}
\end{aligned}$$

Notice this estimate is smaller than the previous upper estimate of 30 miles. We now know that

$$26.25 \text{ miles} \leq \text{Total Distance Traveled} \leq 28.75 \text{ miles.}$$

Now the difference between the upper and lower estimates is 2.5 miles, half of what it was previously. Halving the length of each subinterval has halved the difference between lower and upper estimates.

Visualizing the Distance Traveled

The lower and upper estimates of the distance traveled can be represented on a graph of the velocity. First, a velocity-time graph can be sketched by plotting the 5 original data points (0, 180), (1, 240), (2, 300), (3, 360), (4, 420), (5, 480) and then connecting them.

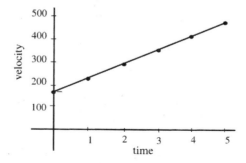

Now with each subinterval we can associate an inscribed and a circumscribed rectangle with its base on the horizontal axis.

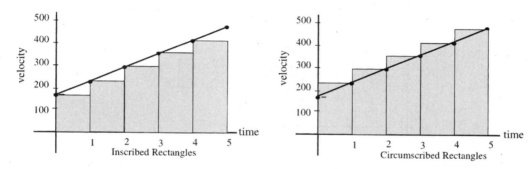

The area of each rectangle gives the distance traveled over that subinterval, assuming that the velocity is constant and equal to the height of the rectangle. Thus, the sum of the areas of the inscribed rectangles gives a lower approximation of the distance traveled, 25 miles; the sum of the areas of the 5 circumscribed rectangles gives an upper approximation of the distance traveled, 30 miles.

       In the figure below, the sum of the areas of the shaded regions represents the difference between the lower and upper estimates. To compute the difference between the two estimates the shaded rectangles have been slid horizontally so that they fit into a single rectangle with width 1 and height 300. The height, 300, is the difference between the initial and final velocity, 300 = 480 − 180; the width, 1, is the time in minutes between velocity measurements. And, the area of the rectangle is

$$\text{width} \cdot \text{height} \ = \ 1 \ \text{min} \cdot 300 \, \frac{\text{mile}}{\text{hour}} \ = \ \frac{1}{60} \, \text{hour} \cdot 300 \, \frac{\text{mile}}{\text{hour}} \ = \ 5 \ \text{miles}.$$

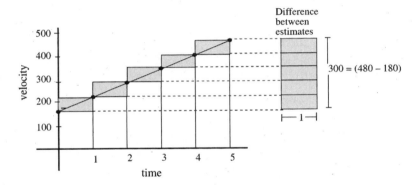

As we saw earlier, measuring the velocity more frequently produces more accurate estimates of the true distance traveled.

Data for the velocities measured every half minute are displayed below.

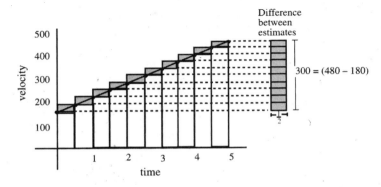

The difference between lower and upper estimates is represented by the shaded rectangles. Once again, we can slide the little rectangles to the right and stack them on top of one another forming a single rectangle with the same height as before, but with half the width. Thus its area is half of what it was before.

$$\text{width} \cdot \text{height} = \frac{1}{2} \text{ min} \cdot 300 \frac{\text{mile}}{\text{hour}} = \frac{1}{120} \text{ hour} \cdot 300 \frac{\text{mile}}{\text{hour}} = 2.5 \text{ miles}.$$

The difference between lower and upper estimates is proportional to the length of time between measurements of the velocity. For example, if the velocity is calculated every 15 seconds (or $\frac{1}{240}$ hours), we have

$$\text{difference} = (300)\left(\frac{1}{240}\right) = 1.25 \text{ miles}.$$

The examples above suggest a general procedure for estimating distance traveled from velocity-time data.

Suppose a function f defines the velocity v of a moving object at a given time t, $v = f(t)$. To estimate the distance traveled by the object from $t = a$ to $t = b$ we subdivide the time interval $a \le t \le b$ into n equal parts, each part with length $\Delta t = \dfrac{b-a}{n}$. Then in each subinterval $\left[t_{j-1}, t_j\right]$ we choose the left-endpoint, t_{j-1}, and form a rectangle of base Δt, and height $f(t_{j-1})$.

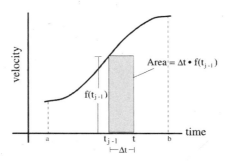

The area of this rectangle, $f(t_{j-1}) \cdot \Delta t$, is an estimate of the distance traveled in the time interval $\left[t_{j-1}, t_j\right]$; and

$$f(t_0) \cdot \Delta t + f(t_1) \cdot \Delta t + f(t_2) \cdot \Delta t + \dots + f(t_{n-1}) \cdot \Delta t,$$

the sum of all these areas, is an approximation of the total distance traveled. This approximating sum is called a **left-hand sum** because the velocity from the left end of each subinterval was used to calculate distance. A **right-hand sum** may be calculated using the value of the velocity at the right end of each subinterval. The n velocities we get are $f(t_1)$, $f(t_2)$, $f(t_3)$, ..., $f(t_n)$; and

$$f(t_1) \cdot \Delta t + f(t_2) \cdot \Delta t + f(t_3) \cdot \Delta t + \dots + f(t_n) \cdot \Delta t$$

is the right-hand approximation of the total distance traveled.

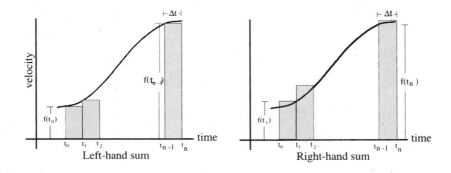

If the velocity function, $v = f(t)$ is an increasing function, the left-hand sum is a lower estimate of the distance traveled and the right-hand sum is an upper estimate. For a decreasing velocity function a left-sum is an upper estimate, a right-sum is an lower estimate.

Whether the velocity function f is increasing or decreasing, the exact value of the distance traveled lies between the left- and right-hand approximating sums. Thus, if a number that lies between the approximating sums, say the average of the two sums, is used as an estimate of the distance traveled, the accuracy of the estimate will depend on closeness of the left and right-hand sums. The following figures suggest that as the number, n, of subintervals increases the length of each subinterval, Δt, approaches 0 and the sums approach the area under the graph of the velocity function which represents the total distance traveled.

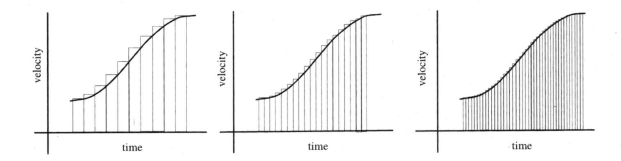

3.1   Exercises

1.   A car is traveling at 36 meters per second when the driver brakes to a stop in 5 seconds. The table shows the velocity-time data.

Time (sec)	0	1	2	3	4	5
Velocity (mps)	36	30	23	15	6	0

   a) Sketch a velocity-time graph by plotting and connecting the data points.

   b) Give lower and upper estimates of the distance the car traveled after the brakes were applied.

   c) On the velocity-time graph show the lower and upper estimates and the difference between them.

2.   The table below shows the velocity of a child sliding down a giant "Water Splash".

Time (sec)	0	2	4	6	8	10
Velocity (ft/sec)	2	3.9	5.4	6.6	7.4	8

   a) Sketch a velocity-time graph by plotting and connecting the data points.

   b) Give lower and upper estimates of the distance traveled in 10 seconds.

   c) On the velocity-time graph show the lower and upper estimates and the difference between them.

3.   The graph below shows an airplane's velocity, v, in miles per hour during a two hour flight.

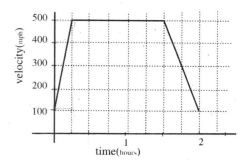

The plane becomes airborne at a velocity of 100 mph and its speed increases rapidly until a cruising velocity of 500 mph is reached. That takes 15 minutes. The plane maintains its cruising speed until 30 minutes before landing, when it starts to slow and begins its descent for landing, which occurs at 100 mph. Find lower and upper estimates to answer the following questions.

   a) What distance did the plane cover in the first 30 minutes of the flight?

   b) What was the average velocity of the plane for the 2 hour flight?

4.   A car is accelerating and the table below gives the velocity every second.

Time (sec)	0	1	2	3	4
Velocity (ft/sec)	2	3.1	4.4	5.9	7.6

   a) Give lower and upper estimates of the distance the car traveled in 4 seconds.

   b) On a sketch of a velocity-time graph, show the lower and upper estimates and the difference between them.

   c) Increased accuracy can be obtained by halving the interval width. Use your graph to estimate the car's speed after 0.5 seconds, 1.5 seconds, etc..

   d) Use the new data to calculate more accurate lower and upper estimates of the distance traveled.

5.   Below is a graph that shows the velocity, v, of an object in meters per second.

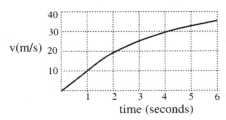

   Estimate the total distance the object traveled between $t = 0$ and $t = 6$.

6.   The velocity of a car is given by

$$v(t) = t^2 + 2t$$

   where $v(t)$ is measured in ft/sec and t is measured in seconds.

   a) Estimate the distance traveled from $t = 1$ to $t = 4$ using 3 subdivisions.

   b) How accurate is your answer?

7.   A car is moving along a straight road and its speed is continually increasing. Speedometer readings are recorded at two second intervals and the results are tabulated below.

Time (sec)	0	2	4	6	8
Velocity (ft/sec)	30	36	38	40	44

   a) In the first two second interval what is the minimum distance the car could have traveled? What is the maximum distance?

   b) Make a lower and upper estimate of the distance the car could have traveled in the 8 second interval.

   c) If you had to guess how far the car went in the 8 second interval, what would you guess? What is the maximum possible difference between your guess and the actual distance?

   d) If the speedometer readings became available for each second, by how much would the upper estimate exceed the lower estimate?

8.   Oil is leaking out of a tank into a water reservoir. The rate of flow is measured every two
       hours for a 12 hour period and the data is listed in the table below.

Time (hr)	0	2	4	6	8	10	12
Amount (gallons/hr)	40	38	36	30	26	18	8

   a) Draw a graph of the rate at which the oil is spilling into the reservoir as a function of
      time.

   b) Estimate the total amount of oil that has spilled during the 12 hour period. Explain
      your method.

9.   Water flows into a large tank at a rate shown in the figure below.

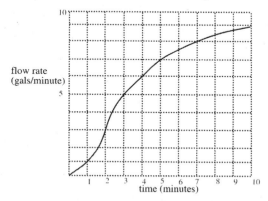

   a) Give a lower and upper estimate of the total number of gallons that have entered the tank
      during the ten minutes shown in the graph.

   b) If the average of the lower and upper estimate in part a) is used to approximate the exact
      number of gallons, how accurate is the approximation?

10.   Suppose a car is moving with increasing speed according to the following table.

time (sec)	0	2	4	6	8	10
speed (ft/sec)	30	36	40	48	54	60

   Approximate the distance traveled in the first 10 seconds and explain your method.

3.2   Calculating Areas; Riemann Sums

In the previous section, distance traveled was approximated by summing areas of rectangular regions. If we take the limit of the approximating sums, as the width of the subintervals gets smaller, we get the exact distance traveled which is also the exact area of the region under the graph.

The general area problem asks how to measure the area of a plane region. For special regions such as triangles, rectangles, trapezoids and circles well known formulas may be employed. Area problems for regions bounded by less familiar curves, such as graphs of functions, are more challenging. This section develops methods for calculating areas of more general regions that are based on the methods used in the previous section for determining distance traveled.

Let us try to approximate the area of the region T in the first quadrant bounded above by the parabola $f(x) = x^2$ and below by the x-axis on the interval $[0, 1]$.

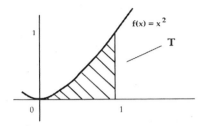

Our strategy is to approximate the area of region T by using inscribed and circumscribed rectangles. First we divide the interval $[0, 1]$ into 4 subintervals of equal length; then we consider the rectangles with bases on the subintervals. Since f is increasing on the interval, the height of each circumscribed rectangle is the value of the function at the right endpoint, whereas the height of each inscribed rectangle is the function value at the left endpoint.

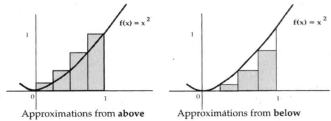

Approximations from **above**   Approximations from **below**

Thus, the sum of the areas of the 4 circumscribed rectangles is

$$\frac{1}{4} \cdot f(\tfrac{1}{4}) + \frac{1}{4} \cdot f\left(\frac{1}{2}\right) + \frac{1}{4} \cdot f\left(\frac{3}{4}\right) + \frac{1}{4} \cdot f(1) = \frac{1}{4}\left[\frac{1}{16} + \frac{4}{16} + \frac{9}{16} + \frac{16}{16}\right] = 0.46875$$

and the sum of the 4 inscribed rectangles is

$$\frac{1}{4} \cdot f(0) + \frac{1}{4} \cdot f\left(\frac{1}{4}\right) + \frac{1}{4} \cdot f\left(\frac{1}{2}\right) + \frac{1}{4} \cdot f\left(\frac{3}{4}\right) = \frac{1}{4}\left[0 + \frac{1}{16} + \frac{4}{16} + \frac{9}{16}\right] = 0.21875.$$

Using the labels U_4 and L_4 to stand for the upper and lower approximations, we have

$$U_4 = 0.46875 \text{ and } L_4 = 0.21875.$$

Since the area of region T is between U_4 and L_4, we could average them to find a single approximation

$$\text{Area}(T) \approx \frac{U_4 + L_4}{2} = 0.34375.$$

If the number of inscribed and circumscribed rectangles is increased, the lower and upper approximations (and the average) converge on a single limiting value. Since the area of region T is always sandwiched between upper and lower approximations, it must be this single value. It can be shown that for the function $f(x) = x^2$ on the interval [0, 1], this unique limiting value is $\text{Area}(T) = \dfrac{1}{3}.$

The approximating sums appearing in the equations above are examples of **Riemann sums.**

> Let f be a function defined on the closed interval $[a, b]$. Subdivide $[a, b]$ into n subintervals with lengths Δx_1, Δx_2, Δx_3, ... , Δx_n. In each subinterval, choose an evaluation point x_1, x_2, ... , x_n. Then
> $$f(x_1) \cdot \Delta x_1 + f(x_2) \cdot \Delta x_2 + ... + f(x_n) \cdot \Delta x_n \text{ is a \textbf{Riemann sum} for } f \text{ on } [a, b].$$

Informally, a Riemann sum is a sum of function values multiplied by the lengths of the corresponding intervals. It is important to note that a Riemann sum is just a real number.

For a particular function f defined on an interval $[a, b]$ there are many possible Riemann sums. First, there is more than one way to divide an interval into subintervals. In addition, there are many ways in which to choose an evaluation point in each subinterval (and each choice determines the value of the function in that subinterval).

In evaluating a Riemann sum with a calculator, it is useful to work with subintervals of equal length and to choose the same point from each interval (for example, left endpoint, right endpoint, or midpoint).

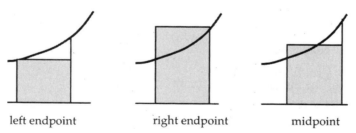

left endpoint   right endpoint   midpoint

The Riemann sums that result from choosing left endpoints and right endpoints are called **left-hand** and **right-hand sums.**

Notice that if a function is increasing over a given interval, then the left-hand sum is a lower estimate of the area under the graph, while the right-hand sum is an upper estimate. In this case, the sum resulting from choosing midpoints will be between the left- and right-hand sums. The situation is reversed when the function is decreasing over an interval.

Riemann Sum Program

The following TI-83 program computes the Riemann sum for n subintervals, evaluating the function stored in Y1 at a given point x in each subinterval.

PROGRAM:RSUM	Comments
:Input "LEFT:",A	A = left endpoint;
:Input "RIGHT:",B	B = right endpoint;
:Input "X CHOICE:",C	C is the choice of X, $0 \le C \le 1$, in each subinterval
:Input "NUMBER:",N	N= number of subintervals
:(B–A)/N→H	Store the subinterval width in H
:0→S	Start the sum at S = 0
:For(I,1,N)	Start a For loop using the counter I
:A+(I–1+C)H→X	Compute the X-value used as input for Y1
:S+Y1(X)H→S	Add the current rectangle area to S
:End	End of the For loop
:Disp "SUM =",S	
:Stop	

The program RSUM has four inputs: A and B are the endpoints of the interval; N is the number of subintervals. The input C, $0 \le C \le 1$ allows us to change the evaluation points from left endpoint (C = 0) to right endpoint (C = 1) to anything between ($0 < C < 1$). For example, to use midpoints in each interval, let C = 0.5.

> **Technology Tip:**   *Repeated Executions of the RSUM program*
> When you run RSUM several times for different values of N, the repeated prompts for the endpoints A and B and the evaluation point C can be tedious. Thus, you may want to insert a Repeat ($N = -1$) command before the prompt for N and add an End command before Stop so that the loop is repeated until the terminating value $N = -1$ is entered.

Example 1:   *Using a Riemann Sum to Appoximate Area*
Use Riemann sums to approximate the area of the region bounded by the graph of
$f(x) = 1 + x^2$ over the interval [0, 1].

Solution:

Set Y1 = 1 + X^2 . Since the function

$f(x) = 1 + x^2$ is strictly increasing on
the interval [0, 1], executing the
program RSUM with C = 0 will
produce a left-hand sum that is a lower
Riemann sum. Executing RSUM with

C = 1 will produce a right-hand sum
that is an upper Riemann Sum.

N	Left-hand sum	Right-hand Sum
4	1.2188	1.4688
6	1.2546	1.4213
10	1.2850	1.3850
20	1.3088	1.3588
40	1.3209	1.3459

$$\text{Area approximation} = \frac{U_{40} + L_{40}}{2} = 1.3334.$$

The function $f(x) = 1 + x^2$ from Example 1 is strictly decreasing on the interval $[-1, 0]$. If we wish to estimate the area under the graph of f on this new interval, upper estimates are now left-hand sums (C = 0) and lower estimates are right-hand sums (C = 1).

3.2   Exercises

In Exercises 1–5, approximate the area of the region bounded by the graph of the function over the indicated interval. Make your approximation by hand calculating left- and right-hand sums for $n = 4$ subdivisions. Draw a sketch that illustrates the approximating sums.

1. $f(x) = x + 1$, $[0, 4]$

2. $f(x) = 4 - 2x$, $[0, 2]$

3. $f(x) = x^2 - 8x + 20$, $[0, 4]$

4. $f(x) = \sin x$, $[0, \frac{\pi}{2}]$

5. $f(x) = \ln x$, $[1, 5]$

In Exercises 6–9, approximate the area of the region bounded by the graph of the given function over the indicated interval. Make your approximation by computing left and right-sums with the program RSUM for the given values of N. Note that each function is either increasing or decreasing on the given interval.

6. $f(x) = 3 - \dfrac{3}{2}x$, $[0, 2]$

N	Left-Sum	Right-Sum
4		
8		
16		

Area approximation = _____

Area of triangular region = _____

Compare your approximations with the actual area obtained by using the formula for the area of a triangle, $A = \dfrac{1}{2}bh$.

7. $f(x) = \sqrt{4 - x^2}$, $[0, 2]$

N	Left-Sum	Right-Sum
4		
8		
16		

Area approximation = _____

Area of quarter-circular region = _____

Compare your approximations with the actual area obtained by using the formula for the area of a quarter-circle, $A = \dfrac{\pi r^2}{4}$.

8. $A = \dfrac{1}{4}x + 2, \quad [0, 4]$

N	Left -Sum	Right-Sum
4		
8		
16		

Area approximation = _____

Area of trapezoidal region = _____

Compare your approximations with the actual area obtained by using the formula for the area of a trapezoid, $A = \dfrac{1}{2}h(b_1 + b_2)$.

9. $f(x) = \sin x, \quad [\,0, \dfrac{\pi}{2}\,]$

N	Left- Sum	Right-Sum
4		
8		
16		

Area approximation = _____

In Exercises 10–11, estimate the area of the region bounded above by the graph of f and below by the interval [–4, 4]. Assume that the grid lines are spaced 1 unit apart.

10. **graph of f**

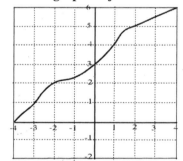

11. **graph of f**

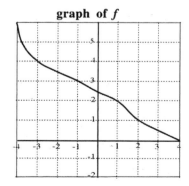

3.3   Definite Integrals

Graphical intuition suggests that all of our approximating sums – left-hand, right-hand and midpoint – should converge to a single limiting value.

If the function f is continuous on a closed interval $[a, b]$, there is one and only one number to which all Riemann sums converge as the number of subdivisions increases and the width of all subdivisions decreases. This number is called the **definite integral** and is denoted by $\int_a^b f(x)\,dx$. In symbols

$$\int_a^b f(x)\,dx = \lim_{n \to \infty} \left[f(x_1) \cdot \Delta x_1 + f(x_2) \cdot \Delta x_2 + \dots + f(x_n) \cdot \Delta x_n \right]$$

The Definite Integral as an Area

If the continuous function f is positive-valued on the interval $[a, b]$, then each term $f(x_1)\Delta x_1$, $f(x_2)\Delta x_2$, ..., in a Riemann sum can be interpreted as the area of a rectangle. As the widths Δx_i of the rectangles approach zero, the rectangles fit the graph of the function more closely, and the sum of their areas gets closer and closer to the area of the region under the graph.

Thus if f is positive-valued, the corresponding definite integral $\int_a^b f(x)\,dx$ is positive. In this case the area of the region bounded above by the graph of f on $[a, b]$ is defined to be the number designated by $\int_a^b f(x)\,dx$.

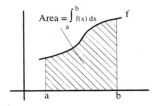

Example 1: *Area Under a Curve*

Given the integral $\int_1^4 (x+1)\,dx$.

a) Interpret the integral as the area of a region and find its exact value.

b) Estimate the integral using left and right sums with 50 subdivisions.

Solution:

a) The integral is the area of the region under the graph of $f(x) = x+1$ from $x = 1$ to $x = 4$, which is a trapezoid with altitude 3 and parallel bases of lengths 2 and 5.

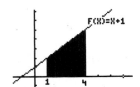

The area of the trapezoid is $\dfrac{1}{2}h(b_1 + b_2) = \dfrac{1}{2}(3)[2 + 5] = 10.5$.

b) With $n = 50$, the left sum = 10.41 and the right sum = 10.59.

What happens if $f(x)$ takes on negative values? If the graph of f lies below the x-axis, then each $f(x)$ is negative, so each term $f(x)\Delta x$ in the Riemann sum is negative, and the definite integral is negative. In this case the definite integral is the negative of the area of the region bounded by the graph of f and the x-axis.

Example 2: *Approximating a Definite Integral*

If $f(x) = x$, approximate the definite integral $\int_a^b f(x)\, dx$ over the interval $[-2, 0]$. Using your approximations, guess the value of the definite integral.

Solution:

The graph $f(x) = x$ lies below the x-axis between $x = -2$ and $x = 0$.

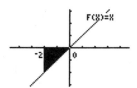

Set Y1 = X. Since $f(x) = x$ is strictly increasing on the interval $[-2, 0]$, executing the program RSUM with $C = 1$ will produce right hand sums that are upper Riemann sums. Executing the program with $C = 0$ will produce left-hand sums that are lower Riemann sums.

n	Right- sum	Left-sum
2	-1	-3
4	-1.5	-2.5
8	-1.75	-2.25
16	-1.875	-2.125
32	-1.9375	-2.0625
64	-1.96875	-2.03125

We guess $\int_{-2}^{0} x\, dx = -2$.

The integral is negative and can not be interpreted as area. However, the region bounded by the graph of $y = x$ and the interval $[-2, 0]$ has an area of 2 and, except for the negative sign, this is the exact value of the integral. In general, if a function is negative-valued, the definite integral is the negative of the area of the region between the graph of f and the x-axis.

If the function f takes on both positive and negative values over the interval [a, b] the definite integral $\int_a^b f(x)\, dx$ is the sum of the areas of the regions above the x-axis minus the total of the areas below the x-axis.

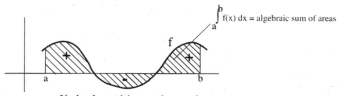

f is both positive and negative

Example 3: *Finding Integrals Using Signed Areas*

Evaluate $\int_0^5 (3-x)\,dx$.

Solution:

The integral is the area of the region above the x-axis, A_1, minus the area of the region below the x-axis, A_2, from $x = 0$ to $x = 5$.

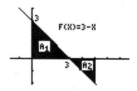

Therefore,

$$\int_0^5 (3-x)\,dx \; = \; Area\,A_1 - Area\,A_2 = \frac{9}{2} - 2 = \frac{5}{2}.$$

Evaluating Definite Integrals Numerically with the TI-83

The TI-83 has a built-in function named **fnInt** for approximating definite integrals. The format of **fnInt** is

fnInt(*expression, variable, lower, upper*)   or

fnInt(*expression, variable, lower, upper, tolerance*)

where **fnInt** returns the numerical approximation of the definite integral of *expression* with respect to *variable,* given *lower* limit and *upper* limit and optional *tolerance* (if no tolerance is specified, .0001 is used).

Example 4: *Using the Calculator to Numerically Approximate a Definite Integral*

Approximate $\int_0^1 \frac{4}{1+x^2}\,dx$.

Solution:

Press ▭Y=▭ and define $Y1 = 4/(1 + X^2)$. Return to the Home screen; press ▭MATH▭ [9:**fnInt**( ] and enter the expression to approximate the definite integral.

It is reassuring to note that the exact value of the integral is π (a fact that you will learn later) and by displaying π on the Home screen we see that the definite integral approximation is correct to nine digits.

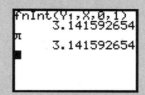

Technology Tip: *Using the Integral Command*

The TI-83 also has a special integral command available from the Calc menu. To see this for the last example, start by graphing $f(x) = \dfrac{4}{1+x^2}$ in the ZDecimal viewing rectangle. Press 2nd [CALC] to display the Calc menu. Select option 7, [7: ∫ f(x) dx]. The Trace cursor will appear at the point (0, 4). Since 0 is the lower bound of integration, press ENTER. Now move the cursor to the point (1, 2). Since 1 is the upper bound of integration, press ENTER. Note that the region between $x = 0$ and $x = 1$ is shaded and an approximation of the definite integral is displayed at the bottom of the screen.

∫f(x)dx=3.1415927

A function that is either increasing or decreasing but not both on an interval [a, b] is called **monotone.** If f is monotone on [a, b], left and right sums always bracket the integral $\int_a^b f(x)\, dx$.

Error Bounds for Monotone Functions

In any approximation process, the problem solver should have some notion of how close an approximation is to the true value. We define the **error** in an approximation to be the magnitude of the difference between the approximate and exact values. Since the exact value of a definite integral is between the lower estimate and the upper estimate for increasing or decreasing functions, the error must be less than the difference between these estimates.

When f is a positive-valued function, by visualizing the upper and lower estimates as areas we can also visualize their difference as an area. In the first figure below, the sum of the shaded regions represents the difference between the upper and lower Riemann sums for an increasing function f.

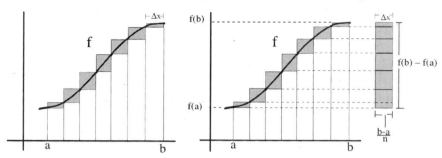

In the second figure above, the shaded rectangles have been slid horizontally so that they fit into a single rectangle. The area of this rectangle is $[f(b) - f(a)] \cdot \Delta x$; therefore, the error is less than the area of our "error rectangle". Thus, on the interval [a, b]

$$\text{Error} \ \leq \ \left| \begin{array}{c} \text{difference between} \\ \text{the lower and upper sum} \end{array} \right| \ = \ |f(b) - f(a)| \cdot \left(\frac{b-a}{n} \right).$$

Notice that the only variable in the expression $|f(b) - f(a)| \cdot \left(\dfrac{b-a}{n} \right)$ is the number of subintervals, n.. Thus, by increasing the number of equal subintervals we can control the size of the error in approximating a definite integral.

With an error bound formula, we can determine in advance how many subdivisions are needed for a prescribed accuracy.

Example 5: *Controlling the Error when Approximating a Definite Integral*

For the function defined by $f(x) = x^2$ on the interval $[0, 2]$, how many intervals of equal

length must be used so that an estimate differs from the value of $\int_0^2 x^2 \, dx$ by less than 0.01 ?

Solution:

If we use n equal subintervals, each of length $\left(\dfrac{b-a}{n}\right) = \dfrac{2}{n}$, then the area of the

error rectangle is $|f(b) - f(a)| \cdot \left(\dfrac{b-a}{n}\right) = |4 - 0| \cdot \left(\dfrac{2}{n}\right) = \dfrac{8}{n}$.

By making $\dfrac{8}{n} < 0.01$ we make the area of the error rectangle less than 0.01, and this ensures

that that the approximation will differ from the actual value by less than 0.01. For $\dfrac{8}{n} < 0.01$,

$n > 800$.

A Reminder: Most functions are not monotone. However, it is often possible to find an error

bound for an integral $\int_a^b f(x) \, dx$ by first splitting the interval $[a, b]$ into

subintervals on which f is monotone.

3.3   Exercises

1.   Sketch a region whose area is given by the following integrals. Then evaluate each integral by using a geometric formula.

a) $\int_1^3 4\,dx$ b) $\int_2^4 (4-x)\,dx$ c) $\int_{-1}^1 \sqrt{1-x^2}\,dx$

2.   The graph of f is shown below.

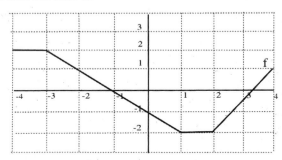

Evaluate each of the following.

a) $\int_{-4}^{-2} f(x)\,dx$ b) $\int_{-2}^1 f(x)\,dx$ c) $\int_{-4}^4 f(x)\,dx$ d) $\int_0^3 f(x)\,dx$

3.   For each function below use a graph of f to evaluate

i) $\int_0^2 f(x)\,dx$ ii) $\int_1^3 f(x)\,dx$ iii) $\int_{-2}^3 f(x)\,dx$

a) $f(x)=3x$ b) $f(x)=x+1$ c) $f(x)=5-2x$

In Exercises 4–7, approximate the definite integral by computing left- and right-hand sums with 4, 8 and 16 subdivisions. Using your approximations, guess the value of the definite integral. [Note that each function is monotone on the interval.]

4. $\int_1^2 (x^2-x)\,dx$ 5. $\int_0^2 \frac{1}{x-3}\,dx$

6. $\int_0^\pi \cos x\,dx$ 7. $\int_{-1}^2 (x^3-1)\,dx$

8.   a) Find a nonzero value of b so that $\int_0^b (x-x^2)\,dx = 0$.

(Hint: Use trial and error with **fnInt** or the Calc option 7, [7: $\int$f(x)dx]).

b) Find an interval $[a, b]$ so that $\int_0^b (\sin x - \cos x)\,dx = 0$.

9.   If the region in the first quadrant bounded by the graph of $y = \cos x$ and the interval $[0, \frac{\pi}{2}]$
        is divided into two regions of equal area by the line $x = c$, guess c and use the TI-83 to
        verify your conjecture.

10.   Find a number k so that the area under the graph of $f(x) = \frac{1}{x}$ on the interval $[1, k]$ is 1.

In Exercises 11–14, use a minimum number of subdivisions to approximate each definite integral
to within .05 of the exact answer (Hint: You need | error | ≤ .05). In each case, specify the
number of subdivisions being used.

11. $\int_1^2 x^2 \, dx$ 12. $\int_1^2 2^x \, dx$

13. $\int_1^e \ln x \, dx$ 14. $\int_0^{\pi/2} \sin x \, dx$

15.   Let f be the function graphed below.

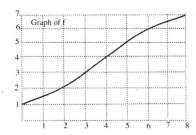

a) Estimate the value of $\int_2^8 f(x) \, dx$ using a right-hand sum with three equal subintervals.

b) Estimate the value of $\int_2^8 f(x) \, dx$ using a left-hand sum with three equal subintervals.

c) Estimate the value of $\int_0^8 f(x) \, dx$ using a right-hand sum with four equal subintervals.

d) Estimate the value of $\int_0^8 f(x) \, dx$ using a left-hand sum with four equal subintervals.

16.   Let f be the function graphed below. Estimate $\int_{-4}^4 f(x) \, dx$ by calculating left- and right-
        hand sums, each with 4 equal subdivisions, then draw sketches that illustrate these sums
        geometrically.

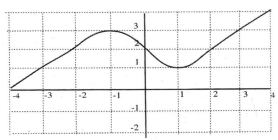

17.   The table below contains values of a continuous function g at several inputs x.

x	0	1	2	3	4	5
$g(x)$	0.15	0.22	0.30	0.38	0.46	0.56

Estimate $\int_1^4 g(x)\, dx$ using a left-hand sum with 3 equal subintervals, then draw a sketch that illustrates this sum geometrically.

18.   It is a fact that $\int_0^\pi \sin x\, dx = 2$. With that information, what you know about integrals, "signed area", and transformations, make a guess for the value of each of the following integrals. A graph will give you clues.

a) $\int_0^{2\pi} \sin x\, dx$ b) $\int_0^{\pi/2} \sin x\, dx$ c) $\int_0^{3\pi/2} \sin x\, dx$

d) $\int_{\pi/2}^{3\pi/2} \sin x\, dx$ e) $\int_0^\pi (\sin x + 2)\, dx$

19.   The table below contains values of a continuous function f at several inputs x.

x	0	0.5	1	1.5	2
$f(x)$	-1	$-.96$	$-.67$	0.13	1.67

Estimate $\int_0^2 f(x)\, dx$ using a right-hand sum with 4 equal subintervals, then draw a sketch that illustrates this sum geometrically.

Exploratory Worksheet

Purpose: Compare Riemann sum approximations of a definite integral.

      So far we have formed Riemann sums using left-hand and right-hand endpoints of an interval. Another Riemann sum uses rectangles whose heights are obtained by taking midpoints of each interval.

      Use the RSUM program to calculate the lefthand (C = 0), righthand (C = 1) and the midpoint (C = 0.5) approximating sums for each of the following definite integrals. Use 20 and 50 equally spaced subdivisions.

1. $\int_0^2 \sqrt{4 - x^2}\ dx = \pi$

N	Left Sum	Right Sum	Midpt Sum
20			
50			

2. $\int_1^2 \frac{1}{x}\ dx = \ln 2$

N	Left Sum	Right Sum	Midpt Sum
20			
50			

3. $\int_0^1 xe^{x^2}\ dx = \frac{e - 1}{2}$

N	Left Sum	Right Sum	Midpt Sum
20			
50			

4.   Which of the two methods, midpoint or endpoint Riemann sums, seems to give better approximations to the definite integral? Try to give a general explanation for your conclusion.

5.   Use the **fnInt(** command on your TI-83 to obtain another approximation of each integral above. Comment on the degree to which the estimates of the definite integral agree.

3.4   The Fundamental Theorem of Calculus

In this section we study the Fundamental Theorem of Calculus. It shows that differentiation and integration are inverse operations, much the same as multiplication and division are inverse operations. The fact that integration and differentiation are inverse operations should not come as a surprise when you think about the way in which velocity-time and position-time functions are related. Given a position-time function, $s = f(t)$, you differentiate to find the velocity, $v = \dfrac{ds}{dt}$; and, given a velocity-time function, you integrate to find the total net distance traveled, $s(b) - s(a) = \int_a^b s'(t)\, dt$.

We now show that, in general, integrating a rate of change function f' over an interval $[a, b]$ gives the total change in f, $f(b) - f(a)$, over the same interval. Suppose $f'(t)$ is the rate of change of some quantity $f(t)$ with respect to time, and we want to determine the net change in f between $t = a$ and $t = b$. We divide the interval $[a, b]$ into n equal subintervals, each of length Δt. For each small interval, we compute the change in f and add them all up. Because the subintervals are small, we can obtain a good approximation to the change in f by assuming that the rate of change remains constant on each subinterval, and we have

$$\text{change in } f = \text{rate of change} \cdot \text{time}$$

For the first subinterval from t_0 to t_1, the rate of change can be approximated as $f'(t_0)$, so

$$\text{change in } f \approx f'(t_0) \cdot \Delta t.$$

Similarly, for the second interval

$$\text{change in } f \approx f'(t_1) \cdot \Delta t.$$

In the figure, we assume f is increasing on $[a, b]$, and each $f'(t_j) \cdot \Delta t$ is the product of a slope and the change in t. This gives the vertical change in f for each subinterval.

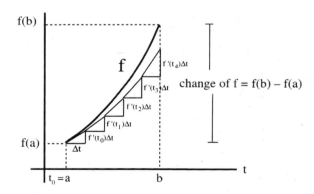

Summing over all subintervals, we have a left-hand sum that gives an approximation of the total change in f.

$$\text{total change in } f = f(b) - f(a) \approx f'(t_0) \cdot \Delta t + f'(t_1) \cdot \Delta t + f'(t_2) \cdot \Delta t + \ldots + f'(t_{n-1}) \cdot \Delta t\ .$$

As n increases and Δt approaches zero, this approximation improves, approaching a definite integral as a limit.

$$\lim_{n \to \infty} \left[f'(t_0) \cdot \Delta t + f'(t_1) \cdot \Delta t + f'(t_2) \cdot \Delta t + \ldots + f'(t_{n-1}) \cdot \Delta t \right] = \int_a^b f'(t)\, dt$$

Thus, we have a result that is called the **Fundamental Theorem of Calculus**.

Let the function f be continuous on $[a, b]$ with derivative f'. Then

$$\text{total change in } f = f(b) - f(a) = \int_a^b f'(t)\, dt.$$

In words: Integrating a rate of change function f' over an interval $[a, b]$ gives the total change in f over the same interval, $f(b) - f(a)$.

     If we know $f(a)$, the Fundamental Theorem enables us to reconstruct the function from a knowledge of its derivative.

Example 1:   *Determing Cost from Changing Price Rate*

     The present price of a new car is \$14,500. The price of a new car is changing at a rate of $\dfrac{dP}{dt} = 120 + 180\sqrt{t}$ dollars per year. To the nearest ten dollars, how much will a new car cost 5 years from now?

Solution:

     Since the new car price is changing at a rate of $\dfrac{dP}{dt} = 120 + 180\sqrt{t}$ dollars per year, we have

$$\text{change in price} = P(5) - P(0) = \int_0^5 P'(t)\, dt = \int_0^5 (120 + 180\sqrt{t})\, dt.$$

The rate function P' is monotone, so the integral will be bracketed between the left and right sum. For $n = 50$

$$\text{left sum} = 1920.37 < \int_0^5 (120 + 180\sqrt{t})\, dt < 1960.62 = \text{right sum}.$$

So

$$\text{change in price} = \int_0^5 (120 + 180\sqrt{t})\, dt \approx \frac{\text{left sum} + \text{right sum}}{2} = \$1940.50.$$

Since

$$P(5) - P(0) = \int_0^5 (120 + 180\sqrt{t})\, dt \text{ and } P(0) = 14,500,$$

the cost of a new car in 5 years will be (to the nearest ten dollars)

$$P(5) = P(0) + \int_0^5 (120 + 180\sqrt{t})\, dt = 14,500 + 1940 = \$16,440.$$

Example 2:   *Determing Volume from Changing Inflation Rate*

     Suppose air is pumped into a balloon at a rate given by $r(t) = \frac{1}{2}t - 1$ ft^3/sec for $t \geq 1$. If the volume of the balloon is 1.3 ft^3 at $t = 1$ sec, what is the volume of the balloon at $t = 5$ sec ?

Solution:

$$v(5) - v(1) = \int_1^5 r(t)\, dt \Rightarrow v(5) = v(1) + \int_1^5 \left(\tfrac{1}{2}t - 1\right) dt = 1.3 + 2 = 3.3\, ft^3$$

Evaluating Definite Integrals

The Fundamental Theorem provides a method for computing definite integrals *exactly*. Given an integral $\int_a^b f(x)\, dx$, we try to recognize f as the derivative F' of some other function F, $F'(x) = f(x)$. If we succeed, then we can evaluate the integral by writing

$$\int_a^b f(x)\, dx = \int_a^b F'(x)\, dx = F(b) - F(a)$$

Example 3: *Using the Fundamental Theorem*

The function $F(x) = x^3$ has the derivative $F'(x) = 3x^2$. Compute $\int_1^2 3x^2\, dx$ using

a) left and right sums with 50 subdivisions;

b) the Fundamental Theorem of Calculus.

Solution:

a) With $n = 50$, the left sum is 6.9102 and the right sum is 7.0902. Since $F'(x) = 3x^2$ is monotone increasing on [1, 2], we have

$$6.9102 < \int_1^2 3x^2\, dx < 7.0902 .$$

b) According to the Fundamental Theorem $\int_a^b F'(x)\, dx = F(b) - F(a)$. So we take $F(x) = x^3$ and $F'(x) = 3x^2$ and obtain

$$\int_1^2 3x^2\, dx = F(2) - F(1) = 2^3 - 1^3 = 7.$$

The result, 7, can be interpreted as the change in $F(x) = x^3$ from $x = 1$ to $x = 2$ and as the area under the curve defined by $F'(x) = 3x^2$ from $x = 1$ to $x = 2$.

Example 4: *Finding Distance Traveled When Velocity Varies*

The velocity of a car in miles per hour is given by $v(t) = 3t^2 + 2t$ for $t \geq 0$. Calculate the distance traveled from $t = 1$ to $t = 4$ hours

a) using left and right sums with $n = 100$ subdivisions;

b) using the Fundamental Theorem and the fact that if $F(t) = t^3 + t^2$, then $F'(t) = 3t^2 + 2t$.

Solution:

a) With $n = 100$, the left sum is 77.2364 and the right sum is 78.7664 . Since $v(t) = 3t^2 + 2t$ is monotone increasing on [1, 4], we have the following lower and upper estimates of the distance traveled

$$77.2364 \text{ miles} < \int_1^4 (3t^2 + 2t)\, dt < 78.7664 \text{ miles}.$$

b) The function $F(t) = t^3 + t^2$ has the derivative $F'(t) = 3t^2 + 2t$, so by the Fundamental Theorem the exact distance is

$$\int_1^4 (3t^2 + 2t)\, dt = F(4) - F(1) = (4^3 + 4^2) - (1^3 + 1^2) = 78 \text{ miles.}$$

Example 5: *Applying the Fundamental Theorem*

Evaluate $\int_0^{\pi/2} \cos x \, dx$ using the fact that $\dfrac{d}{dx}(\sin x) = \cos x$.

Solution:

The function $F(x) = \sin x$ has the derivative $F'(x) = \cos x$ so

$$\int_0^{\pi/2} \cos x \, dx = \int_0^{\pi/2} F'(x) \, dx = F\left(\frac{\pi}{2}\right) - F(0) = \sin\left(\frac{\pi}{2}\right) - \sin(0) = 1.$$

      Evaluating definite integrals by antidifferentiation requires first finding a usable antiderivative. Finding an antiderivative is hard or impossible for a surprising number of integrands; however, given $f(a)$ and the derivative f' , we can use an equivalent form of the Fundamental Theorem, $f(b) = f(a) + \int_a^b f'(x)\, dx$, and a calculator to produce an approximation of $f(b)$.

Example 6: *Approximating a Function Value*

Given a function f with $f'(x) = \sin(x^2)$ and $f(3) = -5$, find $f(1)$.

Solution:

$$f(3) - f(1) = \int_1^3 f'(t) \, dt$$

$$f(1) = f(3) - \int_1^3 f'(t) \, dt$$

$$f(1) = -5 - \int_1^3 \sin(x^2) \, dx = -5.463$$

Example 7: *Approximating a Function Value*

A pizza with a temperature of $92°C$ is put into a $22°C$ room at $t = 0$. The pizza's temperature is decreasing at a rate of $r(t) = 5e^{-0.1t}$ per minute. Estimate the pizza's temperature at $t = 6$.

Solution:

$$\text{Temperature} = 92 - \int_0^6 5e^{-0.1t}\, dt = 69.441°C$$

3.4   Exercises

1.   The function $F(x) = \sqrt{x}$ has the derivative $F'(x) = \dfrac{1}{2\sqrt{x}}$. Compute $\displaystyle\int_1^4 \dfrac{1}{2\sqrt{x}}\,dx$ using

   a) left and right sums with 50 subdivisions;

   b) the Fundamental Theorem of Calculus.

2.   The function $F(x) = \sin(2x)$ has the derivative $F'(x) = 2\cos(2x)$. Compute
   $\displaystyle\int_0^{\pi/2} 2\cos(2x)\,dx$, using

   a) left and right sums with 50 subdivisions;

   b) the Fundamental Theorem of Calculus.

3.   The function $F(x) = x\ln x$ has the derivative $F'(x) = 1 + \ln x$. Compute $\displaystyle\int_1^e (1 + \ln x)\,dx$
   using

   a) left and right sums with 50 subdivisions;

   b) the Fundamental Theorem of Calculus.

4.   The graph below shows the rate in gallons per hour at which oil is leaking out of a tank.

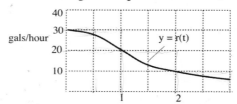

   a) Write a definite integral that represents the total amount of oil that leaks out in the first
      hour.

   b) Shade the region whose area represents the total amount of oil that leaks out in the first
      hour.

   c) Give a lower and upper estimate of the total amount of oil that leaks out in the first hour.

5.   After t hours, a population of bacteria is growing at a rate of 2^t hundred bacteria per hour.

   a) Write a definite integral that measures the total increase in the bacteria population during
      the first 4 hours.

   b) Evaluate the definite integral in part a).

6.   The rate of production for a new motorcycle is $100\left(\dfrac{t}{t+10}\right)^2$ bikes per week t weeks after
   initial production begins. Estimate how many motorcycles are made in the first 5 weeks.

In Exercises 7–10 use the given derivative formula to evaluate the definite integral.

7. $\dfrac{d}{dx}\left(x^2 + x\right) = 2x + 1;\quad \displaystyle\int_0^2 (2x + 1)\,dx$ 8. $\dfrac{d}{dx}(x\ln x - x) = \ln x;\quad \displaystyle\int_1^e \ln x\,dx$

9. $\dfrac{d}{dx}\left(e^{x^2}\right) = 2xe^{x^2}$; $\int_0^2 2xe^{x^2}\,dx$ 10. $\dfrac{d}{dx}\left(\dfrac{1}{2}\sin(2x)\right) = \cos(2x)$; $\int_0^{\pi/4}\cos(2x)\,dx$

11.   A car is moving along a straight road from A to B, starting from A at time $t = 0$. Below is the velocity (positive direction from A to B) plotted against time.

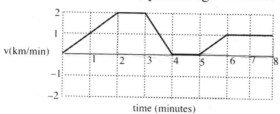

v(km/min)   time (minutes)

     a) How many kilometers away from A is the car at time $t = 2, 4,\ 8$?

     b) Sketch a graph of the acceleration of the car.

12.   Suppose the rate at which ice in a pond is melting is given by $\dfrac{dV}{dt} = 4t + 2$, where V is the volume of the ice in cubic feet, and t is the time in minutes.

     a) Write a definite integral that represents the amount of ice that has melted in the first 4 minutes.

     b) Evaluate the definite integral in part a).

13.   A company purchases a new machine for which the rate of depreciation is

$$\frac{dV}{dt} = 500(t - 6)\ ,\ 0 \le t \le 5$$

where V is the loss in value of the machine in dollars after t years.

     a) Write a definite integral that represents the total loss of value of the machine over the first 3 years.

     b) Evaluate the definite integral in part a).

     c) If the machine was purchased at a price of \$15,500, what is its value after 3 years?

14.   When you are fresh, you find that you can harvest 150 bushels of corn per hour. But as the day wears on, your efficiency decreases. In fact, after t hours from the beginning of the day, you are harvesting at a rate of

$$\frac{dB}{dt} = 150e^{-0.5t}\ ,\ 0 \le t \le 8$$

where B is the number of bushels after t hours.

     a) Write a definite integral that measures how many bushels of corn you harvest after arriving in the corn field and working for 4 consecutive hours.

     b) Evaluate the definite integral in part a).

In Exercises 15–18 use the Fundamental Theorem and your calculator to answer the questions.

15.   Given the function f with $f'(x) = \cos(x^3)$ and $f(0) = 2$. Find $f(1)$.

16.   Let f be the function whose graph passes through the point $(4, 6)$ and whose derivative is given by $f'(x) = \frac{x^3}{1+e^x}$. Find $f(4.1)$.

17.   A particle moves along a line so that at any time $t \geq 0$ its velocity is given by $v(t) = \frac{t}{1+t^2}$. At time $t = 0$, the position of the particle is $s(0) = 5$. Determine the position of the particle at $t = 3$.

18.   Water flows into a tank at a rate of $\frac{dW}{dt} = \sqrt{300 + 20t - t^2}$ where $\frac{dW}{dt}$ is measured in gallons per hour and t is measured in hours. If there are 120 gallons of water in the tank at time $t = 0$, how many gallons of water are in the tank at $t = 24$?

19.   Let $F(x) = \int_0^x f(t)\, dt$, $0 \leq t \leq 4$, where f is the function graphed below.

graph of f

a) Complete the following table of values for F.

x	0	1	2	3	4
$F(x)$				11/4	

b) Sketch a graph of F.

20. $G(x) = \int_{-1}^x g(t)\, dt$ for $-1 \leq t \leq 4$, where g is the function graphed below.

graph of g

a) Complete the following table of values for G.

x	−1	0	1	2	3	4
$G(x)$						

b) Sketch a graph of G.

c) Which is larger, $G(2)$ or $G(3)$? Justify your answer.

d) Where is G increasing?

Chapter 3   Supplementary Problems

1.   The graph of a function f whose domain is the interval $[-4, 4]$ is shown below.

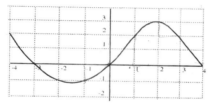

a) Estimate the average rate of change of f over the interval from $x = -2$ to $x = 3$.

b) Estimate the slope of the tangent line at the point where $x = 2$.

c) Evaluate a left sum approximation of $\int_{-1}^{3} f(t)\, dt$ with $n = 4$ subdivisions.

d) Estimate the value of $\int_{-1}^{3} f(t)\, dt$.

2.   The rate at which a popular cola drink is being sold is continuously increasing. Suppose the rate (in bottles per year) is given by $r = f(t)$, where t is measured in years and $t = 0$ is the start of 1993.

a) Write a definite integral that represents the total quantity of cola sold between the start of 1993 and the start of 1998.

b) Suppose $r = 24e^{0.05t}$. Using a left-hand Riemann sum, with 5 subdivisions, find an approximate value for the total quantity of cola sold between the start of 1993 and the start of 1998.

c) Interpret each of the five terms in the sum from part b) in terms of cola sold.

3.   Given the function $f(x) = 2 + \cos x, \ 0 \le x \le 3\pi$.

a) Sketch a graph of f on the interval $[0, 3\pi]$.

b) List in increasing order the following three quantities:

$\int_{0}^{3\pi} (2 + \cos x)\, dx$, the left-hand sum and the right-hand sum with $n = 3$ subdivisions.

4.   The graph of f is shown below.

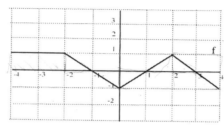

Evaluate each of the following.

a) $\int_{-4}^{-2} f(x)\, dx$ b) $\int_{-2}^{1} f(x)\, dx$ c) $\int_{-4}^{4} f(x)\, dx$ d) $\int_{0}^{3} f(x)\, dx$

5.   For each function below use a graph of f to evaluate

        i) $\int_0^2 f(x)\,dx$ ii) $\int_1^3 f(x)\,dx$ iii) $\int_{-2}^3 f(x)\,dx$

        a) $f(x) = 3 - x$ b) $f(x) = 2x + 1$ c) $f(x) = 1 - |x|$

6.   Calculate left-hand and right-hand sums with $n = 4$ and $n = 10$ for $\int_{-1}^5 \dfrac{12}{x+2}\,dx$.
     How do these sums compare with the true value of the integral?

7.   In a memory experiment the rate of memorization is given by

$$M'(t) = 2 - 0.003t^2$$

     where $M(t)$ is the number of words memorized in t minutes. How many words are memorized in 8 minutes?

8.   The sales of a company are expected to grow at a rate given by $\dfrac{dS}{dt} = 20e^t$

     where $S(t)$ is the sales in dollars in t days. Find the accumulated sales through the first 5 days using

     a) a lower and upper estimate with $n = 10$ subdivisions;

     b) the Fundamental Theorem, if the function $S(t) = 20e^t$ is its own derivative, $S'(t) = 20e^t$.

9.   Buyers' Guide is testing braking performance of two cars: the Maxima and the Lexus. The test consists of using the brakes to bring the car to a complete stop when the car is traveling at 60 mph. The Maxima slows down at the uniform rate of 12 miles per hour per second. The Lexus' velocity is given in the graph below.

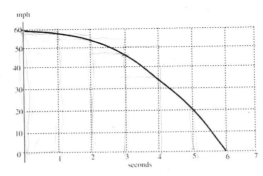

     a) How long will it take each car to stop?

     b) How far will each car travel before it stops?

     c) Write a brief paragraph summarizing the braking performance of each car. If the data indicate that one of the cars is superior explain your conclusion.

10. The population of geese in New Foundland is growing at the rate of 3% per year. The current population of geese is 50,000.

    a) If $P(t)$ is the population of geese t years from now, give an equation of the line tangent to the population graph at time $t = 0$.

    b) Use the tangent line at time $t = 0$ to approximate the population of geese 4 months from now.

11. The graph below gives the velocity of a car traveling along a straight highway. The car is 180 miles from Boston traveling toward Boston at $t = 0$.

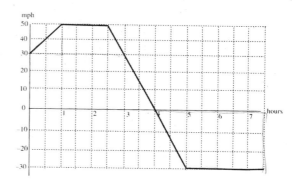

    a) When did the car change direction?

    b) How many miles did the car travel between $t = 0$ and $t = 3$ hours?

    c) When was the car closest to Boston? Explain why.

    d) When was the car farthest from Boston? Explain why.

    e) Did the car reach Boston during the 8 hour trip?

12. The table below contains values of a continuous function f.

x	1.0	1.25	1.5	1.75	2.0	2.25	2.5	2.75	3.0
$f(x)$	2.0	2.5	2.0	1.5	1.0	.5	0	1.0	2.0

    a) Approximate the integral $\int_1^3 f(x)\, dx$.

    b) If a function A is defined by $A(x) = \int_1^x f(t)\, dt$, $1 \le x \le 3$, estimate the average rate of change of A on the interval from $x = 1$ to $x = 3$.

13. The graph of f is shown in the figure.

    a) What can you say about $\lim_{x \to 0} f(x)$?

    b) What can you say about $f'(0)$?

    c) What can you say about $\int_{-1}^0 f(x)\, dx$?

    d) Approximate $\int_{-4}^2 f(x)\, dx$ using a left sum with $n = 6$ subdivisions.

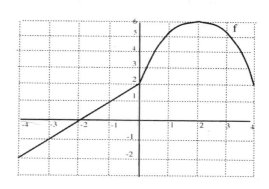

14. The figure at the right shows the graph of f' the derivative of a function f. The domain of f is the interval $[0, 2]$.

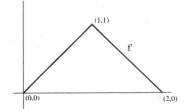

a) Write an expression for $f'(x)$ in terms of x.

b) On what intervals is f increasing?

c) On what intervals is f decreasing?

d) On what intervals is f concave up?

e) If $f(0) = -\dfrac{1}{2}$, sketch a graph of f on the given coordinate axes.

15. Below are the graphs of four functions. On the same axes as each function, sketch a graph of the derivative of each function.

a) 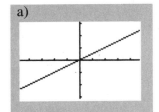 b) 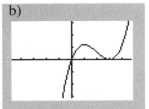 c) 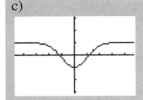 d)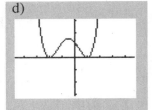

16. To the right is the graph of f', the derivative of a function f with domain $[0, 6]$.

Based on this graph,

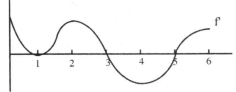

a) where does the graph of f have a horizontal tangent?

b) on what interval(s) is f increasing?

c) on what interval(s) is the graph of f concave up?

d) at what x-coordinate(s) does f have a local minimum value?

e) at what x does f reach its largest value in the interval $[0, 6]$?

17. The table below shows some values of a function g and its first derivative g'.

x	0	1	2	3	4	5
$g(x)$	1	0.60	0.21	−0.17	−0.54	−0.88
$g'(x)$	−0.40	−0.39	−0.38	−0.37	−0.35	−0.33

a) Determine the average rate of change of g from $x = 0$ to $x = 5$.

b) Use the fact that $g(0) = 1$ and $g'(0) = -0.4$ to estimate $g(-0.3)$. (Show your method clearly.)

18. The rate R (in gallons per hour) at which water is leaking from the town water tank is decreasing throughout one night, as shown in the table below.
(t represents the number of hours since 8:00 pm, when the leak started.)

t (hours)	0	2	4	6	8	10
R (gal/hr)	71	53	35	23	15	11

a) In the first two hour period, what is the maximum amount of water that could have leaked out of the tank? What is the minimum?

b) Using all of the available information, make lower and upper estimates for the total amount of water that leaked out during the ten hours.

c) The town engineer estimates the total amount of water that leaked out by using the average of your lower and upper estimates (from part b)). By how many gallons might her estimate differ from the actual, exact amount of water lost?

d) In order to estimate the actual amount of water lost with an error of twenty gallons or less, how often (during the ten hour period) must the rate at which water is leaking be recorded?

19. Each of the following limits is $f'(a)$ for some function f and some number a . Identify f and a, then find the value of the limit.

a) $\displaystyle \lim_{h \to 0} \frac{\cos\left(\dfrac{\pi}{2}+h\right) - \cos \dfrac{\pi}{2}}{h}$

b) $\displaystyle \lim_{x \to 4} \frac{x^{3/2} - 8}{x - 4}$

20. Let the function f be defined by $f(x) = \begin{cases} ax^2 + bx, & \text{if } x \le 3 \\ 3 - 2x, & \text{if } x > 3 \end{cases}$.

a) For what values of a and b is the function f both continuous and differentiable at $x = 3$.

b) Find values of a and b so that the function f is continuous but not differentiable at $x = 3$.

21. Suppose the derivative of the function f is $f'(x) = (x-1)(x+3)$.

a) On what intervals, if any, is f increasing?

b) At which inputs x, if any, does f have a local maximum?

c) On which intervals, if any, is f concave down?

d) Sketch a possible graph of f .

22. The function f is defined on the interval $[-5, 5]$
and its graph is shown to the right. Use the graph
to evaluate the following quantities.

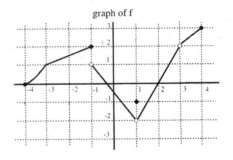
graph of f

a) $\lim_{x \to 1} f(x)$

b) $\lim_{h \to 0} \dfrac{f(2+h) - f(2)}{h}$

c) $\lim_{x \to -1^+} f(x)$

d) $\lim_{x \to 0} f'(x)$

e) $\int_{-3}^{-1} f(x)\, dx$

23. Use your calculator to zoom in on the graph of $f(x) = |x|^{3/2}$ at $x = 0$.

a) Is the function locally linear at $x = 0$?

b) What is the average rate of change of f from $x = 0$ to $x = 0.2$?

c) Can the average rate of change in part b) be used to estimate the rate of change at $x = 0$?

24. The domain of each of the following functions is the interval $[-3, 3]$. Define a
function and sketch its graph if:

a) the function has a limit at each input and is continuous at each input;

b) the function has a limit at each input, but is not continuous at $x = 0$ and at $x = 1$;

c) the function has no limit at $x = 0$ and is not continuous at $x = 0, 1, 2$.

25. Which of the following functions has a derivative at $x = 0$?

a) $f(x) = |x^3 - 3x^2|$ b) $g(t) = \sqrt{t^2 + .01} - |t - 1|$ c) $h(x) = \dfrac{e^x}{\cos x}$

26. Let f be a polynomial function with $f(0) = 2$ and derivative given by
$$f'(x) = (x^2 - 4)(x^2 - 3x + 2).$$

a) Find the slope of the graph of f at the point where $x = 0$.

b) Write an equation for the line tangent to the graph of f at $x = 0$.

c) Determine the intervals on which f is increasing.

d) Determine the x-coordinate of the inflection points on the graph of f.

CHAPTER 4
Differentiation Rules

If a function is defined graphically or numerically, the derivative, or instantaneous rate of change, can be found only approximately. However, if we are given an exact formula for a function, the definition of the derivative can be employed to find an exact formula for the derivative. The process of calculating the derivative formula is called **differentiation**.

The value of the derivative, f', at each x in the domain of f is defined as the limiting value of the average rate of change of f as h approaches zero, provided the limiting value exists.

$$f'(x) = \lim_{h \to 0} \frac{f(x+h) - f(x)}{h}$$

This definition can be used to find the derivative formula of a specified function. The goal of this chapter is to develop a list of rules by which derivatives can be calculated without having to apply the definition each time. Before doing so, however, we introduce an additional symbol for the derivative

$$\frac{d}{dx}[f(x)] = f'(x).$$

For example, $\dfrac{d}{dx}\left(x^2\right) = 2x$ means the derivative of x^2 with respect to x.

4.1 Derivative Rules for Basic Functions

Almost all basic functions have simple formulas for their derivative functions.

Linear Function Rule

The slope of a straight line is a constant. Thus the derivative of a linear function is a constant.

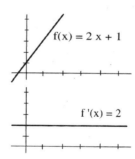

Derivative of a Linear Function

If $f(x) = mx + b$ where m and b are constants, then the derivative of f is $f'(x) = m$.

From the definition of the derivative we have:

$$f'(x) = \lim_{h \to 0} \frac{f(x+h) - f(x)}{h}$$

$$= \lim_{h \to 0} \frac{[m(x+h)+b] - [mx+b]}{h} = \lim_{h \to 0} m$$

$$= m.$$

Example 1: *Using the Linear Function Rule*

Function	Derivative
$y = 3x - 4$	$\dfrac{dy}{dx} = 3$
$f(x) = 7 - 5x$	$f'(x) = -5$
$y = \pi x - 7$	$y' = \pi$

Constant Function Rule

Constant functions are a special case of linear functions. The graph of a constant function is a horizontal line with slope 0, therefore, its derivative is 0.

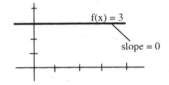

f(x) = 3

slope = 0

Derivative of a Constant Function

If $f(x) = c$, where c is a constant, then $f'(x) = 0$.

For example, $\dfrac{d}{dx}(7) = 0$ and $\dfrac{d}{dx}(\pi) = 0$.

Power Function Rule

The power function, $f(x) = x^n$, where n is a nonnegative integer, can be used as a building block for polynomial functions. Once we know its derivative, it will be easy to find the derivative of a polynomial function.

We know how to compute derivatives for the simpler power functions,

Function	Derivative
$f(x) = x^0 = 1$	$f'(x) = 0$
$f(x) = x$	$f'(x) = 1$
$f(x) = x^2$	$f'(x) = 2x$

For the case $n = 3$ we have

Example 2:   *Using the Definition to Find a Derivative*

If $f(x) = x^3$, find $f'(x)$.

Solution:

$$f'(x) = \lim_{h \to 0} \frac{f(x+h) - f(x)}{h}$$

$$= \lim_{h \to 0} \frac{(x+h)^3 - x^3}{h}$$

$$= \lim_{h \to 0} \frac{(x^3 + 3x^2 h + 3xh^2 + h^3) - x^3}{h}$$

$$= \lim_{h \to 0} (3x^2 + 3xh + h^2)$$

$$= 3x^2$$

There is a pattern and in every case so far ($n = 0, 1, 2, 3$) we have

---

If n is a nonnegative integer and $f(x) = x^n$, then $f'(x) = nx^{n-1}$.

To see that this formula is valid for $n > 3$, we note that with $f(x) = x^n$, from the definition of the derivative we have

$$f'(x) = \lim_{h \to 0} \frac{f(x+h) - f(x)}{h}$$

$$= \lim_{h \to 0} \frac{(x+h)^n - x^n}{h}.$$

If n is a positive integer, then we can expand $(x+h)^n$ by using the binomial theorem, obtaining

$$= \lim_{h \to 0} \frac{(x^n + nx^{n-1} \cdot h + \frac{n(n-1)}{2!} x^{n-2} \cdot h^2 + \dots + nx \cdot h^{n-1} + h^n) - x^n}{h}$$

$$= \lim_{h \to 0} \left[nx^{n-1} + \frac{n(n-1)}{2!} x^{n-2} \cdot h + \dots + nx \cdot h^{n-2} + h^{n-1} \right].$$

Each term within the brackets, except the first, contains a positive power of h; therefore, every term except the first approaches zero as h approaches zero, thus

$$f'(x) = nx^{n-1}.$$

In words: *To differentiate a power function, you multiply by the old exponent, then reduce the exponent by 1 to get the new exponent.*

For example, $\dfrac{d}{dx}\left(x^7\right) = 7x^6$ and $\dfrac{d}{dx}\left(x^{100}\right) = 100x^{99}$.

Constant Multiple Rule

Multiplying the outputs of a function by a constant stretches or shrinks its graph (and reflects it in the x-axis if the constant is negative). If a graph is stretched vertically and possibly reflected about the x-axis, then the tangent line at each point is changed in the same way. In other words, if a function is multiplied by a constant, so is its derivative.

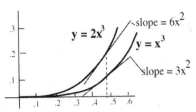

The graphs of $f(x) = x^3$ and the stretched curve $y = 2x^3$. Multiplying the outputs by 2 multiplies the slopes by 2.

For any constant multiple of a differentiable function we have the following result.

> ### The Constant Multiple Rule
>
> If c is any constant and f is a differentiable function, then
> $$\frac{d}{dx}[c \cdot f(x)] = c \cdot f'(x) .$$

In words: *The derivative of a constant times a function is the constant times the derivative of the function.*

Example 3: *Using the Constant Multiple Rule*

Function	Derivative
$y = 3x^7$	$\dfrac{dy}{dx} = 3\left(7x^6\right) = 21x^6$
$f(x) = -4x^4$	$f'(x) = -4\left(4x^3\right) = -16x^3$
$y = \dfrac{1}{2}t^2$	$y' = \dfrac{1}{2}(2t) = t$

The Sum and Difference Rules

The next rule says that the derivative of a sum, or difference, of two differentiable functions, is the sum, or difference, of their derivatives.

> ### The Sum and Difference Rule
>
> $$\frac{d}{dx}[f(x) + g(x)] = [f'(x) + g'(x)]$$
>
> $$\frac{d}{dx}[f(x) - g(x)] = [f'(x) - g'(x)]$$

Using the definition again, we have

$$\frac{d}{dx}[f(x)+g(x)] = \lim_{h\to 0}\frac{[f(x+h)+g(x+h)]-[f(x)+g(x)]}{h}$$

$$= \lim_{h\to 0}\frac{f(x+h)+g(x+h)-f(x)-g(x)}{h}$$

$$= \lim_{h\to 0}\frac{f(x+h)-f(x)}{h}+\lim_{h\to 0}\frac{g(x+h)-g(x)}{h}$$

$$= f'(x)+g'(x).$$

The Sum and Difference rules can be extended to any finite number of functions.

Knowing how to differentiate powers, constant multiples, and sums means that we can differentiate any polynomial function.

Example 4: *Differentiating a Polynomial Function*

If the polynomial function p is defined by $p(x) = 4x^3 - 3x^2 + 7$, find $p'(x)$.

Solution:

$$\frac{d}{dx}\left(4x^3 - 3x^2 + 7\right) = \frac{d}{dx}\left(4x^3\right)-\frac{d}{dx}\left(3x^2\right)+\frac{d}{dx}(7)$$

$$= 4\frac{d}{dx}\left(x^3\right)-3\frac{d}{dx}\left(x^2\right)+\frac{d}{dx}(7)$$

$$= 4\left(3x^2\right)-3(2x)+0$$

$$= 12x^2 - 6x.$$

Negative and Rational Exponents

The fact that the Power rule holds for nonnegative integers n leads to a simple formula for differentiating polynomial functions, but this is only half the story. The Power rule also holds for negative integers.

For example,

$$\frac{d}{dx}\left(\frac{1}{x}\right) = \lim_{h\to 0}\frac{\frac{1}{x+h}-\frac{1}{x}}{h}$$

$$= \lim_{h\to 0}\frac{x-(x+h)}{h\cdot x\cdot(x+h)}$$

$$= \lim_{h\to 0}\frac{-h}{h\cdot x\cdot(x+h)}$$

$$= \lim_{h\to 0}\frac{-1}{x(x+h)} = -\frac{1}{x^2}$$

So,
$$\frac{d}{dx}\left(\frac{1}{x}\right) = -\frac{1}{x^2}.$$

Substituting x^{-1} for $\frac{1}{x}$ and $-x^{-2}$ for $-\frac{1}{x^2}$ in the equation above yields

$$\frac{d}{dx}\left(x^{-1}\right) = -x^{-2}$$

which is the Power Rule with $n = -1$. This example suggests that the rule

---

The Power Rule

If n is any real number, then $\dfrac{d}{dx}\left(x^n\right) = nx^{n-1}$

---

may hold for negative integers also. In fact, the Power Rule holds for any real constant n. Although a proof must wait until a later section, we will assume that the Power Rule holds when the exponent n is any real number.

Example 5: *Using the Power Rule.*

Function	Derivative
$y = x^2 + x^{-5}$	$\dfrac{dy}{dx} = 2x - 5x^{-6}$
$s = t^{3/4} - t^{-3}$	$\dfrac{ds}{dt} = \dfrac{3}{4}t^{-1/4} + 3t^{-4}$
$f(u) = 3u^{1/3} - u^{-1/3}$	$f'(u) = u^{-2/3} + \dfrac{1}{3}u^{-4/3}$

Recall that an object that moves along a line with position $s(t)$ has velocity $v(t) = \dfrac{ds}{dt}$ and acceleration $a(t) = \dfrac{dv}{dt} = \dfrac{d^2s}{dt^2}$. The speed of the object is $\left| v(t) \right|$.

Example 6: *Using Derivatives to Find Velocity and Acceleration*

A marble rolls up a smooth inclined plane. After t seconds its distance s feet from the starting position is given by $s = 24t - 3t^2$. Find the velocity and acceleration of the marble at $t = 3.1$ seconds.

Solution:

The velocity, v, is the derivative of position

$$v = \frac{ds}{dt} = 24 - 6t \ \ \text{ft/sec}$$

and the acceleration, a, is the derivative of velocity

$$a = \frac{dv}{dt} = -6 \ \text{ft/sec}^2.$$

Thus, at $t = 3.1$,

$$v(3.1) = 5.4 \ \text{ft/sec} \ \text{ and } \ a(3.1) = -6 \text{ft/sec}^2.$$

What f' and f'' tell us about f

While the first derivative f' and the second derivative f'' can be interpreted as velocity and acceleration in a motion problem, the graphical interpretation of the first and second derivative is based on the fact that

> *The sign of the derivative of a function tells whether the original function is increasing or decreasing.*

Thus, we know that

- If $f' > 0$ on an interval, then f is increasing on the interval.
  If $f' < 0$ on an interval, then f is decreasing on the interval.

- If $f'' > 0$ on an interval, then f' is increasing and f is concave up on the interval.
  If $f'' < 0$ on an interval, then f' is decreasing and f is concave down on the interval.

Example 7: *Concavity and the Second Derivative*

Find and interpret graphically the second derivative of $f(x) = x^3 - 3x^2$.

Solution:

For $f(x) = x^3 - 3x^2$,

$$f'(x) = 3x^2 - 6x \ \ \text{and} \ \ f''(x) = 6x - 6.$$

At $x = 1$, $f''(1) = 0$. For $x > 1$ we see that $f''(x)$ is positive which means that the f-graph is concave up; for $x < 1$, $f''(x)$ is negative and the f-graph is concave down. Since the concavity of the f-graph changes at the point $(1, -2)$, this is a point of inflection.

4.1   Exercises

1.   Find the slope of the tangent line to the graph of f at the point $(1, 1)$.

   a) $f(x) = x^{1/2}$

   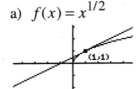

   b) $f(x) = x^3$

   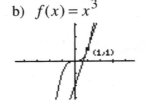

   c) $f(x) = x^{3/2}$

   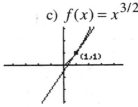

In Exercises 2–17, use the rules of differentiation to find the derivative of each function.

2. $y = 5$

3. $y = 3 - 4x$

4. $g(x) = 3x - 5$

5. $f(x) = x^3 + 3x^2$

6. $f(t) = -2t^2 + 3t - 6$

7. $y = 3x^2 - 8x + 7$

8. $f(x) = 2x^3 - x^2 + 3x$

9. $g(t) = 2t^3 - 4t^2 - 5$

10. $y = 5x^3 - 3x^5$

11. $y = x^5 - \dfrac{x^3}{3} - 2x + 3$

12. $g(x) = x^{3/2} - x^{1/2} - x^{-1/2}$

13. $f(x) = x^{-3} - x^{-2} - x^{-1} - 1$

14. $y = -4.8x^{1/3}$

15. $f(x) = 3x^{-2/3} + x^{3/4}$

16. $s = 4t^{-1/2} + t^{1/2}$

17. $y = -2x^{2/3} + x^{-4/5}$

18.   Find an equation of the tangent line to the graph of the function at the indicated point in two ways: i) algebraically and ii) graphically.

   a) $f(x) = x^4 - 3x^2 + 2$, $P = (1, 0)$

   b) $f(x) = x^3 + x$, $P = (-1, -2)$

   c) $f(x) = x^{-2/3}$, $P = (8, 1/4)$

   d) $f(x) = x^{3/2}$, $P = (4, 8)$

19.   Determine the points at which the function has a horizontal tangent.

   a) $f(x) = x^4 - 3x^2 + 2$

   b) $f(x) = x + \dfrac{1}{x}$

   c) $f(x) = x^3 - 6x^2 + 9x$

   d) $f(x) = x - \sqrt{x}$

20.   A particle moves along a coordinate line according to the position function $s(t) = t^3 + 4t - 3$. Determine

   a) the average velocity of the particle from $t = 2$ to $t = 4$;

   b) the instantaneous velocity of the particle at $t = 2$.

21.   The motion of a particle along a line is given by the position function $s(t) = \dfrac{t^3}{3} - \dfrac{9t^2}{2} + 14t - 6$.
       At what times from $t = 0$ to $t = 10$ is the particle at rest?

22.   For what values of k will the graph of $y = x^3 + kx^2 + x + 2$ have two horizontal tangent lines?

23.   Determine whether the function f is increasing or decreasing at $x = c$, and how fast.

       a) $f(x) = x^{1/2} + 2x - 13$, $c = 4$ b) $f(x) = x^{-2} - 3x + 15$, $c = 1$

24.   Find the x-coordinates of all points on the graph of $y = x^3 - 3x^2$ at which the tangent line
       is parallel to the line $3x - y = 8$.

25.   Find all points on the graph of $y = x^{3/2} - x^{1/2}$ at which the tangent line is parallel to the
       line $y - x = 3$.

26.   Below is the graph of f', the derivative of a function f whose domain is the interval $[-4, 4]$.
       Answer the following questions giving a brief justification in each case.

       a) Find all x at which the graph of f has a horizontal
           tangent.
       b) On what interval(s) is f increasing?
       c) On what interval(s) is the graph of f concave up?
       d) At which x-coordinate does f have its absolute
           maximum value?
       e) On what interval(s) is f increasing at an
           increasing rate?

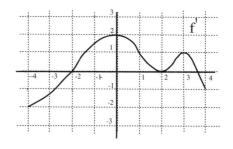

27.   The graph of the second derivative of a function f is shown below.

       a) Where is the graph of f concave up?
       b) Where does f have points of inflection?
       c) Suppose $f'(0) = 0$. Is f increasing or decreasing
           at $x = 2$? Justify your answer.
       d) Suppose $f'(3) = 0$. Is f increasing or decreasing at
           $x = 4$? Justify your answer.

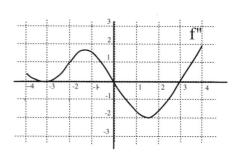

In Exercises 28 – 31 solve each problem in two ways: i) algebraically and ii) graphically.

28.   On what intervals is the function $f(x) = x^4 - 4x^3 + 10$ both concave down and decreasing?

29.   On what intervals is the function $f(x) = 4x^{3/2} - 3x^2$ both concave down and increasing?

30.   A particle moves along a coordinate line so that at $t \geq 0$ its position is given by
$$s(t) = t^4 - 2t^3 - t^2 + 2.$$
a) For what values of t is the particle moving to the right?

b) For what values of t is the velocity of the particle increasing?

31.   The position of a particle moving on a coordinate line is given by $s(t) = 3t^{4/3} - 9t^{7/3}$
Find the velocity and acceleration when $t = 8$.

32.   A particle moves along the x-axis in such a way that its position at time t, where $t \geq 0$, is given by
$$x(t) = \frac{1}{3}t^3 - 3t^2 + 8t.$$
a) Show that at time $t = 0$ the particle is moving to the right.

b) Find all values of t for which the particle is moving to the left.

c) What is the position of the particle at time $t = 3$?

d) What is the total distance the particle travels between times $t = 0$ and $t = 3$?

33.   Two curves are said to be tangent to each other at a point where they intersect if they have the same tangent line at that point. Find a constant k such that the curves $y = 1 + kx - kx^2$ and $y = x^4$ are tangent at the point (1, 1).

34.   Use the definition of the derivative to show that if $g(x) = c \cdot f(x)$ and f' exists, then
$$g'(x) = c \cdot f'(x).$$

[Hint: Start as follows: $g'(x) = \lim_{h \to 0} \dfrac{g(x+h) - g(x)}{h} = \lim_{h \to 0} \dfrac{c \cdot f(x+h) - c \cdot f(x)}{h} = \dots]$

35.   Verify the difference rule for derivatives: $\frac{d}{dx}[f(x) - g(x)] = f'(x) - g'(x)$.

4.2   Differentiating Exponential Functions

The graph of an increasing exponential function $f(x) = b^x$ is shown below.

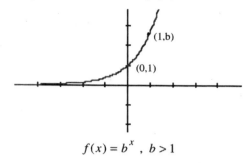

$$f(x) = b^x, \ b > 1$$

We see that slopes of tangent lines to the graph of f are positive. They start small and increase through larger and larger positive values. This means that the derivative function, f', is positive and increasing, which suggests that the graph of f' resembles the graph of f. Below are graphs of $y = 2^x$ and $y = 3^x$ and their derivatives.

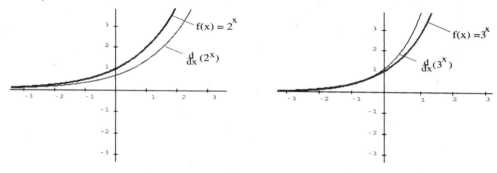

Calculating the Derivative of b^x

For an exponential function, $f(x) = b^x$, the derivative is given by

$$f'(x) = \lim_{h \to 0} \frac{f(x+h) - f(x)}{h}$$

$$= \lim_{h \to 0} \frac{b^{x+h} - b^x}{h}$$

$$= \lim_{h \to 0} \frac{b^x \cdot b^h - b^x}{h} \qquad \text{Using laws of exponents}$$

$$= \lim_{h \to 0} b^x \frac{b^h - 1}{h}$$

$$= b^x \cdot \lim_{h \to 0} \frac{b^h - 1}{h} \qquad \text{The limit involves only the variable h; x and b are fixed.}$$

Hence, $\qquad\qquad f'(x) = b^x \cdot \lim_{h \to 0} \frac{b^h - 1}{h}$.

Thus, the differentiation of b^x depends on $\displaystyle\lim_{h\to 0}\frac{b^h - 1}{h}$.

But this limit is the derivative of the exponential function $f(x) = b^x$ at $x = 0$ because

$$\frac{f(0+h) - f(0)}{h} = \frac{b^{0+h} - b^0}{h} = \frac{b^h - 1}{h}.$$

Assuming that $f(x) = b^x$ is differentiable at $x = 0$, we see that

$$\frac{d}{dx}\left(b^x\right) = \left(\text{Derivative of } b^x \text{ at } x = 0\right)\cdot b^x.$$

In particular, for $f(x) = 2^x$,

$$f'(x) = f'(0)\cdot 2^x .$$

Similarly, for $g(x) = 3^x$,

$$g'(x) = g'(0)\cdot 3^x .$$

In order to approximate the derivatives of 2^x and 3^x at $x = 0$, we can either zoom in on the graphs where they cross the y-axis and use $\boxed{\text{2nd}}$ [calc] 6: dy/dx,

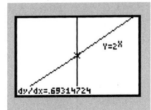 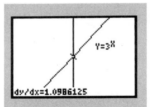

or examine the following table:

h	$\dfrac{2^h - 1}{h}$	$\dfrac{3^h - 1}{h}$
0.1	0.7177	1.1612
0.01	0.6956	1.1047
0.001	0.6934	1.0992
0.0001	0.6932	1.0987
0	undefined	undefined
−0.0001	0.6931	1.0986
−0.001	0.6929	1.0980
−0.01	0.6908	1.0926
−0.1	0.6697	1.0404

In either case, we conclude that

$$\frac{d}{dx}\left(2^x\right) = f'(0)\cdot 2^x \approx (0.693)\cdot 2^x$$

and

$$\frac{d}{dx}\left(3^x\right) = f'(0)\cdot 3^x \approx (1.099)\cdot 3^x.$$

Notice that the derivative of an exponential function is simply the constant $k = f'(0)$ times the function itself,

$$\text{If} \quad y = b^x \text{, then} \quad \frac{dy}{dx} = k \cdot b^x = k \cdot y \ .$$

In other words,

> *The derivative of an exponential function is proportional to the function itself.*

The Derivative of e^x

Since the base $b = 2$ gives a constant multiplier less than 1, while $b = 3$ gives a constant multiplier greater than 1, a question arises: Is there one base b between 2 and 3 for which the constant multiplier is 1? If so, we would have an exponential function that is its own derivative. It turns out that this is true when the base $b = e \approx 2.71828$ is used, and we have the remarkable fact that if $b = e$, the exponential function is its own derivative. Geometrically this means that the slope of the curve is equal to the y-coordinate of the point.

$$\frac{d}{dx}(e^x) = e^x$$

Not only is the function $y = e^x$ equal to its own derivative, it is the only nontrivial function that has this property. (See Exercise 26.)

In addition, the constants connected with the derivatives of 2^x and 3^x are related to e as they are natural logarithms. Since $\ln 2 \approx 0.6931$ and $\ln 3 \approx 1.0986$, this suggests that

$$\frac{d}{dx}\left(2^x\right) = (\ln 2) \cdot 2^x \quad \text{and} \quad \frac{d}{dx}\left(3^x\right) = (\ln 3) \cdot 3^x.$$

In a later section it will be shown that

$$\frac{d}{dx}\left(b^x\right) = (\ln b) \cdot b^x$$

Example 1: *Differentiating Exponential Functions*

If the function f is defined by $f(x) = 2 \cdot 3^x + 4e^x$, find $f'(x)$.

Solution:

$$\frac{d}{dx}\left(2 \cdot 3^x + 4e^x\right) = \frac{d}{dx}\left(2 \cdot 3^x\right) + \frac{d}{dx}\left(4e^x\right)$$

$$= 2\frac{d}{dx}\left(3^x\right) + 4\frac{d}{dx}\left(e^x\right)$$

$$= 2(\ln 3) \cdot 3^x + 4e^x$$

4.2   Exercises

In Exercises 1–16, find the derivative of each given function.

1. $f(x) = 3e^x + 5$

2. $f(x) = e^2 - 7e^x$

3. $f(x) = 2e^x + x^2$

4. $f(x) = 2x^4 - 5e^x$

5. $s(t) = 6^t$

6. $y = 5 \cdot 3^x$

7. $f(x) = 2^x + x^2 + 7$

8. $s(t) = 2.7^t - t^5$

9. $y = e^x + x^e$

10. $f(x) = e^\pi + x^\pi$

11. $f(x) = (\ln 2)e^x$

12. $f(x) = (\ln 6)6^x$

13. $f(x) = x^9 + 9^x$

14. $f(x) = 7^{x+1}$ [Hint: Rewrite using laws of exponents.]

15. $f(x) = 5x^4 - 2^{x+1}$

16. $f(x) = \left(\dfrac{1}{2}\right)^x$

In Exercises 17 and 18, solve each problem in two ways: i) algebraically and ii) graphically.

17.   Find an equation of the tangent line to the graph of $y = e^x$ at $x = 0$.

18.   Find an equation of the tangent line to the graph of $y = 2^x$ at $x = 1$.

19.   Let $f(x) = e^{x+2}$.

   a) Explain how a graph of f can be obtained from a graph of $y = e^x$ using only a vertical
      stretch.
   b) Use part a) to find $f'(x)$.

20.   A particle moves along a line so that at time t, where $t \geq 0$, its position is given by
       $s(t) = 2e^t - t^2$. What is the velocity of the particle when its acceleration is zero?

21.   A particle moves along a line so that at time t, where $0 \leq t \leq 5$, its position is given by
       $s(t) = 2e^t - 2t^3$.

   a) Find the average velocity of the particle over the interval $0 \leq t \leq 1$.
   b) When does the particle change direction?
   c) Find the values of t for which the particle is slowing down. [Hint: Speed will be
      decreasing when the velocity and acceleration have opposite signs.]

22. The position of a particle moving on the x-axis at time $t > 0$ seconds is $x(t) = e^t - \sqrt{t}$ feet.

    a) In what direction and how fast is the particle moving at $t = 1$ second?

    b) When is the particle moving to the right?

    c) Find the position of the particle when its velocity is zero.

23. The table gives the population of a small town over a five year period.

year	1990	1991	1992	1993	1994
population	111	122	134	147	161

    a) Estimate the rate of population growth in 1992 by averaging the slopes of two secant lines.

    b) Find an exponential function, $y = ab^x$, that models the population data.

    c) Use the exponential model to estimate the rate of growth in 1992. Compare your estimate with the result in part a).

24. The **normal line** to a curve C at a point P is, by definition, the line that passes through P and is perpendicular to the tangent line to C at P. Find an equation of the normal line to the curve $y = 3x - e^x$ at the point $(0, -1)$. Sketch the curve and its normal line.

25. The equation $y' - y = -1$ is called a **differential equation** because it involves an unknown function y and its derivative y'. A **solution** of a differential equation is any function that satisfies the differential equation. Show that the function $y = 1 - 2e^x$ is a solution of the given differential equation.

26. a) Show that the zero constant function is a solution of the differential equation $y' = y$.

    b) Find three different solutions of the differential equation $y' = y$. [Hint: Use the Constant Multiple Rule, p. 196]

4.3   The Product and Quotient Rules

Creating new functions from old is a central theme in calculus. In this section and the next we examine the problem of differentiating a function that is built-up from elementary functions using function operations.

The Product Rule

Knowing that the derivative of a sum is the sum of the derivatives, you are probably prepared to guess a rule for the derivative of a product. But first consider the fact that for $f(x) = x^2$ and $g(x) = x^3$, we have $f'(x) = 2x$ and $g'(x) = 3x^2$, so

$$f'(x) \cdot g'(x) = 2x \cdot 3x^2 = 6x^3.$$

But $f(x) \cdot g(x) = x^2 \cdot x^3 = x^5$, and by the Power Rule,

$$\frac{d}{dx}\left(x^5\right) = 5x^4.$$

which is not the product, $6x^3$, of the derivatives. Therefore, in general, the derivative of a product is not the product of the derivatives. The following is the correct Product Rule.

> **The Product Rule**
> $$\frac{d}{dx}[f(x) \cdot g(x)] = f(x) \cdot g'(x) + g(x) \cdot f'(x)$$

In words,

The derivative of a product of two functions is the first function times the derivative of the second, plus the second function times the derivative of the first.

Example 1: *Using the Product Rule*

Find the derivative of $f(x) = (x^2 + 1)(x^3 + 4)$ in two ways: (i) using the Product Rule and (ii) by multiplying out and then differentiating.

Solution:

i) Applying the Product Rule:

$$\frac{d}{dx}\left[(x^2 + 1)(x^3 + 4)\right] = (x^2 + 1)\frac{d}{dx}(x^3 + 4) + (x^3 + 4)\frac{d}{dx}(x^2 + 1)$$

$$= (x^2 + 1) \cdot (3x^2) + (x^3 + 4) \cdot (2x)$$

$$= 5x^4 + 3x^2 + 8x.$$

ii) Multiplying the two binomials and then differentiating:

$$\frac{d}{dx}\left[(x^2 + 1)(x^3 + 4)\right] = \frac{d}{dx}\left[x^5 + x^3 + 4x^2 + 4\right] = 5x^4 + 3x^2 + 8x.$$

A Geometric Proof of the Product Rule

Suppose we know that f and g are two positive-valued, increasing functions, and that the derivatives of functions f and g exist. To calculate the derivative of the product $f \cdot g$, we start with the difference quotient

$$\frac{[f \cdot g](x+h) - [f \cdot g](x)}{h} = \frac{f(x+h) \cdot g(x+h) - f(x) \cdot g(x)}{h} \, ,$$

but it is not clear how to connect this to the difference quotients for f and g. The interpretations of products as areas of rectangles gives us a hint.

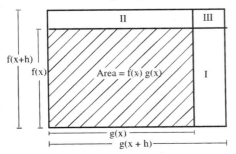

Area of I: $[g(x+h) - g(x)] \cdot f(x)$

Area of II: $[f(x+h) - f(x)] \cdot g(x)$

Area of III: $[g(x+h) - g(x)] \cdot [f(x+h) - f(x)]$

The area of the large rectangle is $f(x+h) \cdot g(x+h)$ and the area of the shaded rectangle is $f(x) \cdot g(x)$. The difference, $f(x+h) \cdot g(x+h) - f(x) \cdot g(x)$, is the sum of the areas of rectangles I, II and III which add up to

$$[g(x+h) - g(x)] \cdot f(x) + [f(x+h) - f(x)] \cdot g(x) + [g(x+h) - g(x)] \cdot [f(x+h) - f(x)].$$

Now, dividing by h, we can write the difference quotient for $f \cdot g$ in terms of the difference quotients for f and g as follows:

$$\frac{[f \cdot g](x+h) - [f \cdot g](x)}{h} = \frac{f(x+h) \cdot g(x+h) - f(x) \cdot g(x)}{h} =$$

$$f(x) \cdot \frac{g(x+h) - g(x)}{h} + g(x) \cdot \frac{f(x+h) - f(x)}{h} + \frac{g(x+h) - g(x)}{h} \cdot \frac{f(x+h) - f(x)}{h} \cdot h$$

The last step in deriving the Product Rule is to take the limit as $h \to 0$.

$$\lim_{h \to 0} \frac{[f \cdot g](x+h) - [f \cdot g](x)}{h} =$$

$$f(x) \cdot \lim_{h \to 0} \frac{g(x+h) - g(x)}{h} + g(x) \cdot \lim_{h \to 0} \frac{f(x+h) - f(x)}{h} + \left[\lim_{h \to 0} \frac{g(x+h) - g(x)}{h} \right] \left[\lim_{h \to 0} \frac{f(x+h) - f(x)}{h} \right] \cdot \lim_{h \to 0} h$$

Therefore,

$$\frac{d}{dx}[f(x) \cdot g(x)] = f(x) \cdot g'(x) + g(x) \cdot f'(x) + g'(x) \cdot f'(x) \cdot 0$$

and so,

$$\frac{d}{dx}[f(x) \cdot g(x)] = f(x) \cdot g'(x) + g(x) \cdot f'(x).$$

Example 2:   *Using the Product Rule*

Differentiate: a) $y = (x^5)e^x$ b) $y = \dfrac{e^x}{x^3}$

Solution:

a) $\dfrac{d}{dx}\left[(x^5)e^x\right] = x^5 \cdot \dfrac{d}{dx}(e^x) \; + \; e^x \cdot \dfrac{d}{dx}(x^5) = x^5 \cdot e^x \; + \; e^x(5x^4) = (x^5 + 5x^4)e^x$

b) $\dfrac{d}{dx}\left[\dfrac{e^x}{x^3}\right] = \dfrac{d}{dx}\left[e^x \cdot x^{-3}\right] = e^x \cdot \dfrac{d}{dx}(x^{-3}) + x^{-3} \cdot \dfrac{d}{dx}(e^x)$

$\qquad\qquad = e^x \cdot (-3x^{-4}) + x^{-3} \cdot (e^x)$

$\qquad\qquad = (-3x^{-4} + x^{-3})e^x$

The Quotient Rule

Suppose we know the derivatives of f and g and we want to calculate the derivative of the quotient $\dfrac{f}{g}$. Starting with the identity

$$q(x) = \frac{f(x)}{g(x)}$$

and assuming that $g(x) \neq 0$, we multiply on both sides by $g(x)$ to get the identity

$$g(x) \cdot q(x) = f(x).$$

We now take the derivative of both sides, applying the Product Rule on the left:

$$g(x) \cdot q'(x) \; + \; q(x) \cdot g'(x) \; = f'(x).$$

Now solving for $q'(x)$ and simplifying we get

$$q'(x) = \frac{f'(x) - q(x) \cdot g'(x)}{g(x)}$$

$$= \frac{f'(x)}{g(x)} - \frac{f(x)}{g(x)} \cdot \frac{g'(x)}{g(x)} \qquad\qquad \text{Substituting } \frac{f(x)}{g(x)} \text{ for } q(x)$$

$$q'(x) = \frac{g(x) \cdot f'(x) - f(x) \cdot g'(x)}{[g(x)]^2}$$

The Quotient Rule

$$\frac{d}{dx}\left[\frac{f(x)}{g(x)}\right] = \frac{g(x) \cdot f'(x) - f(x) \cdot g'(x)}{[g(x)]^2}$$

In words,

The derivative of a quotient is the denominator times the derivative of the numerator minus the numerator times the derivative of the denominator all divided by the square of the denominator.

Example 3: *Using the Quotient Rule*

Find the derivative of $f(x) = \dfrac{3x - 2}{x^2 + 1}$.

Solution:

$$\frac{d}{dx}\left[\frac{3x-2}{x^2+1}\right] = \frac{(x^2+1)\dfrac{d}{dx}(3x-2) - (3x-2)\dfrac{d}{dx}(x^2+1)}{(x^2+1)^2}$$

$$= \frac{(x^2+1)\cdot(3) - (3x-2)\cdot(2x)}{(x^2+1)^2}$$

$$= \frac{-3x^2 + 4x + 3}{(x^2+1)^2}.$$

A calculator can be used to check that the derivative in Example 3 is reasonable. The graphs of f and f' are displayed in the viewing rectangle below. Notice that f' is positive-valued when f is increasing, negative-valued when f is decreasing and $f'(x) = 0$ when f has a horizontal tangent.

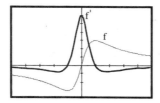

A liberal use of parentheses is recommended for all types of differentiation problems. For example, with the quotient rule, it is a good idea to enclose all factors and derivatives in parentheses and to pay attention to the subtraction in the numerator.

At each input x where functions f and g have derivatives, the functions $f + g$, $f \cdot g$ and f / g have derivatives (in the last case, under the assumption that $g(x) \neq 0$). The derivative formulas are summarized in the following equations.

1) $(f + g)' = f' + g'$

2) $(f \cdot g)' = f \cdot g' + g \cdot f'$

3) $\left(\dfrac{f}{g}\right)' = \dfrac{gf' - fg'}{g^2}$

Derivatives of Trigonometric Functions

For Y1 = sin (X), graphs of Y2 = nDeriv(Y1,X,X) and Y3 = cos (X) in the same viewing rectangle, and tables of inputs and outputs, strongly suggest that the derivative of the sine function is the cosine function.

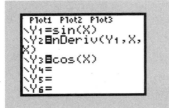

$$\frac{d}{dx}(\sin x) = \cos x$$

Similarly, for Y1 = cos (X), graphs of Y2 = nDeriv(Y1,X,X) and Y3 = –sin (X) in the same viewing rectangle, and tables of inputs and outputs, strongly suggest that the derivative of the cosine function is the negative of the sine function.

$$\frac{d}{dx}(\cos x) = -\sin x$$

We will confirm these results analytically below for $y = \sin x$ and leave $y = \cos x$ for the exercises.

To find $\dfrac{d}{dx}(\sin x)$, we use the addition identity $\sin(x + y) = \sin x \cos y + \cos x \sin y$ and

the limits: $\lim\limits_{h\to 0} \dfrac{\sin h}{h} = 1$ and $\lim\limits_{h\to 0} \dfrac{\cos h - 1}{h} = 0$. From the definition of the derivative we have

$$
\begin{aligned}
\frac{d}{dx}(\sin x) &= \lim_{h\to 0} \frac{\sin(x + h) - \sin(x)}{h}\\[2mm]
&= \lim_{h\to 0} \frac{(\sin x \cos h + \cos x \sin h) - \sin x}{h}\\[2mm]
&= \lim_{h\to 0} \frac{\sin x \cos h - \sin x}{h} + \lim_{h\to 0} \frac{\cos x \sin h}{h}\\[2mm]
&= \sin x \cdot \lim_{h\to 0} \frac{\cos h - 1}{h} + \cos x \cdot \lim_{h\to 0} \frac{\sin h}{h}\\[2mm]
&= \sin x \cdot 0 + \cos x \cdot 1 = \cos x.
\end{aligned}
$$

Example 4: *Differentiating Trigonometric Functions*

Differentiate   a) $y = x^3 \sin x$ b) $y = \dfrac{\cos x}{1 - \sin x}$

Solution:

a) $\dfrac{d}{dx}(x^3 \sin x) = x^3 \cdot \dfrac{d}{dx}(\sin x) + (\sin x) \cdot \dfrac{d}{dx}(x^3) = x^3 \cdot (\cos x) + (\sin x) \cdot 3x^2$

b) $\dfrac{d}{dx}\left(\dfrac{\cos x}{1 - \sin x}\right) = \dfrac{(1 - \sin x) \cdot \dfrac{d}{dx}(\cos x) - \cos x \cdot \dfrac{d}{dx}(1 - \sin x)}{(1 - \sin x)^2}$

$$= \frac{(1 - \sin x) \cdot (-\sin x) - \cos x \cdot (-\cos x)}{(1 - \sin x)^2}$$

$$= \frac{-\sin x + \sin^2 x + \cos^2 x}{(1 - \sin x)^2}$$

$$= \frac{1 - \sin x}{(1 - \sin x)^2} = \frac{1}{1 - \sin x}.$$

Since $\tan x = \dfrac{\sin x}{\cos x}$, we can differentiate $\tan x$ using the Quotient Rule for derivatives.

Example 5: *Derivative of the Tangent Function*

Find $\dfrac{d}{dx}(\tan x)$.

Solution:

$$\frac{d}{dx}(\tan x) = \frac{d}{dx}\left(\frac{\sin x}{\cos x}\right)$$

$$= \frac{\cos x \cdot \dfrac{d}{dx}(\sin x) - \sin x \cdot \dfrac{d}{dx}(\cos x)}{\cos^2 x}$$

$$= \frac{\cos x \cdot \cos x + \sin x \cdot \sin x}{\cos^2 x}$$

$$= \frac{1}{\cos^2 x} = \sec^2 x.$$

We summarize this result as an important fact.

$$\frac{d}{dx}(\tan x) = \sec^2 x$$

The derivatives of the remaining trigonometric functions, csc, sec, and cot, can be found easily by employing the Quotient Rule (see Exercise 39). The differentiation formulas for the basic trigonometric functions are listed below.

Derivatives of Trigonometric Functions

$$\frac{d}{dx}(\sin x) = \cos x \qquad\qquad \frac{d}{dx}(\csc x) = -\csc x \cdot \cot x$$

$$\frac{d}{dx}(\cos x) = -\sin x \qquad\qquad \frac{d}{dx}(\sec x) = \sec x \cdot \tan x$$

$$\frac{d}{dx}(\tan x) = \sec^2 x \qquad\qquad \frac{d}{dx}(\cot x) = -\csc^2 x$$

Notice that the minus signs go with the derivatives of the cofunctions, that is, cosine, cosecant, and cotangent.

Example 6: *Finding a Tangent Line*

Find an equation of the tangent line to the curve $y = \dfrac{-1}{\sec x + \tan x}$ at the point $(0,-1)$.

Solution:

By the Quotient Rule we have

$$\frac{dy}{dx} = \frac{\sec x \cdot \tan x + \sec^2 x}{(\sec x + \tan x)^2} = \frac{\sec x(\tan x + \sec x)}{(\sec x + \tan x)^2} = \frac{\sec x}{\sec x + \tan x}.$$

So the slope of the tangent line at $(0, -1)$ is $y'(0) = 1$. Therefore, an equation of the tangent line is

$$y + 1 = 1 \cdot (x - 0) \ \text{ or } \ y = x - 1.$$

The graph of the curve and its tangent line are shown below.

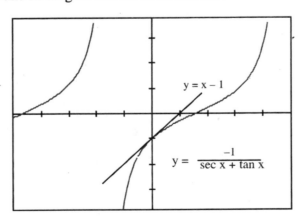

4.3   Exercises

1.   Find the derivative of each of the following functions two ways: (i) using the Product Rule and (ii) by multiplying out and then differentiating.

a) $f(x) = x^2(x^3 - 1)$ b) $f(x) = (x + 3)^2$

c) $f(x) = (4x + 1)(3x - 2)$ d) $f(x) = (x^2 + 2)(x^2 - 2)$

2.   Suppose functions f and g are defined so that $f(1) = 5$, $f'(1) = 10$ and $g(1) = 4$, $g'(1) = 3$.

a) If $h(x) = f(x) \cdot g(x)$, find $h'(1)$.

b) If $k(x) = \dfrac{f(x)}{g(x)}$, find $k'(1)$.

In Exercises 3–24, find a rule for the derivative of the given function.

3. $f(x) = x \sin x$ 4. $f(x) = (x^2 + 1)(x^2 - 1)$

5. $y = (x^2 + 1)\sin x$ 6. $f(x) = e^x \cos x$

7. $f(x) = xe^x$ 8. $s = (t^2 + 1)e^t$

9. $f(x) = -4x + 2 \tan x$ 10. $y = \sin x \cdot \cos x$

11. $y = \dfrac{2x}{x + 1}$ 12. $f(x) = \dfrac{x + 1}{x - 1}$

13. $s(t) = \dfrac{2t + 3}{2t - 2}$ 14. $f(x) = \dfrac{x^2 + 1}{x^2 - 1}$

15. $y = \dfrac{\sin x}{x + 1}$ 16. $f(x) = \dfrac{1}{1 + \sqrt{x}}$

17. $f(x) = \dfrac{x}{\sin x}$ 18. $y = \dfrac{\tan x}{e^x}$

19. $y = \dfrac{1 - \sin x}{1 + \sin x}$ 20. $f(x) = \dfrac{\sin x}{1 + \cos x}$

21. $y = \dfrac{x^2 + 1}{1 - x}$ 22. $y = \dfrac{x \sin x}{1 + \cos x}$

23. $y = \dfrac{2^x}{3^x}$ 24. $y = x \cdot 5^x$

25.   Suppose that f is a differentiable function and $g(x) = [f(x)]^2$. Use the Product Rule to show that $g'(x) = 2f(x) \cdot f'(x)$.

26.   The function f is defined by the rule $f(x) = \dfrac{x^2}{x-1}$.

a) Compute and simplify a formula for $f'(x)$.

b) Determine the points on the graph of f where the tangent lines have zero slope.

c) Compute and simplify a formula for $f''(x)$.

27.   Let $f(x) = \dfrac{x-1}{e^x}$.

a) For what inputs x is f increasing?

b) For what inputs x is the graph of f concave up?

28.   Let $f(x) = \dfrac{x^2 - 3}{e^x}$. Find the x-coordinates of all points at which $f'(x) = 0$.

29.   a) Use the identity $\sin(2x) = 2\sin x \cos x$ to find $\dfrac{d}{dx}(\sin 2x)$.

b) Find $\dfrac{d}{dx}(\cos 2x)$. [Hint: $\cos 2x = 2\cos^2 x - 1$]

30.   A population of bacteria is changing in a culture according to the equation

$$p(t) = 100\left(\frac{1 + 2t}{20 + t^2}\right)$$

where t is measured in hours. Find the rate at which the population is growing at $t = 2$.

31.   Find an exact equation for the line tangent to the curve $y = x\sin x$ at the point $(\frac{\pi}{2}, \frac{\pi}{2})$.

32.   Suppose that $f(0) = 3$ and f' is the function shown below. Let $g(x) = (x^2 + 1) \cdot f(x)$.

a) Evaluate $g'(0)$.

b) Use the Fundamental Theorem of Calculus to estimate $f(1)$.

c) Is g increasing at $x = 1$? Justify briefly.

d) Estimate $g''(0)$.

e) Is g concave up at $x = 1$? Justify briefly.

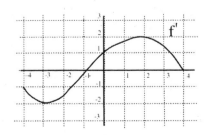

33.   Suppose that $f(0) = 5$ and f' is the function shown below. Let $g(x) = \dfrac{f(x)}{e^x}$.

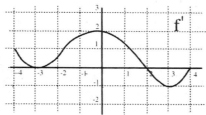

   a) Evaluate $g'(0)$.

   b) Use the Fundamental Theorem of Calculus to estimate $f(1)$.

   c) Is g increasing at $x = 1$? Justify briefly.

   d) Estimate $g''(0)$.

34.   If the graph of $f(x) = \dfrac{ax + b}{(x-1)(x-4)}$ has a horizontal tangent at the point $(2, 1)$ find a and b

35.   The function f is defined by $f(x) = (x - 2)(x^2 + 4x - 7)$. Find all points on the curve where the tangent line is horizontal.

36.   Find an exact equation of the tangent line to each curve at the point where $x = 0$.

   a) $y = \dfrac{8}{x^2 + 4}$

   b) $y = \dfrac{x + \sin x}{\cos x}$

37.   Some sample values of the differentiable functions f and g and their derivatives are shown in the tables below.

x	−1	2	3	4
$f(x)$	0.3	2	−3	−1
$f'(x)$	0.5	−1	0	2

x	−1	2	3	4
$g(x)$	−2	−5	−1	−3
$g'(x)$	−3	0	2	0.25

   If $h(x) = f(x) \cdot g(x)$ and $k(x) = \dfrac{f(x)}{g(x)}$, find the following.

   a) $h'(-1)$ b) $h'(3)$ c) $k'(2)$ d) $k'(-1)$ e) $k'(3)$

38.   Use the definition of the derivative and the identity $\cos(x + y) = \cos x \cdot \cos y - \sin x \cdot \sin y$ to show that $\dfrac{d}{dx}(\cos x) = -\sin x$.

39.   Use the Quotient Rule for derivatives and the definition of the trigonometric functions to find $f'(x)$ if $f(x)$ equals

   a) $\cot x$ b) $\sec x$ c) $\csc x$

4.4   The Derivative of a Composite Function: the Chain Rule

More often that not, the function that you must cope with in a given problem is the composite of two or more simpler functions. In this section we develop the rule for finding the derivative of such a function.

For example, suppose you wanted the derivative of the function $h(x) = (\sin x)^2$. Here h is the composite of an outer function f and an inner function g where

$$f(x) = x^2, \quad g(x) = \sin x, \quad \text{and} \quad h(x) = [f \circ g](x) = (\sin x)^2.$$

In order to find $h'(x)$ we rewrite h as $h(x) = \sin x \cdot \sin x$ and proceed to find $h'(x)$ by using the Product Rule:

$$h(x) = \sin x \cdot \sin x$$
$$h'(x) = \sin x \cdot (\cos x) + \sin x \cdot (\cos x)$$
$$h'(x) = 2 \sin x \cdot \cos x$$

The derivative of the composite $f \circ g$ can be expressed in terms of f' and g' if we note that the derivatives for the outer and inner functions

$$f'(x) = 2x \quad \text{and} \quad g'(x) = \cos x,$$

appear in the formula $[f \circ g]'(x) = 2 \sin x \cos x$ as follows:

The Chain Rule

$$[f \circ g]'(x) = f'(g(x)) \cdot g'(x).$$

This formula, referred to as the Chain Rule, is, in fact, true in general. It asserts that

The derivative of a composite function is the derivative of the outer function composed with the inner, times the derivative of the inner function .

Example 1: *Using the Chain Rule*

If $h(x) = \cos(3x + 1)$, find $h'(x)$.

Solution:

With $f(x) = \cos(x)$ and $g(x) = 3x + 1$, we have

$$h(x) = [f \circ g](x) = \cos(3x + 1).$$

Since $f'(x) = -\sin x$ and $g'(x) = 3$, the Chain Rule gives us

$$h'(x) = [f \circ g]'(x)$$
$$= f'(g(x)) \cdot g'(x)$$
$$= -\sin(g(x)) \cdot g'(x)$$
$$= -\sin(3x + 1) \cdot 3.$$

Example 2: *Using the Chain Rule*

If $h(x) = \sqrt{x^3 + x}$, find $h'(x)$.

Solution:

Let $f(x) = \sqrt{x}$ and $g(x) = x^3 + x$. Then $h(x) = [f \circ g](x) = \sqrt{x^3 + x}$. Since

$f'(x) = \dfrac{1}{2\sqrt{x}}$ and $g'(x) = 3x^2 + 1$, the Chain Rule gives us

$$h'(x) = [f \circ g]'(x)$$
$$= f'(g(x)) \cdot g'(x)$$
$$= \frac{1}{2\sqrt{g(x)}} \cdot g'(x)$$
$$= \frac{1}{2\sqrt{x^3 + x}} \cdot (3x^2 + 1).$$

Example 3: *Using the Chain Rule*

If $y = \left(\dfrac{x}{x+1}\right)^2$, find $\dfrac{dy}{dx}$.

Solution:

For an outer function $f(x) = x^2$ and an inner function $g(x) = \dfrac{x}{x+1}$,
we have

$$\frac{dy}{dx} = [f \circ g]'(x)$$
$$= f'(g(x)) \cdot g'(x)$$
$$= 2g(x) \cdot g'(x)$$
$$= 2\left(\frac{x}{x+1}\right)\left[\frac{(x+1) \cdot 1 - x \cdot 1}{(x+1)^2}\right] = \frac{2x}{(x+1)^3}.$$

As is possible with derivative rules for sums and products, the Chain Rule can be extended to the composite of more than two functions.

Example 4: *Repeated Application of the Chain Rule*

If $h(x) = \sqrt{\sin(2x + 1)}$, find $h'(x)$.

Solution:

Let $h(x) = [f \circ (g \circ k)](x)$, with $f(x) = \sqrt{x}$, $g(x) = \sin x$, $k(x) = 2x + 1$; then

$f'(x) = \dfrac{1}{2\sqrt{x}}$ and $g'(x) = \cos x$, $k'(x) = 2$. And, by the Chain Rule,

$$h'(x) = \left[f' \circ (g \circ k) \right] \cdot \left[g \circ k \right]'(x)$$

$$= \left[f'(g(k(x))) \right] \cdot \left[g'(k(x)) \cdot k'(x) \right]$$

$$= \frac{1}{2\sqrt{g(k(x))}} \cdot \cos(k(x)) \cdot 2$$

$$= \frac{1}{2\sqrt{\sin(2x + 1)}} \cdot \cos(2x + 1) \cdot 2$$

Leibniz Notation for the Chain Rule

The Chain Rule assumes a very suggestive form in Leibniz notation. Suppose differentiable functions f and g are given, and let

$$y = f(u) \text{ and } u = g(x).$$

Then $\dfrac{dy}{du} = f'(u)$, $\dfrac{du}{dx} = g'(x)$, and $y = f(g(x))$. So, by the Chain Rule,

$$\frac{dy}{dx} = f'(g(x)) \cdot g'(x) = f'(u) \cdot g'(x) = \frac{dy}{du} \cdot \frac{du}{dx}$$

More concisely,

$$\frac{dy}{dx} = \frac{dy}{du} \cdot \frac{du}{dx}.$$

Example 5: *Leibniz Notation and the Chain Rule*

If $y = u^6$ and $u = \sin(3x)$, find $\dfrac{dy}{dx}$.

Solution:

The problem can be solved by substituting $\sin(3x)$ for u in the first equation and then computing $\dfrac{dy}{dx}$. Another approach is to compute $\dfrac{dy}{du}$ and $\dfrac{du}{dx}$; $\dfrac{dy}{du} = 6u^5$ and, using the Chain Rule, $\dfrac{du}{dx} = 3\cos(3x)$.

Since $\dfrac{dy}{dx} = \dfrac{dy}{du} \cdot \dfrac{du}{dx} = 6u^5 \cdot 3\cos(3x)$, we substitute $\sin(3x)$ for u to obtain

$$\frac{dy}{dx} = \frac{dy}{du} \cdot \frac{du}{dx} = 6(\sin(3x))^5 \cdot 3\cos(3x) .$$

The Leibniz notation is a convenient aid to the solution of a collection of problems that depend on relating one rate of change to another by the Chain Rule.

Example 6: *Finding the Rate at which an Oil Slick is Spreading*

Oil spilling from a tanker anchored in a harbor spreads out in a circular slick. An hour after the spill started, the radius of the slick is increasing at the rate of 3 feet per hour and the area of the slick is increasing at the rate of 10 square feet for each foot increase in the radius. How fast (in ft^2 per hour) is the area of the slick increasing at this time?

Solution:

We know that at the moment in question

$$\frac{dr}{dt} = 3 \text{ ft/hr} \qquad \text{rate radius is increasing with respect to time}$$

$$\frac{dA}{dr} = 10 \text{ ft}^2/\text{ft} \qquad \text{rate area is increasing with respect to radius}$$

To calculate the rate at which the slick is growing with respect to time, or $\dfrac{dA}{dt}$, we use the fact that A is a function of r, $A = f(r)$, and r is a function of t, $r = g(t)$. Thus $A = f(g(t))$, and by the Chain Rule,

$$\frac{dA}{dt} = \frac{dA}{dr} \cdot \frac{dr}{dt} = 10 \frac{\text{ft}^2}{\text{ft}} \cdot 3 \frac{\text{ft}}{\text{hr}} = 30 \frac{\text{ft}^2}{\text{hr}} \ .$$

Thus, the area is increasing at 30 ft^2/hr.

Derivative of ln x

We estimated from graphs in an earlier section that the derivative of $\ln x$ is $\dfrac{1}{x}$. The following analytical argument to establish this fact is based on the Chain Rule. Since the exponential function $f(x) = e^x$ is the inverse of the natural logarithm function $g(x) = \ln x$, we have (for $x > 0$)

$$f(g(x)) = e^{\ln x} = x.$$

Differentiating both sides and using the Chain Rule on the left side gives

$$\frac{d}{dx}\left(e^{\ln x}\right) = \frac{d}{dx}(x)$$

$$e^{\ln x} \cdot \frac{d}{dx}(\ln x) = 1 \qquad \text{Since } e^x \text{ is the outer function and ln x is the inner function.}$$

$$x \cdot \frac{d}{dx}(\ln x) = 1. \qquad \text{Since } e^{\ln x} = x \ .$$

So,

$$\boxed{\frac{d}{dx}(\ln x) = \frac{1}{x} \ .}$$

Example 7: *Differentiating Logarithmic Functions*

a) Using the Chain Rule: $\dfrac{d}{dx}\left[\ln(2x)\right] = \dfrac{1}{2x}\cdot\dfrac{d}{dx}(2x) = \dfrac{1}{2x}\cdot(2) = \dfrac{1}{x}$.

b) Using the Product Rule: $\dfrac{d}{dx}\left[x\ln x\right] = x\cdot\dfrac{d}{dx}(\ln x) + \ln x\cdot\dfrac{d}{dx}(x) = 1 + \ln x$.

c) Using the Chain Rule: $\dfrac{d}{dx}\left[(\ln x)^3\right] = 3(\ln x)^2\cdot\dfrac{d}{dx}(\ln x) = 3(\ln x)^2\cdot\dfrac{1}{x}$.

Example 8: *Using Logarithmic Properties*

If $y = \ln\left(\dfrac{x^2}{x^3+1}\right)$, find $\dfrac{dy}{dx}$.

Solution:

If we use the Chain Rule immediately, we are faced with the task of finding the derivative of

the inner function $\dfrac{x^2}{x^3+1}$. While we can do this with the Quotient Rule, it is easier to first

use the log properties: $\ln\left(\dfrac{a}{b}\right) = \ln a - \ln b$ and $\ln(a^n) = n\ln a$ and then differentiate.

We have
$$y = \ln(x^2) - \ln(x^3+1) = 2\ln x - \ln(x^3+1).$$

Now we differentiate to obtain
$$\frac{dy}{dx} = 2\cdot\frac{1}{x} - \frac{1}{x^3+1}\cdot\frac{d}{dx}(x^3+1) = \frac{2}{x} - \frac{3x^2}{x^3+1},$$

avoiding the Quotient Rule altogether.

Derivative of b^x

Earlier it was shown that the derivative b^x is proportional to b^x. Now we show that the constant of proportionality is $\ln a$. We use the identity
$$\ln(b^x) = x\ln b.$$

Differentiating, using $\dfrac{d}{dx}(\ln x) = \dfrac{1}{x}$ and the chain rule, and recalling that $\ln b$ is a constant, we obtain:

$$\frac{d}{dx}(\ln b^x) = \frac{1}{b^x}\cdot\frac{d}{dx}(b^x) = \ln b$$

Solving this equation for the derivative of b^x gives the result suggested in Section 4.2.

$$\frac{d}{dx}\left(b^x\right) = (\ln b)\cdot b^x$$

4.4   Exercises

1.   Find the derivative of each of the following functions in two ways: (i) by using the Chain Rule and (ii) by multiplying out and then differentiating.

   a) $h(x) = (5x - 3)^2$ b) $h(x) = (e^x + 1)^2$

   c) $h(x) = (2x + x^2)^2$ d) $h(x) = (\sqrt{x} + 1)^2$

2.   Suppose functions f and g and their derivatives with respect to x have the following values at $x = 0$ and $x = 1$:

x	$f(x)$	$g(x)$	$f'(x)$	$g'(x)$
0	1	1	5	3
1	1	0	$-1/3$	$-5/3$

   a) If $h(x) = [f \circ g](x)$, find   i) $h'(0)$ ii) $h'(1)$

   b) If $k(x) = [g \circ f](x)$, find   i) $k'(0)$ ii) $k'(1)$

In Exercises 3–24, find the derivatives of the given functions.

3. $h(x) = (x^2 - 3)^3$ 4. $h(x) = (3x - x^2)^2$

5. $h(x) = \sqrt{2x}$ 6. $h(x) = \sin(2x)$

7. $f(x) = \sqrt{x^2 + 1}$ 8. $s = (1 + x^2)^{10}$

9. $f(x) = e^{2x}$ 10. $y = e^{-3x}$

11. $f(x) = \cos(\sin x)$ 12. $f(x) = \sin^2 x$

13. $s(t) = \sin(t^3 - 1)$ 14. $f(x) = \cos^3(2x)$

15. $y = (x^3 - 4)^4$ 16. $f(x) = \sqrt{9 - x^2}$

17. $f(x) = \sin(x^2)$ 18. $y = \sqrt{\tan x}$

19. $f(x) = \tan \sqrt{x}$ 20. $f(x) = e^{\sin x}$

21. $y = xe^{2x}$ 22. $y = \sqrt{2x} \cdot \sin(3x)$

23. $y = \left(x + \dfrac{1}{x}\right)^2$ 24. $y = \left(x^2 - \dfrac{1}{x}\right)^3$

In Exercises 25–34, find the derivative of the given function. Where possible, use the properties of logarithm functions first to simplify differentiation.

25. $f(x) = \ln\left(\dfrac{x}{x+1}\right)$ 26. $y = \ln(3x)$

27. $f(x) = \ln \sqrt{x}$ 28. $y = \ln\left(\dfrac{2x+3}{x^2+4}\right)$

29. $f(x) = \ln(x^2 + 1)$ 30. $f(x) = (\ln x)^2$

31. $f(x) = \ln(\ln x)$ 32. $y = \ln(\sin x)$

33. $y = x^2 \ln x$ 34. $y = \ln(e^{\ln x})$

In Exercises 35–37, functions f and g are defined by the graphs below.

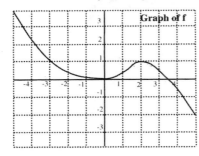
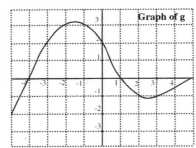

35.   Let h be a function defined by $h(x) = f[g(x)]$.

a) Evaluate $h(3)$.

b) Is $h'(-2)$ positive or negative? Justify your answer.

c) Estimate $h'(-1)$.

d) Determine the value(s) of x which correspond to critical points of h.

36.   Let $h(x) = f(x^2)$. Is h increasing, decreasing or neither at $x = -1$? Justify your answer.

37.   Let $k(x) = g(e^{-x})$. Is k increasing, decreasing or neither at $x = 0$? Justify your answer.

In Exercises 38–41, solve each problem in two ways: i) algebraically and ii) graphically.

38.   Write an equation of the line tangent to the graph of $y = \sqrt{25 - x^2}$ at the point (–3, 4).

39.   Write an equation of the line tangent to the graph of $y = \left(\dfrac{x}{x-1}\right)^3$ at the point (2, 8).

40.   Write an equation of the line tangent to the graph of $y = x\sqrt{16 + x^2}$ at the origin.

41.   Write an equation of the line tangent to the graph of $y = \sqrt[5]{3x^3 + 4x}$ at the point (2, 2).

42.   Find an equation of a straight line tangent to the curve $y = \ln x$ and passing through the origin. [Hint: Let $P(h, k)$ be the point where the tangent line intersects the curve. What can be said about h and k?]

43.   Given $f(3) = -2$, $f'(3) = 5$ and $g(3) = 3$, $g'(3) = -4$. Find $h'(3)$, if possible, for each of the following:

   a) $h(x) = f(x) \cdot g(x)$ b) $h(x) = \dfrac{f(x)}{g(x)}$

   c) $h(x) = f[g(x)]$ d) $h(x) = [f(x)]^3$

44.   The functions whose graphs are shown below give a cross country skier's velocity, $v(t)$ (meters/minute), t minutes after she has begun skiing and her oxygen consumption, $A(v)$ (liters/minute), when she is skiing v meters/minute. At what rate is her oxygen consumption increasing 2 minutes after she starts skiing?

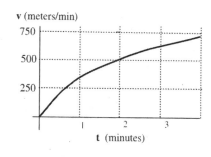
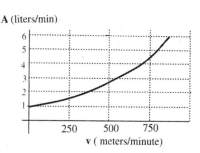

45.   Let $h(x) = f[g(x)]$ and $k(x) = \dfrac{f(x)}{g(x)}$. Fill in the missing entries in the table below.

x	$f(x)$	$f'(x)$	$g(x)$	$g'(x)$	$h(x)$	$h'(x)$	$k(x)$	$k'(x)$
–1	–1	4	1		–1	8	–1	
0	1	0	0	0			2	0
1		–4	1		–1	–8	–1	

4.5   Functions Defined Implicitly

Determining the slope of the tangent line to the circle

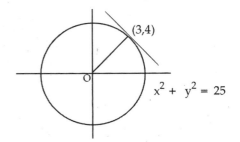

$x^2 + y^2 = 25$ at the point (3, 4) is a problem which can be
solved without calculus. Since the line tangent to the circle
at (3, 4) is perpendicular to the radius to (3, 4), its slope is
the negative reciprocal of $\frac{4}{3}$, or $-\frac{3}{4}$.

However, using calculus to find the slope does illustrate a method that can be generalized and
is used to solve similar but more complex problems. The technique is called **implicit differentiation**.

The circle $C = \left\{(x, y): x^2 + y^2 = 25\right\}$ is not a function.

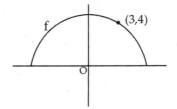

However, there are subsets of C which are functions. For
example, the upper half of the circle is the graph of a
function f which is differentiable at $x = 3$.

The pairs $(x, f(x))$ in the function f satisfy the equation $x^2 + y^2 = 25$, so that $x^2 + \left[f(x)\right]^2 = 25$
with $-5 \le x \le 5$ and $f(x) \ge 0$. We say the function f is defined implicitly by the above equation.
The slope of the tangent at (3, 4) is the value of the derivative f' at $x = 3$. One method of finding
the derivative is to first determine an explicit rule for f by solving $x^2 + \left[f(x)\right]^2 = 25$ for $f(x)$:

$$f(x) = \sqrt{25 - x^2}, \quad -5 \le x \le 5$$

Then, using the Chain Rule,

$$f'(x) = \frac{1}{2\sqrt{25 - x^2}} \cdot (-2x).$$

At $x = 3$, we have $f'(3) = -\frac{3}{4}$.

The solution above can be made easier if $f'(x)$ is determined directly from the equation

$$x^2 + \left[f(x)\right]^2 = 25.$$

Note that the equation above is satisfied by all x in the domain of f. Therefore, the functions h and
g where

$$h(x) = x^2 + \left[f(x)\right]^2 \text{ and } g(x) = 25$$

are equal ; that is, $h(x) = g(x)$. Since equal functions have equal derivatives, we assume that the
implicitly defined function f has a derivative and differentiate both sides of the equation
$x^2 + \left[f(x)\right]^2 = 25$ to obtain

$$2x + 2f(x) \cdot f'(x) = 0.$$

Solving for $f'(x)$, assuming that $f(x) \neq 0$, gives

$$f'(x) = \frac{-x}{f(x)}.$$

At $x = 3$, we evaluate f' to obtain the now familiar result

$$f'(x) = -\frac{3}{4}.$$

The process by which f' was obtained is called **implicit differentiation**. It is usually employed when the equation that defines a function implicitly is so complicated that an explicit rule for $f(x)$ can not be produced.

Example 1:   *Finding the Slope of a Curve*

Assuming $y^2 - 2x = y$ implicitly defines a differentiable function f, find the slope of the tangent line to the graph of f at $(1, -1)$.

Solution:

Replacing y with $f(x)$ we have

$$[f(x)]^2 - 2x = f(x).$$

Differentiating,

$$2f(x) \cdot f'(x) - 2 = f'(x),$$

and solving for $f'(x)$, we get

$$f'(x) \cdot [2f(x) - 1] = 2$$

$$f'(x) = \frac{2}{2f(x) - 1}.$$

Since $f(1) = -1$, the slope of the tangent line at $(1, -1)$ is $f'(1) = -\frac{2}{3}$.

Example 2:   *Finding the Derivative of an Implicitly Defined Function*

Find a rule for the derivative of a function defined implicitly by $x^3 + xy + y^3 = 1$.

Solution:

Assuming y is the name of a differentiable function of x and using y' instead of $f'(x)$, we differentiate and solve for y'. The Product Rule is used to differentiate the xy term.

$$3x^2 + [xy' + y \cdot 1] + 3y^2 \cdot y' = 0$$

$$xy' + 3y^2 y' = -3x^2 - y$$

$$y'\left(x + 3y^2\right) = -3x^2 - y$$

$$y' = \frac{-3x^2 - y}{x + 3y^2}.$$

Example 3:   *Slope of a Tangent to a Curve*

Find the slope of the tangent line to the curve $x^2 y + \sin y = 2\pi$ at the point $(1, 2\pi)$.

Solution:

Differentiating implicitly with respect to x and remembering that y is a function of x, we get

$$x^2 y' + y \cdot 2x + \cos y \cdot y' = 0.$$

Now substitute the coordinates $x = 1$, $y = 2\pi$ and solve the resulting equation for y'.

$$1 \cdot y' + 2\pi \cdot 2 + \cos(2\pi) y' = 0, \text{ so } 2y' = -4\pi \text{ and } y' = -2\pi.$$

Thus, the slope of the tangent line at $(1, 2\pi)$ is -2π.

Although the graph of an equation may be a complicated curve, it is generally true that the curve looks like a straight line locally, just like a differentiable function. Thus, the derivative formula obtained from implicit differentiation may be useful in analyzing a curve locally. For example, implicit differentiation may be used to locate where tangent lines to a curve are horizontal or vertical.

Example 4:   *Horizontal and Vertical Tangent Lines*

Find all points where tangent lines to the graph of $x^2 = 4y - y^2$ are a) horizontal   b) vertical.

Solution:

By implicit differentiation,

$$\frac{dy}{dx} = \frac{x}{2 - y}$$

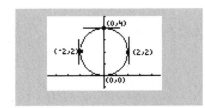

a) A tangent line is horizontal when $\dfrac{dy}{dx} = 0$, thus when $x = 0$. If $x = 0$, then substituting in the original equation, we have $0 = 4y - y^2$ and $y = 0$ or $y = 4$. Thus, there are horizontal tangent lines at the points $(0, 0)$ and $(0, 4)$.

b) A tangent line is vertical when the denominator of the derivative is 0 and the numerator is not 0, thus when $y = 2$. If $y = 2$ then $x^2 = 4$ and $x = 2$ or $x = -2$. Thus, there are vertical tangent lines at the points $(2, 2)$ and $(-2, 2)$.

Example 5:   *Power Rule for Rational Powers of x*

If $y = x^{5/3}$, show that the Power Rule for derivatives gives the correct derivative.

Solution:

$$y^3 = x^5 \qquad\qquad \text{Cubing both sides of the equation.}$$

$$3y^2 \cdot y' = 5x^4 \qquad\qquad \text{Differentiating implicitly with respect to x.}$$

$$y' = \frac{5x^4}{3y^2}$$

$$y' = \frac{5}{3} \cdot \frac{x^4}{(x^{5/3})^2} \qquad \text{Substituting } x^{5/3} \text{ for } y.$$

$$y' = \frac{5}{3} \cdot \frac{x^4}{x^{10/3}} = \frac{5}{3} x^{2/3}$$

4.5   Exercises

In Exercises 1–10, find $\dfrac{dy}{dx}$.

1. $x^2 + y^2 = 4$

2. $x^2 + y^2 = 5x - 4y$

3. $x^2 - xy = 2$

4. $x^2 y + y^2 = 6$

5. $xy + y^2 = x + y$

6. $2x^3 - 3y^2 = 7$

7. $x^2 = 2y - \sin y$

8. $e^y = xy$

9. $\sqrt{x} + \sqrt{y} = 16$

10. $x \ln y = y - \ln x$

In Exercises 11–18, find an equation of the tangent line at the given point.

11. $xy^2 = 16$ at $(4, -2)$

12. $x^2 - y^2 = 1$ at $(\sqrt{5},\, 2)$

13. $\sin y = x$ at $(0, 0)$

14. $y + x \cdot \cos y = 5$ at $(5, 0)$

15. $x^{2/3} + y^{2/3} = 5$ at $(8, 1)$

16. $x^{1/4} + y^{1/4} = 4$ at $(16, 16)$

17. $e^y = \ln x$ at $(e, 0)$

18. $\sqrt{x} + xy^2 = 18$ at $(4, 2)$

19.   A graph of $x^3 + y^3 = 64$ is shown at the right.

   a) Find $\dfrac{dy}{dx}$ at $x = 0$ and $x = 4$.

   Show that your answers are consistent
   with the graph.

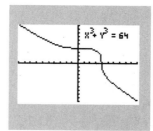

   b) Find $\dfrac{dy}{dx}$ at the point where $y = x$.

20.   A graph of $y^3 + x^2 = 5$ is shown at the right.

   a) Find $\dfrac{dy}{dx}$ at $x = 0$ and $x = 2$.
   Show that your answers are consistent with the graph.

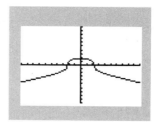

   b) Find all points where tangent lines are horizontal or
      vertical.

21.   A curve is given by the equation $y = \ln(x^2 + y^2)$.

     a) Find $\dfrac{dy}{dx}$.

     b) Write an equation for the line tangent to the curve at the point $(1, 0)$.

22.   A curve is given by the equation $y^3 + 3x^2y + 13 = 0$.

     b) Write an equation for the line tangent to the curve at the point $(2, -1)$.

     c) Find the minimum y-coordinate of any point on the curve. Justify your answer.

23.   A curve is given by the equation $xe^{5y} = 3y$.

     a) Find $\dfrac{dy}{dx}$.

     b) Write an equation for the line tangent to the curve at the point $(0, 0)$.

24.   If $x^3 + y^3 = 8$, show that the second derivative of y with respect to x is $-\dfrac{16x}{y^5}$.

25.   Use implicit differentiation twice to find $\dfrac{d^2y}{dx^2}$ at $(3, 4)$ if $x^2 + y^2 = 25$.

26.   A graph of $4y^2 - xy^2 = x^3$ is shown in the figure.

     a) Find $\dfrac{dy}{dx}$ when $x = 2$.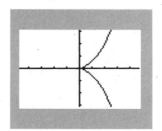

       Show that your answers are consistent
       with the graph.

     b) Find $\dfrac{dy}{dx}$ at $(0, 0)$.

27.   As in Example 5, use implicit differentiation to show that the Power Rule for derivatives gives the correct derivative if $y = x^{2/5}$.

4.6   Related Rates of Change

Frequently, in trying to solve a physical problem we must establish a connection between rates of change (derivatives). This connection is usually made by using implicit differentiation and the Chain Rule.

For example, suppose a spherical balloon is being inflated so that both its radius r and volume V are related to time t by differentiable functions. Suppose we want to know the relationship between the rates of change $\frac{dV}{dt}$ and $\frac{dr}{dt}$ at any instant. First, for any spherical balloon there is a function which relates the volume V to the radius r: $V = \frac{4}{3}\pi r^3$.

Now, differentiating implicitly with respect to t and employing the Chain Rule, we produce an equation which relates the derivatives $\frac{dV}{dt}$ and $\frac{dr}{dt}$:

$$\frac{dV}{dt} = 4\pi r^2 \cdot \frac{dr}{dt}.$$

The general method of determining relationships between derivatives (rates of change) is best explained by examples.

Example 1: *An Inflating Balloon*

A balloon is being inflated at a rate of 10π cu ft/sec. At what rate is the radius increasing when $r = 2$ ft ?

Solution:

Given the constant rate of change of volume with respect to time, $\frac{dV}{dt} = 10\pi$, we want to find $\frac{dr}{dt}$ when $r = 2$. We know that the volume V is related to the radius r by the formula $V = \frac{4}{3}\pi r^3$. If we assume that V and r are related to time t by differentiable functions and differentiate implicitly, we get

$$\frac{dV}{dt} = 4\pi r^2 \cdot \frac{dr}{dt}.$$

Now, letting $\frac{dV}{dt} = 10\pi$ and $r = 2$, we have $10\pi = 4\pi \cdot 4\frac{dr}{dt}$, so that $\frac{dr}{dt} = \frac{5}{8}$. Thus the radius is increasing at a rate of $\frac{5}{8}$ ft/sec at the instant when $r = 2$.

Example 2: *A Sliding Ladder*

A ladder 25 feet long is leaning against a wall. If the foot of the ladder is being pulled away from the bottom of the wall at a rate of 14 feet per second, at what rate is the top of the ladder moving down the wall when it is 7 feet above the ground?

Solution:

If we let y represent the distance from the top of the ladder to the ground, and let x represent the distance from the foot of the ladder to the wall, then by the Pythagorean Theorem:

$$x^2 + y^2 = 625.$$

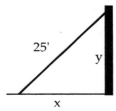

We are given $\frac{dx}{dt} = 14$, and are asked to find $\frac{dy}{dx}$ when $y = 7$. Given $x^2 + y^2 = 625$ we assume x and y are related to t by differentiable functions and differentiate implicitly with respect to time t to obtain

$$2x\frac{dx}{dt} + 2y\frac{dy}{dt} = 0.$$

Now when $y = 7$, $x^2 + 49 = 625$ so that $x = 24$. Substituting $y = 7$, $x = 24$, and $\frac{dx}{dt} = 14$ in the equation above, we have

$$2(24)(14) + 2(7)\frac{dy}{dt} = 0$$

$$14\frac{dy}{dt} = -48 \cdot 14$$

$$\frac{dy}{dt} = -48.$$

The fact that $\frac{dy}{dt}$ is negative is consistent with the fact that, since the top of the ladder is moving down the wall, the distance y is decreasing at a rate of 48 ft/sec when it is 7 feet above the ground. In some problems care in assigning and interpreting signs of the rates of change is important in order to learn whether a quantity is increasing or decreasing.

Example 3:   *Change of Distance Between Two Cars*

At a given moment a car is 30 miles north of an intersection, traveling toward the intersection at 45 mph. At the same time a truck is 40 miles east of the intersection, traveling away from the intersection at 35 mph. Is the distance between the vehicles increasing or decreasing at that moment? At what rate?

Solution:

Begin by drawing a diagram. From the Pythagorean Theorem you know that $x^2 + y^2 = z^2$. Differentiating with respect to time

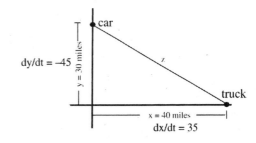

$$2x\frac{dx}{dt} + 2y\frac{dy}{dt} = 2z\frac{dz}{dt}$$

or simply

$$x\frac{dx}{dt} + y\frac{dy}{dt} = z\frac{dz}{dt}.$$

We want to find $\frac{dz}{dt}$ when $x = 40$, $y = 30$, $\frac{dx}{dt} = 35$ and $\frac{dy}{dt} = -45$. At that time,

$$z = \sqrt{x^2 + y^2} = \sqrt{40^2 + 30^2} = 50.$$

The equation relating the rates of change then becomes

$$(40)(35) + (30)(-45) = 50\frac{dz}{dt}, \text{ so } \frac{dz}{dt} = 1 \text{ mph.}$$

At the given moment, the distance between the vehicles is increasing at the rate of 1 mph.

Example 4: *Rising Water Level*

Water is flowing into a cone-shaped tank at the rate of 5 cubic inches per second. If the cone has an altitude of 4 inches and a base radius of 3 inches, how fast is the water level rising when the water is 2 inches deep?

Solution:

If the volume of water in the cone is V cubic inches and the depth is h inches, we are given that $\dfrac{dV}{dt} = 5$, and asked to determine $\dfrac{dh}{dt}$ when $h = 2$. First we find an equation, valid at any time, which relates V to h. The formula for the conical volume is: $V = \dfrac{1}{3}\pi r^2 h$.

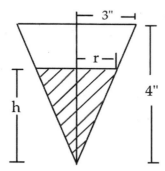

This relates V to both the radius r and the altitude h. But, by similar triangles we have $\dfrac{r}{3} = \dfrac{h}{4}$, or $r = \dfrac{3}{4}h$. Substituting, we have

$$V = \frac{1}{3}\pi\left(\frac{9}{16}h^2\right)\cdot h$$

or

$$V = \frac{3}{16}\pi h^3.$$

Assuming there are differentiable functions which relate V and h to time t, we differentiate implicitly:

$$\frac{dV}{dt} = \frac{3\pi}{16}\cdot 3h^2\frac{dh}{dt}.$$

Given that $\dfrac{dV}{dt} = 5$, when $h = 2$ we have

$$5 = \frac{3\pi}{16}\cdot 3(2)^2\cdot\frac{dh}{dt}$$

and

$$\frac{dh}{dt} = \frac{20}{9\pi}.$$

Thus, the water level is rising at a rate of $\dfrac{20}{9\pi}$ inches per second when the water is 2 inches deep.

Example 5: *Changing Circular Region*

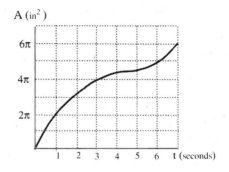

A (in^2)

The function A whose graph is sketched in the figure gives the area $A(t)$ (in^2) of a circular region after t seconds. Approximately how rapidly is the radius changing after 3 seconds?

Solution:

If the radius of the circular region is r inches at t seconds then its area is $A = \pi r^2$ in^2 and the rate of change of the area with respect to time is

$$\frac{dA}{dt} = 2\pi r \frac{dr}{dt} \ \ \text{in}^2/\text{sec}.$$

From the graph above we see that A is approximately 4π at $t = 3$, so that

$$4\pi = \pi r^2.$$

Solving for r we obtain $r = 2$ inches.

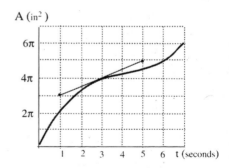

A (in^2)

The slope $\dfrac{dA}{dt}$ of the tangent line in the figure

is approximately $\dfrac{5\pi - 3\pi}{5 - 1} = \dfrac{\pi}{2}$. Therefore, substituting

we obtain

$$\frac{\pi}{2} \approx 2\pi(2)\frac{dr}{dt}, \text{ so that at } t = 3, \ \ \frac{dr}{dt} = \frac{1}{8} \ \text{in/sec}.$$

The strategy used in the examples above can be summarized as follows :

1)   State an equation, valid at any time, which relates the quantities that are changing.

2)   Differentiate with respect to time to obtain a relationship between the rates of change (derivatives) of the quantities.

3)   Substitute the specific values that the variables take on at the instant in question.

4.6   Exercises

In Exercises 1–4, A, R, and h are related to t by differentiable functions.

1.   If $A = \pi r^2$, find $\dfrac{dA}{dt}$ when $r = 2$ and $\dfrac{dr}{dt} = 3$.

2.   If $A = 2\pi rh$, find $\dfrac{dr}{dt}$ when $r = 2$, $h = 4$, $\dfrac{dA}{dt} = 16\pi$ and $\dfrac{dh}{dt} = 2$.

3.   If $\dfrac{r}{3} = \dfrac{h-4}{h}$, find $\dfrac{dh}{dt}$ when $r = 2$, $h = 12$, and $\dfrac{dr}{dt} = \dfrac{1}{2}$.

4.   If $A^2 = R^2 + h^2$, find $\dfrac{dA}{dt}$ when $A = 10$, $R = 8$, $\dfrac{dR}{dt} = \dfrac{1}{2}$, and $\dfrac{dh}{dt} = \dfrac{1}{3}$.

5.   A particle moves along the curve $y = \sqrt{x^2 + 1}$ in such a way that $\dfrac{dx}{dt} = 4$.

   Find $\dfrac{dy}{dt}$ when $x = 3$.

6.   Two automobiles start from a point A at the same time. One travels west at 80 miles per hour; the other travels north at 45 miles per hour. How fast is the distance between them increasing 3 hours after they start?

7.   A spherical balloon is being inflated at the rate of 12 in^3/sec. How fast is the radius r changing at the moment when $r = 2$ in ?

8.   A particle is moving along the curve $y = x^2$ in such a way that when $x = \dfrac{1}{2}$,

   $\dfrac{dx}{dt} = 2$ ft/sec. Determine $\dfrac{dy}{dt}$ at that moment.

9.   A ladder 15 feet tall leans against a vertical wall of a house. If the bottom of the ladder is pulled away horizontally from the house at 4 ft/sec, how fast is the top of the ladder sliding down the wall when the bottom of the ladder is 9 feet from the wall?

10.   A cone (point down) with a height of 10 inches and a radius of 2 inches is being filled with water at the constant rate of 2 in^3/sec. Determine how fast the water surface is rising when the water depth is 6 inches.

11.   A particle is moving along the graph of $y = \sqrt{x}$. At what point on the curve are the x-coordinate and the y-coordinate of the particle changing at the same rate?

12.   A winch at the end of a dock is 10 feet above the level of the deck of a boat. A rope attached to the deck is being hauled in by the winch at a rate of 5 ft/sec. How fast is the boat approaching the dock when 26 feet of rope are out?

13.   The cross-section of a trough 6 feet long is an equilateral triangle with one vertex pointing down.

     a)   Define a function which relates the volume of water in the trough to the depth of the water.

     b)   If water is flowing into the trough at a rate of 10 ft^3/sec, find the rate at which the depth of the water is increasing when the depth is 2.5 feet.

14.   A rope 35 feet long runs over the top of a wall 12 feet high. Each end of the rope is attached to a cement block on the ground. One block is being pulled away from the wall at the rate of 30 ft/min. At the moment that the first block is 16 feet from the foot of the wall, how fast is the other block being dragged toward the wall?

15.   A streetlight is 15 feet above the sidewalk. A man 6 feet tall walks away from the light at the rate of 5 ft/sec.

     a)   Determine a function relating the length of the man's shadow to his distance from the base of the streetlight.

     b)   Determine the rate at which the man's shadow is lengthening at the moment that he is 20 feet from the base of the light.

16.   A ship is anchored 2 miles off a straight shore, and its searchlight is following a car that is moving along the shore at 40 miles per hour. How fast is the light turning (in radians per hour) when the car is 4 miles from the ship?

17.   Two sides of a triangle are equal in length, and each increases at the constant rate of 2 in/min. The angle between these two sides increases at the constant rate of 1 rad/min. Determine the rate of change of the third side at the moment that the other two sides are each 10 inches long and their included angle has a measure of $\frac{\pi}{2}$ radians. [Hint: The Law of Cosines may be useful.]

18.   An observer stands 25 feet from the base of a 50 foot flagpole and watches a flag being lowered at a rate of 5 ft/sec. Determine the rate at which the angle of elevation (from the observer) to the flag is changing at the instant that the flag is 25 feet above eye-level.

19.   A 13 foot ladder is leaning against a wall so that the foot of the ladder is 1 foot from the wall. A gust of wind causes the ladder to begin sliding down the wall. The motion of the top of the ladder as it slides down the wall is described by

$$y = -16t^2 + .05t + \sqrt{168}.$$

where t is measured in seconds.

     a)   When does the top of the ladder reach the ground?

     b)   Determine the velocity of the end of the ladder that is resting on the ground when it is 5 ft from the wall.

20. The length L of a rectangle is decreasing at the rate of 2 cm/sec while the width W is increasing at the rate of 2 cm/sec. When $L = 12$ and $W = 2$, find the rate of change of

     a) the area;

     b) the perimeter;

     c) the length of a diagonal.

21. The function V whose graph is sketched below gives the volume of air $V(t)$ that a man has blown into a balloon after t seconds. Approximately how rapidly is the radius changing after 6 seconds? ($V = \frac{4}{3}\pi r^3$)

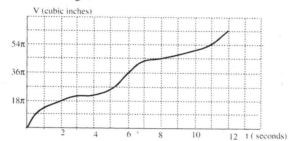

22. One ship traveling west is $W(t)$ nautical miles west of a lighthouse and a second ship traveling south is $S(t)$ nautical miles south of the lighthouse at time t (hours). The graphs of W and S are shown below. At what approximate rate is the distance between the ships increasing at $t = 1$? (nautical miles per hour = knots)

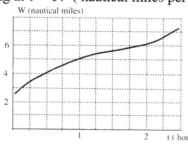

23. Sand is being dumped on a pile in such a way that it forms a cone whose base radius is always 3 times its height. The function V whose graph is sketched in the figure gives the volume of the conical sand pile, $V(t)$, measured in cubic feet, after t minutes. ($V = \frac{1}{3}\pi r^2 h$)

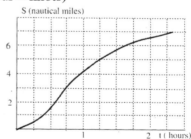

At what approximate rate is the radius of the base changing after 6 minutes.

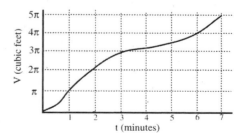

24. A 12 foot ladder is leaning against a house. If the bottom of the ladder is pulled away from the base of the wall at a rate of 3 ft/sec, at what rate is the angle between the ladder and the ground changing when the lower end of the ladder is 6 ft from the house?

4.7   Approximations

Many problems in mathematics are too difficult to solve exactly and all we can hope to do is find approximate solutions that are correct to within some acceptable tolerance. In this section two methods of approximation are described: linear approximation and Newton's method. Each method uses the idea of local linearity, the fact that most curves often look like straight lines when you zoom in closely enough.

Linear Approximation

Suppose the function f is differentiable at $x = a$. Then an equation of the line tangent to the graph of f at the point $(a, f(a))$ is

$$y = f(a) + f'(a)(x - a).$$

Since the curve remains close to its tangent line in the vicinity of the point of tangency, we approximate values of the function by values of the tangent line.

> The tangent line approximation of f at $x = a$ is
> $$L(x) \approx f(a) + f'(a)(x - a).$$
> The tangent line L is called the **local linearization** of f at $x = a$.

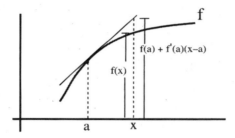

Example 1: *A Tangent Line Approximation*

What is the tangent line approximation of $y = e^x$ near $x = 0$?

Solution:

Since $f'(x) = e^x$, we have $f(0) = 1$ and $f'(0) = 1$. Thus, the tangent line at $x = 0$ is

$$y = f(0) + f'(0)(x - 0) = 1 + x.$$

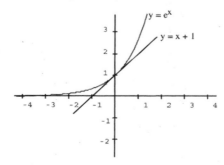

This means that, near 0, the function $f(x) = e^x$ can be approximated by its tangent line $y = x + 1$.

The following example illustrates the use of linearization to find approximate values of functions near points where the values of the functions and their derivatives are known.

Example 2: *A Tangent Line Approximation*

Use the tangent line approximation of $y = \sqrt{x}$ at $x = 25$ to find an approximate value of $\sqrt{26}$.

Solution:

If $f(x) = \sqrt{x}$, then $f'(x) = \dfrac{1}{2\sqrt{x}}$. Since we know that $f(25) = 5$ and $f'(25) = \dfrac{1}{10}$, the linearization of f at $x = 25$ is

$$y = 5 + \frac{1}{10}(x - 25).$$

Substituting x = 26, we get

$$\sqrt{26} = f(26) \approx 5 + \frac{1}{10}(26 - 25) = 5.1.$$

Finding Function Zeros with Newton's method

In 1669, Issac Newton described a method for approximating zeros of differentiable functions. Newton introduced his method to find a zero of the function $f(x) = x^3 - 2x - 5$ between $x = 2$ and $x = 3$. This method makes essential use of the differentiability of the function.

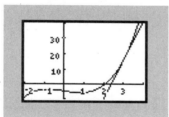

Graph of $f(x) = x^3 - 2x - 5$

We know there is a zero in the interval [2, 3] because $f(2) < 0$ and $f(3) > 0$.

Newton's method begins with a reasonable approximation of the zero, call it x_0. For instance, we take the right-hand endpoint of the interval $x_0 = 3$. Then we draw a tangent line to the graph at the point $(3, f(3))$ and determine the next approximation x_1 from the x-intercept (zero) of this tangent line.

This number x_1 is a better approximation of the actual zero and can be computed easily. Since the tangent line at the point $(x_0, f(x_0))$ has slope $m = f'(x_0)$, its equation is

$$y - f(x_0) = f'(x_0)(x - x_0).$$

We find where it intersects the x-axis by letting $y = 0$ and solving for $x = x_1$:

$$0 - f(x_0) = f'(x_0)(x_1 - x_0)$$

and

$$x_1 = x_0 - \frac{f(x_0)}{f'(x_0)}.$$

Once we have found x_1, the process can be repeated to find a still better approximation of the zero. In general, if we have calculated n approximations, the next approximation, x_{n+1}, is given by the formula

$$x_{n+1} = x_n - \frac{f(x_n)}{f'(x_n)},$$

provided $f'(x_n) \neq 0$. Each successive approximation is called an **iteration**.

Example 3: *Using Newton's Method*

Complete four iterations of Newton's method for the function $f(x) = x^2 - 2$ starting with an initial guess of $x_0 = 1.5$.

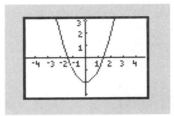

Solution:

To apply Newton's formula $x_{n+1} = x_n - \dfrac{f(x_n)}{f'(x_n)}$ to the function $f(x) = x^2 - 2$, store the function f in Y1 and its numerical derivative in Y2. Next, return to the Home screen and store the initial approximation $x_0 = 1.5$ in X and enter Newton's formula as shown below. Now, pressing the $\boxed{\text{ENTER}}$ key repeatedly produces the following approximations of a zero for $f(x) = x^2 - 2$.

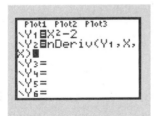
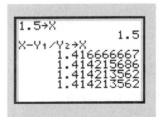

Note that the two zeros of f are $\pm\sqrt{2}$, and to 9 decimal places, $\sqrt{2} = 1.414213562$. Thus, after only three iterations of Newton's method, we have obtained an approximation that is accurate to nine decimal places.

Example 4: *A Graphical Demonstration of Newton's Method*

Graphically demonstrate two iterations of Newton's method for the function $f(x) = \cos x$ on the interval $[-\frac{\pi}{2}, \frac{3\pi}{2}]$. Use an initial guess of $x_0 = 0.5$ to estimate the smallest positive zero.

Solution:

Graph the cosine function in the ZDecimal viewing rectangle. Select Draw option [5:Tangent] and trace to the x-value of .5 and press $\boxed{\text{ENTER}}$ to draw the first tangent line. The x-coordinate where the tangent line crosses the x-axis is x_1, the next estimate of the zero of f.

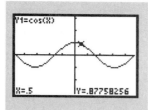
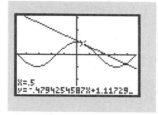

Again, select Draw [5:Tangent] and trace along the cosine curve to the approximate location of $(x_1, \cos(x_1))$ and press **ENTER** to draw the second tangent line. The x-intercept of this tangent line provides a close approximation of the desired zero.

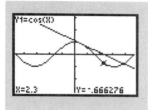
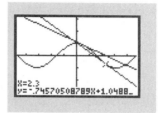

The procedure used in Example 3 for implementing Newton's method on the Home screen can be automated with a short program. Program NEWTON, given below, approximates a zero of a function stored in Y1 given an initial guess z. The program stops calculating new estimates when successive approximations are within a specified tolerance (1E–5 = .00001) of each other.

```
PROGRAM: NEWTON
:ClrHome
:Input "INITIAL GUESS: ",Z
:Repeat (abs (Z-X)<1E–5)
:Z→X
:X-Y1/nDeriv(Y1,X,X,.0001)→Z
:Disp Z
:Pause
:End
:Z→X
:Disp "FUNCTION VALUE",Y1
```

When Does Newton's Method Fail?

When it can be applied, Newton's method is a very efficient algorithm for approximating zeros; however, we can see graphically that Newton's method fails if at any step the approximation term x_n produces a first derivative value $f'(x_n) = 0$ (the tangent line in this case is horizontal and does not cross the x-axis).

Another situation where Newton's method may fail to converge is when the iterations oscillate or cycle periodically through the same set of values. It is also possible for Newton's method to diverge away from a zero.

Example 5: *Approximating a Function Zero with Newton's Method*

Try to find a zero for the function $f(x) = \sqrt[3]{x}$ using Newton's method.

Solution:

The function f certainly has a zero at $x = 0$. However, if we form the iteration equation, we get

$$x_{n+1} = x_n - \frac{f(x_n)}{f'(x_n)} = x_n - \frac{x_n^{1/3}}{\frac{1}{3}x_n^{-2/3}} = x_n - 3x_n = -2x_n.$$

We see that x_{n+1} is always twice as big as x_n in absolute value. The graph of $f(x) = \sqrt[3]{x}$ is shown below. Notice how the shape causes the estimates to diverge.

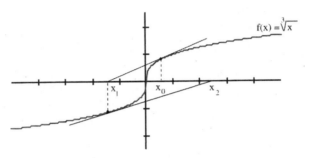

Newton's method is a simple iterative process based on the concept of linear approximations. It is only one of several common zero finding techniques; the Bisection method is another. The zero-finder that is built into the TI-83 (2nd Calc 2:zero) is based on a method similar to Newton's method but uses secant lines instead of tangent lines. In general, the zero-finder on a calculator can not be used effectively without the user knowing how it works and when it might fail.

4.7   Exercises

In Exercises 1–8, find the local linearization, $L(x) = f(a) + f'(a)(x - a)$, of the given function at the given input value a. Use the local linearization to predict the function's output value at $a + 1$. Compare the predicted value $L(a + 1)$ with the actual function value $f(a + 1)$ and calculate the error.

1. $f(x) = x^2$ at $a = 3$

2. $f(x) = \sqrt{x+1}$ at $a = 0$

3. $f(x) = \sin x$ at $a = \pi$

4. $f(x) = \sqrt{4-x}$ at $a = 0$

5. $f(x) = \dfrac{1}{x+1}$ at $a = 0$

6. $f(x) = \ln x$ at $a = 1$

7. $f(x) = \dfrac{1}{x}$ at $a = \dfrac{1}{2}$

8. $f(x) = e^{-x}$ at $a = 0$

9.   Estimate the value of each of the following expressions using a linear approximation. Then compute the difference between your estimate and the value given by your calculator.

   a) $\sqrt{103}$ b) $\sqrt[3]{29}$

In Exercises 10–14, complete five iterations of Newton's method for the given function using the indicated initial guess.

10. $f(x) = 3x^2 - 2x - 4, \quad x_0 = 1$

11. $f(x) = x^3 + x - 1, \quad x_0 = 0$

12. $f(x) = \sin x - x + 1, \quad x_0 = 1$

13. $f(x) = x^5 - x + 1, \quad x_0 = -1$

14. $f(x) = x - \cos x, \quad x_0 = 1$

15.   Graphically demonstrate Newton's method to find the zero of $f(x) = x - \sin(x^2 + 1)$ with $x_0 = 1.9$. What happens when you use $x_0 = 2$? Predict graphically if Newton's method will converge to the zero of f if $x_0 = -0.9$. Confirm your conjectures by completing several iterations of Newton's method on the Home screen with each of these initial values.

16.   Find the largest real zero of the function f defined by $f(x) = x^3 + 6x^2 + 9x + 2$. Sketch a graph of f on the TI-83 to make sure that your first guess leads to the largest zero.

17.   Using $x_0 = 1$, explain why Newton's method fails to converge for the function $f(x) = x^{1/3}$. Is it possible to choose another initial guess and succeed with Newton's method? Why?

18.   Solve the equation $x + \ln x = 2$ by applying Newton's method to the function
       $f(x) = x + \ln x - 2$.

19.   Find the smallest positive solution of the equation $\cos x = e^{-x}$.

20.   Use Newton's method to solve $e^{-x} = \ln x$. List your initial guess and the first five
       approximations that are generated on your calculator.

21.   Using $x_0 = 1.2$, explain why Newton's method fails to converge for the function

       $f(x) = \dfrac{1 + \ln x}{x}$. Why does the initial guess $x_0 = 0.5$ succeed with Newton's method?

22.   Consider the fourth degree polynomial function $f(x) = 4x^4 + 12x^3 + 11x^2 + 3x$.

       a) Show that the zeros of f are in arithmetic progression and find the common
           difference.

       b) Show that the zeros of the derivative of f are also in arithmetic progression and
           estimate the common difference.

Chapter 4   Supplementary Problems

1.   The function $f(x) = \frac{x}{e^x}$ has the derivative $f'(x) = \frac{1-x}{e^x}$. Compute or estimate

$\int_0^1 \frac{1-x}{e^x} \, dx$ in two ways:

   a) using left- and right-sums with $n = 30$ subdivisions, and

   b) using the Fundamental Theorem of Calculus.

2.   Find an equation of the tangent line to the graph of $f(x) = x^{3/2} - x^{1/2}$ at the point
where $x = 4$.

3.   On what intervals is the function $f(x) = 4x^{3/2} - 3x^2$ both concave down and increasing?

4.   Let $f(x) = (x^2 - 1)(x^3 - 1)$.

   a) Find $f'(x)$.

   b) Show that the function f is concave down at the point where $x = 0$.

5.   An object moves along the x-axis in such a fashion that its x-coordinate at any time t
is given by the rule $x(t) = t^2 - 2t^3$. What is the object's velocity at the instant the
acceleration is zero?

6..   Find an equation of the tangent line to the graph of the given function at the given point.

   a) $f(x) = \sqrt{x^2 + 5}$ at $(2, 3)$ b) $f(x) = xe^x$ at $(1, e)$

7.   Let $f(x) = x^x$.

   a) Using your calculator estimate $f'(2)$.

   b) Find the local linearization of f at $x = 2$.

   c) Use the local linearization of f at $x = 2$ to predict the function's output value at $x = 2.2$.

   d) Use your calculator to evaluate $f(2.2)$ and compare the result with the predicted value.

8.   A curve is given by the equation $x^2 - xy + y^2 = 7$.

   a) Find $\frac{dy}{dx}$.

   b) Write an equation for the line tangent to the curve at the point $(-1, 2)$.

9. Let $f(x) = x^2 e^{-x}$ for $-1 \le x \le 3$.

    a) Show that $f'(x) = e^{-x}(2x - x^2)$ and $f''(x) = e^{-x}(x^2 - 4x + 2)$.

    b) For what x is f increasing?

    c) For what x is the graph of f concave down?

10. Find the derivatives of the following functions. Check to see that your answers are reasonable by comparing the graphs of f and f'.

    a) $f(x) = x \ln x - x$ b) $f(x) = \sin(3x) - \cos(2x)$

    c) $f(x) = \dfrac{e^x}{1 + e^x}$ d) $f(x) = \dfrac{2 \sin x}{1 + 2 \cos x}$

11. Let $f(x) = x^4 - 4x + 2$.

    a) How many zeros does f have?

    b) Approximate one of the zeros by first getting an initial estimate and then improving it by using Newton's method once.

    c) An initial estimate of $x = 1$ or very near 1 does not work well for Newton's method. Why?

12. Graphs of functions f and g are shown below.

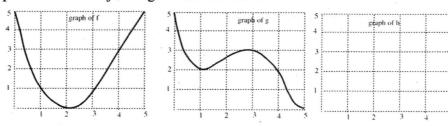

    Let $h(x) = f[g(x)]$. Use the graphs to answer the following questions about the function h.

    a) Approximate the critical points of h and classify them as local maximum, local minimum or neither.

    b) Where is the function h decreasing?

    c) Sketch a graph of h.

13.   a) Find an equation of the tangent line to $f(x) = e^{-3x}$ at $x = 2$.

    b) Use the local linearization of f at $x = 2$ to predict the value of f at $x = 2.2$.

14. Find a point on the graph of $y = e^{3x}$ at which the tangent line passes through the origin.

15. Let $f(x) = \sqrt{1 - \sin x}$

   a) What is the domain of f?

   b) Find $f'(x)$.

   c) What is the domain of f'?

   d) Write an equation for the line tangent to the graph of f at $x = 0$.

16. Find $\dfrac{dy}{dx}$ for each of the following.

   a) $y = x^3 \tan x$ b) $y = \dfrac{\sin x}{x}$ c) $y = (2x^3 - x)^5$ d) $y = \cos^2(3x - \pi)$

17. Suppose the derivative of the function f is $f'(x) = \dfrac{x^2 - 3}{x^4}$.

   a) On what interval(s) is f increasing?

   b) Find the exact x-coordinate of each point of inflection.

18. Suppose the derivative of the function f is $f'(x) = \dfrac{1 - \ln x}{x^2}$.

   a) On what interval(s) is f increasing?

   b) On what interval(s) is the graph of f concave up?

19. a) Show that $y = 1 - x$ is the local linearization of $f(x) = \dfrac{1}{1 + x}$ at $x = 0$.

   b) Find an equation of the line tangent to the curve $x^{2/3} - y^{2/3} - 2y = 2$ at $(1, -1)$

20. Consider the function f defined on the closed interval $[-1, 3]$ by $f(x) = 8x^3 - 3x^4$.

   a) Determine $f'(x)$ and $f''(x)$.

   b) Using the results in part a), determine the x- and y-coordinates of all points where the tangent lines are horizontal.

   c) Using the results in part a), determine the x- and y-coordinates of all inflection points of f on the interval $[-1, 3]$.

21. Let f be the function defined by $f(x) = x^{2/3}(5 - 2x)$ with derivatives

   $f'(x) = \dfrac{-10(x - 1)}{3x^{1/3}}$ and $f''(x) = \dfrac{-10(2x + 1)}{9x^{4/3}}$.

   a) Find the intervals on which f is increasing and those on which it is decreasing.

   b) Find the intervals on which f is concave up and where it is concave down.

22. For each of the following functions:
    i) solve $f'(x) = 0$;
    ii) solve $f''(x) = 0$.

    a) $f(x) = x^3 - 5x^2 - 8x + 12$

    b) $f(x) = 2x^3 - 7x^2 + 9$

    c) $f(x) = 3x^3 - 4x^2 - 17x + 6$

23. Sketch a graph of f' and f'' given the graph of f.

    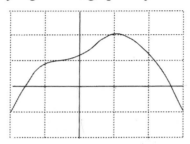

24. Determine the x-coordinates of all points on the graph of $f(x) = (x+1)(x^2 - 1)$ at which the slope of the tangent line is 2.

25. Find all inputs x for which the graph of $f(x) = 4x^3 - 21x^2 + 36x - 4$ is decreasing and concave up.

26. A graph of the function f is shown below

    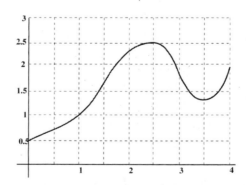

    Arrange the following numbers in ascending order.

    $$f(1), \quad f'(3), \quad \int_1^2 f(x)\,dx, \quad f'(3.5), \quad \lim_{h \to 0} \frac{f(2+h) - f(2)}{h}, \quad \frac{f(2.5) - f(2)}{2.5 - 2}$$

27.   If $g(x) = x^3 + 5x^2 - 3x + 1$ and $h(x) = x^2 - 5x + 2$,

   a) determine all inputs x such that $g'(x) = h'(x)$.

   b) Describe the graphical/geometric significance of the result in part a).

28.   A variable line through the point $(1, 2)$ intersects the x-axis at the point $A(x, 0)$ and the y-axis at the point $B(0, y)$, where x and y are positive. If O is the origin, how fast is the area of triangle AOB changing at the instant when $x = 5$, if x is increasing at a constant rate of 116 units per sec?

29.   The graph of f', the derivative of the function f, is shown below for $-1 \le x \le 8$. The graph of f' has horizontal tangents at $x = 1, 3.5$ and 6.4.

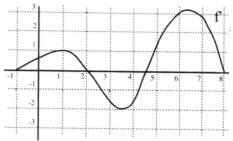

   a) On what interval(s) is f decreasing?.

   b) On what interval(s) is f increasing and concave down?

   c) Where does f have a local minimum?

   d) Where does f have points of inflection?

30.   The graph of the function f is shown below.

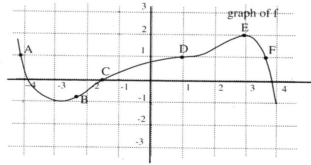

   Which of the labeled points on the graph have

   a) $f(x) > 0$ and $f'(x) < 0$?

   b) $f'(x) > 0$ and $f''(x) < 0$?

   c) $f'(x) = 0$ and $f''(x) < 0$?

   d) $f'(x) = 0$ and $f''(x) = 0$?

   e) $f'(x) < 0$ and $f''(x) < 0$?

31.   a) If $g(x) = \dfrac{\cos x}{1 - \sin x}$ find and simplify $g'(x)$.

b) Find the exact slope of the tangent line to the graph of $y = \ln \sqrt{x}$ at $(e^2, 1)$.

32.   A graph of $y^3 + 7x^2 = x^3$ is shown below.

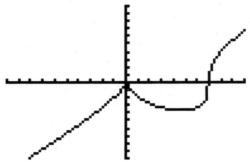

a) Using the graph, estimate where $\dfrac{dy}{dx}$ might be zero or undefined.

b) Find $\dfrac{dy}{dx}$ using implicit differentiation.

c) Find an equation of the tangent line to the curve at the point $(8, 4)$. Sketch the tangent line.

d) Find the x-coordinates of all points where tangent lines are horizontal or vertical. Justify your answer.

33.   A graph of the function g is shown below.

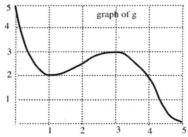

grap$b$ of g

Use the graph to approximate each of the following.

a) $g'(2)$

b) If $h(x) = g(x^2)$, find $h'(2)$.

c) If $k(x) = g^2(x)$, find $k'(4)$.

d) If $f(x) = g[g(x)]$, find $f'(4)$.

34.   The economists at the Bank of Boston have discovered a new cyclic law governing
        mortgage interest rates given by

$$r(t) = .03\cos(.53t - 1.57) + .09$$

where $t = 0$ on Jan 1, 1990 and t is measured in years.

On the other hand, the Boston Board of Realtors has noticed that housing starts vary with
the mortgage rate according to the formula

$$N(r) = 42772(.45)^r.$$

At what rate was the number of housing starts changing 5 years after Jan. 1, 1990?

35.   Evaluate the following limits by recognizing each as the definition of a derivative.

a) $\lim\limits_{h \to 0} \dfrac{\tan\left(\dfrac{\pi}{4} + h\right) - \tan\left(\dfrac{\pi}{4}\right)}{h}$

b) $\lim\limits_{h \to 0} \dfrac{\ln(1 + h) - \ln 1}{h}$

36.   Administrators at Massachusetts General Hospital believe that the hospital's daily
        expenditures $E(B)$, measured in dollars, are a function of how many beds B are in use,
        with

$$E(B) = 14000 + (B + 1)^2.$$

On the other hand, the number of beds B is a function of time t, measured in days, and it
is estimated that

$$B(t) = 20\sin\left(\frac{t}{10}\right) + 50.$$

At what rate are the expenditures changing at $t = 100$?

37.   The function h is defined by $h(x) = f[g(x)]$, where f and g are functions whose graphs
        are shown below.

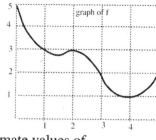
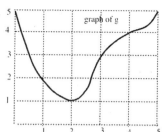

Give approximate values of

a) $h(1)$ b) $h(3)$ c) $h'(1)$ d) $h'(3)$

38.   The function F is defined by $F(x) = G[3x + G(x)]$ where the graph of the function G is shown below.

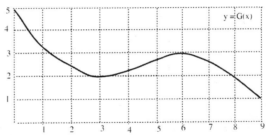

Give the approximate values of

a) $F(2)$ b) $F'(2)$

39.   Oil spilled from a tanker anchored in a harbor spreads out in a circular slick. When the radius of the slick is 100 ft., the area of the slick is increasing at 50 ft²/sec. Find the rate of increase of the radius of the slick at that moment.

40.   The length of a rectangle is increasing at 6 ft/min while its width is decreasing at a rate of 4 ft/min. Find the rate of change of the area when the length is 100 ft and the width is 30 ft.

41.   The functions f and g are defined on the interval $[-4, 4]$ by the graphs below.

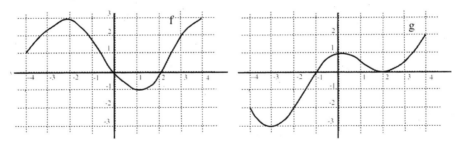

a) Find $f[g(0)]$.

b) Where does the graph of $y = f[g(x)]$ have horizontal tangent lines?

c) Is $y = f[g(x)]$ increasing or decreasing at $x = 3$?

d) Estimate the rate of change of the composite function $y = f[g(x)]$ at $x = 1$.

42.   Let the function f be defined by $f(x) = \begin{cases} kx - x^2, & \text{if } x \le 2 \\ 4 - x, & \text{if } x > 2 \end{cases}$

a) For what values of k will f be continuous at $x = 2$? Justify your answer.

b) Using the value of k found in part a), determine whether f is differentiable at $x = 2$.

c) Let $k = 4$. Determine whether f is differentiable at $x = 2$. Justify your answer.

CHAPTER 5
More Applications of the Derivative

The derivative of a function gives us much useful information about the graph of a function. As we have already seen in Chapter 2, the interpretation of the first and second derivative is based on the fact that

The sign of the derivative of a function tells whether the original function is increasing or decreasing.

Thus, we know that

- If $f' > 0$ on an interval, then f is increasing on the interval.
  If $f' < 0$ on an interval, then f is decreasing on the interval.

- If $f'' > 0$ on an interval, then f' is increasing and f is concave up on the interval.
  If $f'' < 0$ on an interval, then f' is decreasing and f is concave down on the interval.

We can now apply these principles to functions that are defined by formulas. In particular, we will be interested in using the derivative to solve optimization problems – finding the maximum and minimum value of a function.

5.1 Increasing and Decreasing Functions

When sketching the graph of a function by hand or with a calculator you cannot normally show the entire graph. The choice of which part of the function to display is often crucial. For example, which of the following viewing rectangles better represents the graph of

$$f(x) = x^3 + 3x^2 - 24x + 18?$$

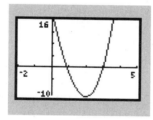 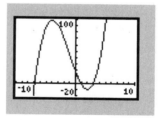

The viewing rectangle on the right gives a more complete representation of the graph; however, would a third rectangle display other interesting features of the graph? This is an example of a situation where knowing how to find and interpret derivatives is useful. We begin by using the first derivative to decide where f is increasing and where f is decreasing.

The derivative of f is

$$f'(x) = 3x^2 + 6x - 24 = 3(x+4)(x-2).$$

Using the factored form of the derivative

$$f'(x) = 3(x+4)(x-2),$$

we see that $f'(x) = 0$ at $x = -4$ and $x = 2$. These points divide the x-axis into three intervals

$$(-\infty, -4), \quad (-4, 2) \text{ and } (2, \infty),$$

and since f' is continuous , f' cannot change sign within any of these intervals. We determine the sign of the derivative by calculating its value at one point in each of the three intervals. For example, we select –5, 0, and 5. Since $f'(-5) = 21 > 0$ we know that f' is positive for $x < -4$, so f is increasing on $(-\infty, -4)$. Similarly, since $f'(0) = -24 < 0$ and $f'(5) = 81 > 0$, we know that f is decreasing on $(-4, 2)$ and increasing on $(2, \infty)$. This information is summarized on the following coordinate line.

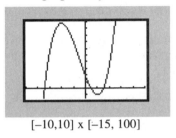

At the zeros of the derivative we have $f(-4) = 98$ and $f(2) = -10$. Thus, on the interval $(-4, 2)$ the function decreases from a high of 98 to a low of –10. By setting a viewing rectangle to $[-10,10] \times [-15, 100]$ we are guaranteed a graph of f that includes all of its interesting features.

$[-10,10] \times [-15, 100]$

Critical Points

Recall that at inputs $x = c$ in the domain of a function f where the derivative f' is neither positive nor negative there are two possibilities – either $f'(c) = 0$ or $f'(c)$ is not defined. In either case, $x = c$ is called a **critical number** for the function f.

> A number c in the interior of the domain of a function f is called a **critical number of f** if either $f'(c) = 0$ or $f'(c)$ is not defined. The point $(c, f(c))$ is called a **critical point** of the graph of f.

Geometrically, a critical point occurs where a tangent to the graph is horizontal or the graph has a sharp corner, cusp or vertical tangent. The following graph has several different critical points.

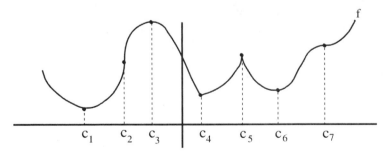

The numbers c_1, c_3, c_6, and c_7 are critical because the tangent line is horizontal and the derivative is zero. The number c_2 is critical because the tangent is vertical and $f'(c_2)$ does not exist. The numbers c_4 and c_5 are critical because a sharp change of direction of the graph occurs and, again, the derivative does not exist.

Example 1:   *Finding the Critical Numbers*

Find the critical numbers for the given functions.

a) $f(x) = x^5 - 5x^4 + 5x^3 + 20$ b) $g(x) = \dfrac{e^x}{x-2}$ c) $h(x) = x^{3/5}(4-x)$

Solution:

a) $f'(x) = 5x^4 - 20x^3 + 15x^2$, which is defined at every input x, so the critical numbers occur only where $f'(x) = 0$. We have

$$5x^4 - 20x^3 + 15x^2 = 0$$

$$5x^2(x^2 - 4x + 3) = 0$$

$$5x^2(x-1)(x-3) = 0$$

$$x = 0, 1, 3 \text{ are the critical numbers.}$$

b) The Quotient Rule gives $g'(x) = \dfrac{(x-2)e^x - e^x(1)}{(x-2)^2} = \dfrac{e^x(x-3)}{(x-2)^2}$.

The derivative is not defined at $x = 2$, but g is not defined at 2 either, so $x = 2$ is not a critical number. The actual critical numbers are found by solving $g'(x) = 0$:

$$\frac{e^x(x-3)}{(x-2)^2} = 0$$

$$x = 3$$

Since $e^x > 0$, 3 is the only critical number.

c) The Product Rule gives $h'(x) = x^{3/5}(-1) + \dfrac{3}{5}x^{-2/5}(4-x)$

$$= \frac{3(4-x) - 5x}{5x^{2/5}} = \frac{12 - 8x}{5x^{2/5}}.$$

The derivative is not defined at $x = 0$; however, we have $h(0) = 0$, so we see that the function h is defined at $x = 0$, which means $x = 0$ is a critical number. For other critical numbers, solve $h'(x) = 0$:

$$\frac{12 - 8x}{2x^{2/5}} = 0$$

$$12 - 8x = 0$$

$$x = \frac{3}{2}$$

So, the critical numbers are $\dfrac{3}{2}$ and 0.

The figure shows a graph of the function $h(x) = x^{3/5}(4-x)$. It supports our answer because there is a horizontal tangent at $x = \dfrac{3}{2}$ and a vertical tangent at $x = 0$.

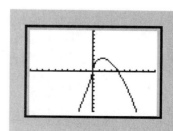

Maximum and Minimum Points

Here is the graph of a differentiable function.

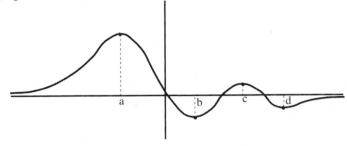

Its most interesting features occur at the places where the graph levels off and the derivative is zero. It is necessary to distinguish between **local maximums** and **local minimums** like those occurring at $x = c$ and $x = d$ and the **global maximums** and **global minimums**, like those occurring at $x = a$ and $x = b$.

> Suppose that c is a critical number of a function f. Then
>
> - f has a **local minimum** at $x = c$ if, for all x near c, $f(c) \leq f(x)$;
>
> - f has a **local maximum** at $x = c$ if, for all x near c, $f(c) \geq f(x)$.

The graph below illustrates the local extrema for our critical number example.

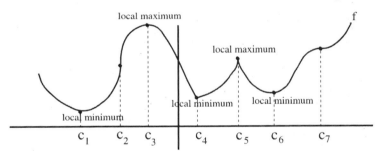

Notice in the graph above that every local minimum and local maximum occurred at a critical point. On the other hand, there is a critical point at $x = c_7$ and another at $x = c_2$ where there is neither a local minimum nor a local maximum. In other words, a function does not necessarily have a local extrema at every critical point.

The derivative can be employed to determine whether a critical point is the location of a local minimum, local maximum or neither.

The First Derivative Test

> **The First Derivative Test**
>
> If c is a critical number and if f' changes sign at $x = c$, then
>
> - f has a local minimum at $x = c$ if f' is negative to the left of c and positive to the right of c;
>
> - f has a local maximum at c if f' is positive to the left of c and negative to the right of c.

If f' is positive (or negative) on both sides of a critical point, then f has neither a maximum nor a minimum value at that point. In this case the critical point will usually (but not always) be an inflection point, a concept that will be discussed further in the next section.

The First Derivative Test is illustrated below.

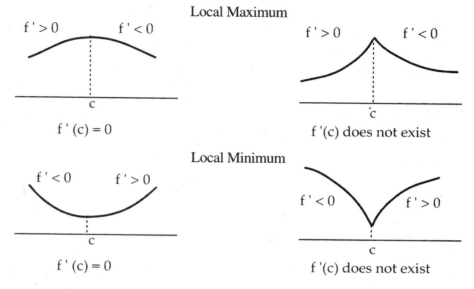

Local Maximum

f ' > 0   f ' < 0   f ' > 0   f ' < 0

c   c

f ' (c) = 0   f '(c) does not exist

Local Minimum

f ' < 0   f ' > 0   f ' < 0   f ' > 0

c   c

f ' (c) = 0   f '(c) does not exist

Example 2: *Using the First Derivative Test*

Use the First Derivative Test to find the local maximum and minimum values of f if
$$f(x) = x^4 + 2x^3 + x^2 - 8.$$

Solution:

With $f'(x) = 4x^3 + 6x^2 + 2x = 2x(2x+1)(x+1)$, the critical numbers of f are -1, $-\frac{1}{2}$, and 0. Checking the sign of the derivative on either side of the critical numbers we have

$$f'(x) < 0 \text{ for } x \text{ in } (-\infty, -1)$$
$$f'(x) > 0 \text{ for } x \text{ in } (-1, -\frac{1}{2})$$
$$f'(x) < 0 \text{ for } x \text{ in } (-\frac{1}{2}, 0)$$
$$f'(x) > 0 \text{ for } x \text{ in } (0, \infty).$$

Therefore, by the First Derivative Test, f has local minimums at -1 and 0 and a local maximum at $-\frac{1}{2}$.

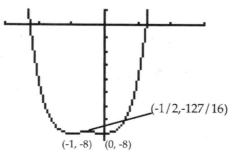

(-1/2,-127/16)

(-1, -8)   (0, -8)

Global Maxima and Minima

In many important applications of calculus the goal is to find the global maximum or the global minimum value of a function.

> A function f has
>
> - a **global maximum value** $f(c)$ at the input c if $f(x) \leq f(c)$ for every x in the domain of f;
>
> - a **global minimum value** $f(c)$ at the input c if $f(x) \geq f(c)$ for every x in the domain of f.

A function can have at most one global maximum and one global minimum value, though this value can be assumed at many points. For example, $f(x) = \sin x$ has a global maximum value of 1 at many inputs x. Maximum and minimum values of a function are collectively referred to as **extreme values.**

   A function does not necessarily need to have either a global maximum or a global minimum value. The reciprocal function, $f(x) = \dfrac{1}{x}$, is an example of such a function. However, if a continuous function f is defined on a closed interval, it is guaranteed by the Extreme Value Theorem that the function f will have a global maximum and minimum value on the closed interval.

> **The Extreme Value Theorem**
>
> If a function f is continuous on a closed interval $[a, b]$, then f has a global maximum and a global minimum value on $[a, b]$.

These extreme values can only occur at the critical points or at one or more endpoints. In any case, there is a simple four-step strategy for locating the global extrema of a continuous function f on a closed interval $[a, b]$.

> Step 1.   Find the critical numbers that lie in the interval (a, b).
>
> Step 2.   Calculate $f(c)$ for each critical number c.
>
> Step 3.   Calculate endpoint values $f(a)$ and $f(b)$.
>
> Step 4.   Compare outputs at all critical numbers and endpoints to identify the global maximum and minimum values.

Example 3: *Finding Global Maximum and Minimum Values*

   Find the global maximum and minimum values of $f(x) = 3x - x^2$ on the interval $[0, 4]$.

Solution:

   $f'(x) = 3 - 2x$. $f'(x) = 0$ at $x = \dfrac{3}{2}$. Thus, $\dfrac{3}{2}$ is the only critical number and the candidates for global maximum and minimum values on $[0, 4]$ are the values of f at the critical number $\dfrac{3}{2}$ and at the two endpoints 0 and 4. These are

$$f(0) = 0, \qquad f\left(\frac{3}{2}\right) = \frac{9}{4}, \qquad f(4) = -4.$$

The global maximum value of f on the interval $[0, 4]$ is $\frac{9}{4}$, and occurs at the critical number $\frac{3}{2}$. The global minimum value is -4, and occurs at the endpoint 4.

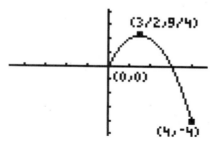

Example 4: *Finding Global Maximum and Minimum Values*

Find the global maximum and minimum values of $f(x) = 5x^{2/3} - 2x^{5/3}$ on the interval $[-1, 2]$.
Solution:

$$f'(x) = \frac{10}{3}x^{-1/3} - \frac{10}{3}x^{2/3} = \frac{10}{3}x^{-1/3}(1-x). \quad f'(x) = 0 \text{ at } x = 1 \text{ and } f'(x) \text{ is}$$

undefined at $x = 0$. Thus, 1 and 0 are critical numbers. The candidates for global maximum and minimum values on $[-1, 2]$ are the values

$$f(-1), \ f(0), \ f(1), \ f(2).$$

Values at Endpoints: $f(-1) = 7$

          $f(2) = 2^{2/3} \approx 1.587$

Critical Values: $f(0) = 0$

          $f(1) = 3$

The global maximum value occurs at $x = -1$ and is $f(-1) = 7$; the global minimum occurs at $x = 0$ and is $f(0) = 0$.

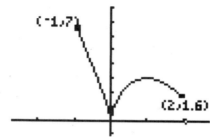

5.1   Exercises

In Exercises 1–12, determine the critical numbers of each function, then use the First Derivative Test to determine whether the function has a local maximum, a local minimum, or neither at each critical number. Confirm your answers graphically.

1. $f(x) = \frac{1}{3}x^3 - x^2 - 3x + 3$

2. $f(x) = 3x^4 - 4x^3 - 6x^2 + 12x$

3. $f(x) = x^4 - 4x^3$

4. $f(x) = \sin(2x)$

5. $f(x) = x^{2/3}(x - 5)$

6. $f(x) = 2x\sqrt{3 - x}$

7. $f(x) = x^{4/3} + 4x^{1/3}$

8. $f(x) = \dfrac{x}{x^2 + 1}$

9. $f(x) = xe^x$

10. $f(x) = \dfrac{\ln x}{x}$

11. $f(x) = x^2 e^{-x}$

12. $f(x) = \ln x - \ln(x^2 + 1)$

In Exercises 13–18, use your calculator to plot the graph of f. Approximate within 0.1 all values of c such that $f(c)$ is a local extreme value, and identify each as a local maximum or a local minimum point.

13. $f(x) = x^5 - x^4 + x^2 - 1$

14. $f(x) = \dfrac{10x + 2}{x^4 + 1}$

15. $f(x) = 3\sin x - x$ for $-2\pi \le x \le 2\pi$

16. $f(x) = e^x + \sin x$ for $-6 \le x \le 3$

17. $f(x) = \dfrac{x^2 + x - 6}{x^4 + 1}$

18. $f(x) = (5x^3 - 1)\sqrt{4 - x^2}$

19.   For each of the following graphs, mark any critical points or extremes. Identify which extreme is local and which is global.

a)

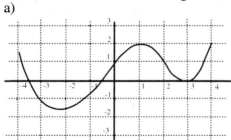

b)

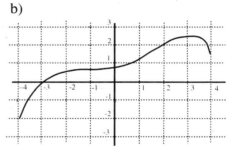

c)

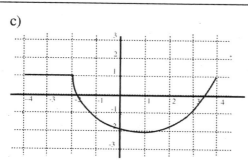

d)

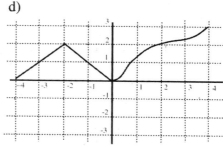

In Exercises 20–31, find the global maximum and minimum function values on the given interval.

20. $f(x) = x^3 - 3x + 1$, $[1, 3]$

21. $f(x) = x^4 - 2x^2 + 3$, $[-1, 2]$

22. $f(x) = x^3 - 3x^2 - 9x + 5$, $[-2, 4]$

23. $f(x) = \dfrac{x-1}{x^2 + 8}$, $[-3, 6]$

24. $f(x) = x - 2\sin x$, $[0, \frac{\pi}{2}]$

25. $f(x) = \cos x(1 + \sin x)$, $[0, 2\pi]$

26. $f(x) = x^{2/3}$, $[-8, 8]$

27. $f(x) = x(x-1)^{1/3}$, $[-7, 2]$

28. $f(x) = x\sqrt{4 - x^2}$, $[0, 2]$

29. $f(x) = x^3 - e^x$, $[-3, 2]$

30. $f(x) = x - \ln x$, $[0.1, 2]$

31. $f(x) = \sin^2 x - \cos x$, $[0, \pi]$

In Exercises 32–35, the derivative of a function f is given. Use the graph of the derivative to answer the following questions.

   a) On which intervals is the function f increasing?

   b) At what values of x, if any, does f have a local maximum? A local minimum?

   c) Sketch a possible graph of f.

32. $f'(x) = (x - 2)(x + 4)$

33. $f'(x) = (x - 1)^2 (x - 4)$

34. $f'(x) = \dfrac{x + 1}{x - 2}$

35. $f'(x) = (x - 1)^2 (x - 2)(x - 4)$

36.   The graph of g', the derivative of a function g, is shown below for $-4 \le x \le 4$.
       The graph of g' has horizontal tangents at x = -3. 0.5, and 3.5.

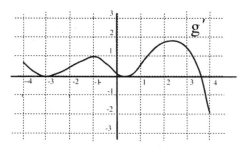

a) On which intervals is g increasing?

b) Where does g have critical numbers?

c) Where does g have a local maximum?

d) Suppose $g(-4) = 0$. Sketch a possible graph of g.

37.   The graph of f', the derivative of f, is shown below for $-4 \le x \le 4$.

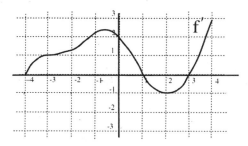

a) Suppose that $f(2) = -1$. Find an equation of the line tangent to the graph of f at the
   point $(2, -1)$.

b) Where does f have critical numbers?

c) On what intervals is f decreasing?

d) Where does f have a local maximum? Explain briefly.

e) Suppose $f(0) = 0$. Sketch a possible graph of f.

In Exercises 38–39 sketch the graph of a function f that is continuous on the interval [0, 4] and
has the given properties.

38.   Global maximum at 0, global minimum at 4, local minimum at 2 and local maximum at 3.

39.   2 is a critical number, but no local maximum or minimum.

40.   Find the x-coordinate at which the absolute maximum value of $f(x) = x^3 - 3x^2 + 12$
       occurs on the closed interval [-2, 4].

41.   Find the total number of local maximum and minimum points of the function whose derivative, for all x, is given by $f'(x) = x(x-3)^2(x+1)^4$.

42.   If the function f is defined for all positive numbers by $f(x) = x \ln(2x)$, find its local minimum value.

43.   Determine constants a and b so that the function $f(x) = x^3 + ax^2 + b$ will have a critical point $(1, 2)$.

44.   Determine constants a and b so that the function $f(x) = a \sin x + bx$ will have a critical point (π, π).

45.   How many critical points does the function $f(x) = (x+2)^5(x^2-1)^4$ have?

46.   Give an example and sketch the graph of a function f that satisfies all of the following conditions.

     i) f satisfies the hypothesis of the Extreme Value Theorem on the interval $[0, 4]$.

     ii) f has two critical numbers in the interval $(0, 4)$.

     iii) Neither the global maximum nor the global minimum of f on the interval $[0, 4]$ occurs at either critical number.

5.2   Applications of the Second Derivative

The second derivative f'' of a function f is simply the derivative of f' and it tells us how fast the derivative f' is changing with respect to changes in the input variable.

One physical interpretation of the second derivative is provided by the phenomenon of motion. Recall that when an object moves along a straight path, its instantaneous acceleration is defined as the derivative or rate of change of the velocity function. But velocity, in turn, is defined as the rate of change of the object's distance or displacement from some fixed point of reference along the path.

Therefore, given a distance-time function $y = s(t)$, velocity is obtained by differentiating, $v(t) = \dfrac{ds}{dt}$,

which means that acceleration is the second derivative, $a(t) = \dfrac{d}{dt}\left(\dfrac{ds}{dt}\right) = \dfrac{d^2s}{dt^2}$, of the distance function.

Example 1: *Modeling Vertical Motion*

The vertical distance, in feet, above the ground of a golf ball thrown from the top of a tall building is given by

$$s(t) = -16t^2 + 36t + 50$$

where time t is measured in seconds. Find the golf ball's position, velocity, and acceleration at $t = 3$ seconds.

Solution:

$s(3) = 14$ feet above the ground. Since $v(t) = \dfrac{ds}{dt} = -32t + 36$, we have $v(3) = -60$ ft/sec.

For the acceleration we have, $a(t) = \dfrac{d^2s}{dt^2} = -32$ ft/sec^2 , so $a(3) = -32$ ft/sec^2.

Notice that the acceleration is a familiar constant which, of course, is due to the force of gravity.

Concavity and Points of Inflection

While the second derivative f'' can be interpreted in a physical sense as acceleration, its graphical interpretation is related to both the graph of the original function and the graph of the first derivative.

The second derivative f'' is related to the first derivative f' in the same way that the first derivative f' is related to the function f. For example, if the second derivative is positive for each input x in an interval, this means that the first derivative is increasing on the interval. Geometrically this means that the slopes of the tangent lines to the graph of f are increasing, and consequently the graph of f should bend upward on the interval. When this occurs the graph of f is called **concave up** on the interval.

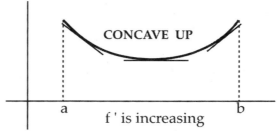

> The graph of a differentiable function f is called **concave up**
> on an interval if the derivative f' is increasing on the interval. Note that if
> $f''(x) > 0$ for all x in an interval, then f is concave up on the interval.

On the other hand, if for each input x in an interval the second derivative f'' is negative valued, then the first derivative is decreasing on the interval. Geometrically this means that the slopes of the tangent lines to the graph of f are decreasing and the graph of f should bend downward on the interval. When this occurs the graph of f is called **concave down** on the interval.

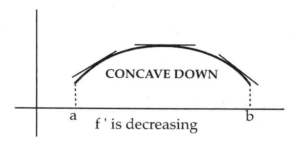

> The graph of a differentiable function f is called **concave down**
> on an interval if the derivative f' is decreasing on the interval. Note that if
> $f''(x) < 0$ for all x in an interval, then f is concave down on the interval.

According to the above definitions, concavity is defined only on intervals where the derivative of the given function exists and is not constant. Thus, a function is neither concave up nor concave down on an interval where its graph is a straight line segment.

In sketching the graph of a function, the points where the concavity of the graph changes should be plotted. Such points are called **points of inflection**.

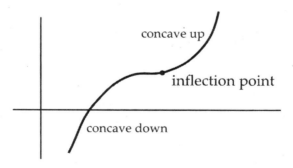

> A point on the graph of a function where there is a tangent line and the
> concavity changes is called **a point of inflection.**

If $(c, f(c))$ is an inflection point on the graph of f, then the second derivative must change sign at $x = c$. There are two situations under which this can happen: a) $f''(c) = 0$ or b) $f''(c)$ is undefined. We look at the places where $f''(x)$ is zero or undefined, and check the concavity on both sides of such inputs x.

As an example of the four possible cases, we consider:

a) $f(x) = x^3$; $f''(x) = 6x$, so $f''(0) = 0$.   b) $f(x) = x^4$; $f''(x) = 12x^2$, so $f''(0) = 0$.

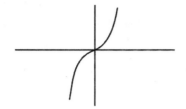 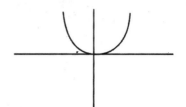

c) $f(x) = x^{1/3}$; $f''(x) = -\frac{2}{9}x^{-5/3}$, so   d) $f(x) = x^{2/3}$; $f''(x) = -\frac{2}{9}x^{-4/3}$, so

    $f''(0)$ does not exist. $f''(0)$ does not exist.

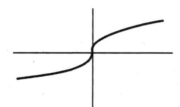 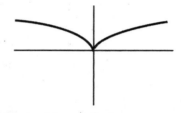

Cases a) and c) have inflection points at $x = 0$, while cases b) and d) do not.

In view of these examples, we can consider only the critical points of f' (where $f''(c)$ is zero or is undefined) as candidates for inflection points on the graph of f. A sign change in the second derivative must occur before we can be sure that a particular candidate really does give an inflection point.

Example 2: *Finding Inflection Points*

Find the inflection points of $f(x) = x^4 + 2x^3 - 1$.

Solution:

From the graph of f in the figure, we see that f has inflection points.

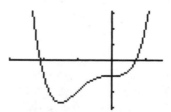

We calculate $f'(x) = 4x^3 + 6x^2$ and $f''(x) = 12x^2 + 12x = 12x(x+1)$, so $f''(x) = 0$ when $x = 0$ or when $x = -1$. Since, $f''(x)$ is positive for $x < -1$, negative for $-1 < x < 0$, and positive again for $x > 0$, changes in concavity occur at $x = 0$ and at $x = -1$. Therefore, points $(0, -1)$ and $(-1, -2)$ are inflection points.

The Second Derivative Test

It is often possible to classify a critical point on the graph of f by examining the sign of the second derivative. Suppose that c is a critical number of f such that $f'(c) = 0$ and the graph of f has a horizontal tangent at c. If the graph of f is concave up at c, the derivative f' is increasing, so f' must be negative to the left of c and positive to the right of c. Thus, by the First Derivative Test, f has a local minimum at $x = c$. Similarly, if the graph of f is concave down, f has a local maximum at $x = c$.

The Second Derivative Test

- If $f'(c) = 0$ and $f''(c) > 0$ then f has a local minimum at c.

- If $f'(c) = 0$ and $f''(c) < 0$ then f has a local maximum at c.

Example 3:   *Applying the Second Derivative Test*

If the function f is defined by $f(x) = 2x^3 - 9x^2 + 27$, find the local maximum and minimum values by applying the Second Derivative Test.

Solution:

With $f'(x) = 6x^2 - 18x = 6x(x - 3)$, the critical numbers for f are 0 and 3.
Since $f''(x) = 12x - 18$, $f''(0) = -18$ and $f''(3) = 18$.
So, by the Second Derivative Test,

$\quad\quad f(0) = 27$ is a local maximum and
$\quad\quad f(3) = 0$ is a local minimum.

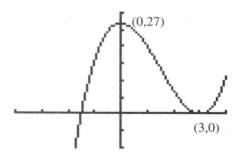

It should be noted that if both $f'(c) = 0$ and $f''(c) = 0$, then the Second Derivative Test is useless as a test for a local maximum or minimum point. For example, if $f(x) = x^3$ and $g(x) = x^4$, both $f'(0) = g'(0) = 0$ and $f''(0) = g''(0) = 0$. However, the point $x = 0$ is a minimum for $g(x) = x^4$ but is neither a maximum or minimum for $f(x) = x^3$. When the Second Derivative Test fails to give information about a critical point c because $f''(c) = 0$, the First Derivative Test can still be useful. Also, for many functions the second derivative is more complicated to calculate than the first derivative, so the First Derivative Test is likely to be of more use in classifying critical points than the Second Derivative Test.

5.2   **Exercises**

1.   Use the graph below to determine the inputs x that lead to inflection points. Assume the curve in the figure is the graph of

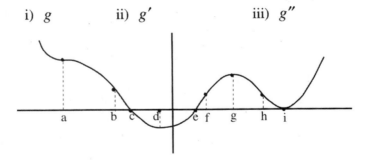

       i) g ii) g' iii) g''

2.   The graph of the second derivative of a function g is shown in the figure.

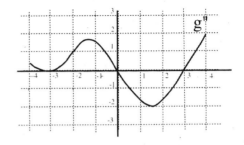

   a) On which intervals is g concave down?

   b) Where does g have inflection points? Justify your answer.

   c) Where does g' have a local maximum?

   d) Suppose $g'(0) = 0$. Is g increasing or decreasing at $x = 2$? Justify your answer.

3.   The graph of the derivative of a function f is shown in the figure.

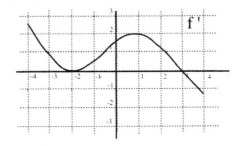

   a) Where is the graph of f concave up?

   b) Where does f have points of inflection?

   c) Sketch a graph of f''.

In Exercises 4–13, find $f''(x)$ and use it to locate any inflection points and determine the intervals on which the graph of the function is concave up or concave down. Confirm your answers graphically.

4. $f(x) = -\dfrac{3}{2}x^2 + x$ 5. $f(x) = x^3 - 3x^2$

6. $f(x) = x^3 - 6x^2 + 12x - 4$ 7. $f(x) = x^4 - 6x^2 + 8$

8. $g(x) = \dfrac{x}{x^2 + 1}$ 9. $g(x) = x\sqrt{x - 1}$

10. $f(x) = xe^x$

11. $g(x) = x^2 e^{-x}$

12. $f(x) = x \ln x$

13. $f(x) = x^2 + \ln x$

In Exercises 14–21, classify the critical points of the function using the Second Derivative Test whenever possible.

14. $f(x) = 3x^3 - 36x - 3$

15. $f(x) = x(x-2)^2 + 1$

16. $f(x) = x^3 + \dfrac{1}{x}$

17. $f(x) = \sin x - \cos x, \quad 0 \le x \le 2\pi$

18. $f(x) = 5x^3 - 3x^5$

19. $f(x) = (x^2 - 4)^3$

20. $f(x) = \dfrac{x}{e^{2x}}$

21. $f(x) = x^{2/3}(6-x)^{1/3}$

In Exercises 22–27, you are given a position function s for an object moving along a coordinate line. ($s(t)$ is measured in feet and t is measured in seconds). For each function s find the following:

   a) the velocity of the object at $t = 1$;

   b) the acceleration of the object at $t = 1$;

   c) the object's velocity at the instant when the acceleration is 0.

22. $s(t) = 2t^3 - 6t^2 + 18t$

23. $s(t) = t^3 - 3t^2$

24. $s(t) = \dfrac{2}{t^2 + 1}$

25. $s(t) = \ln(1 + t^2)$

26. $s(t) = \dfrac{t}{e^t}$

27. $s(t) = \cos(\pi t) - \sin(\pi t), \quad 0 \le t \le 2\pi$

28. Given the function f defined by $f(x) = \dfrac{x^2}{2x^2 + 1}$ with derivatives

   $$f'(x) = \frac{2x}{(2x^2 + 1)^2} \quad \text{and} \quad f''(x) = \frac{2 - 12x^2}{(2x^2 + 1)^3}.$$

   a) Show that f is an even function.

   b) Find an equation of the tangent line to the graph of f at the point $(1, f(1))$.

   c) Give the coordinates of all local maximum and minimum points.

   d) Find the x-coordinates of all inflection points.

29.   Given the function f defined by $f(x) = \dfrac{x(3-x)}{(3+x)^2}$ with derivatives

$f'(x) = \dfrac{9(1-x)}{(3+x)^3}$ and $f''(x) = \dfrac{18(x-3)}{(3+x)^4}$.

a) Show that f is not an even function.

b) Find an equation of the tangent line to the graph of f at the point $(1, f(1))$.

c) Give the coordinates of all local maximum and minimum points.

d) Find the x-coordinates of all inflection points.

30.   Let g be the function defined by $g(x) = \dfrac{4x}{3+x^4}$ with derivatives

$g'(x) = \dfrac{12(1-x^4)}{(3+x^4)^2}$ and $g''(x) = \dfrac{-48x^3(5-x^4)}{(3+x^4)^3}$.

a) Show that g is an odd function.

b) Give the coordinates of all local maximum and minimum points.

c) Find an equation of a line normal to the graph of g at the point $(0, g(0))$.

d) Find the x-coordinates of all inflection points.

31.   Given the function f defined by $f(x) = xe^{-3x}$.

a) Find $f'(x)$ and $f''(x)$.

b) Determine the interval(s) on which f is increasing.

c) Determine the interval(s) on which the graph of f is concave up.

32.   Sketch a graph of one continuous function f with all of the following properties:

i) $f'(x) > 0$ when $x < 2$ and when $2 < x < 5$;

ii) $f'(x) < 0$ when $x > 5$;

iii) $f'(2) = 0$;

iv) $f''(x) < 0$ when $x < 2$ and when $4 < x < 7$;

v) $f''(x) > 0$ when $2 < x < 4$ and when $x > 7$.

5.3   Limits Involving Infinity

The information that can be obtained from examining the first and second derivative of a given function is a great help in producing an accurate graph of a function. To complete our discussion of graphing we extend our notion of limit so that we will have some basis for deciding the behavior of a graph when inputs become very large, very small, or close to points where the function is undefined.

The graph of the function $f(x) = \dfrac{1}{x^2 + 1}$ is shown below.

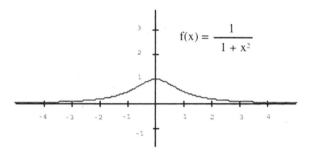

Note that for very large x or for very small x, the corresponding outputs are close to 0. In fact, we can make the outputs $f(x)$ as close to 0 as we choose by making x sufficiently large. In such a case we write

$$\lim_{x \to +\infty} \frac{1}{x^2 + 1} = 0.$$

In general we have

The function f has a limit L as x increases without bound, written

$$\lim_{x \to +\infty} f(x) = L \ ,$$

if the outputs $f(x)$ can be made as close to L as we like by taking x sufficiently large.

In the example above it is also true that for sufficiently small inputs x, the outputs $f(x)$ can also be made close to 0 and we write

$$\lim_{x \to -\infty} \frac{1}{x^2 + 1} = 0.$$

The symbol ∞ does not represent a real number and the symbols $x \to +\infty$ and $x \to -\infty$ are used to indicate the behavior of the function outputs when x is very large or very small.

Example 1: *Determining Limits at Infinity*

Find $\displaystyle\lim_{x \to +\infty} \frac{1}{x}$ and $\displaystyle\lim_{x \to -\infty} \frac{1}{x}$.

Solution:

Note that when x is large, $\frac{1}{x}$ is small. For instance,

$$\frac{1}{100} = 0.01 \qquad \frac{1}{1000} = 0.001 \qquad \frac{1}{10000} = 0.0001$$

In fact, by taking x large enough, we can make $\frac{1}{x}$ as close to 0 as we please. Therefore, we have

$$\lim_{x \to +\infty} \frac{1}{x} = 0.$$

By similar reasoning we see that when is x very large negatively, $\frac{1}{x}$ is small negatively, so we also have

$$\lim_{x \to -\infty} \frac{1}{x} = 0.$$

Most of the limit rules that were given in Section 2.7 also hold for limits at infinity. In particular, the rules on the limit of a constant, sum, and product, and quotient remain unchanged when "$x \to a$" is replace with $x \to +\infty$ or $x \to -\infty$.

Example 2: *Determinng Limits at Infinity*

Find $\displaystyle\lim_{x \to +\infty} \frac{1 + x^2}{2 - x^2}$.

Solution:

The trick that will enable us to use the results of Example 1 is to divide the numerator and denominator by x^2 and then apply limit rules.

$$\lim_{x \to +\infty} \frac{1 + x^2}{2 - x^2} \;=\; \lim_{x \to +\infty} \frac{\dfrac{1}{x^2} + 1}{\dfrac{2}{x^2} - 1}$$

$$= \frac{\displaystyle\lim_{x \to +\infty} \frac{1}{x^2} + 1}{2\,\displaystyle\lim_{x \to +\infty} \frac{1}{x^2} - 1}$$

$$= \frac{\left(\displaystyle\lim_{x \to +\infty} \frac{1}{x}\right)\left(\displaystyle\lim_{x \to +\infty} \frac{1}{x}\right) + 1}{2\left(\displaystyle\lim_{x \to +\infty} \frac{1}{x}\right)\left(\displaystyle\lim_{x \to +\infty} \frac{1}{x}\right) - 1}$$

$$= \frac{0 \cdot 0 + 1}{2 \cdot 0 \cdot 0 - 1} = -1.$$

Asymptotes

The graph of the function $f(x) = \dfrac{1}{x}$ is shown below.

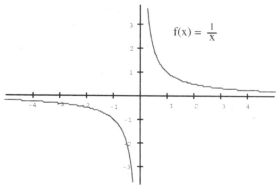

From the results of Example 1, we know that $\lim\limits_{x \to +\infty} f(x) = 0$. Geometrically this means that the graph of f gets close to the line $y = 0$ when x is large. Also $\lim\limits_{x \to -\infty} f(x) = 0$ which means that the graph of f gets close to the line $y = 0$ when x is a large negative number. We call the line $y = 0$ a horizontal asymptote of the graph. In general, if either

$$\lim_{x \to +\infty} f(x) = b \quad \text{or} \quad \lim_{x \to -\infty} f(x) = b,$$

the line $y = b$ is called a **horizontal asymptote** of the graph of f.

Example 3: *Finding Horizontal Asymptotes*

Find the horizontal asymptotes for the graph of f if $f(x) = \dfrac{x^4 + x^2}{x^4 + 1}$.

Solution:

Since
$$\lim_{x \to +\infty} \frac{x^4 + x^2}{x^4 + 1} = \lim_{x \to +\infty} \frac{1 + \dfrac{1}{x^2}}{1 + \dfrac{1}{x^4}} = 1,$$

and
$$\lim_{x \to -\infty} \frac{x^4 + x^2}{x^4 + 1} = \lim_{x \to -\infty} \frac{1 + \dfrac{1}{x^2}}{1 + \dfrac{1}{x^4}} = 1,$$

the line $y = 1$ is a horizontal asymptote.

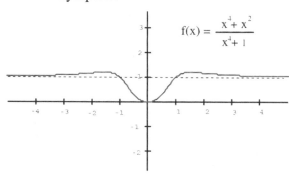

Notice that the graph of f crosses its asymptote twice. There is a misconception among some students that curves cannot cross their asymptotes.

Returning to the graph of $f(x) = \dfrac{1}{x}$ we note that

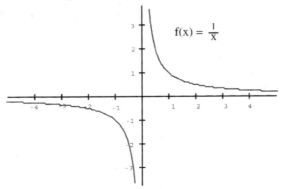

when inputs are chosen just to the right of the origin, the corresponding outputs are large and positive. And when inputs are chosen just to the left of the origin, the corresponding outputs are large and negative. In such a case we call the line $x = 0$ a **vertical asymptote** of the graph.

In general, the vertical asymptotes will be vertical lines at points where the function is undefined because of a zero in the denominator. To show that a vertical asymptote occurs at an input $x = a$ it is sufficient to show that the reciprocal $\dfrac{1}{|f(x)|}$ is close to zero when x is close to a.

Example 4: *Finding Vertical Asymptotes*

Find the vertical asymptotes of the graph of f if $f(x) = \dfrac{1}{x^2 - 1}$.

Solution:

We have $\dfrac{1}{f(x)} = x^2 - 1$ with $\lim\limits_{x \to 1}(x^2 - 1) = 0$ and $\lim\limits_{x \to -1}(x^2 - 1) = 0$. Therefore, the lines $x = 1$ and $x = -1$ are vertical asymptotes. Also $\lim\limits_{x \to +\infty} f(x) = 0$ and $\lim\limits_{x \to -\infty} f(x) = 0$ so that the line $y = 0$ is a horizontal asymptote.

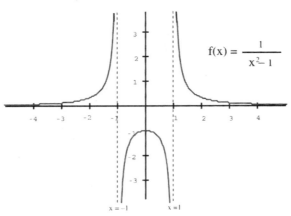

5.3   Exercises

In Exercises 1–10, evaluate the limits algebraically if possible. Support your answers numerically or graphically.

1. $\displaystyle\lim_{x\to+\infty}\frac{3x+1}{5x-3}$

2. $\displaystyle\lim_{x\to+\infty}\frac{x-1}{2x+1}$

3. $\displaystyle\lim_{x\to-\infty}\frac{2x^2}{x^2+1}$

4. $\displaystyle\lim_{x\to-\infty}\frac{1}{x^2+1}$

5. $\displaystyle\lim_{x\to-\infty}\frac{x^2}{x+1}$

6. $\displaystyle\lim_{x\to-\infty}\frac{x+4}{3x^2-5}$

7. $\displaystyle\lim_{x\to+\infty}\frac{3x^2-2x+1}{2x+1}$

8. $\displaystyle\lim_{x\to+\infty}\frac{3x^2-2x+1}{2x^2+1}$

9. $\displaystyle\lim_{x\to+\infty}\frac{3x^3+25x^2-x+1}{4x^3-7x^2+2x+2}$

10. $\displaystyle\lim_{x\to+\infty}\frac{2x^5-3x^2+15}{3x^2+4x-5}$

In Exercises 11–20, find the horizontal and vertical asymptotes for the graph of f. Check your work by graphing the curve.

11. $f(x)=\dfrac{1}{x^2-4}$

12. $f(x)=\dfrac{1}{x^2-x}$

13. $f(x)=\dfrac{4x}{x^2-1}$

14. $f(x)=\dfrac{x^2}{4-x^2}$

15. $f(x)=\dfrac{x^2+3x+2}{x^2+2x-3}$

16. $f(x)=\dfrac{x^2-5x}{x^2-25}$

17. $f(x)=\ln(1+e^x)$

18. $f(x)=\dfrac{(x-1)^2(x+1)^2}{x^4}$

19. $f(x)=\dfrac{2}{\sqrt{x^2-4}}$

20. $f(x)=\dfrac{x}{\sqrt{x^2-9}}$

21.   Sketch the graph of a function f having all of the following properties:

i)   Domain of f is all real numbers;

ii)   Range of f is $(-\infty, 2]$;

iii) $\displaystyle\lim_{x\to-\infty}f(x)=-1$;

iv) $\displaystyle\lim_{x\to+\infty}f(x)=1$;

v) $\displaystyle\lim_{x\to2}f(x)$ does not exist.

Can you find a formula for such a function?

5.4   Using Calculus to Solve Optimization Problems

Finding a maximum or minimum is often an important part of problem solving. For example, manufacturers want to maximize their profits, contractors want to minimize their costs, and a physician would like to select the smallest dosage of a drug that will cure a disease.

The process of maximizing or minimizing is called **optimization.** Entire courses in mathematic are dedicated to this topic, and in this section we will develop procedures involving calculus to solve problems that seek a maximum or minimum value.

In Section 5.1, we learned that the absolute (global) extrema of a continuous function defined on a closed interval must occur at a critical point or at an endpoint of the domain of the function. Using this information we can develop a general strategy for solving optimization problems. The following guidelines are often helpful.

---

1. Read and understand the problem. Identify the given quantities and those you must find.

2. Sketch a diagram and label it appropriately, introducing variables for unknown quantities.

3. Decide which quantity is to be optimized and express this quantity as a function f of one or more other variables.

4. Using available information, express f as a function of just one variable.

5. Determine the domain of f and draw its graph.

6. Find the global extrema of f, considering any critical points and endpoints.

7. Convert the result obtained in Step 6 back into the context of the original problem. Be sure you have answered the question asked.

Example 1:   *Finding Maximum Volume*

An open box with a rectangular base is to be constructed from a rectangular piece of cardboard 16 inches wide and 21 inches long by cutting congruent squares from each corner and then bending up the sides. Find the size of a corner square that will produce an opened-top box with the largest possible volume.

Solution:

1. Read the problem more than once.

2. Sketch a diagram of the cardboard and the box, letting x represent the side of a square   and $21 - 2x$ and $16 - 2x$ represent the dimensions of the base.

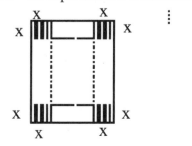
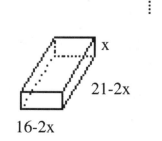

3. The quantity to be optimized is the volume V of the box, and

$$\text{Volume} = \text{length} \cdot \text{width} \cdot \text{height}$$

4. $V = x(16 - 2x)(21 - 2x) = 2(168x - 37x^2 + 2x^3)$.

5. Since $0 \le 2x \le 16$, the domain of V is $0 \le x \le 8$.

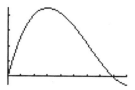

6. To find the critical numbers, differentiate the function V.

$$\frac{dV}{dx} = 2(168 - 74x + 6x^2) = 4(3x - 28)(x - 3).$$

The only critical number in the domain $[0, 8]$ is 3; the endpoints $x = 0$ and $x = 8$ yield the minimum value $V = 0$. For the critical number $x = 3$ we have $V(3) = 450$, which is the maximum value.

7. Thus, a 3-inch square should be cut from each corner of the cardboard in order to maximize the volume of the box.

There are a variety of ways in which a graphing calculator might be employed in the process of solving an optimization problem. The following example illustrates some of the many possible methods that could be used.

Example 2: *Finding the Minimum Time*

A hiker finds herself in a forest 2 miles from the nearest point on a straight road. She wants to wal to her car which is parked 10 miles down the road. If she can walk 4 mph along the road, but only mph through the forest, toward what point on the road should she walk in order to reach her car in the least time?

Solution:

Read the problem more than once. Sketch a diagram, letting x represent the distance of the aiming point P along the road.

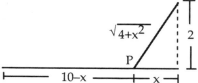

From the diagram, the distance to the point P is $\sqrt{x^2 + 4}$ and the distance from P to the car is $10 - x$. Thus the total time to reach the car is

$$T = \frac{\sqrt{4 + x^2}}{3} + \frac{10 - x}{4} \quad \text{with } 0 \le x \le 10.$$

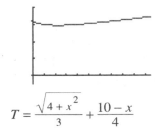

$$T = \frac{\sqrt{4 + x^2}}{3} + \frac{10 - x}{4}$$

We can find the zero of the derivative of T by using the **solve** command in the MATH menu (option 0), with some value in the domain of T, such as 5, as the guess for the solution. The critical number for T is approximately $x = 2.26779$.

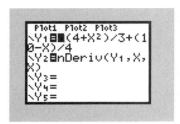
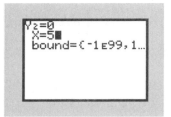
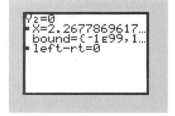

Thus, to find the minimum value of the continuous function T on the closed interval $[0, 10]$ we evaluate the function at the critical number, the endpoints, and then select the smallest T value as the minimum time. Using the Table Setup **Ask** option for the independent variable allows us to evaluate the function only at the x values we specify.

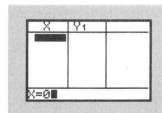
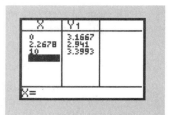

Therefore, the hiker should walk towards a point approximately 2.2678 miles down the road to reach her car in the minimum time of about 2.941 hours.

The critical number could also be found by graphing the derivative of T and using the **zero** command in the CALC menu (option 2). To verify the solution, use the **minimum** command in the CALC menu (option 3) with the graph of T above on the interval $[0, 10]$.

Another method for finding the minimum value is to compute the derivative of T and search for a zero of the derivative near $x = 2.26$. We will use Newton's Method, storing the derivative function in Y1 of the Y= menu.

$$\text{Y1} = \frac{x}{3\sqrt{4 + x^2}} - \frac{1}{4}.$$

On the Home screen, we store the initial guess 2 in X; then we enter and execute Newton's formula.

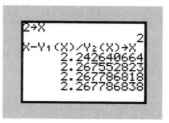

Since numerical differentiation can be inaccurate, we check the value of the derivative at $x = 2.267786838$ to confirm that we are approaching a zero.

$$T'(2.267786838) = -2.67\text{E}-12 \approx 0.$$

In conclusion, the hiker should walk toward a point approximately 2.27 miles down the road in order to reach her car in the least time.

Example 3:   *Finding the Largest Area*

Suppose you would like to construct a rain gutter from a sheet of metal which is 12 inches wide. One third of the metal's width on each side is bent up through an angle θ in order to form the gutter's side. How large should the angle θ be made in order to maximize the area of the trapezoidal cross section?

Solution:

Letting h denote the altitude of the trapezoid, and x denote a leg of the right triangle, we have
$\sin\theta = \dfrac{h}{4}$ and $\cos\theta = \dfrac{x}{4}$, so $h = 4\sin\theta$ and $x = 4\cos\theta$.

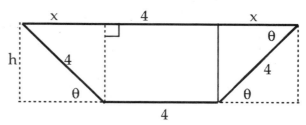

The area of the trapezoid, $A = \dfrac{1}{2}h(b_1 + b_2)$, is related to the angle θ by the equation,

$$A = \frac{1}{2}(4\sin\theta)[4 + (4 + 8\cos\theta)], \text{ with } 0 \le \theta \le \frac{\pi}{2}$$

$$= 2(\sin\theta)(8 + 8\cos\theta)$$

$$A = 16(\sin\theta)(1 + \cos\theta), \quad 0 \le \theta \le \frac{\pi}{2}.$$

Computing the derivative, we have

$$\frac{dA}{d\theta} = 16\left[-\sin^2\theta + \cos\theta + \cos^2\theta\right] = 16\left[2\cos^2\theta + \cos\theta - 1\right]$$

$$\frac{dA}{d\theta} = 16(2\cos\theta - 1)(\cos\theta + 1), \quad 0 \le \theta \le \frac{\pi}{2}.$$

The derivative $\frac{dA}{d\theta}$ is defined everywhere in the interval $[0, \frac{\pi}{2}]$ and is zero at $\theta = \frac{\pi}{3}$. Checking the value of A at the critical number $\frac{\pi}{3}$ and at the endpoints, we have

$$A = 0 \ \text{ at } \theta = 0,$$

$$A = 12\sqrt{3} \ \text{ at } \ \theta = \frac{\pi}{3},$$

$$A = 16 \ \text{ at } \ \theta = \frac{\pi}{2}.$$

Thus, the largest area, $12\sqrt{3}$, occurs when $\theta = \frac{\pi}{3}$.

5.4   **Exercises**

1.   Each side of a cardboard square is 12 inches. What is the maximum volume of a box that can be made from the square by cutting congruent squares from the corners and turning up the four sides?

2.   A long rectangular sheet of metal, 12 inches wide, is to be made into a rain gutter by turning up two sides so that they are perpendicular to the sheet. How many inches should be turned up to give the gutter its greatest capacity?

3.   The sum of the two bases and the altitude of a trapezoid is 16 ft.

   a) Define the area A of the trapezoid as function of its altitude.

   b) Find the altitude for which the trapezoid has the largest possible area.

4.   A rectangle has its base on the x-axis and its upper vertices on the parabola $y = 6 - x^2$, as shown in the figure. Find the maximum possible area of the rectangle.

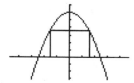

5.   A rectangle is inscribed between the graphs of $y = 4 - x^2$ and $y = \frac{1}{4}x^4 - 1$.

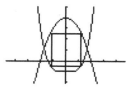

   Find the width of the rectangle that has the largest area.

6.   Find the maximum distance, measured vertically, between the graphs of $y = x$ and $y = x^{1/3}$ for $0 \le x \le 1$.

7.   A cardboard box of 108 in^3 volume with a square base and no top is to be constructed. Find the minimum area of the cardboard needed.

8.   Mr. Phillips decides to have a completely fenced-in rectangular garden. It will be laid out so that one side is adjacent to his neighbor's property. The neighbor agrees to pay for half of that part of the fence which will border his property. The garden is to contain 432 square yards. What dimensions should Mr. Phillips select for his garden so that his own cost will be a minimum?

9.   A rectangle is inscribed in a right triangle with sides 6, 8, and 10. Two sides of the rectangle lie along legs of the triangle, and the remaining vertex of the rectangle is on the hypotenuse of the triangle. What dimensions for the rectangle will maximize its area?

10.   What is the x-coordinate of the point on the curve $y = e^x$ that is closest to the origin? Give your result accurate to the nearest thousandth.

11.   A rectangle is inscribed under one "bump" of the cosine curve, with two vertices on the x-axis and the other two vertices on the curve $y = \cos x$ where $-\dfrac{\pi}{2} \le x \le \dfrac{\pi}{2}$.

       a) What is the area of the rectangle, given in terms of the x-coordinate of one of the vertices on the x-axis?

       b) What is the maximum possible area of the rectangle?

12.   A boater finds herself 4 miles from the nearest point to a straight shore line which is 6 miles from a shoreside motel. She plans to row to shore and then walk to the motel. If she can walk at 3 mph, what speed is the minimum she must be able to row so that the quickest way to get to the motel is to row directly?

13.   A rectangle is drawn with sides parallel to the coordinate axes and with its upper two vertices on the parabola $f(x) = 4 - x^2$ and its lower two vertices on the parabola $g(x) = \dfrac{x^2}{2} - 2$. What is the maximum possible area of the rectangle?

14.   An isosceles triangle is to be drawn with its two equal sides of length 10. What is the maximum possible area for the triangle? [Note: The area of a triangle is given by half the product of two sides with the sine of the included angle.]

15.   Let $A(x)$ be the area of the rectangle inscribed under the curve $y = e^{-2x^2}$ with vertices at $(-x, 0)$ and $(x, 0)$, where $x \ge 0$.

       a) Determine $A(1)$.

       b) Determine the greatest possible value for $A(x)$. Justify your answer.

16.   A particle moves along a line so that at any time t its position is given by $x(t) = 2\pi t + \cos(2\pi t)$. Determine the particle's maximum velocity.

17.   Estimate the shortest distance from the point $(5, 3)$ to the nearest point on the graph of $y = \sin x - \sin(x^2)$, $0 \le x \le 10$.

18.   A company is designing shipping crates and wants the volume of each crate to be 4 cubic feet, and the crate's base to be a square between 1 and 1.5 feet per side. The material for the bottom costs $5, the sides $3 and the top $1 per square foot. What dimensions will give the minimum cost?

5.5   The Mean Value Theorem

     The derivative was used in the previous section to help locate maximum and minimum values. We have learned that many other properties can be deduced from a close examination of the derivative. The justification for most of these conclusions is an interesting theorem, the Mean Value Theorem, which is stated below.

The Mean Value Theorem

If the function f is continuous on the closed interval $[a, b]$ and differentiable on the open interval (a, b), then there exists at least one number c in the open interval (a, b) such that

$$f'(c) = \frac{f(b) - f(a)}{b - a}$$

     We can see that this theorem is reasonable by interpreting f' as a slope. The figure below shows the points $(a, f(a))$ and $(b, f(b))$ on the graph of a differentiable function f.

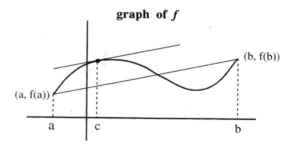

graph of f

     Each side of the equation in the Mean Value Theorem represents a slope. The quotient $\dfrac{f(b) - f(a)}{b - a}$ is the slope of the secant segment connecting the endpoints $(a, f(a))$ and $(b, f(b))$ and $f'(c)$ is the slope of the tangent line at $x = c$ The theorem asserts that, for at least one number c between a and b, these two slopes are equal (the tangent line and the secant line are parallel). In the case shown graphically above, one such c has been identified. Is there another number between a and b where the tangent line and secant line are parallel?

     Interpreting f' as an instantaneous rate we see that if an object moves in a straight line with position function $s = f(t)$, then the average velocity between $t = a$ and $t = b$ is $\dfrac{f(b) - f(a)}{b - a}$ and the velocity at $t = c$ is $f'(c)$. Thus, the Mean Value Theorem says that at some time $t = c$ between a and b the instantaneous velocity $f'(c)$ is equal to the average velocity. For example, if a car averages 50 mph over an interval of time, then, at some instant the speedometer must read exactly 50 mph.

     Up to now we have stressed but not proven that the sign of f' determines whether f increases or decreases over an interval. We can now employ the Mean Value Theorem to prove the following theorem.

Theorem   If $f'(x) > 0$ for all x in an interval I, then f is increasing on I.

Proof: We must show that, for any a and b in I with $a < b$, $f(a) < f(b)$.

Let a and b be any two numbers in the interval I. Then f is continuous on $[a, b]$ and differentiable on (a, b). Applying the Mean Value Theorem, there is some number c in (a, b) such that

$$f'(c) = \frac{f(b) - f(a)}{b - a} \quad \text{or} \quad f(b) - f(a) = f'(c) \cdot (b - a).$$

Since both factors on the right side are positive, it follows that
$$f(b) - f(a) > 0 \quad \text{and} \quad f(b) > f(a).$$

The Mean Value Theorem is an existence theorem. It tells us the number c exists, without telling us how to find it. The next example illustrates how to locate the number guaranteed by the Mean Value Theorem.

Example 1: *Applying the Mean Value Theorem*

If the function f is defined on $[1, 3]$ by $f(x) = 4 - \dfrac{3}{x}$, show that the Mean Value Theorem can be applied to f and find a number c which satisfies the conclusion.

Solution:

$f'(x) = \dfrac{3}{x^2}$ so that f is both continuous and differentiable on $[1, 3]$.

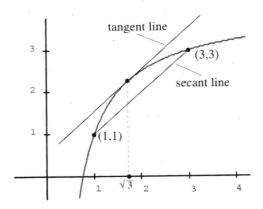

The slope of the secant line through $(1, f(1))$ and $(3, f(3))$ is

$$\frac{f(3) - f(1)}{3 - 1} = \frac{3 - 1}{3 - 1} = 1.$$

With $f'(c) = 1$, we have $\dfrac{3}{c^2} = 1$, $c^2 = 3$ and $c = \pm\sqrt{3}$. Thus, in the interval $(1, 3)$ we have $c = \sqrt{3}$.

5.5   **Exercises**

1.   Use the graph of f to estimate the numbers
      in [0, 8] that satisfy the conclusion of the
      Mean Value Theorem.

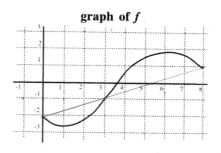

graph of f

2.   Use the graph of g to estimate the numbers
      in [0, 8] that satisfy the conclusion of the
      Mean Value Theorem.

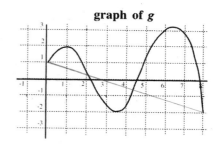

graph of g

In Exercises 3–10, determine whether f satisfies the hypotheses of the Mean Value Theorem on
the interval [a, b]. If it does, find all numbers c in (a, b) such that $f'(c) = \dfrac{f(b)-f(a)}{b-a}$.

3. $f(x) = 3x^2 + x - 4$;   [1, 5]   4. $f(x) = x^{2/3}$;   [−1, 2]

5. $f(x) = \cos x - \sin x$; $[\frac{\pi}{4}, \frac{5}{4}\pi]$ 6. $f(x) = 1 - |x|$;   [−1, 2]

7. $f(x) = \dfrac{x-1}{x+1}$;   [0, 3]   8. $f(x) = x^{1/3}$;   [0, 6]

9. $f(x) = \sqrt{x}$;   [4, 9]   10. $f(x) = 2\sin x + \sin(2x)$;   [0, π]

11.   Sketch a graph of the function f if $f(x) = \begin{cases} x+2, & \text{for } x \le 1; \\ x^2, & \text{for } x > 1. \end{cases}$

      Show that f fails to satisfy the hypothesis of the Mean Value Theorem on the interval
      [−2, 2] but the conclusion of the theorem is still valid.

12.   Suppose that $s(t) = t^2 - t + 4$ is the position function of the motion of a particle moving in
      a straight line.

      a) Explain why the function s satisfies the hypothesis of the Mean Value Theorem.

      b) Find the value of t in [0, 3] where the instantaneous velocity is equal to the average
      velocity.

5.6   Antiderivatives

Finding derivative functions is one skill we need to develop. However, there are many instances in which we need to solve the opposite problem. Given the function f, determine a function F whose derivative is f. That is, $F' = f$. If $F' = f$, then the function F is called an **antiderivative** of f.

A scientist who knows the acceleration of an object may want to determine its velocity or its position at a particular time. A coast guard official who knows the rate at which oil is leaking out of a grounded vessel might want to know the actual amount of oil that has spilled. In each of these examples, a derivative f is given and the problem is to find an antiderivative F, so that $F' = f$.

For example, if $f(x) = 2x$ and $F(x) = x^2 + 4$, then F is an antiderivative of f, since $F' = f$. Also, if $G(x) = x^2 + 2$, then $G'(x) = 2x$, so that G is another antiderivative of f. In fact, any function F with $F(x) = x^2 + C$, and C a constant, is an antiderivative of the function defined by $f(x) = 2x$.

Example 1:   *Finding an Antiderivative*

Find at least two antiderivatives of the function $f(x) = x^3 + x$.

Solution:

By reversing the differentiation rule for polynomials we find the function F defined by $F(x) = \frac{1}{4}x^4 + \frac{1}{2}x^2$ is one antiderivative. And, since the derivative of a constant function is zero, it follows that adding a constant to a function does not change its derivative; hence, the function G defined by $G(x) = \frac{1}{4}x^4 + \frac{1}{2}x^2 + C$ with C any real constant is also an antiderivative of f.

The above examples show that if the function F is an antiderivative of the function f, then so is $F + C$ for any constant C.

If two functions F and G have the same derivative, then the tangent lines to their graphs have the same slope at each input x. Thus the graphs have the same shape, and the equation $G(x) = F(x) + C$ holds because each graph is obtained from the other by shifting it up or down.

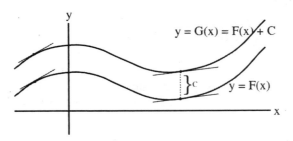

It follows that if a function f has an antiderivative it will have a whole family of them and each member of the family can be obtained from one of them by the addition of an appropriate constant. The family of solutions is called the **general antiderivative of** f.

For example, given the squaring function $f(x) = x^2$, we see that the general antiderivative of f is $F(x) = \dfrac{x^3}{3} + C$. By assigning different values to the constant C we obtain a family of curves whose graphs are vertical translates of each other.

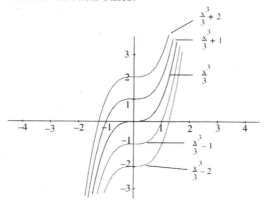

The differentiation rules of the last chapter provide formulas for antiderivatives. The table below lists some basic antiderivatives. We use the notation $F' = f$ and $G' = g$. Each formula in the table is true because the derivative of the function in the right column appears in the left column. Note that to anti-differentiate a power of x, you must increase the exponent by 1 and divide by the new exponent.

Function	Antiderivative
$kf(x)$	$kF(x) + C$
$f(x) + g(x)$	$F(x) + G(x) + C$
$x^n \ (n \neq -1)$	$\dfrac{x^{n+1}}{n+1} + C$
$\cos x$	$\sin x + C$
$\sin x$	$-\cos x + C$
$\sec^2 x$	$\tan x + C$
e^x	$e^x + C$

Example 2: *Finding Antiderivatives*

Find the antiderivatives of the following functions.

$$\text{a) } f(x) = x^2 - 4x \qquad\qquad \text{b) } g(x) = 4\sin x + \sqrt{x}$$

Solution:

Using the formulas in the table we obtain

$$\text{a) } F(x) = \frac{x^3}{3} - 4\left(\frac{x^2}{2}\right) + C = \frac{x^3}{3} - 2x^2 + C$$

$$\text{b) } G(x) = 4(-\cos x) + \frac{2}{3}x^{3/2} + C = -4\cos x + \frac{2}{3}x^{3/2} + C.$$

Applications

Antidifferentiation is useful in analyzing the motion of an object moving in a straight line. Recall that if an object has position function $s = f(t)$, then the velocity function is $v(t) = \dfrac{ds}{dt}$. This means that the position function is an antiderivative of the velocity function. Similarly, the acceleration function is $a(t) = \dfrac{dv}{dt}$, so the velocity function is an antiderivative of the acceleration function. Thus if the acceleration function and initial values $s(0)$ and $v(0)$ are known, then the position function can be found by antidifferentiating.

Example 3:   *Modeling the Motion of a Car*

The brakes on a certain car can produce a constant deceleration of 22 ft/sec^2. If the car is traveling at 60 mph (88 ft/sec) when the brakes are applied, how far will it travel before coming to a stop?

Solution:

Let $s(t)$, $v(t)$ and $a(t)$ represent the position, velocity and acceleration of the car t seconds after the brakes are applied. Since $v'(t) = -22$, antidifferentiation gives

$$v(t) = -22t + C.$$

Note that $v(0) = C$. But we are given that $v(0) = 88$, so $C = 88$ and

$$v(t) = -22t + 88.$$

Since $v(t) = s'(t)$, s is an antiderivative of v and antidifferentiation gives

$$s(t) = -11t^2 + 88t + C.$$

Assuming that s is measured from the point where the brakes are applied we have $s(0) = 0$, so $0 = -11(0) + 88(0) + C$ and $C = 0$. Thus,

$$s(t) = -11t^2 + 88t.$$

Finally, the car comes to a stop when its velocity is 0, so solving $v(t) = 0$ for t,

$$0 = -22t + 88$$

$$t = 4 \text{ seconds.}$$

This means that the car decelerates for 4 seconds before coming to a stop, and therefore it travels

$$s(4) = 176 \text{ feet.}$$

In the next example we employ the fact that an object near the surface of the earth is acted upon by the force of gravity that is 9.8 m/sec^2 or 32 ft/sec^2.

Example 4:   *Modeling the Motion of a Ball Thrown Vertically Upward*

A ball is thrown vertically upward from a tall building 154 feet above the ground with an initial velocity of 64 feet/sec. When does it reach its maximum height? When does it hit the ground?

Solution:

The motion is vertical and we assume the positive direction to be upward. At time t the distance above the ground is $s(t)$ and the velocity $v(t)$ is decreasing. Thus acceleration is negative and we have

$$a(t) = -32.$$

Taking antiderivatives, we have

$$v(t) = -32t + C$$

But since $v(0) = 64$, we have $64 = -32(0) + C$, so $C = 64$ and

$$v(t) = -32t + 64.$$

The maximum height is reached when $v(t) = 0$, that is, after 2 seconds. Since $s'(t) = v(t)$, we antidifferentiate again and obtain

$$s(t) = -16t^2 + 64t + C.$$

Since $s(0) = 154$, we have $154 = -16(0)^2 + 64(0) + C$, so $C = 154$ and

$$s(t) = -16t^2 + 64t + 154.$$

The ball will strike the ground when $s(t) = 0$, that is, when

$$0 = -16t^2 + 64t + 154.$$

Using the Quadratic Formula to solve this equation, we get

$$t \approx 5.691 \text{ seconds.}$$

In Chapter 2, we used differentiation to compute the slope at each point on the graph of a function. That procedure is reversed in the next example.

Example 5:   *Finding an Antiderivative*

The graph of a function F has slope $2x^3 - 4$ at each point (x, y) and contains the point $(2, 1)$. Find the function F.

Solution:

Because the slope of the tangent line at each point (x, y) is given by the derivative $F'(x)$, we have

$$F'(x) = 2x^3 - 4.$$

Antidifferentiating, we get

$$F(x) = \frac{1}{2}x^4 - 4x + C.$$

Using the fact that $F(2) = 1$, we have

$$1 = \frac{1}{2}(2)^4 - 4(2) + C, \text{ so } C = 1 \text{ and}$$

$$F(x) = \frac{1}{2}x^4 - 4x + 1.$$

Families of Curves

From our previous work we know the graph of $y = ax^2$ is a parabola with its vertex at the origin and that the coefficient a determines the shape of the parabola. In particular, the smaller the magnitude of a, the wider the parabola. Also, if a is positive the parabola opens upward and when a is negative the parabola opens downward.

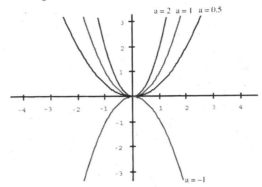

The family of curves defined by $y = ax^2$

We say that all functions of the form $y = ax^2$ form a **family of curves**; their graphs are similar to $y = x^2$, except for stretches or compressions determined by the value of a. The constant a is called a **parameter**. The word parameter is used to denote a constant in the formula for a function when specification of the constant completely determines the function. Since different values of the parameter give rise to different members of the family it is useful to understand how the graph of the function changes as the parameter changes. Often the first step in modeling or curve fitting is recognizing the family of functions that might fit the data. The next step is to determine the values of the parameter.

Example 6: *Investigating a Family of Curves*

Investigate the family of curves given by $f(x) = x^2 + bx + 1$. What features do the members of the family have in common? How do they differ?

Solution:

This is a family of parabolas with $\dfrac{dy}{dx} = 2x + b$ and $\dfrac{d^2y}{dx^2} = 2$.

The formula for $\dfrac{dy}{dx}$ is a linear function, so there is only one critical number at $x = -\dfrac{b}{2}$. The second derivative is the positive constant 2, so the graph is always concave up and the point $(-\dfrac{b}{2}, f(-\dfrac{b}{2}))$ is a relative minimum point. The effect of varying the parameter b is to move the vertex of the parabola. The figure below shows several curves in the family on the same axes.

$Y_1 = x^2 + \{0,1,2,3,4,5,6\}x + 1$

5.6   Exercises

In Exercises 1–10, find the general antiderivative of the given function.

1. $f(x) = 4$ 2. $f(x) = 2x + 3$

3. $f(x) = 3x^2 + 2x + 1$ 4. $f(x) = 3x^4 + 5x + 6$

5. $f(x) = e^x + 2x^2$ 6. $f(t) = \sin t + \sqrt{t}$

7. $f(x) = x^3 - \dfrac{1}{x^2}$ 8. $f(x) = \sin x - 2\cos x$

9. $f(x) = \sec^2 x + 1$ 10. $f(x) = \sqrt{x} - \dfrac{1}{\sqrt{x}}$

11.   Find three different antiderivatives for the function $f(x) = x^3 - 2x^2$.

12.   Find a quadratic function f whose derivative f' is defined by $f'(x) = 5x + 7$.

13.   The graph of the function g passes through the point $(1, -2)$ and $g'(x) = 2x - 3$. Find such a function g.

14.   The graph of the function g passes through the point $(0, 1)$ and $g'(x) = 3 - 2\sin x$. Find such a function g.

15.   A stone is thrown vertically upward from a position 144 feet above the ground with an initial velocity of 96 ft/sec. Find

     a) the stone's distance above the ground after t seconds;

     b) the length of time the stone rises;

     c) when and with what velocity the stone strikes the ground.

16.   A stone is thrown directly downward from a height of 96 feet with a velocity of 16 ft/sec. Find

     a) the stone's height above the ground after t seconds;

     b) when it strikes the ground;

     c) the velocity at which it strikes the ground.

17.   A car is traveling at a speed of 60 mph, what constant (negative) acceleration will enable it to stop in 9 seconds?

18.   With what initial velocity must an object be thrown upward (from ground level) to reach a maximum height of 50 feet.

19.   A train is moving at the rate of 45 mph (66 ft/sec) when the engineer sees a truck stalled at the railroad crossing, blocking the tracks 275 feet from the train. The train's brakes are fully applied, producing a constant deceleration of 8 ft/sec^2. Will the train be able to stop in time to avoid hitting the truck?

In Exercises 20–24 investigate the family of curves. Explain the effect that the parameter has on each curve and confirm your conclusions graphically.

20. $y = x^2 + ax$ for $a = -2, -1, 0, 1, 2$.

21. $y = x^3 - 3ax$ for $a = -2, -1, 0, 1, 2$.

22. $y = x^3 - 3ax^2$ for $a = -1, 0, 1, 2, 3$

23. $y = e^{-\frac{x^2}{a}}$

24. $y = x + \dfrac{a}{x}$

25.   a) Use your calculator to plot $y = x^2 + bx + 1$ for at least 5 positive values of b and five negative values of b.

        b) The vertices of the family of graphs in part a) appear to lie on a single parabola. Find a quadratic function that goes through all their vertices and plot it along with the family.

26.   Consider the function $f(x) = x - k\sqrt{x}$, where k is a positive constant and $x \geq 0$.

        a) Find all critical points of f.

        b) On the same axes, sketch and label the graphs of f for three positive values of k.

        c) What effect does increasing the value of k have on the position of the local minimum points?

27.   Consider the function $f(x) = e^x - kx$, where k is a positive constant and $x \geq 0$.

        a) Find all critical points of f.

        b) On the same axes, sketch and label the graphs of f for three positive values of k.

        c) What effect does increasing the value of k have on the position of the local minimum points?

28.   Consider the function f given by $f(x) = \dfrac{ax}{x^2 + 1}$.

        a) Find all critical points of f.

        b) On the same axes, sketch and label the graphs of f for three positive values of a.

        c) What effect does increasing the value of a have on the position of the local minimum and local maximum points?

Chapter 5   Supplementary Problems

1.   a) Determine the critical numbers of the function f defined by $f(x) = xe^{-2x}$

   b) Determine an equation of the line tangent to the graph of $x^2 + xy + y^3 = 3$ at the
      point $(-2, 1)$.

2.   a) The function f has a derivative defined by the rule $f'(x) = x^3 - 4.5x^2 - 12x - 5$.
      Determine the intervals on which the graph of the original function f is concave up.

   b) Determine the value of k so that $f(x) = x^3 - kx^2 + 2$ will have a point of inflection
      at $x = 1$.

3.   The graph of f', the derivative of a function f, is shown below on the closed interval
      $[-3, 5]$. The graph of f' has horizontal tangents at $x = -1$ and 3.

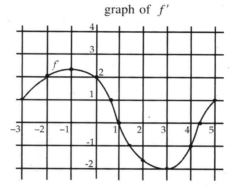

graph of f'

   a) What are the critical numbers of f ?

   b) Determine the x-coordinate(s) of any local minimum point(s) of f.

   c) On what interval is f both increasing and concave down?

   d) Let another function g be defined by $g(x) = x^2 - 3x - 1$. With this information
      and the graph of f above, determine the value of the derivative of the composition
      $y = f(g(x))$ at $x = 3$.

4.   An open-topped box with a square base is to be constructed from 192 square inches of
      material. What should the dimensions of the box be in order to obtain a maximum volume?

5.   Let $f(x) = x^3 + px^2 + qx$.

   a) Find the values of p and q so that $f(-1) = -8$ and $f'(-1) = 12$.

   b) Find the value of p so that the graph of f changes concavity at $x = 2$.

   c) Under what conditions on p and q will the graph of f be increasing everywhere?

6.   Let g be the function defined so that it has derivatives

$$g'(x) = \frac{12(1-x^4)}{(3+x^4)^2} \quad \text{and} \quad g''(x) = \frac{-48x^3(5-x^4)}{(3+x^4)^3}$$

a) Show that g' is an even function.

b) Give the x-coordinates of all local maximum and minimum points for g. Show your reasoning.

c) If $g(1) = 1$, find an equation of the line normal to the graph of g at the point $(1, 1)$.

d) Find the x-coordinates of all inflection points of g. Show your reasoning.

7.   Two points, A and B, are 275 ft apart. At a given instant, a balloon is released at B and rises vertically at a constant rate of 2.5 ft/sec, and, at the same instant, a cat starts running from A to B at a constant rate of 5 ft/sec.

a) After 40 seconds, is the distance between the cat and the balloon decreasing or increasing? At what rate?

b) Describe what is happening to the distance between the cat and the balloon at $t = 50$ seconds.

8.   If $f(x) = 4x^3 - 21x^2 + 36x - 4$, find the interval(s) on which the graph of f is both decreasing and concave down.

9.   Find the coordinates of the absolute maximum point on the curve $y = xe^{-kx}$, where k is a fixed positive number.

10. .   The function h is defined by $h(x) = f[g(x)]$, where f and g are functions whose graphs are shown below.

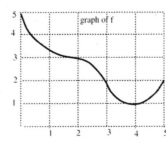
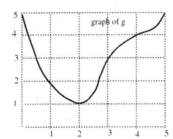

a) Evaluate $h(2)$.

b) Estimate $h'(1)$.

c) Is the graph of the composite function h increasing or decreasing at $x = 3$? Show your reasoning.

d) Find all inputs x for which the graph of h has horizontal tangents. Show your reasoning.

11.   The graph of the derivative of the function f is defined on the interval $[-4, 4]$. The graph of f' has horizontal tangents at $x = -3, -1$ and 2.

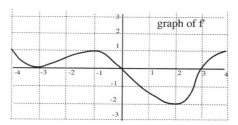

a) Suppose $f(-4) = 0$. Find an equation for the tangent line to the graph of f at $(-4, 0)$.

b) Where does the graph of f have a local minimum? Justify briefly.

c) On what intervals, if any, is the graph of f concave down?

d) If $f(-4) = 0$, sketch a possible graph of f.

12.   The graph of $y = x^2(x - k)$ for a positive constant k looks something like the picture below.

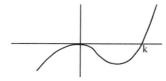

a) Find the coordinates of the local minimum point in terms of k.

b) If the constant k is a negative real constant what might the graph of f look like?

13.   Let c and d be positive constants. Let the function f be defined for all positive numbers by

$$f(x) = (c + x)(d + \frac{1}{x}).$$

a) Determine, in terms of c and d, the critical numbers of f.

b) Does f have a local maximum or minimum at its critical numbers? Justify briefly.

14.   Let f be the function defined so that it has derivatives

$$f'(x) = \frac{1 - x^2}{(1 + x^2)^2} \quad \text{and} \quad f''(x) = \frac{2x(x^2 - 3)}{(1 + x^2)^3}.$$

a) Show that the derivative is an even function.

b) Give the x-coordinates of all local maximum and minimum points of f. Show your reasoning.

c) Find the x-coordinates of all inflection points of f. Show your reasoning.

d) If $f(0) = 0$, sketch a possible graph of f.

15.   The following is the graph of the derivative f'. The domain of f is the interval $[-3, 4]$.

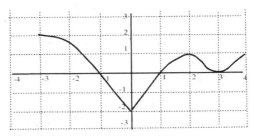

      a) Determine the critical numbers of f.

      b) On what interval(s) is f decreasing?

      c) At which x-coordinates does f have a local maximum? Justify your answer.

      d) On which interval(s) is f concave down?

      e) If $f(-1) = 2$, $f(1) = -1$, and $f(3) = 0$, sketch a possible graph of f.

16.   Given an equilateral triangle with all three sides of length $2s$, as drawn. A rectangle is inscribed in this triangle so that one side of the rectangle lies along a side of the triangle. Determine the area of the largest such rectangle. Your answer will be in terms of the positive constant s.

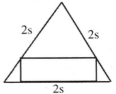

17.   An eight foot fence is located 1 foot from the base of a wall of a building. An extension ladder is to be placed to lean against the wall in such a way that the ladder rests on the fence. Find the length of the shortest ladder that can be used.

18.   Show that the graph of $f(x) = \dfrac{x+1}{x^2 + 1}$ has three inflection points and that these points are collinear.

19.   Find a and b so that the graph of $y = x^3 + ax^2 - 3x + b$ has an inflection point at $(1, 1)$.

20.   Determine the maximum and minimum values of the function f on the interval $[0, 13]$ if $f(x) = \sqrt{x} \cdot (x - 5)^{1/3}$.

21.   A tank with a rectangular base and rectangular sides is to be open at the top. It is to be constructed so that its width is 4 meters and its volume is 36 cubic meters. If building the tank costs \$10 per square meter for the base and \$5 per square meter for the sides, what is the cost of the least expensive tank?

22.   Let f be the function defined by $f(x) = 2\ln(x^2 + 3) - x$ with the domain $-3 \le x \le 5$.

   a) Use your calculator to graph f on the
      viewing window [-3, 5] by [0, 8].
      Carefully reproduce this graph on the
      graph window to the right.

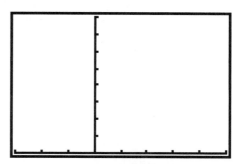

   b) Find the exact x-coordinate of each relative maximum point and each relative minimum
      point of f. Justify your answer by using the first derivative test.

   c) Find the exact x-coordinate of each inflection point of f.

   d) Find the absolute maximum value of $f(x)$ exactly.

23.   Let f and g be functions defined by the graphs below. For each function, the domain
      is the set of real numbers where $-3 \le x \le 4$. The graph of f has a horizontal tangent at
      $x = 1$; the graph of g has horizontal tangents at $x = -1$ and $x = 3$.

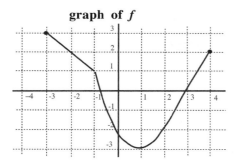

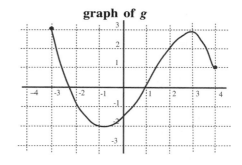

   Furthermore, let h be a function defined by $h(x) = f[g(x)]$. With this information,

   a) evaluate $h(-2)$ and $h(1)$;

   b) approximate $h'(-1)$ and $h'(0)$;

   c) determine what values of x correspond to critical points of h.

24.   Consider the family of curves $f(x) = \dfrac{1}{x^n}$, where n is a positive integer.

   a) Graph the function for several values of the parameter n.

   b) How does the graph change when n changes ?

25.   Suppose f is a continuous and differentiable function on the interval $[0, 1]$ and $g(x) = f(3x)$. The table below gives some values of f.

x	0.1	0.2	0.3	0.4	0.5	0.6
$f(x)$	1.01	1.042	1.180	1.298	1.486	1.573

What is the approximate value of $g'(0.1)$? Show your reasoning.

26.   The graph of the function g passes through the point $(1, 2)$ and $g'(x) = 2x - 3$. Find such a function g.

27.   A particle moves along the x-axis in such a way that its acceleration at time $t \geq 0$ is given by $a(t) = 12t - 6$. When $t = 0$ the position of the particle is 0 and the velocity is -12.

a) Write an equation for the velocity, $v(t)$, of the particle for all $t \geq 0$.

b) Find the value of t for which the particle is at rest.

c) Write an equation for the position, $s(t)$, of the particle for all $t \geq 0$.

28.   Consider the family of curves $f(x) = x^4 + bx^2 + x$.

a) Graph the function for several values of the parameter b.

b) How does the graph change when b changes ?

29.   Two particles move along the x-axis and their positions at time $0 \leq t \leq 2\pi$ are given by $x_1 = \cos t$ and $x_2 = e^{(t-2)/3} - 0.75$. For what values of t do the two particles have the same velocity?

30. · Find an equation of the tangent line to the curve $y^2 = 5x^4 - x^2$ at the point $(1, 2)$.

31.   A conical drinking cup is made from a circular piece of paper of radius 8 centimeters by cutting out a sector and joining the edges CA and CB. Find the maximum capacity of such a cup.

32. Let f be a function defined on the closed interval $-6 \leq x \leq 6$ with $f(0) = 5$. The graph of f', the derivative of f, consists of three line segments and a semicircle .

graph of f'

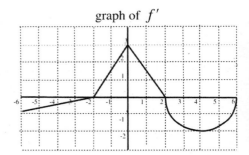

a) Write an equation for the tangent line to the graph of f at $x=0$.

b) Find all values of x on the interval $-6 \leq x \leq 6$ at which f has a relative maximum.

c) For $-6 \leq x \leq 6$, find all values of x at which the graph of f has a point of inflection. Justify.

d) Find the absolute maximum value of $f(x)$ over the closed interval $-6 \leq x \leq 6$. Explain your reasoning.

CHAPTER 6
Integrals

6.1   The Definite Integral Again

In Chapter 3, distance traveled over a given time interval was determined by finding the area under the graph of the velocity function. For example, if $v(t)$ is the velocity of a car at time t, then the net distance covered by the car over the interval $1 \leq t \leq 3$ hours corresponds to the definite integral $\int_1^3 v(t)\, dt$.

Recall that a definite integral is defined to be the limiting value of Riemann sums

$$\int_a^b f(x)\, dx = \lim_{n \to \infty} \left[f(x_1) \cdot \Delta x + f(x_2) \cdot \Delta x + \ \ldots\ + f(x_n) \cdot \Delta x \right]$$

where the interval $[a, b]$ is subdivided into equal subintervals of length Δx, one input x_n is chosen from each subinterval, and we sum the products $f(x_n) \cdot \Delta x$ for all subintervals.

A definite integral can be evaluated numerically or by the Fundamental Theorem. For a numerical approximation of an integral, we can evaluate a Riemann sum for a particular subdivision size and choice of input from that subdivision. The left- and right-hand sums introduced in Chapter 3 are special types of Riemann sums.

On the other hand, if F is an antiderivative of the continuous function f, that is $F' = f$, then the Fundamental Theorem of Calculus states that the exact value of the integral is given by

$$\int_a^b f(x)\, dx = F(b) - F(a).$$

Example 1: *Evaluating a Definite Integral*

The function $F(x) = x \ln x - x$ has the derivative $F'(x) = f(x) = \ln x$.

Compute $\int_1^2 \ln x\ dx$ using

a) left- and right-hand Riemann sums with 50 subdivisions;

b) the Fundamental Theorem of Calculus.

Solution:

a) With $n = 50$, the left-sum is approximately 0.379 and the right-sum is approximately 0.393. Since $F(x) = x \ln x - x$ is monotone increasing on $[1, 2]$, we have

$$0.379 < \int_1^2 \ln x\ dx < 0.393$$

b) According to the Fundamental Theorem of Calculus $\int_a^b F'(x)\, dx = F(b) - F(a)$.

So with $F(x) = x \ln x - x$ and $F'(x) = \ln x$ we have

$$\int_1^2 \ln x \, dx \; = F(2) - F(1) = (2 \cdot \ln 2 - 2) - (1 \cdot \ln 1 - 1) = 2\ln 2 - 1, \text{ or}$$

$$\int_1^2 \ln x \, dx \; \approx 0.386.$$

This result can be interpreted as the change in $F(x) = x \ln x - x$ from $x = 1$ to $x = 2$ and as the area under the curve defined by $F'(x) = \ln x$ from $x = 1$ to $x = 2$.

Rules for Definite Integrals

Just as there are rules that tell us how to find the derivative of various combinations of functions, there are integral rules that tell us how to find the integral. Each integral rule follows from its interpretation as area. Here are two that are analogous to differentiation rules.

If f and g are any continuous functions on the interval $[a,b]$ and c is any constant, then

1) $\displaystyle\int_a^b c \cdot f(x) \, dx = c \int_a^b f(x) \, dx$ (Constant Multiple Rule)

2) $\displaystyle\int_a^b [f(x) \pm g(x)] \, dx = \int_a^b f(x) \, dx \pm \int_a^b g(x) \, dx$ (Sum and Difference Rule)

The following picture illustrates the constant multiple rule for integrals. Multiplying a positive-valued function by a positive constant c results in a stretching of all the outputs of f by a factor of c, which results in an increase in the area by a factor of c. For example, if $f(x) = x$ and $c = 2$, then we double outputs and hence double the area.

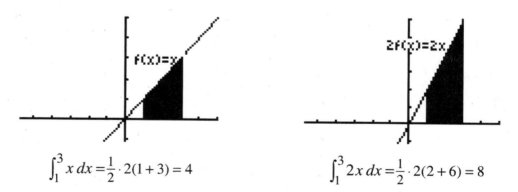

$$\int_1^3 x \, dx = \frac{1}{2} \cdot 2(1+3) = 4 \qquad\qquad \int_1^3 2x \, dx = \frac{1}{2} \cdot 2(2+6) = 8$$

Here is an example where we can use both integral rules. **Bracket notation** is also introduced as a convenient shorthand for calculating integrals. For any function f,

$$[f(x)]_a^b \quad \text{means} \quad f(b) - f(a).$$

Example 2: *Integrating a Polynomial Function*

Evaluate $\int_1^3 (2x^2 + 3x + 1)\, dx$.

Solution:

$$\int_1^3 (2x^2 + 3x + 1)\, dx = 2\int_1^3 x^2\, dx + 3\int_1^3 x\, dx + \int_1^3 1\, dx$$

$$= 2 \cdot \left[\frac{x^3}{3}\right]_1^3 + 3 \cdot \left[\frac{x^2}{2}\right]_1^3 + [x]_1^3$$

$$= 2 \cdot \left[\frac{3^3 - 1^3}{3}\right] + 3 \cdot \left[\frac{3^2 - 1^2}{2}\right] + [3 - 1]$$

$$= \frac{94}{3}$$

There are two additional properties that are unique to integrals. If you visualize an integral as area, it is clear why these properties hold. The first says that the area of a region is the sum of its parts. The following picture illustrates the

Additivity of Integrals over Intervals

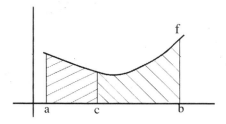

Additive Interval Rule

If $a < c < b$, then $\int_a^b f(x)\, dx = \int_a^c f(x)\, dx + \int_c^b f(x)\, dx$.

Example 3: *Using the Additivity of the Definite Integral*

If $\int_{-2}^1 f(x)\, dx = 4$ and $\int_{-2}^6 f(x)\, dx = -3$, what is $\int_1^6 f(x)\, dx$?

Solution:

From the additivity of integrals, we have

$$\int_{-2}^6 f(x)\, dx = \int_{-2}^1 f(x)\, dx + \int_1^6 f(x)\, dx \text{ ; thus}$$

$$\int_1^6 f(x)\, dx = \int_{-2}^6 f(x)\, dx - \int_{-2}^1 f(x)\, dx = (-3) - 4 = -7.$$

Another property of integrals involves inequalities. If $f(x) \le g(x)$ for all x on some interval, the area of the region under the graph of f must be less than, or equal to, the area of the region under the graph of g on that interval. The following picture illustrates the

Inequality of Integrals

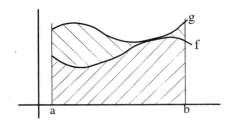

> **Inequality Rule**
> If $f(x) \le g(x)$ for every input x in the interval $[a, b]$ then
> $$\int_a^b f(x)\, dx \;\le\; \int_a^b g(x)\, dx \;.$$

The Definite Integral of a Rate Gives Total Change

If F' is the rate of change of F, then the Fundamental Theorem of Calculus says that the definite integral $\int_a^b F'(x)\, dx$ represents the total change in F over the interval $[a, b]$.

> $$\text{total change in } F \;=\; F(b) - F(a) \;=\; \int_a^b F'(x)\, dx.$$

In other words, the integral of a rate of change of any quantity gives the total change in that quantity. Here are some examples.

Example 4: *Finding Total Sales Given Rate of Sales*

A company estimates the rate of sales of a given item is given by $\dfrac{dS}{dt} = -3t^2 + 36t$, where t is the number of weeks after an advertising campaign has begun. What is the total number of items sold in the first five weeks?

Solution:

Since the rate at which the sales are growing is $\dfrac{dS}{dt} = -3t^2 + 36t$, we have

$$\text{total sales} = S(5) - S(0) = \int_0^5 (-3t^2 + 36t)\, dt$$

$$= -3\int_0^5 t^2\, dt + 36\int_0^5 t\, dt$$

$$= -3 \cdot \left[\frac{t^3}{3}\right]_0^5 + 36 \cdot \left[\frac{t^2}{2}\right]_0^5$$

$$= -3 \cdot \left[\frac{5^3 - 0^3}{3}\right] + 36 \cdot \left[\frac{5^2 - 0^2}{2}\right]$$

$$= 325$$

Example 5: *Estimating Total Change Given Rate of Change*

The rate at which a natural resource is extracted often increases at first until the easily accessible part of the resource is exhausted. Then the rate of extraction tends to decline. The figure below shows the rate, $A'(t)$, measured in millions of cubic feet per year, at which natural gas is being extracted from a new field. Estimate the total amount A of natural gas extracted in the first four years.

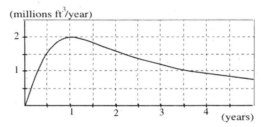

Solution:

Since the rate at which natural gas is being extracted is $A'(t)$,

$$A(4) - A(0) = \int_0^4 A'(t)\, dt$$

is the total volume of natural gas extracted in the first four years. We approximate the value of the integral using the Midpoint rule with four subdivisions and $\Delta t = 1$.

$$\int_0^4 A'(t)\, dt \approx \left[A'(0.5) + A'(1.5) + A'(2.5) + A'(3.5)\right] \Delta t$$

$$\approx \left[1.6 + 1.8 + 1.4 + 1\right] \cdot 1 = 5.8.$$

Approximately 5.8 million cubic feet of natural gas is extracted in the first 4 years.

6.1   Exercises

1.   Evaluate $\int_1^4 (3x^2 + 2)\, dx$ using

    a) left and right sums with 30 subdivisions;

    b) the Fundamental Theorem of Calculus.

2.   Evaluate $\int_1^e \frac{1}{x}\, dx$ using

    a) left and right sums with 30 subdivisions;

    b) the Fundamental Theorem of Calculus.   [Hint: Recall that $\frac{d}{dx}(\ln x) = \frac{1}{x}$.]

3.   Evaluate $\int_{\pi/2}^{\pi} \sin x\, dx$ using

    a) left and right sums with 30 subdivisions;

    b) the Fundamental Theorem of Calculus.

4.   Evaluate $\int_1^2 (2^x \ln 2)\, dx$ using

    a) left and right sums with 30 subdivisions;

    b) the Fundamental Theorem of Calculus.

5.   If the velocity of an object at time t is given by $v(t) = \cos t$, find the total distance traveled by the object between $t = 0$ and $t = \frac{\pi}{2}$.

6.   A company estimates that its rate of sales of an item is given by $\frac{dS}{dt} = 10\sqrt{t}$, where t is measured in years. Find the total sales during the years $2 \le t \le 4$.

7.   Suppose the rate at which oil is leaking out of a small storage tank is $\frac{dA}{dt} = 20 + 4t$ gallons per hour.

    a) Write a definite integral that represents the amount of oil that has leaked in the first 5 hours.

    b) Evaluate the integral in part a).

8.   A car travels on a straight highway with velocity given by $v(t) = 5 + 2t$ ft/sec.

    a) Write a definite integral that represents the distance traveled by the car between $t = 0$ and $t = 6$ seconds.

    b) Evaluate the integral in part a).

    c) Find the function s that gives the position of the car as a function of t if $s(0) = 0$.

    d) Use the function s to find the total distance traveled by the car for $0 \le t \le 6$. Compare your answer with part b).

9.   The figure shows the velocity of two marathon runners racing against each other. They start together at the same time on the same course.

    a) Which runner is ahead after the first minute? Explain your answer.

    b) Which runner is ahead after the second minute? Explain your answer.

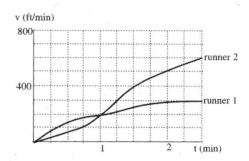

10.   An object moves along a line so that its velocity, measured in meters per second, at various times is listed in the following table.

t	1	3	5	7	9
$v(t)$	1	2	5	18	39

    Use a Riemann sum of your own choosing to approximate the distance traveled by the object from $t = 1$ to $t = 9$ seconds.

11.   Suppose $\int_{-1}^{3} f(x)\, dx = 7$ and $\int_{-1}^{3} g(x)\, dx = 4$. Evaluate each of the following integrals.

    a) $\int_{-1}^{3} [f(x) + g(x)]\, dx$ b) $\int_{-1}^{3} -2g(x)\, dx$

    c) $\int_{-1}^{3} [2f(x) - 3g(x)]\, dx$ d) $\int_{-1}^{3} [f(x) + 1]\, dx$

12.   Let $\int_{1}^{6} f(x)\, dx = 23$, $\int_{1}^{4} f(x)\, dx = 11$ and $\int_{1}^{4} g(x)\, dx = 17$. Evaluate the following integrals.

    a) $\int_{1}^{4} [f + g](x)\, dx$ b) $\int_{4}^{6} f(x)\, dx$

    c) $\int_{1}^{4} 3 \cdot f(x)\, dx$ d) $\int_{1}^{4} [g(x) + 2]\, dx$

13.   Verify the following inequalities without evaluating integrals.

a) $\int_0^1 (3x+1)\, dx \geq \int_0^1 (x+1)\, dx$ b) $\int_1^2 (x^2+1)\, dx \geq \int_1^2 (x+1)\, dx$

14.   Suppose that f is a nonnegative-valued function on $[0, 2]$ such that $\int_0^2 f(x)\, dx = 3$.

a) If f is an *even* function, explain why $\int_{-2}^2 f(x)\, dx = 6.$.

b) If f is an *odd* function, explain why $\int_{-2}^2 f(x)\, dx = 0$.

15.   Suppose f is an odd function and nonnegative on $[0, 2]$. If $\int_0^2 f(x)\, dx = 5$, evaluate each of the following integrals. [Hint: See Exercise 14]

a) $\int_{-2}^0 f(x)\, dx$ b) $\int_{-2}^2 f(x)\, dx$

c) $\int_0^2 [2f(x)+3]\, dx$ d) $\int_{-2}^2 |f(x)|\, dx$

16.   If g is an even function and $\int_{-3}^3 g(x)\, dx = 16$, evaluate each of the following integrals. [Hint: See Exercise 14]

a) $\int_{-3}^0 g(x)\, dx$ b) $\int_0^3 g(x)\, dx$

c) $\int_0^3 [3g(x)+4]\, dx$ d) $\int_{-3}^3 [g(x)+x]\, dx$

17.   The graph below shows the rate in gallons per hour at which water is leaking out of a tank.

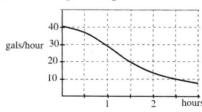

a) Express the total amount of water that leaked out during the first three hours as a definite integral.

b) Use a Riemann sum of your own choosing to approximate the integral in part a).

6.2   Functions Defined by Integrals;   Accumulation Functions

In this section we will define and study a new kind of function, an **accumulation function**, that is created from a continuous function f using the definite integral.

We start by assuming that f is defined so that the integral $\int_a^x f(t)\, dt$ exists for each x in the interval $[a, b]$. If the lower limit a and the function f are fixed, then for each input x in the interval $[a, b]$ the corresponding output is the definite integral $\int_a^x f(t)\, dt$. If we choose A as the name of the function that pairs each x with a definite integral, we have

$$A(x) \; = \int_a^x f(t)\, dt\,.$$

This process enables us to construct a new function A from the given function f, the value of A at each x in $[a, b]$ being determined by the equation above. The function A is called an **accumulation function** since it is evaluated by accumulating the "area" under the graph of the integrand function f.

There is a simple geometric relationship between the function f and its corresponding accumulation function A. If f is nonnegative, then $A(x)$ is the area under the graph of f from a to x. If f takes on both positive and negative values, then $A(x)$ gives the sum of the areas of the regions above the x-axis minus the total of the areas below the x-axis.

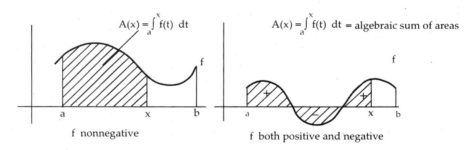

If the integrand f is a continuous function, then the corresponding accumulation function $A(x) \; = \int_a^x f(t)\, dt$ has the same domain as f itself. In particular, $A(x)$ makes sense even if x is to the left of a; in this case we use the sign convention

$$\int_a^x f(t)\, dt \; = \; -\int_x^a f(t)\, dt$$

NOTE: If $x = a$, we have $\int_a^a f(t)\, dt \; = 0.$

An Accumulation Function and its Graph

We have seen that if $v(t)$ represents the velocity of a moving object and $v(t) \geq 0$, then the distance covered by the object in traveling from $t = a$ to $t = b$ corresponds to the area under the graph of the velocity function.

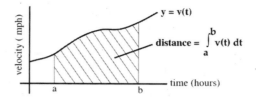

In symbols,

$$\text{Distance} = \int_a^b v(t)\, dt.$$

The graph below shows an object's velocity, v, in meters per second.

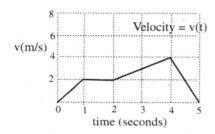

Suppose $D(T)$ meters is the distance traveled in the first T seconds. Then the accumulation function illustrates how the distance depends on velocity.

$$D(T) = \int_0^T v(t)\, dt \quad \text{for } 0 \leq T \leq 5.$$

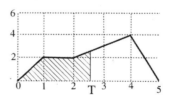

It is easy to find values for the accumulation function D because we know how to find the area of the regions under the velocity graph. For example,

$$D(1) = \int_0^1 v(t)\, dt \ = 1 \text{ meter}$$

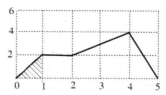

$$D(2) = \int_0^2 v(t)\, dt \ = 3 \text{ meter.}$$

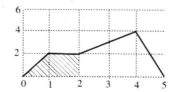

We can make a table of these values, plot and connect them with a continuous curve.

T (sec)	D(meters)
0	0
1	1
2	3
3	5.5
4	9
5	11

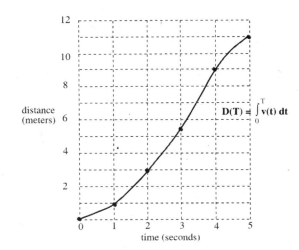

In comparing the graphs of the velocity function and the distance functions notice that the height of the accumulation function graph at any time equals the area under the velocity graph up to that time. For example, when $T = 0$, no area has accumulated, so $D(0) = 0$.

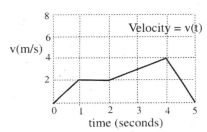

The built-in function **fnInt** in the Math menu can be used to graph an accumulation function.

Example 1:   *Using the Calculator to Graph an Accumulation Function*

Let $f(t) = t$ and $a = 0$. Using the function **fnInt**, graph the accumulation function

$A(x) = \int_a^x f(t)\, dt$. Does $A(x)$ have a simple formula?

Solution:

Press ⬜ **Y=** and enter the numerical integral Y1 = **fnInt**(T, T, 0, X). Press ⬜ **ZOOM** 4 to select the ZDecimal viewing rectangle. Note that the busy indicator displays while the graph below is being plotted.

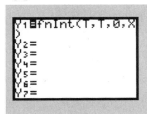

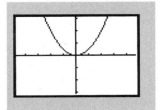

As the graph below shows, the region defining $A(x)$ for a positive input is a triangle, with base x and height x. The formula for $A(x)$, therefore, is

$$A(x) = \frac{\text{base} \cdot \text{height}}{2} = \frac{x^2}{2}.$$

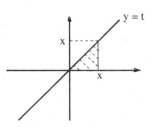

Thus, if $f(x) = x$, then $A(x) = \dfrac{x^2}{2}$. Notice that $A'(x) = f(x)$ so that the accumulation function A is an antiderivative of f.

Example 2: *Graphing an Accumulation Function*

Let $f(x) = 2 - x$ and $a = 0$. Using the function **fnInt**, graph the accumulation function $A(x) = \int_a^x f(t)\, dt$. Does $A(x)$ have a simple formula?

Solution:

Press $\boxed{\text{Y=}}$ and enter the numerical integral Y1 = **fnInt**(2–T, T, 0, X). Press $\boxed{\text{ZOOM}}$ 4 to select the ZDecimal viewing rectangle.

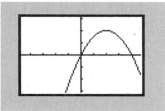

$y = A(x)$ appears to a quadratic function; we will use the Stat menu to find the coefficients a, b and c in the equation $A(x) = ax^2 + bx + c$. Enter the list editor by pressing $\boxed{\text{STAT}}$ and selecting [1:Edit]. After inputting three arbitrary x values in L1, place the cursor on the heading for the list L2 and enter Y1(L1). Pressing $\boxed{\text{ENTER}}$ will generate L2, a list of corresponding function values for $A(x)$. To calculate a quadratic regression equation, key in the sequence $\boxed{\text{STAT}}$ CALC [5:QuadReg] L1, L2, Y2 $\boxed{\text{ENTER}}$. The third argument, Y2, causes the regression equation to be stored in Y2. The equation is also automatically stored in the variable RegEQ. So another way to enter the equation in Y2 is as follows: press $\boxed{\text{Y=}}$, place the cursor on the Y2 line and then press $\boxed{\text{VARS}}$ [5:Statistics...] EQ [1:RegEQ].

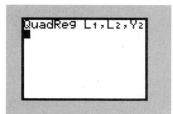 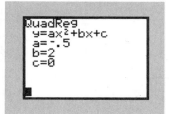 

Thus if $f(x) = 2 - x$, then $A(x) = 2x - \dfrac{x^2}{2}$. Again, notice that $A'(x) = f(x)$.

This result is no accident, as we shall soon see.

Graphing an Accumulation Function in Parametric Mode

The graph of an accumulation function can be plotted more quickly in parametric mode.

Example 3: *Graphing an Accumulation Function in Parametric Mode*

Graph $F(x) = \int_0^x (2-t)\, dt$ in parametric mode.

Solution:

Select parametric mode and enter

$$X_1T = T$$
$$Y_1T = \textbf{fnInt}(2-T,\ T,\ 0,\ T).$$

Enter the following window values and graph.

Tmin = –2	Xmin = –2	Ymin = –6
Tmax = 6	Xmax = 6	Ymax = 4
Tstep = 1	Xscl = 1	Yscl = 1

 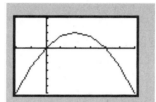

The value of Tstep controls how often X_1T and Y_1T are calculated. For the above graph with Tstep = 1, the parametric equations were evaluated 9 times (for T = –2, –1, 0, 1, ..., 5, 6). Changing Tstep to .5 will increase the number of points plotted to 17. This improves the quality of the graph, but also increases the time required to draw the graph. Whenever the accumulation function is graphed in Function mode, the **fnInt** command is executed 95 times! The advantage of graphing an accumulation function in parametric mode is that you control the speed and accuracy of the graphing.

6.2   Exercises

1.   Let $F(x) = \int_{-1}^{x} f(t)\ dt,\ -1 \le x \le 4$, where

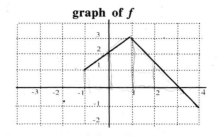

graph of f

f is the function graphed in the figure.

a) Complete the following table of
values for F.

x	−1	0	1	2	3	4
F(x)						

b) Sketch a graph of F.

2. Let $G(x) = \int_{-3}^{x} g(t)\ dt,\ -3 \le x \le 3$, where g is
the function graphed in the figure.

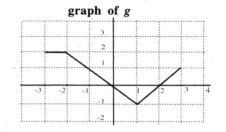

graph of g

a) Complete the following table of values
for G.

x	−3	−2	−1	0	1	2	3
G(x)							

b) Sketch a graph of G.

c) Which is larger $G(0)$ or $G(1)$)?
Justify your answer.

d) Where is G increasing?

3.   Sketch a graph of $f(t) = |t|$.

a) With F defined by $F(x) = \int_{-1}^{x} f(t)\ dt$ for $-1 \le x \le 2$, evaluate $F(-1)$, $F(0)$ and $F(2)$.

b) Sketch a graph of F.

4.   Sketch a graph of $f(t) = 6 - 2t$.

a) With F defined by $F(x) = \int_{0}^{x} f(t)\ dt$ for $0 \le x \le 3$, evaluate $F(0)$, $F(1)$ and $F(3)$
using the area formula for a trapezoid.

b) Sketch a graph of F.

5.   Sketch a graph of $f(t) = \sqrt{4 - t^2}$.

         a) With F defined by $F(x) = \int_{-2}^{x} f(t)\ dt$ for $-2 \le x \le 2$, evaluate $F(-2)$, $F(0)$ and
             $F(2)$ using the area formula for a circle.

         b) Sketch a graph of F.

6.   Let $A(x) = \int_{-3}^{x} f(t)\ dt$, $-3 \le x \le 4$, where f is the function graphed below.

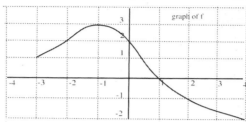

         a) Which is larger $A(-1)$ or $A(1)$? Justify your answer.

         b) Which is larger $A(2)$ or $A(4)$? Justify your anwer.

         c) Where is A increasing?

         d) Explain why A has a local maximum at $x = 1$.

7.   The accumulation function A defined by $A(x) = \int_{0}^{x}(4 - 2t)\ dt$ is a quadratic

         function of the form $A(x) = ax^2 + bx + c$.

         a) Use **fnInt** to define and graph the function $A(x)$.

         b) Using the Stat menu, fit a quadratic regression equation to 3 points on the graph of A.

         c) Find $A(0)$, $A(2)$, and $A(4)$ using the regression equation.

         d) Find $A(0)$, $A(2)$, and $A(4)$ using geometry formulas.

8.   The accumulation function A defined by $A(x) = \int_{0}^{x}(2 - t^2)\ dt$ is a cubic function

         of the form $A(x) = ax^3 + bx^2 + cx + d$.

         a) Use **fnInt(** to define and graph the function.

         b) Using the Stat menu, fit a cubic regression equation to 4 points on the graph of $A(x)$.

         c) Find $A(0)$ $A(\sqrt{2})$, and $A(3)$ using the regression equation.

9. Consider the function f defined by $f(t) = \cos(t^2)$.

a) Plot a graph of f and the accumulation function $A(x) = \int_0^x f(t)\, dt$ in the viewing rectangle $[0, 4]$ x $[-4, 4]$.

b) Make a table for the accumulation function showing the intervals where it is increasing, the intervals where it is decreasing, and its maximum and minimum points. In another row of this table describe where f is positive, negative and zero. Explain any relations you observe. Which relations do you think will hold between any function and its accumulation function?

10. Many important functions are defined as accumulation functions. Consider the function

$$L(x) = \int_1^x \frac{1}{t}\, dt \quad \text{for} \quad x > 0.$$

The function L is undefined for $x \le 0$ since $f(t) = \frac{1}{t}$ is not continuous at $t = 0$.

a) Set $Y_1 = \mathbf{fnInt}(1/T, T, 1, X)$ and graph Y_1 in the viewing rectangle

Xmin = .01	Ymin = −1.5
Xmax = 10	Ymax = 2.5
Xscl = 1	Yscl = 1

b) For what values of x is $L(x) = 0$?

c) Use the TI-83 to make a table of approximate values for $L(x)$ using 6 equally spaced values starting with $x = 1$. [Hint: set **TblMin** $= 1$ and **ΔTbl** $= 1$.]

d) Verify from the table created in c) that:

$L(6) = L(2) + L(3)$

$L(4) = 2 \cdot L(2)$

$L(8) = 3 \cdot L(2)$

e) Using the Stat menu, fit a LnReg curve to the data for $L(x)$.

11. Let $f(x) = \int_{-2}^x t^2 dt$ $\qquad g(x) = \int_0^x t^2 dt$ $\qquad h(x) = \int_2^x t^2 dt$

a) Using a list, define and graph the accumulation functions f, g and h simultaneously with

$$Y_1 = \mathbf{fnInt}(T^2, T, \{-2, 0, 2\}, X).$$

b) Describe the relationship between the graphs of f, g and h.

c) Define and graph the numerical derivatives of f, g and h with

$$Y_2 = \mathbf{nDeriv}(Y_1, X, X).$$

How do the derivatives of f, g and h compare?

d) Repeat steps a) b) c) using parametric mode.

X₁T = T	X₂T = T
Y₁T = **fnInt**(T² , T, {−2, 0, 2}, T).	Y₂T = **nDeriv**(Y₁T, T, T).

6.3   The Fundamental Theorem Again

In the previous section we defined and studied accumulation functions built from an original continuous function f and a fixed number $x = a$,

$$A(x) = \int_a^x f(t)\,dt.$$

The examples and exercises in Section 6.2 showed the accumulation function A to be an antiderivative of f, $A'(x) = f(x)$. Here is a formal statement of this fact which is sometimes called the **Second Fundamental Theorem of Calculus**.

> If f is a continuous function on some interval $[a, b]$ and an
>
> accumulation function A is defined by $A(x) = \int_a^x f(t)\,dt$, then
> $$A'(x) = f(x).$$

A geometric argument that $A'(x) = f(x)$ (when $f(x) \geq 0$) follows.

$A(x + h) = \int_a^{x+h} f(t)\,dt$ is the area under the graph of f from a to $x + h$.

$A(x) = \int_a^x f(t)\,dt$ is the area from a to x.

$A(x + h) - A(x)$ is the area from x to $x + h$.

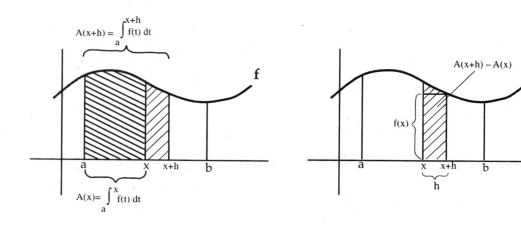

If f is continuous and h is small, then the area of the shaded region should be close to the area of the rectangle with height $f(x)$ and width h. That is

$$A(x + h) - A(x) \approx f(x) \cdot h \qquad \text{or} \qquad \frac{A(x + h) - A(x)}{h} \approx f(x).$$

The closer h is to zero, the better this approximation will be. Therefore

$$A'(x) = \lim_{h \to 0} \frac{A(x+h) - A(x)}{h} = f(x).$$

Again, we see a connection between derivatives and integrals. In this case if we integrate a continuous function f, then differentiate the result, we return to the original function f. In the (first) Fundamental Theorem of Calculus in Chapter 3, we started with a function f, differentiated (obtaining f'), and then integrated to return to the original function.

$$\int_a^b f'(t) \, dt = f(b) - f(a).$$

In either version of the Fundamental Theorem we see that differentiation and integration are inverse processes.

The fact that the derivative of an accumulation function $A(x) = \int_a^x f(t) \, dt$ is the original function f, $A'(x) = f(x)$, allows us to produce an antiderivative for any continuous function f by simply building an accumulation function for it. However, this fact is not necessarily helpful in getting a simple formula.

Example 1: *Finding the Derivative of an Accumulation Function*

If $F(x) = \int_1^x t^2 \, dt$, find $F'(x)$.

Solution:

The formula for $F'(x)$ can be found in two ways.

First we evaluate the integral and then the derivative.

$$F(x) = \int_1^x t^2 \, dt = \left[\frac{t^3}{3} \right]_1^x = \frac{x^3}{3} - \frac{1}{3}.$$

Thus, $F'(x) = \frac{d}{dx} \left(\int_1^x t^2 \, dt \right) = \frac{d}{dx} \left(\frac{x^3}{3} - \frac{1}{3} \right) = x^2$

Or, using the Second Fundamental Theorem,

$$F'(x) = \frac{d}{dx} \left(\int_1^x t^2 \, dt \right) = x^2 \ .$$

Example 2: *Finding an Equation of a Tangent Line*

If $G(x) = \int_1^x \dfrac{1}{t^2 + 1}\, dt$, determine an equation of the tangent line to the graph of G at the point where $x = 1$.

Solution:

$G'(x) = \dfrac{1}{x^2 + 1}$, so we have $G'(1) = \dfrac{1}{2}$. Since $G(1) = 0$, an equation of the tangent line is

$y = \dfrac{1}{2}(x - 1)$.

Example 3: *Finding where the Minimum Value of a Function Occurs*

Find the number x for which the function F defined on the interval $[-\frac{\pi}{2}, \frac{\pi}{2}]$ by the equation

$F(x) = \int_0^x \sin t\, dt$ takes its minimum value.

Solution:

With $F'(x) = \sin x$, there is a critical number at $x = 0$, and we have $F'(x) < 0$ for $-\frac{\pi}{2} \le x < 0$, and $F'(x) > 0$ for $0 < x \le \frac{\pi}{2}$. Thus the function F takes its global minimum value in the interval $[-\frac{\pi}{2}, \frac{\pi}{2}]$ at $x = 0$.

Now consider the accumulation function defined by $F(x) = \int_a^x f(t)\, dt$ as an outer function with a differentiable inner function g. Then by the Chain Rule the composite function H defined by

$$H(x) = F[g(x)] = \int_a^{g(x)} f(t)\, dt$$

has a derivative

$$H'(x) = F'[g(x)] \cdot g'(x) = f[g(x)] \cdot g'(x).$$

Example 4: *Using the Chain Rule*

If $y = \int_0^{x^2} \sin t\, dt$, find $\dfrac{dy}{dx}$.

Solution:

The outer function $F(x) = \int_0^x \sin t\, dt$ has a derivative $F'(x) = \sin x$, and the derivative of the inner function $g(x) = x^2$ is $2x$. Then by the Chain Rule, $\dfrac{dy}{dx} = \sin(x^2) \cdot 2x$.

6.3   Exercises

1.   Determine $\dfrac{dy}{dx}$ if:

a) $y = \int_3^x (t^2 - \sqrt{t})\, dt$

b) $y = \int_1^x \dfrac{1}{t^2 + 4}\, dt$

c) $y = \int_{-1}^x \sqrt{u^2 - 1}\, du$

d) $y = \int_x^0 \sqrt{3s + 1}\, ds$

2.   Let the function F be defined by $F(x) = \int_0^x (t^2 - 4t^3)\, dt$.

a) If they exist, determine critical numbers for F.

b) Discuss the concavity of F.

3.   Let the function F be defined on the interval $[0, 2\pi]$ by $F(x) = \int_0^x \sin t \cdot \cos t\, dt$.

a) If they exist, determine critical numbers for F.

b) Discuss the concavity of F.

4.   Let the function F be defined by $F(x) = \int_0^x \dfrac{(1-t)}{e^t}\, dt$.

a) If they exist, determine critical numbers for F.

b) Discuss the concavity of F.

5.   Find the global maximum value for the function F on the interval $[0, 1]$ if $F(x) = \int_0^x \dfrac{1}{t+1}\, dt$.

6.   Show that the graph of F is concave up everywhere if $F(x) = \int_0^x \dfrac{t}{\sqrt{t^2 + 1}}\, dt$.

7.   A particle moves along the x-axis so that its position at time $t > 0$ is given by

$s(t) = \int_0^t \dfrac{x}{x+1}\, dx$. Find the velocity and acceleration at $t = 2$.

8.   A particle moves along the x-axis so that its position at time $t > 0$ is given by

$s(t) = \int_3^t x \cdot \sin x\, dx$.

a) Find the velocity and acceleration of the particle at $t = \dfrac{\pi}{2}$.

b) What is the particle doing at $t = \dfrac{\pi}{2}$?

9.   Find the interval on which the graph of $y = \int_1^x \frac{1+t}{1+t^2} dt$ is concave up.

10.   Use the Chain Rule to determine $\frac{dy}{dx}$ if:

a) $y = \int_2^{x^2} (t^2 + 1) \, dt$

b) $y = \int_0^{2x+1} \sin t \, dt$

c) $y = \int_{-1}^{\sqrt{x}} \tan u \, du$

d) $y = \int_0^{x^2} \frac{1}{1+t^3} dt$

Exercises 11–15 are Multiple Choice.

11.   The approximate average rate of change of the function $f(x) = \int_0^x \sqrt{1 + \sin(t^2)} \, dt$ over the interval [1, 3] is

A) 0.493   B) 1.025   C) 1.139   D) 1.277   E) 2.051

12.   Let $F(x) = \int_0^x e^{-t^2} dt$. Of the following, which is the best approximation of $F(1)$?

A) 0.333   B) 0.724   C) 0.747   D) 1.457   E) 1.676

13.   If $F(x) = \int_0^x \sqrt{\tan t} \, dt$, then $F''(0.2) =$

A) 0.059   B) 0.148   C) 0.450   D) 1.156   E) 1.594

14.   The tangent line approximation to the graph of $f(x) = 2 + \int_0^x \frac{10}{1+t} dt$ at $x = 0$ is

A) $y = 10x$ B) $y = 12x$ C) $y = 12x + 2$ D) $y = 10x + 2$ E) $y = 10x - 20$

15.   The graph of f', the derivative of the function f, is shown on the right for $-2 \le x \le 6$. The graph consists of two line segments and a quarter circle. If, $f(2) = 3$ which of the following statements must be true?

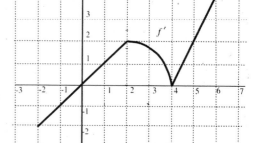

I $f(6) = 5 + \pi$

II   The function f has a relative minimum at $x = 0$.

III The graph of f has points of inflection at $x = 0$ and $x = 4$.

A) I only   B) II only   C) III only   D) I and II only   III) I and III only

6.4   Areas of Plane Regions

The simplest plane regions are bounded above by the graph of one function, below by the x-axis, and on the left and right by vertical lines.

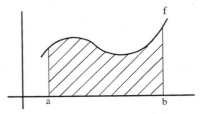

In this simplest case, the definite integral $\int_a^b f(x)\, dx$ measures the shaded area.

Example 1: *Finding the Area of a Region Using the Fundamental Theorem*

Find the area of the region R bounded by the x-axis, the vertical lines $x = \ln 2$ and $x = \ln 4$, and the graph of $y = e^x$.

Solution:

Graphing $y = e^x$ between the vertical lines $x = \ln 2$ and $x = \ln 4$, we see that the area of R is given by the definite integral $\int_{\ln 2}^{\ln 4} e^x\, dx$. Using the Fundamental Theorem we find an exact answer.

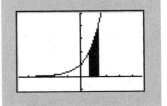

$$\int_{\ln 2}^{\ln 4} e^x\, dx = e^{\ln 4} - e^{\ln 2} = 4 - 2 = 2.$$

Area is a nonnegative number. If the graph of the function f lies below the x-axis over an interval $[a, b]$, then the definite integral is negative and therefore cannot be an area. However, it is just the negative of the area of the region.

Example 2: *Finding the Area of a Region Between a Graph and the x-Axis*

Find the total area enclosed between the graph of $f(x) = x^3 - 3x^2 + 2x$ and the x-axis for $x = 0$ to $x = 2$. Use your calculator to confirm the result.

Solution:

Integrating over the interval $[0, 2]$ we have

$$\int_0^2 (x^3 - 3x^2 + 2x)\, dx = \left[\frac{x^4}{4} - x^3 + x^2 \right]_0^2 = (\tfrac{16}{4} - 8 + 4) - (0) = 0.$$

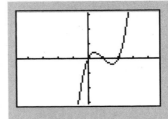

Looking at the graph we see that the net area, counting area below the x-axis as negative, over the interval is 0 square units; however, if we want the area enclosed by the curve and the x-axis over the interval $[0, 2]$, we need to take the absolute value of the definite integral over the interval $[1, 2]$.

$$\int_0^1 (x^3 - 3x^2 + 2x)\, dx = \left[\frac{x^4}{4} - x^3 + x^2\right]_0^1 = \frac{1}{4} - 1 + 1 = \frac{1}{4}, \text{ while}$$

$$\int_1^2 (x^3 - 3x^2 + 2x)\, dx = \left[\frac{x^4}{4} - x^3 + x^2\right]_1^2 = (\frac{16}{4} - 8 + 4) - (\frac{1}{4}) = -\frac{1}{4}.$$

Thus the total area is $\frac{1}{4} + \left|-\frac{1}{4}\right| = \frac{1}{2}$.

A numerical approximation to the area just found can be calculated on the TI-83 by using either **fnInt** or option 7 in the Calc menu to evaluate $\int_0^2 \left|x^3 - 3x^2 + 2x\right| dx$.

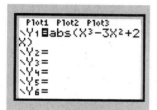 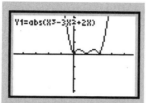  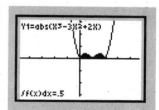

The definite integral can also be employed to measure the area of a region bounded by two or more graphs. Suppose we want to find the area of a region bounded by the graphs of two continuous functions f and g on the interval $[a, b]$.

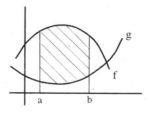

If, as in the figure above, the graphs of f and g both lie above the x-axis, and the graph of g is below the graph of f, the area of the region between the graphs is the area under the graph of g subtracted from the area under the graph of f.

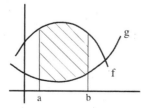 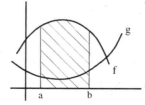 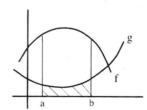

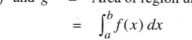

Area of region between f and g $=$ Area of region under f $-$ Area of region under g

$$\int_a^b [f(x) - g(x)]\, dx \qquad = \int_a^b f(x)\, dx \qquad - \int_a^b g(x)\, dx$$

In general, we have the following,

> **Area of a Region Between Two Curves**
> If f and g are continuous functions and $f(x) \geq g(x)$ for all x in the interval $[a, b]$, then the area of the region between the graphs of f and g from $x = a$ to $x = b$ is
> $$\int_a^b [f(x) - g(x)]\, dx.$$

If we seek the area of the region between two graphs that intersect exactly twice but we are not given the interval over which to integrate, we must first determine where the graphs intersect and which graph lies above the other, and then integrate over the corresponding interval.

Example 3:   *Finding the Area of a Region Between Two Curves*

Compute the area of the region X between the graphs of f and g if $f(x) = 4x - x^2$ and $g(x) = 5 - 2x$.

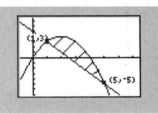

Solution:

Graph the functions and note which is the upper graph on the interval where X lies. In this case $4x - x^2 \geq 5 - 2x$. The graphs intersect where $4x - x^2 = 5 - 2x$. Solving this equation,

$$4x - x^2 = 5 - 2x$$
$$x^2 - 6x + 5 = 0$$
$$(x - 1)(x - 5) = 0$$
$$x = 1 \quad \text{or} \quad x = 5.$$

We substitute in either formula and find that the graphs intersect at $(1, 3)$ and $(5, -5)$.

Then by the method described above,

$$\text{Area } (X) = \int_1^5 \left[(4x - x^2) - (5 - 2x) \right] dx$$

$$= \int_1^5 (6x - x^2 - 5)\, dx$$

$$= \left[3x^2 - \frac{1}{3}x^3 - 5x \right]_1^5$$

$$= (75 - \frac{125}{3} - 25) - (3 - \frac{1}{3} - 5) = \frac{32}{3}.$$

The next example illustrates how the intersection and integration features of a calculator may be employed to find the area of a region.

Example 4:   *Using the Calculator to Find Area*

Find the area of the region bounded by the graphs of f and g if $f(x) = (5x^3 - 3)\sqrt{4 - x^2}$ and $g(x) = 15x^2 - x - 12$.

Solution:

First define the functions as Y1 and Y2. Then graph both functions in the same viewing rectangle.

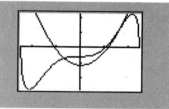

By using the TI-83's intersect command, we find the two points of intersection occur at $A \approx -0.531$ and $B \approx 1.313$. To calculate the area of the region between these curves, we use the TI-83's numerical integrator to calculate

$$\int_{-.531}^{1.313} (Y_1 - Y_2)\, dx \approx 6.429 \; .$$

6.4   Exercises

In Exercises 1–13, sketch graphs of the functions f and g and compute the area of the region between the graphs. Some exercises may require the help of the TI-83.

1. $f(x) = x^2$ and $g(x) = x^3$ 2. $f(x) = x^2$ and $g(x) = x + 2$

3. $f(x) = x^2 - 2x$ and $g(x) = 2x$ 4. $f(x) = x^3 - 12x$ and $g(x) = x^2$

5. $f(x) = x^3 - 4x$ and $g(x) = 3x + 6$ 6. $f(x) = 2x - x^2$ and $g(x) = -x$

7. $f(x) = e^{x^2}$ and $g(x) = x + 2$ 8. $f(x) = \sin x$ and $g(x) = x^2$

9. $f(x) = e^{-x^2}$ and $g(x) = x^3 - x + 1$ 10. $f(x) = \cos x$ and $g(x) = \frac{x}{3} + \frac{1}{3}$

11. $f(x) = \sin(x^2)$ and $g(x) = 1 - x^2$ 12. $f(x) = \sqrt[3]{x}$ and $g(x) = x^2 - 2$

13. $f(x) = x^2 - 7x + 10$ and $g(x) = \ln(x - 1)$

14.   Compute the area of the region in the first quadrant that is bounded above by the graph of $y = 6x - x^2$ and below by the graph of $y = x^2 - 2x$.

15.   Shade the region that is bounded above by the graph of $f(x) = x + 6$ and below by the graphs of $g(x) = 6 - x$ and $h(x) = x^2$. Find the area of this region.

16.   Compute the area of the first quadrant region bounded by the graph of $y = x + \sin x$ from $x = 0$ to $x = \frac{\pi}{2}$.

17.   Find the area of the region between the graphs of $y = \sin x$ and $y = \cos x$ from $x = 0$ to $x = \frac{\pi}{2}$.

18.   Find the area of the region bounded above by the graph of $f(x) = |x|$ and below by the graph of $g(x) = x^3$.

19.   Find the constant k so that the area of the region bounded by $y = x^2 - k^2$ and $y = k^2 - x^2$ is exactly 72.

20.   Find the area of the region enclosed by the graph of $y = e^{x-1}$ and the x-axis between $x = 0$ and $x = 3$.

21.   Find the total area of the region between the curve $y = x^3 - 4x$ and the x-axis.

Chapter 6   Supplementary Problems

1. Find the area of the region bounded by the curves $y = \sin x$ and $y = \cos x$ from $x = 1$ to $x = \dfrac{\pi}{2}$.

2. Evaluate the following exactly:

    a) $\displaystyle\int_{\pi/3}^{\pi/6} \sin x \ dx$

    b) $\displaystyle\lim_{h \to 0} \frac{\sqrt{4+h}-2}{h}$

3. Find the equations for the two tangent lines to the circle $x^2 + y^2 - 2x = 24$ at the points where $x = 4$.

4. Estimate the global maximum value for the function F on the interval $[0, 1]$ if
    $$F(x) = \int_0^x \frac{1}{1+t} \ dt.$$

5. Let the function f be defined on the interval $[0, \pi]$ by $f(x) = \int_0^x (1 + \cos t) \ dt$.

    a) If they exist, determine the critical numbers of f.

    b) If possible, determine the minimum value of f on the interval $[0, \pi]$.

6. Show that the graph of the function $F(x) = \int_0^x \dfrac{1}{t^2 + 1} \ dt$ has an inflection point at the origin.

7. Find the area of the region bounded by the parabolas $y = x^2$ and $y = 2x - x^2$.

8. Approximate the average rate of change of the function $f(x) = \int_{1/2}^x \ln t \ dt$ from $x = 1$ to $x = 3$.

9. The figure shows the graph of f. Determine a so that $\int_{-3}^a f(t) \ dt$ will be as small as possible.

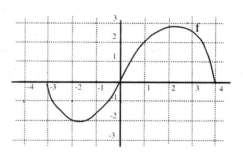

10. Find the area of the region bounded by the curves $y = \cos x$ and $y = \sec^2 x$ from $x = -\dfrac{\pi}{4}$ to $x = \dfrac{\pi}{4}$.

11.   Fuel oil is pumped into a tank at a rate of $r(t) = (150 - 0.9t)$ gallons per minute, where t is the time in minutes. The tank is empty at the begining of the filling process. How much fuel oil is in the tank after 10 minutes?

12.   If $F(x) = \int_0^x \sqrt{\tan t}\; dt$, approximate $F''(0.3)$.

13.   The graph of f is shown in the figure.

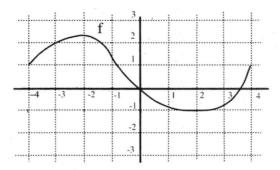

Estimate each of the following:

a) $f'(-1)$ b) $\int_{-4}^{1} f(x)\, dx$ c) $\dfrac{d}{dx}\left[\int_{-4}^{x} f(t)\, dt\right]$ at $x = -2$.

14.   Sketch the derivative and an antiderivative of the function in the figure. Make the antiderivative go through the origin.

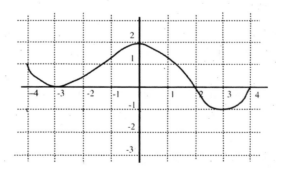

15.   An object is thrown vertically upward with a speed of 10 ft/sec from a height of 6 feet. Find the highest point it reaches and when it hits the ground.

16.   Suppose that f is a continuous function defined on the interval [2, 5] and that $\int_2^5 f(x)\, dx = 7$. Use this information to evaluate each of the following:

a) $\int_2^5 3 \cdot f(x)\, dx$ b) $\int_2^5 [f(x) + 3]\, dx$ c) $\int_{-1}^{2} f(x + 3)\, dx$

17.　A particle moves along the *x*-axis so that at any time $t \geq 0$ its acceleration is given by $a(t) = 18 - 12t$. At time $t = 1$ the velocity of the particle is 36 m/sec and its position is $x = 21$.

　　a) Find the velocity function for $t \geq 0$.

　　b) Find the position function for $t \geq 0$.

　　c) What is the position of the particle when it is farthest to the right?

18.　Evaluate the definite integrals.

　　a) $\int_0^{\pi/4} \dfrac{1 - \sin^2 x}{\cos x} \, dx$

　　b) $\int_0^{2\pi} |\sin x| \, dx$

　　c) $\int_0^{10} 2[f(x) + g(x)] \, dx$, given that $\int_0^4 f(x) \, dx = 8$, $\int_4^{10} f(x) \, dx = 20$, and $\int_{10}^0 g(x) \, dx = -5$.

19.　The graph of f', the derivative of f shown below is defined for all x in the interval $[-2, 4]$. The graph of f' has horizontal tangents at $x = -0.5$ and 2.

　　a) On what interval(s) is f increasing? Why?

　　b) What are the possible *x*-coordinates for the global minimum value of f on $[-2, 4]$? Why?

　　c) On what interval(s) is the graph of f concave up? Why?

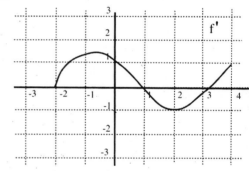

20.　a) Suppose $F(x) = \int_0^{x^2} \dfrac{1}{1 + t^3} \, dt$ for all real *x*; find $F'(1)$.

　　b) Suppose $F(x) = \int_0^{\cos x} \sqrt{1 + t^3} \, dt$ for all real x; find $F'(\frac{\pi}{2})$.

21.　A function G is defined by $G(x) = \int_0^x \sqrt{1 + t^2} \, dt$ for all real numbers *x*. Determine whether the following statements are true or false. Justify your answers.

　　a) G is continuous at $x = 0$.

　　b) $G(3) > G(1)$

　　c) $G'(2\sqrt{2}) = 3$

　　d) The graph of G has a horizontal tangent at $x = 0$.

　　e) The graph of G has an inflection point at $(0, 0)$.

22.   Let $G(x) = \int_{-3}^{x} f(t)\, dt$, where the graph of the function f is given below for $-3 \leq t \leq 3$.
The graph of f has horizontal tangents at $x = -1$ and $x = 1$.

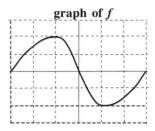

graph of f

     a) Evaluate $G(3)$ and $G(-3)$.

     b) On what interval is G increasing?

     c) On what interval is G concave down?

     d) Where does G have a maximum value?

     e) Sketch a graph of G.

23.   Let f be the function defined by $f(x) = \ln(x + 1) - \sin^2 x$ for $0 \leq x \leq 3$.

     a) Find the x-intercepts of the graph of f.

     b) Find the intervals on which f is increasing.

     c) Find the global maximum and the global minimum value of f. Justify your
        answer.

24.   Given: $f(t) = 2\pi t + \sin(2\pi t)$

     a) Find the value of t in the open interval $(0, 2)$ for which the line tangent at $(t, f(t))$ is
        parallel to the line through $(0, 0)$ and $(2, 4\pi)$.

     b) Suppose the given function describes the position of a particle on the x-axis for time
        $0 \leq t \leq 2$. What is the average velocity of the particle over that interval?

     c) Determine the velocity and the acceleration of the particle at $t = 1$.

25.   Let $f(x) = x^3 + px^2 + qx$.

     a) Find the values of p and q so that $f(-1) = 8$ and $f'(-1) = 12$.

     b) Find the value of p so that the graph of f changes concavity at $x = 2$.

     c) Under what conditions on p and q will the graph of f be increasing everywhere.

26.   Let $f(x) = x^3 + 3x^2 - x + 2$.

     a)   The tangent to the graph of f at the point $P = (-2, 8)$ intersects the graph of f again at
          the point Q. Find the coordinates of the point Q.

     b)   Find the coordinates of point R, the inflection point on the graph of f.

     c)   Show that the segment QR divides the region between the graph of f and its tangent at
          P into two regions whose areas are in the ratio of $\dfrac{16}{11}$.

CHAPTER 7
Finding Antiderivatives

The Fundamental Theorem reduces the problem of evaluating the definite integral $\int_a^b f(x)\,dx$ to that of finding an antiderivative G of f. Once this is accomplished, then

$$\int_a^b f(x)\,dx = G(b) - G(a).$$

In this chapter we begin a discussion of the basic techniques for finding antiderivatives.

7.1   Antiderivatives

If the function f is continuous on the interval $[a, b]$, we know that the function $G(x) = \int_a^x f(t)\,dt$ is one antiderivative for f. And, if C is any real number constant, the function $H(x) = G(x) + C$ is also an antiderivative of f, since $H'(x) = G'(x) + 0 = f(x)$. Thus a given function f has infinitely many different antiderivatives that differ only by an additive constant. It is useful to have a special symbol for the collection of all antiderivatives for a given function f. Because antiderivatives are closely related with the evaluation of definite integrals, Leibniz used the symbol $\int f(x)\,dx$ to denote the set of all antiderivatives for a function f. For example, since $\dfrac{d}{dx}\left[\dfrac{1}{3}x^3\right] = x^2$, we have

$$\int x^2\,dx = \frac{1}{3}x^3 + C \text{ where } C \text{ is any real number.}$$

Also, $\dfrac{d}{dx}[\sin x] = \cos x$, so that

$$\int \cos x\,dx = \sin x + C \quad \text{where } C \text{ is any real number.}$$

In general,

> If G is any particular antiderivative of f, then we write
>
> $$\int f(x)\,dx = G(x) + C.$$

A note about terminology. Since finding an antiderivative is the key step in using the Fundamental Theorem to evaluate a definite integral, the term "integration" is used to refer to the process of finding antiderivatives. So, for example, integrating $\int x^2\,dx$ is understood to mean finding the most general antiderivative $\dfrac{1}{3}x^3 + C$.

Example 1. *Reversing a Derivative Formula*

Integrate $\int \dfrac{1}{2\sqrt{x}}\, dx$.

Solution:

Since $\dfrac{d}{dx}(\sqrt{x}) = \dfrac{1}{2\sqrt{x}}$, $\int \dfrac{1}{2\sqrt{x}}\, dx = \sqrt{x} + C$.

Each of the differentiation formulas we have learned can be turned around to give us a formula for an antiderivative. The formulas in the following list can be verified by differentiating the function on the right side.

$$\int k\, dx = kx + C$$

$$\int x^n\, dx = \frac{x^{n+1}}{n+1} + C \qquad (n \neq -1)$$

$$\int e^x\, dx = e^x + C$$

$$\int \frac{1}{x}\, dx = \ln|x| + C$$

$$\int b^x\, dx = \frac{b^x}{\ln b} + C \qquad (b > 0,\ b \neq 1)$$

$$\int \cos x\, dx = \sin x + C$$

$$\int \sin x\, dx = -\cos x + C$$

$$\int \sec^2 x\, dx = \tan x + C$$

Notice that the formula $\int \dfrac{1}{x}\, dx = \ln|x| + C$ includes an unexpected absolute value. Since the function $f(x) = \dfrac{1}{x}$ is defined for all $x \neq 0$, we would expect an antiderivative to have the same domain. We know that $\int \dfrac{1}{x}\, dx = \ln x + C$ for $x > 0$. If $x < 0$, then $\ln(-x)$ is well defined and $\dfrac{d}{dx}[\ln(-x)] = (-1)\left(\dfrac{1}{-x}\right) = \dfrac{1}{x}$. Thus, $\int \dfrac{1}{x}\, dx = \ln(-x) + C$ for $x < 0$. Using absolute value allows us to say $\int \dfrac{1}{x}\, dx = \ln|x| + C$.

Also, the following general formulas are consequences of the corresponding differentiation results. If f and g are continuous functions and k is any constant, then

$$\int [f(x) + g(x)]\, dx = \int f(x)\, dx + \int g(x)\, dx$$

$$\int k \cdot f(x)\, dx = k \int f(x)\, dx$$

These equations assert that an antiderivative of $f + g$ can be produced by adding an antiderivative of f to an antiderivative of g; and an antiderivative of $k \cdot f$ can be produced by multiplying an antiderivative of f by k.

Example 2: *Applying an Integral Formula*

Integrate $\int \sqrt[3]{x^2}\, dx$.

Solution:

$$\int \sqrt[3]{x^2}\, dx = \int x^{2/3}\, dx = \frac{x^{(2/3+1)}}{\frac{2}{3}+1} = \frac{3}{5}x^{5/3} + C$$

Example 3: *Intergrating Term by Term*

Integrate $\int (2\sin x - 3e^x)\, dx$.

Solution:

$$\int (2\sin x - 3e^x)\, dx = 2\int \sin x\, dx - 3\int e^x\, dx = -2\cos x - 3e^x + C$$

Example 4: *Rewriting Before Integrating*

Evaluate $\int_1^e \frac{x^2 + 1}{x}\, dx$.

Solution:

A technique is needed. We first rewrite the integrand: $\dfrac{x^2+1}{x} = x + \dfrac{1}{x}$.

$$\int_1^e \frac{x^2+1}{x}\, dx = \int_1^e \left(x + \frac{1}{x} \right) dx$$

$$= \left[\frac{x^2}{2} + \ln|x| \right]_1^e$$

$$= \left(\frac{e^2}{2} + \ln e \right) - \left(\frac{1}{2} + \ln 1 \right)$$

$$= \frac{e^2 + 1}{2}.$$

7.1   Exercises

In Exercises 1–18, determine the antiderivatives. In some exercises it will be helpful to rewrite the function to be integrated using algebraic or trigonometric identities. Check your answers by differentiation.

1. $\int (3x+5)\, dx$

2. $\int 3\cos x\, dx$

3. $\int (x - \frac{1}{\sqrt{x}})\, dx$

4. $\int x^{3/5}\, dx$

5. $\int \frac{x^3 - 1}{x^2}\, dx$

6. $\int (\sec^2 x - 1)\, dx$

7. $\int \frac{\sqrt{x}}{x}\, dx$

8. $\int (2x^{5/2} + x^3)\, dx$

9. $\int \frac{x-1}{x}\, dx$

10. $\int \frac{x+2}{x^2}\, dx$

11. $\int (e^x + 5)\, dx$

12. $\int (x+1)(2x-1)\, dx$

13. $\int (\cos^2 x + \sin^2 x)\, dx$

14. $\int (3x^2 - \frac{1}{x^2})\, dx$

15. $\int (3\sin x - 5\cos x)\, dx$

16. $\int \tan^2 x\, dx$

17. $\int \frac{1}{\cos^2 x}\, dx$

18. $\int (2e^x - 6\cos x)\, dx$

In Exercises 19–28, evaluate, if possible, the given definite integrals.

19. $\int_{-1}^{2} (3x-4)\, dx$

20. $\int_{0}^{\pi} (\sin x - 8x^2)\, dx$

21. $\int_{1}^{2} \frac{1+x^2}{x}\, dx$

22. $\int_{0}^{2} (\frac{x^3}{3} + 2x)\, dx$

23. $\int_{0}^{1} 2e^x\, dx$

24. $\int_{0}^{\pi/4} \frac{1}{\cos^2 x}\, dx$

25. $\int_{-1}^{-1/2} \frac{1}{x^2}\, dx$

26. $\int_{0}^{4} (x^{1/2} - x^2)\, dx$

27. $\displaystyle\int_4^9 \frac{1}{\sqrt{x}}\, dx$
                             28. $\displaystyle\int_1^2 \frac{x+1}{x^3}\, dx$

29.   What is the area of the region bounded by the graph of $y = \dfrac{x-1}{x^2}$ and the x-axis from $x = 1$ to $x = e$?

30.   At time $t = 0$ an object starts at rest and moves along a line in such a way that at time t its acceleration is $24t^2$ feet per second per second. Through how many feet does the particle move during the first 2 seconds?

31.   The acceleration at time t of a particle moving along the x-axis is $a(t) = \cos t - \sin t$. At time $t = 0$ the velocity of the particle is $v(0) = 0$.

    a) Find the velocity of the particle at any time t.

    b) Find an equation for the position $x(t)$ if $x(0) = 0$.

    c) What is the first time $t > 0$ that the particle returns to the origin?

32.   What is the area of the closed region bounded by the curve $y = e^x$ and the lines $x = 1$ and $y = 1$?

33.   What is the area of the region bounded by the graph of $y = \sqrt{x} + 1$ and the x-axis from $x = 1$ to $x = 3$?

34.   Use the identity $\sin^2 x = \dfrac{1 - \cos 2x}{2}$ to find $\int \sin^2 x\, dx$.

35.   Use the identity $\cos^2 x = \dfrac{1 + \cos 2x}{2}$ to find $\int \cos^2 x\, dx$.

7.2   Integration Using the Chain Rule

The problem of finding antiderivatives is generally more difficult than the problem of determining derivatives. One reason is the absence of a rule for integrating the product of two functions. In this section we discuss an integration method that can be used on some products and makes use of the Chain Rule for differentiation.

The Chain Rule says that a composite function $y = F[g(x)]$ has the following derivative.

$$\frac{d}{dx}[F[g(x)]] = F'[g(x)] \cdot g'(x).$$

If we call f the derivative of the outer function F, $F' = f$, then

$$\frac{d}{dx}[F[g(x)]] = f[g(x)] \cdot g'(x).$$

Therefore, $F[g(x)]$ is an antiderivative of $f[g(x)] \cdot g'(x)$ and we have

$$\int f[g(x)] \cdot g'(x) \, dx = F[g(x)] + C$$

Thus when trying to integrate a product or a function that can be expressed as a product, we should consider the possibility that the integrand is the derivative of a composite function. Try to break the integrand into two factors: a composite $f[g(x)]$ and the derivative $g'(x)$ of the inner function in that composite. If that can be done, and if an antiderivative of f, the outer function in the composite is known, then the integration can be completed as $F[g(x)] + C$. We call this method of integration "Reversing the Chain Rule".

Example 1:   *Reversing the Chain Rule*

Integrate $\int \sin^2 x \cdot \cos x \, dx$.

Solution:

The integrand is the product of two functions. If we choose $g(x) = \sin x$ as the inner function, then we discover that $g'(x) = \cos x$ is a factor of the integrand. The outer function f in the composite $f(g(x))$ is $f(x) = x^2$. Fortunately, we know an antiderivative for f, namely $F(x) = \frac{1}{3}x^3$. Thus the integrand is the derivative of $F[g(x)] = \frac{1}{3}\sin^3 x$. Hence,

$$\int \sin^2 x \cdot \cos x \, dx = \frac{1}{3}\sin^3 x + C.$$

Example 2:   *Recognizing the $f[g(x)] \cdot g'(x)$ Pattern*

Integrate $\int \sqrt{x^2 + 1} \cdot 2x \, dx$.

Solution:

If we let $g(x) = x^2 + 1$, then g appears as the inner function in the factor $\sqrt{x^2 + 1}$ and $g'(x) = 2x$ is also a factor in the integrand. Using $f(x) = x^{1/2}$ as the outer function, we have $\sqrt{x^2 + 1} \cdot 2x = f(g(x)) \cdot g'(x)$. Hence the integral will be $F(g(x)) + C$ where F is an antiderivative of f. Such a function is known, namely, as $F(x) = \frac{2}{3}x^{3/2}$. Thus

$$\int \sqrt{x^2 + 1} \cdot 2x \, dx = \frac{2}{3}(x^2 + 1)^{3/2} + C.$$

Example 3: *Multiplying and Dividing by a Constant*

Integrate $\int (1+\sin 2x)^3 \cdot \cos 2x \; dx$

Solution:

If we choose $g(x) = 1 + \sin 2x$ as our inner function, then the first factor of the integrand is $f[g(x)]$ where $f(x) = x^3$. Unfortunately, the second factor, $\cos 2x$, is not the same as $g'(x)$, but differs from it by a factor of 2. What we would really like as the integrand is $(1+\sin 2x)^3 \cdot (2 \cos 2x)$. The general formula $\int k \cdot f(x) \, dx = k \cdot \int f(x) \, dx$ allows us to write

$$\int (1+\sin 2x)^3 \cdot \cos 2x \; dx = \frac{1}{2} \int (1+\sin 2x)^3 \cdot \cos 2x \cdot 2 \; dx$$

$$= \frac{1}{2} \int F'[g(x)] \cdot g'(x) \, dx$$

$$= \frac{1}{2} F[g(x)] + C.$$

Since $F(x) = \frac{1}{4} x^4$ is an antiderivative of $f(x) = x^3$, we have

$$\int (1+\sin 2x)^3 \cdot \cos 2x \; dx = \frac{1}{2} \cdot \frac{1}{4} (1+\sin 2x)^4 + C = \frac{1}{8}(1+\sin 2x)^4 + C \; .$$

The TI-83 can be used to provide a graphical check of antiderivatives. For instance, to check the result in Example 3 above, enter the antiderivative $\frac{1}{8}(1+\sin(2x))^4$ as Y1 (we take the case $C = 0$) and the original integrand as Y2 $= (1+\sin(2x))^3 \cdot \cos(2x)$. Then the graph of the numerical derivative Y3 $=$ nDeriv(Y1,X,X) should appear to be virtually the same as the graph of Y2.

If an integral is a quotient with the numerator being the derivative of the denominator, $\int \frac{g'(x)}{g(x)} \, dx$, we can treat the integrand as a product $f(g(x)) \cdot g'(x)$ where $f(x) = \frac{1}{x}$ is the reciprocal function.

Example 4: *Integrating a Quotient*

Integrate $\int \frac{2t}{t^2+1} \, dt$.

Solution:

If we let $g(t) = t^2 + 1$, with $f(t) = \frac{1}{t}$, the integrand is the product $f(g(t)) \cdot g'(t)$ where $g'(t) = 2t$. Since the antiderivative of $f(t) = \frac{1}{t}$ is $F(t) = \ln|t|$ we have

$$\int \frac{2t}{t^2+1} \, dt = F[g(t)] + C = \ln\left|t^2+1\right| + C$$

Summarizing the integration method in the examples above, we see that the Chain Rule can be reversed if the integrand is the product of two functions, one factor being the composite $F'[g(x)]$ and the other factor being $g'(x)$ or a constant multiple of $g'(x)$.

7.2   Exercises

In Exercises 1–22, find the antiderivatives.

1. $\int 2x\sqrt{x^2+1}\ dx$

2. $\int (x^3+1)^{17}\cdot 3x^2\ dx$

3. $\int (2x+3)^3\ dx$

4. $\int \sqrt{1-x}\ dx$

5. $\int \sec^2(2x)\ dx$

6. $\int \sin x\cdot\cos x\ dx$

7. $\int \sqrt{x+2}\ dx$

8. $\int \cos(4x)\ dx$

9. $\int e^{2x-1}\ dx$

10. $\int x\cdot e^{x^2}\ dx$

11. $\int e^{\sin x}\cdot\cos x\ dx$

12. $\int (\ln x)^2\cdot\frac{1}{x}\ dx$

13. $\int (\tan x)^2 \sec^2 x\ dx$

14. $\int \cos^3 x\cdot\sin x\ dx$

15. $\int \frac{\cos\sqrt{x}}{\sqrt{x}}\ dx$

16. $\int x^2\cos(x^3)\ dx$

17. $\int x(x^2-1)^3\ dx$

18. $\int \tan^2(2x)\ dx$

19. $\int \sqrt{\cos x}\cdot\sin x\ dx$

20. $\int \frac{x}{x^2+1}\ dx$

21    $\int \frac{\cos x}{\sin x+1}\ dx$

22. $\int \frac{\sec^2 x}{\tan x}\ dx$

In Exercises 23–30, use the Fundamental Theorem to evaluate each definite integral.

23. $\displaystyle\int_e^{e^3} \frac{2}{x}\, dx$ 24. $\displaystyle\int_1^3 x^3\sqrt{x^4+1}\, dx$

25. $\displaystyle\int_0^{\pi/6} \sin(2x)\, dx$ 26. $\displaystyle\int_0^1 e^{2x}\, dx$

27. $\displaystyle\int_0^2 \frac{1}{x+1}\, dx$ 28. $\displaystyle\int_0^{\sqrt{3}} x\sqrt{1+x^2}\, dx$

29. $\displaystyle\int_0^1 \frac{1+e^x}{1+x+e^x}\, dx$ 30. $\displaystyle\int_0^1 \frac{x}{(1+2x^2)^3}\, dx$

31.   The velocity, in meters per second, of an object moving along the x-axis is given by $v(t) = \dfrac{t}{t^2+1}$. Find the total distance that the object travels between $t=0$ and $t=2$.

32.   Find the total area enclosed by the graph of $y = e^{\sin x} \cdot \cos x$ and the x-axis from $x=0$ to $x=\pi$.

33.   Find the area of the region bounded by the curve $y = x\sin(x^2)$ and the lines $y=0$, $x=0$, and $x=\sqrt{\pi}$.

34.   Find the total distance traveled by an object from $t=1$ to $t=2$ if its velocity is given by $v(t) = (t^2+2t-3)^3(t+1)$.

In Exercises 35–38, find the antiderivative.

35. $\displaystyle\int \tan x\, dx$ [Hint: Rewrite as $\displaystyle\int \frac{\sin x}{\cos x}\, dx$ and use the method of Example 4.]

36. $\displaystyle\int \cot x\, dx$

37. $\displaystyle\int \tan(3x)\, dx$

38. $\displaystyle\int \tan(x^2)\cdot x\, dx$

7.3　　The Method of Substitution

A problem-solving strategy that is used throughout mathematics is that of applying a transformation to a given expression to change it into an expression that is easier to work with, that is better understood, or that is in standard form. Examples of this strategy include the expression of a quadratic as the sum or difference of squares using the completing the square transformation and the expression of multiplication problems as addition problems using a logarithmic transformation.

We now consider a transformation technique – called **substitution** – that is used to transform one antiderivative into another, the second antiderivative being simpler than the first. The key steps in using substitution to find antiderivatives are:

1) **Substitute**	Choose a substitution $u = g(x)$, and write $du = g'(x)\,dx = \dfrac{du}{dx}\,dx$.
	Then substitute both u and du into the original integral
	to produce a new integral of the form $\int f(u)\,du$.
2) **Antidifferentiate**	Antidifferentiate in terms of u; i.e., find a function F for which $F'(u) = f(u)$.
3) **Resubstitute**	Substitute $g(x)$ for u to obtain an antiderivative in terms of x.

Substitution provides a more formal approach to the "Reversing the Chain Rule" technique. We will redo the last three examples in Section 7.2 to illustrate how substitution works with integrals of the form $\int f'[g(x)] \cdot g'(x)\,dx$.

Example 1:　*Using Substitution*

Use substitution to integrate $\int \sqrt{x^2 + 1} \cdot 2x \, dx$.

Solution:

First, let u be the inner function, $u = g(x) = x^2 + 1$. Next we use the derivative of the inner function,

$$\frac{du}{dx} = g'(x) = 2x,$$

to write the symbolic equation

$$du = \frac{du}{dx}\,dx = 2x\,dx,$$

where $\dfrac{du}{dx}$ denotes the derivative of u with respect to x and du and dx are part of the notation for integrals.

The two substitutions

$$u = x^2 + 1 \qquad \text{and} \qquad du = 2x\,dx$$

result in the transformation

$$\int \sqrt{x^2 + 1} \cdot 2x \, dx \rightarrow \int \sqrt{u}\,du.$$

Antidifferentiating the simpler integral with respect to u, we obtain

$$\int \sqrt{u} \, du = \frac{2}{3} u^{3/2} + C.$$

Our final step is to replace u with $x^2 + 1$ so that the final answer is in terms of the original variable x,

$$\int \sqrt{x^2 + 1} \cdot 2x \, dx = \frac{2}{3} (x^2 + 1)^{3/2} + C.$$

We can check our answer by using the Chain Rule to differentiate:

$$\frac{d}{dx} \left[\frac{2}{3} (x^2 + 1)^{3/2} + C \right] = \frac{2}{3} \cdot \frac{3}{2} (x^2 + 1)^{1/2} \cdot \frac{d}{dx} (x^2 + 1) = \sqrt{x^2 + 1} \cdot 2x$$

This use of the substitution technique is outlined in the following diagram where f is an antiderivative of f', $\frac{d}{dx} [f(x)] = f'(x)$.

$$\int f'[g(x)] \cdot g'(x) \, dx = f[g(x)] + C$$

$$\downarrow \quad u = g(x)$$
$$\qquad\qquad\qquad \uparrow \quad u = g(x)$$
$$du = g'(x) \, dx$$

$$\int f'(u) \, du \qquad = \qquad f(u) + C$$

Example 2: *Using Substitution*

Use substitution to integrate $\int (1 + \sin(2x))^3 \cdot \cos(2x) \, dx$.

Solution:

If we let $u = 1 + \sin(2x)$ be the inner function, then $\frac{du}{dx} = 2 \cos(2x)$ and

$\frac{1}{2} du = \cos(2x) \, dx$. Now we substitute to obtain:

$$\int (1 + \sin(2x))^3 \cdot \cos(2x) \, dx \qquad = \qquad \frac{1}{8} (1 + \sin(2x))^4 + C$$

$$\downarrow \quad u = 1 + \sin(2x)$$
$$\qquad\qquad\qquad \uparrow$$
$$\frac{du}{dx} = \cos(2x) \cdot 2 \Rightarrow \frac{1}{2} du = \cos(2x) \, dx \qquad u = 1 + \sin(2x)$$

$$\frac{1}{2} \int u^3 \, du \qquad\qquad\qquad = \qquad \frac{u^4}{8} + C$$

Now differentiate this answer to check that it is correct.

Example 3: *Using Substitution*

Use substitution to integrate $\int \dfrac{2t}{t^2+1}\, dt$.

Solution:

If we let $u = t^2 + 1$ be the inner function, then $\dfrac{du}{dt} = 2t$ and $du = 2t\, dt$. Now we substitute to obtain:

$$\int \frac{2t}{t^2+1}\, dt \qquad = \quad \ln\left| t^2 + 1\right| + C$$

$$\downarrow \quad u = t^2 + 1$$

$$\frac{du}{dt} = 2t \Rightarrow du = 2t\, dt.$$

$$\uparrow \quad u = t^2 + 1$$

$$\int \frac{1}{u}\, du \qquad = \quad \ln|u| + C$$

The method of substitution may also be used when the integrand contains radicals. In such cases it is often helpful to solve for x in terms of u and then find the derivative of x with respect to u, $\dfrac{dx}{du}$. The next two examples illustrate this strategy.

Example 4: *Using Substitution*

Use the substitution $u = x+1$ to integrate $\int x\sqrt{x+1}\, dx$.

Solution:

If we let $u = x+1$, then solving for x gives $x = u-1$. Then $\dfrac{dx}{du} = 1$ and $dx = du$. Now substituting for x, $x+1$ and dx we obtain:

$$\int x\sqrt{x+1}\, dx \qquad = \tfrac{2}{5}(x+1)^{5/2} - \tfrac{2}{3}(x+1)^{3/2} + C$$

$$\downarrow \quad u = x+1, \text{ so } \boxed{x = u-1}$$

$$\frac{dx}{du} = 1 \Rightarrow dx = du$$

$$\uparrow \quad u = x+1$$

$$\int \overset{x}{(u-1)}\,\sqrt{u}\,\underset{dx}{du} \quad = \int (u^{3/2} - u^{1/2})\, du = \tfrac{2}{5}u^{5/2} - \tfrac{2}{3}u^{3/2} + C$$

The answer to Example 4 could be checked by differentiation; however, we will check it visually. In the figure, we have a TI-83 graph of both the integrand $f(x) = x\sqrt{x+1}$ and the its antiderivative $g(x) = \tfrac{2}{5}(x+1)^{5/2} - \tfrac{2}{3}(x+1)^{3/2} + C$ (We let C = 0). Notice that f is negative when g decreases, positive when g increases and g has its minimum value when $f(x) = 0$.

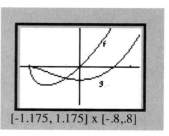

[-1.175, 1.175] x [-.8,.8]

Example 5: *Using Substitution*

Use the substitution $u = \sqrt{x-1}$ to integrate $\int \dfrac{x}{\sqrt{x-1}}\, dx$

Solution:

If we let $u = \sqrt{x-1}$, then solving for x gives $x = u^2 + 1$. Then $\dfrac{dx}{du} = 2u$ so that $dx = 2u \cdot du$. Now we substitute to obtain:

$$\int \frac{x}{\sqrt{x-1}}\, dx \qquad\qquad = \qquad \frac{2}{3}(x-1)^{3/2} + 2\sqrt{x-1} + C$$

$$u = \sqrt{x-1} \text{ , so } \boxed{x = u^2 + 1}$$

$$\frac{dx}{du} = 2u \Rightarrow dx = 2u\, du \qquad\qquad u = (x-1)^{1/2}$$

$$\int \overset{x}{\overbrace{(u^2+1)}} \cdot \frac{1}{u} \cdot \underbrace{2u\, du}_{dx} \;\; = \;\; 2\int (u^2 + 1)\, du \;\; = \;\; \frac{2}{3}u^3 + 2u + C$$

Substitution in Definite Integrals

The method of substitution can also be applied to definite integrals.

Example 6: *Evaluating Definite Integrals*

Evaluate $\displaystyle\int_0^{\pi/2} \sin^2 x \cdot \cos x \; dx$.

Solution:

Let $u = \sin x$. Then $\dfrac{du}{dx} = \cos x$ and $du = \cos x\, dx$. Before substituting, determine new lower and upper limits of integration in terms of u.

Lower limit: When $x = 0$, $u = \sin 0 = 0$.

Upper limit: When $x = \dfrac{\pi}{2}$, $u = \sin \dfrac{\pi}{2} = 1$.

Now we substitute to obtain:

$$\int_0^{\pi/2} \sin^2 x \cdot \cos x \; dx$$

$$u = \sin x$$

$$\frac{du}{dx} = \cos x \Rightarrow du = \cos x\, dx$$

$$\int_{u=0}^{u=1} u^2 \, du \qquad\qquad = \;\; \frac{1}{3}\Big[u^3\Big]_0^1 \;\; = \;\; \frac{1}{3}[1-0] \;\; = \frac{1}{3} \; .$$

7.3   Exercises

In Exercises 1–10, find the antiderivatives by using the indicated substitution. Check your answer by differentiation. (Note: each of these problems can also be done by reversing the Chain Rule)

1. $\int \frac{\ln x}{x} \, dx;$ $u = \ln x$

2. $\int \left[x^3 \cos(x^4 + 2) \right] dx;$ $u = x^4 + 2$

3. $\int \frac{2x}{5x^2 + 7} \, dx;$ $u = 5x^2 + 7$

4. $\int \frac{x^2}{\sqrt{2 + x^3}} \, dx;$ $u = 2 + x^3$

5. $\int \frac{e^{1/x}}{x^2} \, dx;$ $u = \frac{1}{x}$

6. $\int \frac{1}{(2x + 1)^2} \, dx;$ $u = 2x + 1$

7. $\int x^2 \sec^2(x^3) \, dx;$ $u = x^3$

8. $\int \frac{\left(1 + \sqrt{x} \right)^3}{\sqrt{x}} \, dx;$ $u = 1 + \sqrt{x}$

9. $\int \frac{x^3}{(5 - x^4)^2} \, dx;$ $u = 5 - x^4$

10. $\int \left[\csc^2(\sin x) \right] \cdot \cos x \, dx;$ $u = \sin x$

In Exercises 11 –16, find the antiderivative by using the indicated substitution. Check that your answer is reasonable by graphing both the function and its antiderivative (let C = 0).

11. $\int x\sqrt{x + 2} \, dx;$ $u = x + 2$

12. $\int x\sqrt{2x - 1} \, dx;$ $u = 2x - 1$

13. $\int \frac{x}{\sqrt{2x + 3}} \, dx;$ $u = \sqrt{2x + 3}$

14. $\int \frac{x}{\sqrt{x - 1}} \, dx;$ $u = \sqrt{x - 1}$

15. $\int \frac{2x - 1}{\sqrt{x + 3}} \, dx;$ $u = \sqrt{x + 3}$

16. $\int \frac{1}{1 + \sqrt{x}} \, dx;$ $u = \sqrt{x}$

In Exercises 17–26, evaluate the definite integrals exactly using the indicated substitution. Use your calculator to check your work.

17. $\displaystyle\int_0^4 \sqrt{2x+1}\ dx; \quad u = 2x+1$

18. $\displaystyle\int_0^{\sqrt[3]{\pi}} x^2 \sin(x^3)\ dx; \quad u = x^3$

19. $\displaystyle\int_1^e \frac{\ln x}{x}\ dx; \quad u = \ln x$

20. $\displaystyle\int_0^{\pi/4} \tan^3\theta\ \sec^2\theta\ d\theta; \quad u = \tan\theta$

21. $\displaystyle\int_1^2 x\sqrt{x-1}\ dx; \quad u = x-1$

22. $\displaystyle\int_0^4 \frac{x}{\sqrt{1+2x}}\ dx; \quad u = \sqrt{1+2x}$

23. $\displaystyle\int_1^4 \frac{1}{\sqrt{x}(\sqrt{x}+1)}\ dx; \quad u = \sqrt{x}+1$

24. $\displaystyle\int_0^2 \frac{x^2}{(9-x^3)^{3/2}}\ dx; \quad u = 9-x^3$

25. $\displaystyle\int_0^1 \frac{x^3}{\sqrt{x^2+1}}\ dx; \quad u = \sqrt{x^2+1}$

26. $\displaystyle\int_1^4 \frac{\ln\sqrt{x}}{x}\ dx; \quad u = \ln\sqrt{x}$

In Exercises 27–34, choose a suitable substitution function, $u = g(x)$, to find the indicated antiderivative.

27. $\displaystyle\int \frac{(\ln x)^7}{x}\ dx$

28. $\displaystyle\int \frac{e^{2\sqrt{x}}}{\sqrt{x}}\ dx$

29. $\displaystyle\int \frac{\sec(\ln x)\tan(\ln x)}{x}\ dx$

30. $\displaystyle\int \sec^2(e^x)\cdot e^x\ dx$

31. $\displaystyle\int x\sqrt{3x-6}\ dx$

32. $\displaystyle\int \frac{x}{\sqrt{x+1}}\ dx$

33. $\displaystyle\int \frac{4e^{3x}}{3+5e^{3x}}\ dx$

34. $\displaystyle\int (x+5)\sqrt{2x+7}\ dx$

7.4   Integration by Parts

Not all products can be integrated by reversing the Chain Rule. For example, consider the function $f(x) = x \cdot \cos x$. To integrate such functions we need to develop a new technique.

If we consider the derivative of the product $f(x) \cdot g(x)$, we have

$$\frac{d}{dx}[f(x)g(x)] = f(x) \cdot g'(x) + g(x) \cdot f'(x).$$

Thus

$$f(x) \cdot g'(x) = \frac{d}{dx}[f(x)g(x)] - g(x) \cdot f'(x).$$

Taking the antiderivative of both sides we have

$$\int f(x) \cdot g'(x)\, dx = \int \frac{d}{dx}[f(x)g(x)]\, dx - \int g(x) \cdot f'(x)\, dx.$$

Since the first integrand on the right is the derivative of $f(x) \cdot g(x)$, we obtain

$$\int f(x) \cdot g'(x)\, dx = f(x) \cdot g(x) - \int g(x) \cdot f'(x)\, dx$$

This equation is called the **rule of integration by parts** and can be written in more condensed form:

$$\int f \cdot g' = f \cdot g - \int g \cdot f'$$

The word **parts** refers to the functions f and g' in the integrand on the left side.

According to the rule, the integral of a product of two functions may be written as a difference, one term of which is a new integral. Fortunately, in many cases the new integral is simpler than the original integral. Thus integration by parts is a technique by which we replace one integration problem by a simpler one. An important aspect of employing integration by parts is designating the factors $f(x)$ and $g'(x)$ of the original integrand in such a way that $f(x)$ simplifies under differentiation while the factor $g'(x)$ does not become more complicated under integration.

Example 1: *Integration by Parts*

Find $\int x \cdot \cos x\, dx$.

Solution:

If we take $f(x) = x$ and $g'(x) = \cos x$ (so that $f'(x) = 1$ and $g(x) = \sin x$), the parts formula,

$$\int f \cdot g' = f \cdot g - \int g \cdot f', \text{ yields}$$

$$\int x \cdot \cos x\, dx = x \cdot \sin x - \int \sin x\, dx = x \cdot \sin x + \cos x + C.$$

Example 2: *Integration by Parts*

Find $\int x^2 \cdot \ln x \; dx$.

Solution:

If we take $f(x) = \ln x$ and $g'(x) = x^2$, then $f'(x) = \frac{1}{x}$ and $g(x) = \frac{1}{3} x^3$ By the parts

formula, $\int f \cdot g' = f \cdot g - \int g \cdot f'$, we have

$$\int x^2 \cdot \ln x \; dx = \frac{x^3}{3} \cdot \ln x - \int \frac{x^2}{3} \, dx$$

$$= \frac{x^3}{3} \cdot \ln x - \frac{x^3}{9} + C$$

It can happen that a particular integral may require repeated applications of integration by parts. This is illustrated in the following example.

Example 3: *Repeated Integration by Parts*

Find $\int x^2 \cdot e^x \; dx$.

Solution:

Let $f(x) = x^2$ and $g'(x) = e^x$, then $f'(x) = 2x$ and $g(x) = e^x$. Then

$$\int x^2 \cdot e^x \; dx = x^2 \cdot e^x - 2 \int x \cdot e^x \; dx.$$

The integral on the right-side of the equation can be evaluated by a second integration by parts. Let $f(x) = x$, $g'(x) = e^x$, so that $f'(x) = 1$ and $g(x) = e^x$. Then

$$\int x \cdot e^x \; dx = x \cdot e^x - \int e^x \; dx = x \cdot e^x - e^x + C.$$

Substituting above we have

$$\int x^2 \cdot e^x \; dx = x^2 \cdot e^x - 2 \Big[x \cdot e^x - e^x + C \Big] = x^2 \cdot e^x - 2x \cdot e^x + 2e^x + D,$$

where $D = -2C$ still represents any real number constant.

A trick that is useful when integrating a single function by parts is to consider the function $g'(x)$ to be the factor 1. This is illustrated in the next example.

Example 4: *An Integrand with a Single Term*

Find $\int \ln x \; dx$.

Solution:

Take $f(x) = \ln x$ and $g'(x) = 1$, so that $f'(x) = \frac{1}{x}$ and $g(x) = x$. Then

$$\int \ln x \; dx = x \cdot \ln x - \int \frac{1}{x} \cdot x \, dx$$

$$= x \cdot \ln x - \int 1 \, dx$$

$$= x \cdot \ln x - x + C.$$

7.4   Exercises

In Exercises 1–14, evaluate the given integral. Solve by the simplest method – not all require integration by parts. Check your answers by differentiation.

1. $\int x \cdot e^x \, dx$ 2. $\int x \cdot \ln x \, dx$

3. $\int \ln(2x) \, dx$ 4. $\int x \cdot e^{2x} \, dx$

5. $\int x \cdot \sin x \, dx$ 6. $\int (\ln x)^2 \, dx$

7. $\int \ln \sqrt{x} \, dx$ 8. $\int \frac{\ln x}{x^2} \, dx$

9. $\int x \cdot \sec^2 x \, dx$ 10. $\int x^3 \cdot \ln x \, dx$

11. $\int (\ln x)^2 \cdot \frac{1}{x} \, dx$ 12. $\int x^2 \cdot \sin x \, dx$

13. $\int x \cdot \sqrt{1+x} \, dx$ 14. $\int \sqrt{x} \cdot \ln x \, dx$

In Exercises 15–20, evaluate the definite integral exactly. Check your answer by comparing it to a midpoint estimate computed with $n = 10$.

15. $\int_0^1 x \cdot e^{-2x} \, dx$ 16. $\int_0^\pi x \cdot \cos(2x) \, dx$

17. $\int_0^{\pi/2} x \cdot \sin(2x) \, dx$ 18. $\int_0^1 x\sqrt{1-x} \, dx$

19. $\int_2^8 \ln \sqrt{2x} \, dx$ 20. $\int_0^1 \ln(1+x) \, dx$

21.   Find the area of the region bounded by the graph of $y = \ln x$, the x-axis and the line $x = e^2$.

22.   Find the area of the region in the first quadrant bounded by the graph of $y = xe^{-x}$, the x-axis and the line $x = 5$.

7.5   The Trigonometric Functions and Their Inverses

The trigonometic functions, sine, cosine and tangent do not have inverses because they are periodic and therefore not one-to-one. However, by suitably restricting the domain of each of these functions we produce new functions that are one-to-one and that also take on all of the possible function values of the corresponding trigonometric function.

Sine

There are many intervals we could choose to consider as restricted domains on which sine would be one-to-one. Any closed interval of length π and with left endpoint at an odd multiple of $\frac{\pi}{2}$ works. The usual choice is the interval $[-\frac{\pi}{2}, \frac{\pi}{2}]$.

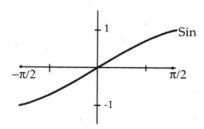

On this interval sine is increasing and takes on all the possible function values for sine; that is, all real numbers between -1 and 1.

We call the new function with this restricted domain **Sine** (pronounced "Cap-Sine")

> The function Sine, denoted by Sin, is defined only on the interval $[-\frac{\pi}{2}, \frac{\pi}{2}]$ by $\text{Sin}(x) = \sin(x)$.

Since the function Sine behaves exactly like sine except for the restriction on its domain, its derivative is still given by cosine. That is for all x in $[-\frac{\pi}{2}, \frac{\pi}{2}]$, $\frac{d}{dx}(\text{Sin}\, x) = \cos(x)$. The function cosine is nonnegative on the interval $[-\frac{\pi}{2}, \frac{\pi}{2}]$, so Sine is an increasing function on its whole domain. Hence Sine has an inverse, which we denote in either of two ways: Sin^{-1} or arcsin.

> The inverse of Sine is denoted by **Sin^{-1} or arcsin**.

The graph of arcsin is obtained from that of Sin by reflecting in the line $y = x$.

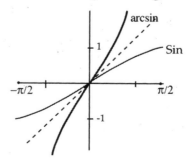

The domain of arcsin is the interval $[-1, 1]$, and its range is $[-\frac{\pi}{2}, \frac{\pi}{2}]$

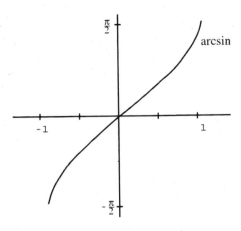

Example 1: *Evaluating Inverse Trigonometric Functions*

Find arcsin (.707), arcsin ($\frac{1}{2}$) and arcsin ($\frac{-\sqrt{3}}{2}$) .

Solution:

With a calculator in radian mode, we find arcsin (.707) $\approx$ .785. Approximations to

arcsin ($\frac{1}{2}$) and arcsin ($\frac{-\sqrt{3}}{2}$) may also be found with a calculator. Or you may use the fact

that Sin ($\frac{\pi}{6}$) = $\frac{1}{2}$ to obtain arcsin ($\frac{1}{2}$) = $\frac{\pi}{6}$. Similarly, arcsin ($\frac{-\sqrt{3}}{2}$) = $-\frac{\pi}{3}$.

Example 2: *Using Right Triangles*

Simplify cos (arcsin (x)).

Solution:

If we let $y = \arcsin(x)$, then $x = \sin y$. Form a right triangle with hypotenuse of length 1
that has an acute angle y such that $y = \arcsin(x)$. Then

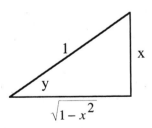

$$\cos (\arcsin x) = \cos y \ = \ \frac{\sqrt{1-x^2}}{1}.$$

So,

$$\cos (\arcsin (x)) = \sqrt{1-x^2} \ .$$

The Derivative of Arcsin x

Let $y = \arcsin x$, $-\frac{\pi}{2} \le y \le \frac{\pi}{2}$. Hence, $\sin y = x$, and if $-\frac{\pi}{2} \le y \le \frac{\pi}{2}$, then $\cos y \ge 0$, so that $\cos y = \sqrt{1 - \sin^2 y}$. Using implicit differentiation it follows that

$$\cos y \cdot \frac{dy}{dx} = 1$$

$$\frac{dy}{dx} = \frac{1}{\cos y} = \frac{1}{\sqrt{1 - \sin^2 y}} = \frac{1}{\sqrt{1 - x^2}}$$

$$\frac{d}{dx}[\arcsin x] = \frac{1}{\sqrt{1 - x^2}}, \quad -1 < x < 1$$

The derivative for the function arcsin gives us a new integration formula.

$$\int \frac{1}{\sqrt{1 - x^2}}\, dx = \arcsin x + C$$

Example 3: *Evaluating the Derivative of Arcsin*

If $f(x) = \arcsin x$, find $f'(\frac{1}{2})$.

Solution:

$$f'(x) = \frac{1}{\sqrt{1 - x^2}}, \text{ so } f'(\tfrac{1}{2}) = \frac{1}{\sqrt{1 - 1/4}} = \frac{2}{\sqrt{3}}.$$

Example 4: *Using the Chain Rule*

If $f(x) = \arcsin \sqrt{1 - x}$, find $f'(x)$.

Solution:

By the Chain Rule, we obtain

$$f'(x) = \frac{1}{\sqrt{1 - \left(\sqrt{1 - x}\right)^2}} \cdot \frac{-1}{2\sqrt{1 - x}} = \frac{-1}{2\sqrt{x - x^2}}$$

Example 5: *Integration Involving the Arcsin Function*

Evaluate $\int \dfrac{2x}{\sqrt{1-x^4}}\,dx$.

Solution:

If we take $f(x) = \dfrac{1}{\sqrt{1-x^2}}$ and $g(x) = x^2$, then the integrand can be considered to be

$f(g(x)) \cdot g'(x) = \dfrac{1}{\sqrt{1-(x^2)^2}} \cdot 2x$. Hence its antiderivative is $F(g(x)) + C$ where F is an

antidervative of f. We therefore have

$$\int \frac{2x}{\sqrt{1-x^4}}\,dx = \int \frac{1}{\sqrt{1-(x^2)^2}} \cdot 2x \; dx = \arcsin(x^2) + C.$$

Cosine

The interval $[-\frac{\pi}{2}, \frac{\pi}{2}]$ is not suitable as a restricted domain for cosine if our goal is to get a one-to-one function. Instead, we take the interval $[0, \pi]$ and define **Cosine** (pronounced "Cap Cosine") to be a function that behaves like cosine on this restricted interval.

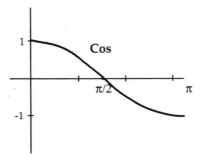

> The function Cosine, denoted by Cos, is defined on the domain $[0, \pi]$ by $\text{Cos}(x) = \cos x$.

Cosine is a one-to-one function, its derivative, $-\sin x$, is nonpositive on $[0, \pi]$, and, hence, it has an inverse.

> The inverse of Cosine is denoted by **Cos⁻¹** or **arccos**.

The domain of arccos is the interval $[-1, 1]$ and its range is $[0, \pi]$.

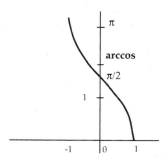

By an argument similar to that for arcsin the derivative of arccos is

$$\frac{d}{dx}[\arccos x] = \frac{-1}{\sqrt{1-x^2}}, \; -1 < x < 1 \; .$$

Example 6: *Evaluating a Definite Integral*

Evaluate $\displaystyle\int_0^{1/2} \frac{-1}{\sqrt{1-x^2}} \; dx$.

Solution:

$$\int_0^{1/2} \frac{-1}{\sqrt{1-x^2}} \; dx = [\arccos x]_0^{1/2} = \arccos(\tfrac{1}{2}) - \arccos(0) = \frac{\pi}{3} - \frac{\pi}{2} = -\frac{\pi}{6}$$

Tangent

The interval used to produce a one-to-one function for tangent is the open interval $(-\frac{\pi}{2}, \frac{\pi}{2})$.

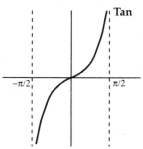

The function defined by $f(x) = \tan x$ but restricted to this interval has an inverse, which we denote Tan^{-1} or arctan. Since the range of the tangent function is the set of all real numbers, the domain of Tan^{-1} is all reals.

> **Tan^{-1}** (or **arctan**) is the inverse of the function f
> defined on the domain $(-\frac{\pi}{2}, \frac{\pi}{2})$ by $f(x) = \tan x$.

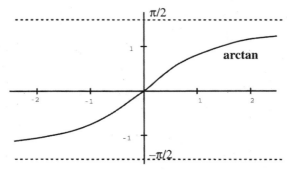

Tan^{-1} has a derivative for all x. To derive the formula for this derivative let $y = \arctan x$, with $-\frac{\pi}{2} < y < \frac{\pi}{2}$. Hence, $\tan y = x$ and differentiating implicitly, it follows that

$$\frac{d}{dx}(\tan y) = \frac{d}{dx}(x)$$

$$\sec^2 y \cdot \frac{dy}{dx} = 1 \ .$$

Using the Chain Rule on the left side,

$$\frac{dy}{dx} = \frac{1}{\sec^2 y} = \frac{1}{1 + \tan^2 y} = \frac{1}{1 + x^2} \ .$$

$$\boxed{\frac{d}{dx}[\arctan x] = \frac{1}{1 + x^2}}$$

Example 7: *Evaluating the Derivative of Arctan*

Find $\frac{dy}{dx}$ at $(1, \frac{\pi}{4})$ if $y = \arctan(x^2 - 2)$.

Solution:

Using the Chain Rule, $\frac{dy}{dx} = \frac{1}{1 + (x^2 - 2)^2} \cdot 2x$. When $x = 1$, we obtain $\frac{dy}{dx} = \frac{2}{2} = 1$.

Example 8: *Integration Involving the Arctan Function*

Evaluate $\displaystyle\int_0^1 \frac{x}{1 + x^4} \, dx$.

Solution:

If we let $f(x) = \frac{1}{1 + x^2}$ and $g(x) = x^2$, then the integrand differs from

$$f(g(x)) \cdot g'(x) = \frac{2x}{1 + (x^2)^2}$$ only by the factor 2. Thus we obtain

$$\int_0^1 \frac{x}{1 + x^4} \, dx = \frac{1}{2} \int_0^1 \frac{2x}{1 + (x^2)^2} \, dx = \frac{1}{2} \left[\arctan(x^2) \right]_0^1$$

$$= \frac{1}{2} \left[\arctan(1) - \arctan(0) \right] = \frac{1}{2} \cdot (\frac{\pi}{4} - 0) = \frac{\pi}{8} \ .$$

Here is a summary of the differentiation and integration formulas that have been derived in this section.

1) $\dfrac{d}{dx}[\arcsin x] = \dfrac{1}{\sqrt{1-x^2}}$ $-1 < x < 1$

2) $\dfrac{d}{dx}[\arccos x] = \dfrac{-1}{\sqrt{1-x^2}}$ $-1 < x < 1$

3) $\dfrac{d}{dx}[\arctan x] = \dfrac{1}{1+x^2}.$

Each of the differentiation formulas above gives rise to an integration formula. The two most useful are the following.

1) $\displaystyle\int \dfrac{1}{\sqrt{1-x^2}}\, dx = \arcsin x + C$

2) $\displaystyle\int \dfrac{1}{1+x^2}\, dx = \arctan x + C$

Completing the Square

Completing the square sometimes helps when quadratic functions are involved with the integrand.

Example 9: *Completing the Square*

Evaluate $\displaystyle\int \dfrac{1}{x^2 + 6x + 13}\, dx$.

Solution:

The denominator can be written as the sum of two squares as follows:

$$x^2 + 6x + 13 = (x^2 + 6x + 9) + 4 = (x+3)^2 + 4$$

Now, $\displaystyle\int \dfrac{1}{x^2 + 6x + 13}\, dx = \int \dfrac{1}{(x+3)^2 + 4}\, dx = \dfrac{1}{4}\int \dfrac{1}{\left(\dfrac{x+3}{2}\right)^2 + 1}\, dx$.

Now we reverse the Chain Rule to integrate. Letting $g(x) = \dfrac{x+3}{2}$ and $f(x) = \dfrac{1}{1+x^2}$, we

have $g'(x) = \dfrac{1}{2}$ and $F(x) = \arctan x$. Thus

$$2\cdot\dfrac{1}{4}\int \dfrac{1}{\left(\dfrac{x+3}{2}\right)^2 + 1}\cdot \dfrac{1}{2}\, dx = \dfrac{1}{2}\int f[g(x)]\cdot g'(x)\, dx = \dfrac{1}{2}\arctan(\dfrac{x+3}{2}) + C.$$

7.5   Exercises

1.   Evaluate each of the following using a calculator as needed:

a) arcsin 1   b) arctan 1   c) arccos 0

d) $\arcsin\left(-\frac{1}{2}\right)$ e) $\arccos\left(\frac{\sqrt{2}}{2}\right)$ f) $\arctan\left(-\sqrt{3}\right)$

g) $\arccos\left(-\frac{1}{2}\right)$ h) $\sin\left(\arccos\frac{1}{2}\right)$ i) $\arctan\left(\sin\left(-\frac{\pi}{2}\right)\right)$

j) arcsin (0.851)   k) arctan (50)   l) arccos (−.755)

2.   Simplify each of the following:

a) sin (arccos x)   b) sin (2 arcsin x)   c) tan (arcsin x)

d) sec (arctan x)   e) cos (2 arcsin x)

In Exercises 3–16, determine $f'(x)$.

3. $f(x) = \arcsin(2x)$ 4. $f(x) = \arccos(-3x)$

5. $f(x) = \arctan(x+1)$ 6. $f(x) = x\arcsin x$

7. $f(x) = \ln(\arctan x)$ 8. $f(x) = \arctan(\sqrt{x})$

9. $f(x) = e^{\arcsin x}$ 10. $f(x) = (\arccos x)^2$

11. $f(x) = \sqrt{\arctan(2x)}$ 12. $f(x) = \ln(\arcsin x)$

13. $f(x) = \arctan(\sin x)$ 14. $f(x) = \arccos\left(\frac{x}{2}\right)$

15. $f(x) = e^x \arcsin x$ 16. $f(x) = \arctan\left(\frac{2x}{1-x^2}\right)$

17.   Show that $\displaystyle\int \frac{1}{\sqrt{a^2 - x^2}}\, dx = \arcsin\frac{x}{a} + C.$ [Hint: Differentiate the right-hand side.]

18.   Show that $\displaystyle\int \frac{1}{a^2 + x^2}\, dx = \frac{1}{a}\arctan\frac{x}{a} + C.$

19.   Show that $\displaystyle\int \frac{1}{a^2 + (x+b)^2}\, dx = \frac{1}{a}\arctan\left(\frac{x+b}{a}\right) + C.$

In Exercises 20–27, evaluate the given integral.

20. $\displaystyle\int_{-1}^{1} \frac{1}{1+x^2}\,dx$

21. $\displaystyle\int_{0}^{\sqrt{2}/2} \frac{1}{\sqrt{1-x^2}}\,dx$

22. $\displaystyle\int_{0}^{4} \frac{1}{16+x^2}\,dx$

23. $\displaystyle\int_{0}^{1/4} \frac{1}{\sqrt{1-4x^2}}\,dx$

24. $\displaystyle\int_{-2}^{0} \frac{1}{25+(x+2)^2}\,dx$

25. $\displaystyle\int_{0}^{1} \frac{1}{x^2+4x+8}\,dx$

26. $\displaystyle\int_{0}^{2} \frac{1}{\sqrt{16-x^2}}\,dx$

27. $\displaystyle\int_{1}^{e} \frac{1}{x[1+(\ln x)^2]}\,dx$

28.   Sketch a graph of the function $y = \arcsin x + \arccos x$.

29.   Determine $\dfrac{dy}{dx}$ if $x\sin y = \arctan y$.

30.   The acceleration of an object moving along the x-axis is related to time by $a(t) = \dfrac{\arctan t}{t^2+1}$.
Determine the velocity at time $t = 0$ given that $v(t) = 1$ when $t = 1$.

31.   Find the area of the region bounded by the graph of $y = \dfrac{8}{x^2+4}$, the x-axis, the y-axis
and the line $x = 2$.

32.   Sketch the region whose area is represented by each integral. Use the integration
capabilities of your calculator to approximate area of each region.

a) $\displaystyle\int_{0}^{1} \arcsin x\,dx$ b) $\displaystyle\int_{0}^{1} \arccos x\,dx$ c) $\displaystyle\int_{0}^{1} \arctan x\,dx$

33.   a) Show that $\displaystyle\int_{0}^{1} \frac{4}{1+x^2}\,dx = \pi$.

b) Approximate the nmber π by using the integration capabilities of your calculator.

7.6   Numerical Integration

Most applications of integration involve the definite integral, $\int_a^b f(x)\,dx$. Thanks to the Fundamental Theorem, we can evaluate such definite integrals provided we can find an antiderivative for the integrand, $\int_a^b f(x)\,dx = F(b) - F(a)$, where $F'(x) = f(x)$. And this is why a considerable amount of time has been spent developing strategies for finding antiderivatives.

However, there are two problems that can prevent our using the Fundamental Theorem.

1) There are some continuous functions for which simple antiderivatives cannot be found; the innocent looking $y = e^{-x^2}$ is an example.

2) We may not be given a formula for the continuous function f. f may be an unknown function whose values at certain points of the interval $[a, b]$ have been determined by experimental measurements.

Nevertheless, the definite integral of a continuous function does exist and is defined to be the limiting value of Riemann sums. Thus, we can approximate the integral's value numerically using left- and right-hand Riemann sums. Recall that a left-hand sum of f over an interval $[a, b]$ with n subdivisions is a sum of terms like $f(x_i)\Delta x$, where the input value x_i is chosen to be the left-hand endpoint in each subinterval. The right-hand Riemann sum is the same except that we choose x_i to be the right-hand endpoint in each subinterval. The purpose of this section is to formally introduce two other approximation methods, the **Midpoint Rule** and the **Trapezoid Rule**. Both methods require only simple calculations and can be easily implemented on a graphing calculator.

Midpoint Rule

One method of approximating a definite integral $\int_a^b f(x)\,dx$ is to use a **midpoint Riemann sum** (program RSUM with C = .5), taking midpoints of equal subdivisions for the input x_i where $f(x_i)$ is calculated.

Example 1: *Approximating a Definite Integral*

Approximate $\int_1^3 x^2\,dx$ using a

a) left-hand sum   b) right-hand sum   c) midpoint sum

with $n = 4$ equal subdivisions. Compare your estimates with the exact value of the integral.

Solution:

Here $a = 1$, $b = 3$, $n = 4$ and $f(x) = x^2$. The subdivision size is $\Delta x = \dfrac{b-a}{n} = \dfrac{3-1}{4} = 0.5$

a) The left-hand sum gives the approximation

$$L_4 = (.5) \cdot f(1) + (.5) \cdot f(1.5) + (.5) \cdot f(2) + (.5) \cdot f(2.5) = .5(1 + 2.25 + 4 + 6.25) = 6.75.$$

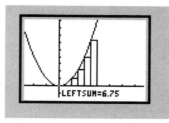

b) The right-hand sum gives the approximation

$$R_4 = (.5) \cdot f(1.5) + (.5) \cdot f(2) + (.5) \cdot f(2.5) + (.5) \cdot f(3) = .5(2.25 + 4 + 6.25 + 9) = 10.75.$$

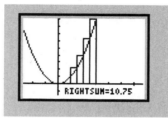

c) The midpoint sum gives the approximation

$$M_4 = (.5) \cdot f(1.25) + (.5) \cdot f(1.75) + (.5) \cdot f(2.25) + (.5) \cdot f(2.75)$$
$$= .5(1.5625 + 3.0625 + 5.0625 + 7.5625) = 8.625.$$

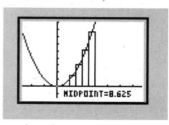

All three Riemann sums are approximating

$$\int_1^3 x^2 \, dx = \left[\frac{1}{3} x^3 \right]_1^3 = \frac{1}{3}(27 - 1) = \frac{26}{3} \approx 8.667$$

Notice that since the function $f(x) = x^2$ is increasing over the interval [1, 3], the left-hand sum is an underestimate of the integral, while the right-hand sum is an overestimate. Similar to the left-hand sum, the midpoint sum underestimates the value of the integral in this case. The error, however, is not as great. Note that each midpoint rectangle is partly above and partly below the graph, so the errors tend to cancel each other.

It is not always clear from a sketch of midpoint rectangles whether the Midpoint Rule overestimates or underestimates the exact value of a definite integral. However, if we consider the Midpoint Rule as fitting a tangent line to the curve at the midpoint of each subinterval, we quickly see whether we have an overestimate or an underestimate.

In the diagrams below, line BF is tangent to the graph of f at the midpoint M. Note that the midpoint rectangle $ACED$ and the trapezoid $ABFD$ have the same area, because the shaded right triangles are congruent and have the same area. Thus, if the graph of the function is concave up, the Midpoint Rule underestimates; if the graph is concave down the Midpoint Rule overestimates.

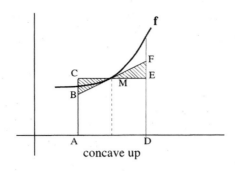

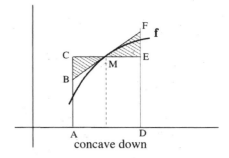

concave up   concave down

Trapezoid Rule

Recall that for a monotone function (either increasing or decreasing) one endpoint rule will give an overestimate and the other endpoint rule will give an underestimate. Thus a Midpoint Rule was introduced as a way of balancing out the errors in the left- and right-hand sums. But instead of using a midpoint sum, we can simply average the left-and right-hand sums. This approximation is called the **Trapezoid Rule**.

Example 2: *Using the Trapezoid Rule*

Approximate $\int_1^3 x^2\, dx$ using the Trapezoid Rule with $n = 4$ equal subdivisions.

Solution:

The left- and right-hand sums for this definite integral were computed in Example 1; therefore, the Trapezoid Rule approximation is

$$T_4 = \frac{L_4 + R_4}{2} = \frac{6.75 + 10.75}{2} = 8.75.$$

The Trapezoid Rule averages the values of f at the left and right endpoints of each subinterval and multiples by Δx. The reason for the name *Trapezoid Rule* can be seen from the figure below which illustrates the case where f is nonnegative, $f(x) \geq 0$. The area of the trapezoid that lies above the kth interval is

$$\frac{1}{2}\,(\text{altitude})(\text{sum of bases}) = \frac{1}{2}\,\Delta x\big[f(x_{k-1}) + f(x_k)\big],$$

and adding up all of these trapezoids gives an approximation of the area under the graph of f.

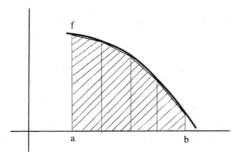

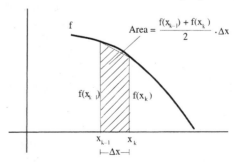

The easiest way to remember the Trapezoid Rule is as the average of the left- and right-hand sums.

In practice, we may not have a rule or formula for the function we are trying to integrate. For example, in experiments in the physical and biological sciences, it is usually the case that the only information available about a function comes from measurements made at a finite collection of points.

Example 3: *Approximating Distance Traveled*

The velocity of an object moving in a straight line was measured at regular intervals with the following results. Use the Trapezoid Rule to estimate how far the object traveled from $t = 0$ to $t = 1$.

t (sec)	0	0.25	0.5	0.75	1.0
v (m/sec)	1	1.3	1.8	1.6	1.4

Solution:

We start by plotting the points and connecting them with straight line segments. Notice that the region created is composed of trapezoids.

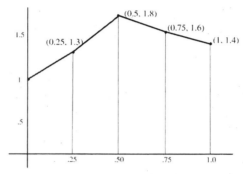

Recall that the area of a trapezoid with sides h_1 and h_2 and base b is given by $b \cdot (h_1 + h_2) / 2$, so the total area is

$$.25 \frac{f(0) + f(.25)}{2} + .25 \frac{f(.25) + f(.5)}{2} + .25 \frac{f(.5) + f(.75)}{2} + .25 \frac{f(.75) + f(1)}{2}$$

$$= \frac{.25}{2} [f(0) + 2f(.25) + 2f(.5) + 2f(.75) + f(1)] = 1.475.$$

the distance traveled is the value of the integral $\int_0^1 v(t)\, dt$, and, for $n = 4$ subdivisions, the trapezoid approximation is 1.475 meters.

Because the trapezoid approximation simply sums the areas of trapezoids, it is possible to give a formula for the approximation. Given the interval $[a, b]$ is divided into n subintervals of equal length $\Delta x = \frac{b-a}{n}$ by points $a = x_0 < x_1 < ... < x_n = b$ the Trapezoid Rule approximation T_n of $\int_a^b f(x)\, dx$ is

$$T_n = \frac{\Delta x}{2} [f(x_0) + 2f(x_1) + 2f(x_2) + ... + 2f(x_{n-1}) + f(x_n)]$$

This formula is frequently used in hand calculations for small values of n.

Comparison of the Midpoint and Trapezoid Rules

It is helpful to know when a rule is producing an underestimate and when it is producing an overestimate. The figures below illustrate the midpoint and trapezoid approximations over subintervals where the function graph is concave down and concave up.

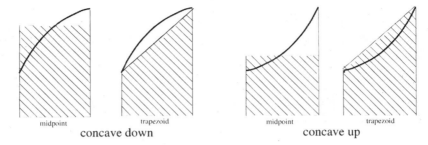

concave down   concave up

Notice that for the concave down graphs the Midpoint Rule overestimates the area while the Trapezoid Rule underestimates the area. Note also that the error for the trapezoid estimate appears to be about twice that for the midpoint estimate. For the concave up graph, the Midpoint Rule now underestimates the area while the Trapezoid Rule overestimates the area. Again, the error for the trapezoid estimate of the area appears to be approximately twice that for the midpoint estimate.

Example 4: *Comparing the Midpoint and Trapezoid Approximations*

For $\int_0^1 3x^2 dx$ compare the midpoint and trapezoid approximations with $n = 4$, $n = 8$ and $n = 16$.

Solution:

For the midpoint approximation, we use the program RSUM with $C = .5$. To compute the trapezoid approximation we use the program RSUM with $C = 0$ for the left-hand sum, $C = 1$ for the right-hand sum and average the sums to get the trapezoid approximation. The results are listed below.

N	Midpoint	Error	Trapezoid	Error
4	.984375	.015625	1.03125	.03125
8	.99609375	.00390625	1.0078125	.0078125
16	.999023435	.000976565	1.00195315	.001953125

From the Fundamental Theorem we have $\int_0^1 3x^2 dx = \left[x^3\right]_0^1 = 1$ exactly. The errors for the two rules here are small, although the Midpoint Rule is slightly more accurate. The geometry of the two rules and the fact that $f(x) = 3x^2$ is concave up on [0, 1] should explain why the midpoint approximations are too low and the trapezoid approximations are too high. And the two Error columns show that the trapezoid error is consistently twice the midpoint error in magnitude.

Error Estimates for the Midpoint and Trapezoid Rules

An approximation process is only valuable if you have some idea about the size of the possible error. The following inequalities, which are studied in numerical analysis, give formulas for estimating the errors involved in the use of the Midpoint Rule, Mid(n), and the Trapezoid Rule, Trap(n).

If $f''(x)$ exists in $[a, b]$ and M_2 is the maximum of $|f''(x)|$ on the interval $[a, b]$, then

$$\left| \int_a^b f(x)\, dx - \text{Mid}(n) \right| \le \frac{M_2 (b-a)^3}{24n^2} \qquad \text{Midpoint Rule}$$

$$\left| \int_a^b f(x)\, dx - \text{Trap}(n) \right| \le \frac{M_2 (b-a)^3}{12n^2} \qquad \text{Trapezoid Rule}$$

There are three factors that contribute to the size of the error when using a midpoint or trapezoid approximation. These are, $b - a$, the width of the interval; n, the number of subdivisions, and M_2, the maximum value of $|f''(x)|$ on the interval $[a, b]$. Knowing that the first derivative measures slope and the second derivative measures concavity, it is reasonable that the size of the second derivative influences the error for both rules. Recall that the Midpoint Rule can be thought of as fitting a tangent line to the curve at the midpoint of each subinterval, thereby completing a trapezoid instead of a rectangle.

Example 5:   *Controlling the Error in the Midpoint Rule*

     Determine how many subdivisions are required with the Midpoint Rule to approximate $\int_1^2 \frac{1}{x}\, dx$ with an error of less than 0.001.

Solution:

     From the graph of $f''(x) = \frac{2}{x^3}$ it can be determined that the largest value of $|f''(x)|$ on the interval $[1, 2]$ is 2. Thus by the error formula above,

$$\left| \int_1^2 \frac{1}{x}\, dx - \text{Mid}(n) \right| \le \frac{M_2 (b-a)^3}{24n^2} = 2 \cdot \frac{1}{24n^2} = \frac{1}{12n^2}.$$

To ensure that the error is less than 0.001 we solve the following inequality:

$$\left(\frac{1}{12n^2} \right) < 0.001$$

and get

$$n > \sqrt{\frac{1000}{12}} \approx 9.13.$$

Thus, we choose $n = 10$ (n must be at least 9.13) and apply the Midpoint Rule to obtain

$$\int_1^2 \frac{1}{x}\, dx \approx .6928353604 \qquad (\text{ Exact value of } \int_1^2 \frac{1}{x}\, dx = \ln 2 \approx .6931471806)$$

Example 6: *Controlling the Error in the Trapezoid Rule*

Use the Trapezoid Rule to determine an approximation of $\int_0^1 \cos(x^2)\, dx$ with an error of no more than .0001. Confirm your results with the calculator approximation **fnInt**$(\cos(X^2), X, 0,1)$.

Solution:

For the function $y = \cos(x^2)$, and we have

$$\frac{dy}{dx} = -2x\sin(x^2),$$

$$\frac{d^2y}{dx^2} = -4x^2 \cos(x^2) - 2\sin(x^2).$$

From a graph of $\frac{d^2y}{dx^2} = -4x^2 \cos(x^2) - 2\sin(x^2)$ it can be determined that the largest value of $|f''(x)|$ on the interval $[0, 1]$ is approximately 3.84488 at $x = .9941$. Applying the error formula for the Trapezoid Rule with $M_2 = 3.84488$,

$$\left| \int_0^1 \cos(x^2)\, dx - \text{Trap}(n) \right| \le \frac{M_2 (b-a)^3}{12n^2} = 3.84488 \cdot \frac{1}{12n^2}.$$

To ensure that the error is less than 0.0001 we solve the following inequality:

$$\left(\frac{3.84488}{12n^2} \right) < 0.0001$$

and get

$$n > \sqrt{\frac{38448.8}{12}} \approx 56.60.$$

Thus we choose $n = 57$ (since n must be at least 56.60).

Choosing $n = 57$, we apply the Trapezoid Rule to obtain

$$\int_0^1 \cos(x^2)\, dx \approx .9044810722.$$

The value of **fnInt**$(\cos(X^2), X, 0,1) = .9045242379$.

7.6   Exercises

In Exercises 1–4, use a) the Midpoint Rule and b) the Trapezoid Rule to approximate the given integral for $n = 10$. Round answers to three decimal places. Then compare with the exact value of the definite integral and the calculator approximation **fnInt(f(X),X, A, B)**.

1. $\int_0^1 \sqrt{1 - x^2} \, dx$

2. $\int_2^4 \frac{x}{x^2 + 1} \, dx$

3. $\int_0^1 \sec^2 x \, dx$

4. $\int_0^1 x \cdot \cos(x^2) \, dx$

None of the definite integrals in Exercises 5–8 can be evaluated using the Fundamental Theorem, because the necessary antiderivative cannot be expressed exactly in terms of elementary functions. Use the Midpoint Rule with $n = 10$ to approximate each integral. Round off each approximation to three decimal places.

5. $\int_0^1 \sqrt{1 + x^3} \, dx$

6. $\int_0^1 \sqrt{\sin x} \, dx$

7. $\int_1^2 \frac{\sin x}{x} \, dx$

8. $\int_0^1 \sin(x^2) \, dx$

9.   Using the Trapezoid Rule with $n = 8$, approximate the area of the first quadrant region bounded by the graph of $y = \frac{x}{e^x} - \frac{1}{10}$ and the x-axis.

10.   a) Estimate $\int_0^1 \frac{1}{1 + \cos x} \, dx$ with $n = 10$ using

   i) a left-hand Riemann sum   ii) a right-hand Riemann sum   iii) the Trapezoid Rule

   b) Explain why the first estimate is too small and the second estimate is too large.

11.   Use the Trapezoid Rule to approximate $\int_1^2 \frac{1}{x} \, dx$ and $\int_1^8 \frac{1}{x} \, dx$ with n = 8. Compare your results. Guess what $\int_1^{16} \frac{1}{x} \, dx$ might be. Check it with the Trapezoid Rule.

12.   a) Verify that $\int_0^1 4\sqrt{1-x^2}\, dx = \pi$.

       b) Using the Midpoint Rule, approximate the integral in part a) for $n = 4, 8, 16$.

       c) Verify that $\int_0^1 \frac{4}{1+x^2}\, dx = \pi$.

       d) Using the Trapezoid Rule, approximate the integral in part c) for $n = 4, 8, 16$.

       e) Which method of approximating π converges faster? How do you explain this?

13.   An experiment was performed in which oxygen was produced at a continuous rate. The rate at which oxygen was produced was measured each minute and the results tabulated.

minutes	0	1	2	3	4	5	6
oxygen (cu ft/min)	0	1.4	1.8	2.2	3.0	4.2	3.6

       Use the Trapezoid Rule to estimate the total amount of oxygen produced in 6 minutes.

14.   The following inputs and outputs for a continuous function f are known.

x	1	1.2	1.4	1.6	1.8	2.0
$f(x)$	7.3	6.8	4.9	5.4	6.0	5.8

        Use the Trapezoid Rule to approximate the area of the region S bounded above by the graph of f and below on the x-axis on the interval $[1, 2]$.

15.   Use the Trapezoid Rule with $n = 10$ to approximate the distance traveled in the first 2 seconds by an object moving along a straight line with velocity given by $v(t) = \sqrt{\sin t}$.

16.   Determine which rules (left-hand, midpoint, trapezoid) give the exact integral for $n = 4$. Explain your answers geometrically.

       a) $\int_0^1 4x\, dx$ b) $\int_0^1 |x - 1|\, dx$ c) $\int_0^1 (\sin^2 x + \cos^2 x)\, dx$

17. The graph of the function f over the interval $[1, 7]$ is shown below. Using values from the graph, find Trapezoid Rule estimates for the integral $\int_1^7 f(x)\, dx$ using the indicated number of subintervals.

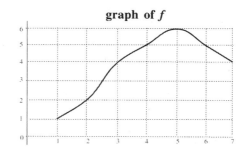

graph of f

    a) $n = 3$

    b) $n = 6$

18. The graph of f over the interval $[1,9]$ is shown in the figure. Using the data in the figure, find a midpoint approximation with 4 equal subdivisions for $\int_1^9 f(x)\, dx$.

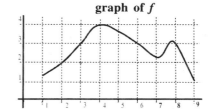

graph of f

19. The graph of g is shown in the figure.

Estimates of $\int_0^2 g(x)\, dx$ were computed using the left, right, trapezoid and midpoint rules, each with the same number of subintervals. The answers recorded were 1.562, 1.726, 1.735, 1.908.

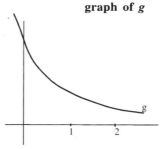

graph of g

a) Match each approximation with the corresponding rule.

b) Between which two approximations does the true value of the integral lie?

20. The graph of f is shown in the figure.

Estimates of $\int_0^2 f(x)\, dx$ were computed using the left, right, trapezoid and midpoint rules, each with the same number of subintervals. The answers recorded were 1.750, 1.9343, 1.9440, 2.1378.

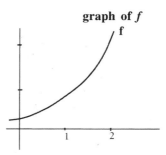

graph of f

a) Match each approximation with the corresponding rule.

b) Between which two approximations does the true value of the integral lie?

21. Let $f(x) = \sin(x^2)$.

    a) Show that $f''(x) = -4x^2 \sin(x^2) + 2\cos(x^2)$.

    b) Use the graph of f'' to verify that the maximum value of $|f''(x)|$ on [0, 1] is 2.2853

    c) What is the smallest value of n for which the Midpoint Rule approximates

    $\int_0^1 \sin(x^2)\, dx$ within 0.001? Justify your answer.

    d) Confirm your results with the calculator approximation **fnInt**$(\sin(X^2), X, 0, 1)$.

22. Let $f(x) = e^{x^2}$.

    a) Show that $f''(x) = (2 + 4x^2)e^{x^2}$.

    b) Use the graph of f'' to verify that the maximum value of $|f''(x)|$ on [0, 1] is 16.3097.

    c) What is the smallest value of n for which the Trapezoid Rule approximates $\int_0^1 e^{x^2}\, dx$ within 0.001? Justify your answer.

    d) Confirm your results with the calculator approximation **fnInt**$(e^\wedge(X^2), X, 0, 1)$.

In Exercises 23–26, determine a value of n such that the midpoint approximation M_n differs from the true value of the definite integral by less than the indicated allowable error E. (Hint: Use the graph of the second derivative to find the maximum value M_2 of $|f''(x)|$)

23. $\int_1^2 \ln x\, dx$;   E = 0.001

24. $\int_0^1 e^{-x}\, dx$;   E = 0.0001

25. $\int_0^1 \dfrac{1}{x+1}\, dx$; E = 0.001

26. $\int_0^4 \cos(x^2)\, dx$;   E = 0.001

In exercises 27-28, determine a value of n such that the trapezoid approximation T_n differs from the true value of the definite integral by less than the indicated allowable error E. (Hint: Use the graph of the second derivative to find the maximum value M_2 of $|f''(x)|$)

27. $\int_0^1 e^{-x}\, dx$;   E = 0.001

28. $\int_0^1 \dfrac{1}{\sqrt{x+1}}\, dx$;   E = 0.0001

Chapter 7   Supplementary Problems

In Exercises 1–20, find the antiderivatives.

1. $\displaystyle\int x^{-5}\,dx$

2. $\displaystyle\int \sec^2(2x)\,dx$

3. $\displaystyle\int \frac{10}{t^6}\,dt$

4. $\displaystyle\int (x-7)^{-2}\,dx$

5. $\displaystyle\int (3x+2)^4\,dx$

6. $\displaystyle\int xe^{3x}\,dx$

7. $\displaystyle\int \left(\frac{2}{\sqrt{x}}-3\sqrt{x}\right)dx$

8. $\displaystyle\int 2x(x^2+4)^5\,dx$

9. $\displaystyle\int x^2(x^3+1)^2\,dx$

10. $\displaystyle\int \frac{1}{\sqrt{2x+1}}\,dx$

11. $\displaystyle\int x^2 \ln x\,dx$

12. $\displaystyle\int \tan^5 x\,\sec^2 x\,dx$

13. $\displaystyle\int x^3\sqrt{2+x^4}\,dx$

14. $\displaystyle\int \frac{x^2}{\sqrt{2+x^3}}\,dx$

15. $\displaystyle\int x\sqrt{1-x}\,dx$

16. $\displaystyle\int \frac{x}{\sqrt{1+2x}}\,dx$

17. $\displaystyle\int \frac{2}{\sqrt{1-x^2}}\,dx$

18. $\displaystyle\int \frac{1}{x^2+4}\,dx$

19. $\displaystyle\int \frac{6}{\sqrt{1-9x^2}}\,dx$

20. $\displaystyle\int \frac{3}{1+4x^2}\,dx$

In Exercises 20–28, evaluate each definite integral using the Fundamental Theorem of Calculus.

21. $\displaystyle\int_1^2 5\,dx$

22. $\displaystyle\int_{-1}^{-1/2} \frac{1}{x^2}\,dx$

23. $\displaystyle\int_1^3 \ln x \, dx$

24. $\displaystyle\int_1^8 2x^{1/3} \, dx$

25. $\displaystyle\int_{-1}^1 \frac{4}{(t+2)^3} \, dt$

26. $\displaystyle\int_{-3}^0 \sqrt{25+3t} \, dt$

27. $\displaystyle\int_{\sqrt{2}}^{\sqrt{3}} \frac{x^3}{(x^4-1)^2} \, dx$

28. $\displaystyle\int_1^{\sqrt{2}} x\sqrt{x^2-1} \, dx$

29. $\displaystyle\int_4^9 \frac{1}{\sqrt{x}} \, dx$

30. $\displaystyle\int_1^4 \frac{1-t}{\sqrt{t}} \, dt$ [Hint: Divide first.]

31. $\displaystyle\int_0^\pi \cos^2 x \sin x \, dx$

32. $\displaystyle\int_0^{\pi/4} \tan^2 x \, dx$

33. The table below shows the velocity of a model train engine moving along a track for 5 seconds. Use the Trapezoid Rule to estimate the distance traveled by the engine.

time (sec)	0	1	2	3	4	5
velocity (in/sec)	0	12	16	18	22	26

Find the total area of each shaded region in Exercises 34 and 35.

34. $y = x\sqrt{4-x^2}$)

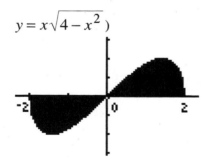

35. $y = (1-\cos x) \cdot \sin x$

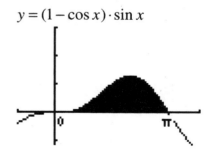

36.   Estimate $\int_0^1 \cos(x^3)\,dx$ using (a) the Trapezoid Rule and (b) the Midpoint Rule with
       $n = 4$. From a graph of the integrand, decide whether your answers are underestimates or
       overestimates.

37.   The left, right, trapezoidal and midpoint approximations

       were used to estimate $\int_0^2 f(x)\,dx$ where f is the function

       whose graph is shown in the figure. The estimates were
       10.284, 9.410, 11.158 and 10.366.

       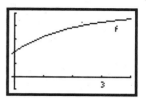

       a) Which rule produced which estimate?

       b) Between which two approximations does the true value
           of the integral lie?

38.   Use a trapezoid approximation with four subdivisions to obtain the area of the first

       quadrant region bounded by the graph of $f(x) = \dfrac{1}{1+x^2}$ on the interval $[1, 3]$.

39.   Determine the exact area of the region bounded between the graph of $y = 1 - x^2$ and the
       x-axis for $0 \le x \le 3$.

40.   Find the exact area of the first quadrant region bounded by the graph of $y = \dfrac{4x}{1+x^2}$,

       the x-axis and the line $x = 2$.

41.   Use a) the Trapezoid Rule and b) the Midpoint Rule with 4 equal subdivisions to
       approximate the definite integral

$$\int_{-1}^3 |2x - 3|\,dx \,.$$

42.   The function f is defined by $f(x) = \dfrac{2x}{\sqrt{x+1}}$.

       a) Approximate $\int_1^4 f(x)\,dx$ with a left-hand Riemann sum, using three subdivisions
       of equal length.

       b) Check your answer in part a) by finding an antiderivative of $f(x) = \dfrac{2x}{\sqrt{x+1}}$ and
       using the Fundamental Theorem of Calculus to evaluate a definite integral.

43. The figure represents an observer at point A watching balloon B as it rises from point C. The balloon is rising at a constant rate of 2 meters per second and the observer is 48 meters from point C.

Find the rate of change in θ at the instant when $y = 24$.

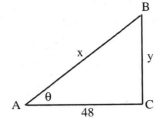

44. Find the point of inflection of the curve $y = (x+1)\arctan x$.

45. Given the function $y = 3x^{2/3} - 2x$ on the interval $[-1, 8]$.

a) Find $f'(x)$.

b) Find all critical values of f on the interval $[-1, 8]$.

c) Find the absolute maximum and minimum values of f on $[-1, 8]$.

46. Determine the radius and height of the cylinder with maximum volume that can be obtained by revolving a rectangle with perimeter 20 about one of its sides. $(V = \pi r^2 h)$

47. Find the area of the region bounded by the graph of $y = x^2 - 2x$ and the x-axis from $x = 0$ to $x = 3$.

48. A storage shed with a volume of 150 cubic feet is to be built in the shape of a rectangular box with a square base. The concrete for the base costs \$4.00 per square foot, the material for the sides and roof costs \$2.50 per square foot. Find the dimensions of the most economical shed.

49. If $G(x) = \displaystyle\int_0^x \frac{t^2}{t^2 + 1}\, dt$, find all intervals on which the graph of G is

a) increasing;   b) concave up

50. The function g is defined by $g(x) = \dfrac{x}{x^2 + 9}$.

a) Find $g'(x)$.

b) Determine the global maximum and minimum values of g on the interval $[0, 4]$.

51. On a distant planet, somewhat smaller than earth, the acceleration due to gravity is -12 ft/sec^2. If a rock is thrown upward from a height of 10 ft with an initial velocity of 30 ft/sec, what is the maximum height reached by the rock?

52.   Given the function g defined by $g(x) = 3x^5 - 10x^4 + 7$, use derivatives to find

a) the intervals on which g is increasing;

b) the x-coordinates of any inflection points.

53.   Use Newton's method to approximate a zero of $y = x^3 - 3x^2 + 1$. Let the initial
approximation be $x_0 = 1$ and determine x_1 and x_2 to three decimal places.

54.   The graph of a differentiable function f on the closed interval $[-4, 4]$ is shown below.
The graph of f has horizontal tangents at $x = -3, -1$ and 2.

Let $G(x) = \int_{-4}^{x} f(t)\, dt$ for $-4 \le x \le 4$.

a) Find $G(-4)$.

b) Find $G'(-4)$.

c) On which interval or intervals is the graph
of G concave down? Justify your answer.

d) Find the value of x at which G has its
maximum on the closed interval $[-4, 4]$.
Justify your answer.

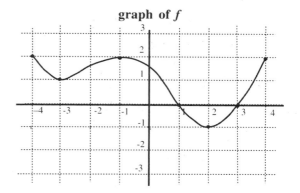

graph of f

55.   Consider a polynomial function f with the following properties:

i) $\int_{1}^{3} f(x)\, dx = \frac{5}{2}$ ii) $\int_{1}^{5} f(x)\, dx = 10$

a) Find the value of $\int_{3}^{5} [2f(x) + 6]\, dx$.

b) If $f(x) = ax + b$, determine a and b.

56.   Let f be a function whose graph is concave down on the the closed interval $[1, 2]$, with
selected values shown in the table below.

x	1.1	1.3	1.5	1.7	1.9
$f(x)$	12	18	21	23	24

If f' and f'' are defined for all x in $[1, 2]$, which of the following statements are true?

a) $f'(1.5) < f'(1.7)$

b) $10 < f'(1.5) < 30$

c) $f(1.7) > f''(1.7)$

57.   Find the nth derivative of the function by calculating the first few derivatives and observing the
pattern that occurs.

a) $f(x) = xe^x$ b) $f(x) = \frac{1}{x}$

58.   Let f be the function given by $f(x) = \dfrac{x^2 + 3x}{x^2 + 1}$.

a) In the viewing rectangle provided below, sketch a graph of f.

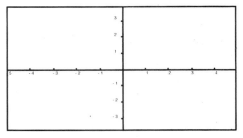

b) Write an equation for each horizontal asymptote of the graph of f.

c) Find the range of f. Use $f'(x)$ to justify your answer. [Note: $f'(x) = \dfrac{3 + 2x - 3x^2}{(x^2 + 1)^2}$]

59.   An object moves along the parabola $y = x^2$ in the first quadrant in such a way that its x-coordinate increases at a steady 10 cm/sec. How fast is the angle of inclination θ of the line joining the object to the origin changing when $x = 3$ cm? What is the limiting value of $\dfrac{d\theta}{dt}$ as $x \to \infty$?

60.   The function f is defined on $[-1, 4]$ by $f(x) = \begin{cases} |x - 2| & \text{for } 1 \le x \le 4 \\ -x & \text{for } -1 \le x < 1 \end{cases}$

a) Sketch the graph of f.

b) If possible, determine values of x on $[-1, 4]$ for which f is *not* continuous. Explain briefly.

c) If possible, determine values of x on $[-1, 4]$ for which the derivative does not exist.

d) Evaluate: $\displaystyle\int_1^4 f(x)\,dx$

61.   Let f be a function defined and differentiable on the closed interval $0 \le x \le 7$ with $f(3) = 5$. The graph of f', the derivative of f, consists of two line segments and a semicircle.

graph of f'

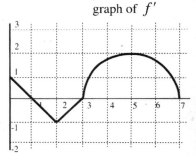

a) Find $f(1)$ and $f(7)$.

b) For what inputs x does f have a relative minimum. Justify.

c) For $0 \le x \le 7$, find all inputs x at which the graph of f has a point of inflection. Justify.

d) Find the absolute maximum value of $f(x)$ over the closed interval $0 \le x \le 7$.. Explain your reasoning.

CHAPTER 8
Using the Definite Integral

In this chapter we explore more applications of the definite integral by using it to measure distance traveled, volume of solids, and the average value of a function.

The problem solving strategy employed in these applications is similar to that used to find areas under curves. Given a continuous function f on an interval $[a, b]$, we start by dividing the interval into n equal subdivisions, each of length $\Delta x = (b-a)/n$. Then we choose a single input x from the subinterval and use the corresponding output $f(x)$ as the approximate constant value of f over each subinterval. If the quantity we are measuring is approximated by the product $f(x) \cdot \Delta x$ for each subinterval, then we add up all of the products to form an approximating Riemann sum. The limiting value of this Riemann sum is a definite integral that measures the desired quantity.

8.1 Net and Total Distance Traveled

In Chapter 3 we discovered that distance traveled is the definite integral of velocity. Consider an object moving along a straight line with velocity $v(t)$ at time t. If $v(t) \geq 0$, then $\int_a^b v(t)\, dt$ gives the distance traveled during the time interval $a \leq t \leq b$. In the figure below the area of each rectangle is $v(t_i)\Delta t$. If the velocity were constant this area would represent the distance traveled by the object during the time period Δt. Because the velocity is not constant, the shaded area only approximates the distance traveled for this time period.

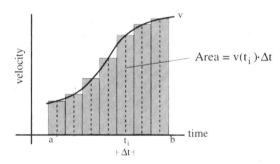

Summing the areas of all the rectangles we get $v(t_1) \cdot \Delta t + v(t_2) \cdot \Delta t + \ldots + v(t_n) \cdot \Delta t$, an approximation to the total distance traveled between $t = a$ and $t = b$. As the number, n, of subintervals increases, the length of each subinterval, Δt, approaches 0 and the sums approach the area under the graph of the velocity function, which represents the total distance traveled.

Example 1: *Finding Distance Traveled When Velocity Varies*

During a trip a car's speedometer reading, in miles per hour, is given by the function v where

$$v(t) = 25 - 25\cos(2\pi t)$$

from $t = 0$ to $t = 1$ hours. Sketch a graph of the velocity function and determine how far the car traveled during the trip.

Solution:

The total distance traveled during the trip is given by the area under the curve from $t = 0$ to $t = 1$. Since $v(t) \geq 0$ for $0 \leq t \leq 1$, we have

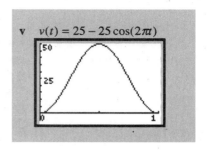

$$\int_0^1 v(t) \, dt = \int_0^1 [25 - 25\cos(2\pi t)] \, dt$$

$$= \left[25t - \frac{25}{2\pi}\sin(2\pi t) \right]_0^1$$

$$= (25) - (0) = 25.$$

The car travels 25 miles from $t = 0$ to $t = 1$.

Integrating velocity over a time interval $[a, b]$ gives the change in position of the moving object over the time interval. We should think of this as the **net distance** traveled. For example, suppose you walk 5 steps forward and then take 3 steps backward; the net change in your position is 2 steps forward, even though you have traveled a total distance of 8 steps.

Start　　　　　　　　　5 steps

　　　　　　　　　　　　　　3 steps

Net Distance
= 2 steps

The direction of movement is indicated by the sign of the velocity. A negative velocity indicates movement in the opposite direction of a positive velocity. Thus, if $v(t)$ is sometimes negative, then $\int_a^b v(t) \, dt$ measures the net distance from the starting position to the ending position. To get the **total distance** traveled by an object during a time interval $[a, b]$, we must calculate $\int_a^b |v(t)| \, dt$.

$$\text{net distance traveled} = \int_a^b v(t) \, dt$$

$$\text{total distance traveled} = \int_a^b |v(t)| \, dt$$

Example 2: *Finding Net and Total Distance Traveled*

An object moves along a coordinate line with distance to the right considered positive. If the velocity of the object is given by $v(t) = t^2 - 7t + 10$ over the time interval $1 \leq t \leq 3$, find the net and the total distance traveled.

Solution:

The graph of the velocity function is shown in the figure

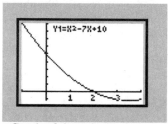

Graph of the velocity function

The direction the object is moving at any time t is determined by the sign of the velocity $v(t)$. In this example, the velocity is positive for $1 \leq t < 2$ and the object is moving to the right during this time. The velocity is negative for $2 < t \leq 3$ and the object is moving to the left during this time. The object reverses direction at $t = 2$ when the velocity is zero. The net distance traveled by the object is measured by the net area of the region bounded by the graph and the t–axis over the time interval $[1, 3]$.

$$\int_1^3 (t^2 - 7t + 10)\, dt \ = \left[\frac{t^3}{3} - \frac{7}{2}t^2 + 10t\right]_1^3 = (9 - \frac{63}{2} + 30) - (\frac{1}{3} - \frac{7}{2} + 10) = \frac{2}{3}$$

This means the object moves a net distance of 2/3 units to the right.

The total distance traveled is

$$\int_1^3 |t^2 - 7t + 10|\, dt \ = \int_1^2 (t^2 - 7t + 10)\, dt \ + \ \int_2^3 (-t^2 + 7t - 10)\, dt$$

$$= \left[\frac{t^3}{3} - \frac{7}{2}t^2 + 10t\right]_1^2 \ + \ \left[-\frac{t^3}{3} + \frac{7}{2}t^2 - 10t\right]_2^3$$

$$= \frac{11}{6} + \frac{7}{6} = 3.$$

We see from this calculation that the object moved $\frac{11}{6}$ units to the right, reversed direction, and then moved $\frac{7}{6}$ units to the left for a total distance of 3 units and a net distance of $\frac{2}{3}$ units to the right.

8.1   Exercises

In Exercises 1–6 the velocity function (in feet per second) is given for an object moving along a line. Find a) the net distance traveled and b) the total distance traveled by the object during the given time interval.

1. $v(t) = 4 - 2t,$ $0 \le t \le 3$

2. $v(t) = t^2 - 6t - 8,$ $1 \le t \le 5$

3. $v(t) = 2t^2 - 6t + 4,$ $0 \le t \le 4$

4. $v(t) = t \cos t,$ $0 \le t \le \pi$

5. $v(t) = \sec^2 t - 2,$ $0 \le t \le \dfrac{\pi}{3}$

6. $v(t) = 5 - e^{0.5t},$ $1 \le t \le 4$

In Exercises 7–10, first find an equation for the velocity of a moving object from the equation of the acceleration. Then find the net and total distance traveled by the moving object in the given time interval.

7. $a(t) = 6t - 18$ ft/sec^2, $v(0) = 24$ ft/sec, from $t = 1$ to $t = 5$ sec

8. $a(t) = \sqrt{t}$ ft/sec^2, $v(0) = -6$ ft/sec, from $t = 0$ to $t = 9$ sec

9. $a(t) = -2\pi \cos(2\pi t)$ ft/sec^2, $v(0) = 1$ ft/sec, from $t = 0$ to $t = 1$ sec

10. $a(t) = \dfrac{3}{t+1}$ cm/sec^2, $v(0) = 2$ cm/sec, from $t = 0$ to $t = 1$ sec

11.   The velocity of a car was read from its speedometer at ten second intervals and recorded in the table. Use the Trapezoid Rule with six equal subdivisions to estimate the distance traveled by the car.

t (sec)	0	10	20	30	40	50	60
v (ft/sec)	0	38	42	48	51	50	45

12.   A particle moves along a line so that at any time $t > 0$ its acceleration is given by $a(t) = \ln t$ ft/sec^2. At time $t = 1$ sec the velocity of the particle is $v(1) = -2$ ft/sec.

a) Write an expression for the velocity of the particle.

b) For what values of t is the particle moving to the right?

c) What is the minimum velocity of the particle?

d) Find the total distance traveled by the particle from $t = 2$ to $t = 4$ seconds.

13. A car is moving forward and backward along a straight road from A to B, starting from A at time $t = 0$. The car's velocity is given by $v(t) = 1 + 2\sin(\frac{\pi t}{6})$ where t is in minutes and v is in km/min. The graph of the velocity function is given below.

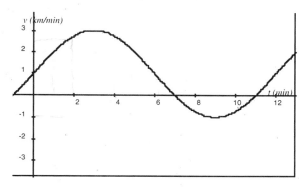

a) What is the velocity of the car at $t = 0$?

b) At what time(s) t does the car change direction?

c) Use a rectangle summing program (e.g., RSUM) to estimate the car's distance from A at $t = 9$.

d) Find the average velocity of the car between $t = 0$ and $t = 9$.

14. A particle moves along a line so that at any time $t \geq 0$ its velocity is given by $v(t) = \dfrac{t}{1 + t^2}$. At time $t = 0$ the position of the particle is $s(0) = 5$.

a) Determine the maximum velocity of the particle. Justify your answer.

b) Determine the position of the particle at $t = 3$.

c) What is the total distance traveled by the particle from $t = 1$ to $t = 3$?

d) Find the limiting value of the velocity as t increases without bound.

15. A car is moving along a straight road from A to B, starting from A at time $t = 0$. Below is a graph of the car's velocity (positive direction from A to B), plotted against time.

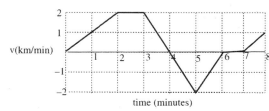

a) How many kilometers away from A is the car at time $t = 6$?

b) At what time does the car change direction? Explain briefly.

c) Sketch a graph of the acceleration of the car.

8.2   Volumes by Slicing

We begin this section with a strategy for using the definite integral to compute the volume of certain solids. The method we use to make these computations is to divide the solid into small pieces whose volumes we can easily approximate. Then we add the volumes of the smaller pieces, creating a Riemann sum that approximates the total volume. As the number of terms in the sum grows larger and larger, the approximation improves, and the limiting value is a definite integral.

If you slice a rectangular block of cheese parallel to a face, the area A of a cross-section does not vary. The same is true for a cylinder of liverwurst.

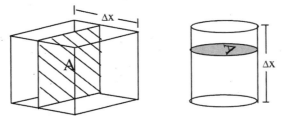

volume = area of cross-section x thickness = $A \cdot \Delta x$

For any solid that has a constant cross-sectional area, its volume is the product of its cross-section area A and its thickness Δx. Most solids do not have regular shapes and the task of computing their volumes requires calculus methods.

Imagine a loaf of bread lying along the x-axis in the xy-plane between $x = a$ and $x = b$.

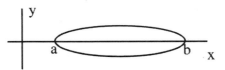

Slicing a loaf of bread- an overhead view

The loaf can be divided into several slices by making cuts perpendicular to the x-axis. The volume of the loaf is the sum of the volumes of all of the individual slices. In general, since the shape of the loaf varies, different slices have different volumes even if they all have the same thickness.

For any x between a and b, let $A(x)$ represent the cross-sectional area created by the cut at x. Suppose we divide the loaf into n slices of thickness $\Delta x = \dfrac{b-a}{n}$. The cross-sectional area of a typical slice is not constant. However, if we replace it with a slice of the same thickness and constant cross-sectional area given by one face of the slice, then the volume of this slice is

$$\text{volume of one slice} \approx A(x) \cdot \Delta x.$$

To estimate the volume of the whole loaf, divide it into n pieces of equal thickness $\Delta x = \dfrac{b-a}{n}$ by making cuts at $a = x_0 < x_1 < x_2 < ... < x_n = b$. Using the cross-sectional areas $A(x_1)$, $A(x_2)$, ..., $A(x_n)$ for the individual slices leads to an estimate for total volume.

$$\text{Total volume} \approx A(x_1)\Delta x + A(x_2)\Delta x + ... + A(x_n)\Delta x.$$

This is a Riemann sum for the cross-sectional area function $A(x)$ on the interval $[a, b]$. Thus, the limiting value as n grows larger and larger is the definite integral $V = \int_a^b A(x)\, dx$. So we conclude:

> The volume of a solid between $x = a$ and $x = b$ having a cross-sectional area $A(x)$ at input x is
>
> $$V = \int_a^b A(x)\, dx.$$

Example 1: *Finding the Volume of a Cone*

Find the volume of a right circular cone with base radius 1 and height 4.

Solution:

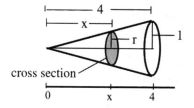

Sketch the cone. A typical cross-section is a circle of radius r; its area is πr^2. But the cross-sectional area function should be a function of the input variable that determines the cross-section, in this case x. Using similar triangles in the figure we have

$$\frac{r}{1} = \frac{x}{4} \quad \text{and} \quad r = \frac{x}{4}.$$

Thus, the cross-sectional area function is $A(x) = \dfrac{\pi x^2}{16}$, and the volume of the cone is

$$V = \int_a^b A(x)\, dx \;\; = \frac{\pi}{16} \int_0^4 x^2\, dx \;\; = \frac{\pi}{16}\left[\frac{x^3}{3}\right]_0^4 = \frac{4\pi}{3} \text{ units}^3.$$

Recall that the volume formula for a right circular cone is $V = \frac{1}{3}\pi r^2 h$. So, for a cone with base radius 1 and height 4, we have $V = \frac{1}{3}\pi(1)^2 \cdot 4 = \frac{4\pi}{3} \text{units}^3$, confirming the calculus result above.

Example 2: *Finding the Volume of a Square-Based Pyramid*

Find the volume of a pyramid with a square base 4 inches on a side and height 6 inches.

Solution:

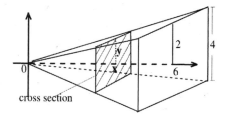

Sketch the pyramid so that the origin is at the vertex and the x-axis is along the central axis. For any x in the interval $[0, 6]$, the cross-section is a square $2y$ units on a side. To express y in terms of x, we use similar triangles.

$$\frac{y}{2} = \frac{x}{6} \quad \text{or} \quad y = \frac{x}{3}.$$

Thus, $2y = \dfrac{2x}{3}$ and $A(x) = \dfrac{4x^2}{9}$.

The volume of the pyramid is

$$V = \frac{4}{9}\int_0^6 x^2 \, dx = \frac{4}{9}\left[\frac{x^3}{3}\right]_0^6 = 32 \text{ in}^3.$$

Solids of Revolution ; Disks and Washers

Another way to create a solid having known cross-sections is to revolve a region in the plane about a line producing a **solid of revolution**; the line is called the **axis of rotation**. The advantage of solids of revolution is that their cross-sections perpendicular to the axis of rotation are circular.

Example 3: *A Solid of Revolution*

Find the volume of a solid of revolution obtained by rotating the first quadrant region S bounded by the curve $y = x^2$ and the lines $x = 1$ and $x = 2$ about the x-axis.

Solution:

The region S, with a typical cross-section displayed, is sketched in the figure. For any x in the interval $[1, 2]$, the cross-section at x is a circle with radius $y = x^2$. The area of each circle is

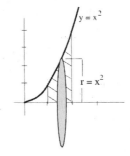

$$A(x) = \pi r^2 = \pi (x^2)^2 = \pi x^4.$$

So the volume of the solid of revolution is

$$V = \int_1^2 \pi x^4 \, dx = \pi \left[\frac{x^5}{5}\right]_1^2 = \frac{31\pi}{5} \text{ units}^3.$$

Rotation about the x-axis is not the only possibility. Whatever the axis of rotation, the volume is found by integrating the cross-sectional area function.

Example 4: *Using the Washer Method*

Find the volume of a solid of revolution obtained by rotating the first quadrant region R bounded by the curve $y = x^2$ and the lines $x = 1$ and $x = 2$ about the line $y = -1$.

Solution:

The region S is the same as in the previous example; however, the axis of rotation is the line $y = -1$. Rotating region S about a line which is 1 unit from the nearest edge of the region produces a solid with a hole in the middle. Each cross section is a "washer" having outer radius $R(x) = x^2 + 1$ and inner radius $r(x) = 1$ for $1 \le x \le 2$. The area of each washer is

$$A(x) = \pi(\text{outer radius})^2 - \pi(\text{inner radius})^2 = \pi\left[(x^2+1)^2 - (1)^2\right] = \pi(x^4 + 2x^2).$$

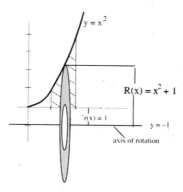

The volume of the solid of revolution is

$$V = \int_1^2 \pi(x^4 + 2x^2)dx = \pi\left[\frac{x^5}{5} + \frac{2x^3}{3}\right]_1^2 = \pi\left[(\tfrac{32}{5} + \tfrac{16}{3}) - (\tfrac{1}{5} + \tfrac{2}{3})\right] \approx 34.14 \ \text{units}^3.$$

The method for finding the volume of a solid of revolution can be extended to find the volumes of more complicated solids. For example, suppose two nonnegative functions f and g satisfy $0 \le g(x) \le f(x)$ on the interval $[a,b]$. When the region between their graphs is rotated about the x-axis, it generates a solid such that each cross-section created by a plane perpendicular to the axis is an annulus (the region between two concentric circles). The area of the cross-section is

$\pi[f(x)]^2 - \pi[g(x)]^2$. Integrating the cross-sectional area function we have

$$V = \pi \int_a^b ([f(x)]^2 - [g(x)]^2) \, dx.$$

Example 5: *Washer Cross-Sections*

The region bounded by the line $f(x) = x + 2$ and the parabola $g(x) = x^2$ is rotated about the
x-axis. Find the volume of the solid generated.

Solution:

We first determine the interval of interest by finding the x-coordinates of the intersection points of the two functions. Letting $x^2 = x + 2$ we find that the graphs intersect at $x = -1$ and $x = 2$.

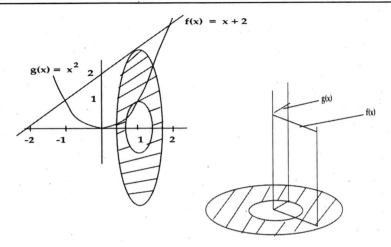

A cross-section of the solid is the region between two concentric circles (a washer) with

$$\text{Area of a cross-section} = \pi(x+2)^2 - \pi x^4.$$

Thus,

$$V = \pi \int_{-1}^{2} \left[(x+2)^2 - x^4 \right] dx$$

$$= \pi \int_{-1}^{2} \left[x^2 + 4x + 4 - x^4 \right] dx$$

$$= \pi \left[\frac{x^3}{3} + 2x^2 + 4x - \frac{x^5}{5} \right]_{-1}^{2}$$

$$= \frac{72\pi}{5} \text{ units}^3.$$

8.2   Exercises

1. A solid is 12 inches high. The cross-section of the solid at height x above its base has area $3x$ square inches. Find the volume of the solid.

2. A solid extends from $x = 1$ to $x = 3$. The cross-section of the solid in the plane perpendicular to the x-axis at x is a square of side x. Find the volume of the solid.

3. A solid is 6 ft high. Its horizontal cross-section at height x ft above the base is a rectangle with length $2 + x$ ft and width $8 - x$ ft. Find the volume of the solid.

4. A solid extends along the x-axis from $x = 1$ to $x = 4$. Its cross-section at any point x is an equilateral triangle with edge $\sqrt{x}$. Find the volume of the solid.

In Exercises 5–10, find the volume of the solid formed by revolving the given region about the x-axis.

5. $y = 1 - x$

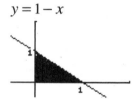

6. $y = 4 - x^2$

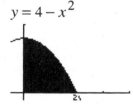

7. $y = \sqrt{x}$

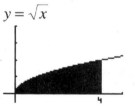

8. $y = \sqrt{\sin x}$

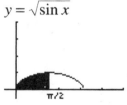

9. $y = x^2, y = x^3$

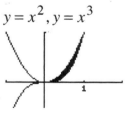

10. $y = e^x$

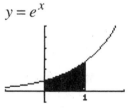

In Exercises 11–14, sketch the region R bounded by the graphs of the given equations. After showing a typical cross-section, find the volume of the solid generated by revolving R about the x-axis.

11. $y = \dfrac{x^4}{4}, x = 4$ and $y = 0$.

12. $y = e^{2x}, x = 0$ and $y = e^2$.

13. $y = \dfrac{1}{x}, x = 1, x = 4,$ and $y = 0$.

14. $y = \tan x, y = 0$ and $x = \dfrac{\pi}{4}$.

15. Find the volume of the solid generated by revolving about the x-axis the region bounded by the line $y = 4x$ and the parabola $y = 4x^2$.

16. Find the volume of the solid generated by revolving about the x-axis the region bounded by the line $y = 2x + 4$ and the curve $y = e^x$.

17. The region bounded by the graph of $y = x^2$ and the line $y = 4$ generates various solids when rotated as follows:

    a) about the x-axis;   b) about the line $y = -1$;   c) about the line $y = 4$.

    Find the volume in each case.

18. The region bounded by the graph of $y = x^2 + 1$ and the line $y = x + 1$ generates various solids when rotated as follows:

    a) about the x-axis;   b) about the line $y = -1$;   c) about the line $y = 3$.

    Find the volume in each case.

19. A CAT scan produces equally spaced cross-sectional views of a human organ. Suppose a CAT scan of a human liver shows cross-sections spaced 1 cm apart. The liver is 10 cm long and its cross-sectional areas A at a distance x from one end are recorded in the following table.

x (cm)	0	1	2	3	4	5	6	7	8	9	10
A (cm²)	0	16	42	60	72	94	82	64	38	24	0

    Use the Midpoint Rule with $n = 5$ to estimate the volume of the liver.

20. The following table lists the known values of a continuous function f.

x	1	1.2	1.4	1.6	1.8	2
$f(x)$	7.3	6.8	4.9	5.4	6.0	5.8

    Let S be the region bounded above by the graph of f and below by the x-axis on the interval $[1, 2]$. Use the Trapezoid Rule with $n = 5$ to approximate the volume of the solid generated when the region S is rotated about the x-axis.

21. The region R shaded in the figure at the right is rotated about the x-axis.

    a) Use the Trapezoid Rule with 5 equal subdivisions to approximate the volume of the resulting solid.

    b) Use the regression capabilities of your TI-83 to find a third-degree polynomial that fits the curve in the figure. Plot the data and graph the model.

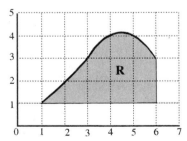

    c) Use your TI-83 to approximate a definite integral whose value is the volume of the solid that is generated when R is rotated about the x-axis. Compare the result with the answer in part a).

8.3   Volume by the Shell Method

So far finding the volumes of solids has depended on our skill at computing cross-sectional areas of the solid. There is an alternate technique that can be applied when it is difficult to determine the cross-sectional area or if the integration is too complicated. It is called the **shell method** and is based on subdividing the solid into shells rather than slicing it into disks or washers.

Let f be a nonnegative, continuous function on the interval $[a, b]$ with $a \geq 0$ and consider the solid of revolution generated by rotating the region under the graph of f about the y-axis.

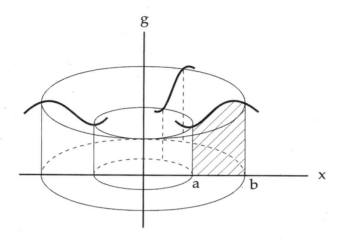

To obtain an approximation for the volume of the solid subdivide the interval $[a, b]$ into n equal subdivisions. Then choose any input x_k^* in each of the n subdivisions, $[x_{k-1}, x_k]$; consider the rectangular regions whose bases are the subintervals $[x_{k-1}, x_k]$ and whose heights are the outputs $f(x_k^*)$. Now if each rectangular region is rotated about the y-axis, it generates a so-called cylindrical shell which approximates a portion of our solid. (Note: A cylindrical shell is the solid that remains when a cylinder of radius r is removed from a cylinder of radius $r + p$ where $p > 0$. For example, a roll of paper towels and a piece of copper water pipe have this shape.)

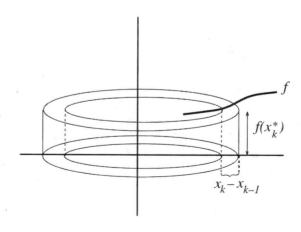

If x_{k-1} is the inner radius for a shell, x_k the outer radius, and $f(x_k^*)$ its height, then the volume V_k of each shell is the difference between the volumes of the inner and outer cylinders.

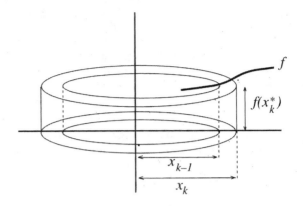

$$V_k = \pi(x_k)^2 \cdot f(x_k^*) - \pi(x_{k-1})^2 \cdot f(x_k^*)$$

$$= \pi \, f(x_k^*) \, (x_k^2 - x_{k-1}^2)$$

Now if the number of subdivisions is made larger, the length of each subinterval gets smaller and the sum of the volumes of the shells

$$\pi f(x_1^*)(x_1^2 - x_0^2) + \pi f(x_2^*)(x_2^2 - x_1^2) + \pi f(x_3^*)(x_3^2 - x_2^2) + \ldots + \pi f(x_n^*)(x_n^2 - x_{n-1}^2)$$

should be a good approximation of the volume of our solid of revolution. We now show that the sum of the volumes of the shells can be expressed as a Riemann sum.

Since each x_k^* may be any input in the subinterval $\left[x_{k-1}, x_k\right]$, we choose it to be the midpoint. Thus $x_k^* = \dfrac{x_k + x_{k-1}}{2}$. Then

$$x_k^2 - x_{k-1}^2 = (x_k + x_{k-1})(x_k - x_{k-1}) = 2x_k^*(x_k - x_{k-1}).$$

Substituting $2x_k^*(x_k - x_{k-1})$ for $x_k^2 - x_{k-1}^2$ in the sum above we have

$$2\pi x_1^* f(x_1^*)(x_1 - x_0) + 2\pi x_2^* f(x_2^*)(x_2 - x_1) + \ldots + 2\pi x_n^* f(x_n^*)(x_n - x_{n-1}).$$

This is a Riemann sum for the function G defined on the interval $[a, b]$ by $G(x) = 2\pi x f(x)$, and the limiting value of the Riemann sum is the definite integral $2\pi \displaystyle\int_a^b x \cdot f(x) \, dx$.

If V is the volume of the solid of revolution obtained by rotating the region under the graph of f between $x = a$ and $x = b$ about the y-axis,

$$V = 2\pi \int_a^b x \cdot f(x) \, dx.$$

Example 1: *Using the Shell Method to Find Volume*

A solid is generated by rotating the region in the first quadrant bounded by the graph of $f(x) = \sqrt{1-x^2}$ about the y-axis. Use the shell method to find its volume.

Solution:

The graph of $f(x) = \sqrt{1-x^2}$ is the upper half of the circle $x^2 + y^2 = 1$ By revolving the quarter circle in the first quadrant about the y-axis we produce a hemisphere of radius 1.

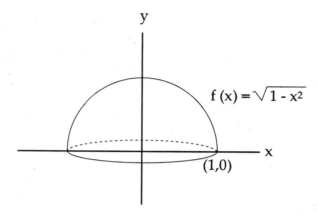

The volume of the hemisphere is

$$V = 2\pi \int_0^1 x\sqrt{1-x^2}\ dx \ = \ -\pi \int_0^1 \sqrt{1-x^2}\,(-2x)\ dx$$

$$= -\pi\left[\frac{2}{3}(1-x^2)^{3/2}\right]_0^1 = -\pi\left[0 - \frac{2}{3}\right] = \frac{2}{3}\pi \text{ units}^3.$$

The integral used to find the volume by the shell method can be thought of as

$$V = 2\pi \int_a^b (\text{radius})(\text{height})(\text{thickness})$$

$$= \ 2\pi \int_a^b x \cdot f(x)\ dx.$$

8.3   Exercises

In Exercises 1–14, the graphs of the given equations bound a region in the plane. Use the shell method to find the volume of the solid generated when the region is rotated about the y-axis.

1. $y = \sqrt{x}, \ x = 4, \ y = 0$.   2. $y = \sin x, \ y = 0, \ x = 0, \ x = \pi$.

3. $y = e^x, \ x = 0, \ y = 0, \ x = 2$.   4. $y = \dfrac{1}{1 + x^2}, \ x = 0, \ y = 0, \ x = 1$.

5. $y = \dfrac{1}{\sqrt{x}}, \ x = 1, \ x = 4, \ y = 0$.   6. $y = \ln x, \ y = 0, \ x = 1, \ x = e$.

7. $y = x^3, \ y = 0, \ x = 2$.   8. $xy = 4, \ x + y = 5$.

9. $y = \sqrt{x}, \ y = x^3$.   10. $y = \sqrt{1 - x^2}, \ y = 0$ on the interval $[\frac{1}{2}, \ 1]$.

11. $y = e^{-x^2}$ from $x = 0$ to $x = \dfrac{\sqrt{2}}{2}$.   12. $y = \sqrt{1 - (x - 2)^2}, \ y = 0$.

13. $y = \dfrac{1}{x\sqrt{1 - x^2}}, \ y = 0$ on the interval $[\frac{1}{2}, \ \dfrac{\sqrt{2}}{2}]$.

14. $\sqrt{x} + \sqrt{y} = 1, \ x + y = 1$.

15.   The region S shaded in the figure at the right is rotated about the y-axis.

a) Use the Trapezoid Rule with 8 equal subdivisions to approximate the volume of the resulting solid.

b) Use the regression capabilities of your TI-83 to find a fourth-degree polynomial that fits the curve in the figure. Plot the data and graph the model.

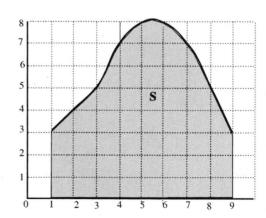

c) Use your TI-83 to approximate a definite integral whose value is the volume of the solid that is generated when S is rotated about the y-axis. Compare the result with the answer in part a).

8.4   Average Value of a Function over an Interval

Often one is interested in the average value of a continuous function on an interval, such as the average temperature over a 24 hour period, the average speed of a car on a 2 hour trip or the average level of carbon dioxide in a given year.

To compute the average of n numbers x_1, x_2, ... x_n you add them up and divide by n. For example, if the function f is defined by the following table

x	1	2	3	4
$f(x)$	5	−1	4	8

the average of the function values is

$$\frac{f(1) + f(2) + f(3) + f(4)}{4} = \frac{5 + (-1) + 4 + 8}{4} = 4$$

But how do you find the average value of f if there are infinitely many numbers? Specifically, what is the average value of $f(x)$ on the interval $a \leq x \leq b$? As an example of how we might approximate the average of a large number of function values $f(x)$ on the interval $a \leq x \leq b$, imagine that the interval is divided into n equal subintervals, each of width

$$\Delta x = \frac{b - a}{n}.$$

For $k = 1, 2, ... , n$, let x_k be any point chosen from the kth subinterval. Then the average value M of $f(x)$ on the interval $[a, b]$ is estimated by the sum

$$S_n = \frac{f(x_1) + f(x_2) + \; ... \; + f(x_n)}{n}.$$

Because $\Delta x = \frac{b - a}{n}$, we know that $\frac{1}{n} = \frac{1}{b - a} \cdot \Delta x$, and substituting in the equation above

$$S_n = \frac{1}{b - a} \cdot \left[f(x_1) + f(x_2) + \; ... \; + f(x_n) \right] \Delta x$$

and

$$S_n = \frac{1}{b - a} \cdot \left[f(x_1) \Delta x + f(x_2) \Delta x + \; ... \; + f(x_n) \Delta x \right].$$

The sum on the right is a Riemann sum and its limiting value is a definite integral, so we have

$$\lim_{n \to \infty} S_n = \frac{1}{b - a} \int_a^b f(x) \, dx.$$

Thus, we define **the average value of a function** as follows:

If f is a continuous function on the interval $[a, b]$, then the average value of f on $[a, b]$ is

$$M(f) = \frac{1}{b - a} \int_a^b f(x) \, dx.$$

There is a nice geometric interpretation of the average value of a function. If the function f is non-negative valued on an interval, then $\int_a^b f(x)\,dx$ is the area under the graph of f on the interval $[a, b]$.

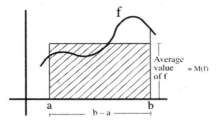

Since

$$M(f)(b-a) = \int_a^b f(x)\,dx,$$

the area under the graph of f is equal to the product of the interval's length, $(b-a)$, and the number $M(f)$. Thus, the average value is the height of a rectangle whose base is $(b-a)$ and whose area is equal to the area of the region under the graph of f. In fact, if f is continuous over $[a, b]$, then this height must be achieved at least once by the function over the interval. In other words, there exists a number c in $[a, b]$ such that $f(c) = M(c)$. The Mean Value Theorem for integrals states that any continuous function must achieve its average value at least once over a closed interval.

> **Mean Value Theorem for Integrals:** If f is continuous on $[a, b]$, then there exists a number c in $[a, b]$ such that
>
> $$\int_a^b f(x)\,dx = f(c)(b-a)$$

Example 1: *Finding the Average Value of a Function*

Find the average value of the function $f(x) = x^2 - 2x + 1$ on the interval $[0, 2]$. Sketch a graph of f and shade the rectangle with base $[0, 2]$ and area equal to $\int_0^2 f(x)\,dx$.

Solution:

$$M(f) = \frac{1}{2-0}\int_0^2 (x^2 - 2x + 1)\,dx$$

$$= \frac{1}{2}\left[\frac{x^3}{3} - x^2 + x\right]_0^2$$

$$= \frac{1}{2}\left[(\tfrac{8}{3} - 4 + 2) - 0\right]$$

$$= \frac{1}{3}$$

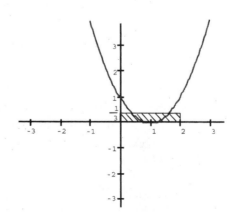

8.4   Exercises

In Exercises 1 – 10, find the average value of the function f on the given interval. If $f(x) \geq 0$ on the given interval, sketch a graph of f and shade the rectangle with base $[a, b]$ and area equal to $\int_a^b f(x)\, dx$.

1. $f(x) = x^2$ on $[-1, 1]$
                                 2. $f(x) = x^2 - x + 1$ on $[-1, 2]$

3. $f(x) = x^3 - 3x^2$ on $[-2, 1]$
                            4. $f(x) = \sin x$ on $[0, \frac{\pi}{4}]$

5. $f(x) = 2\sin x - \cos x$ on $[0, \frac{\pi}{2}]$
            6. $f(x) = \sqrt{4 - x}$ on $[0, 4]$

7. $f(x) = \sqrt[3]{1 - x}$ on $[-7, 0]$
                      8. $f(x) = (2x - 3)^3$ on $[0, 1]$

9. $f(x) = e^{\sin x}$ on $[-\pi, \pi]$
                        10. $f(x) = \sqrt{5x^2 + 4}$ on $[0, 2]$

11.   A continuous function f is defined on the interval $[0, 3]$ and some values of f are given in the following table.

x	0	.5	1.0	1.5	2.0	2.5	3.0
$f(x)$	1.2	1.4	2.2	2.6	3.0	2.8	2.4

     a)   Use the Trapezoid Rule with $n = 6$ to estimate $\int_0^3 f(x)\, dx$.

     b)   Estimate the average value of f over the interval $[0, 3]$.

12.   A continuous function f is defined on the interval $[0, 6]$ and its graph is given in the figure.

     a)   Use the Midpoint Rule with $n = 3$ to estimate $\int_0^6 f(x)\, dx$.

     b)   Estimate the average value of f over the interval $[0, 6]$.

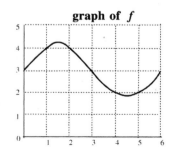

graph of f

13. Suppose that the speed of traffic on Boston's Southeast Expressway is modeled by the function

$$s(t) = 2t^3 - 21t^2 + 60t + 25$$

miles per hour, where t is the number of hours past noon. Compute the average speed between 4 pm and 6 pm.

14. Recorded data indicates that on a recent date t hours past midnight, the temperature at Logan Airport was

$$f(t) = 14\sin(\frac{1}{3}t - 2) + 60$$

degrees Fahrenheit. What was the average temperature between 9 am and noon?

15. The functions f and g are defined on the interval $[0, b]$ by $f(x) = \cos(2x)$ and $g(x) = e^x - 1$. For what value of b will functions f and g have the same average value?

16. a) Find the average value of $f(x) = e^{-x}$ over the interval $[0, k]$.

    b) What happens to the average value of f on $[0, k]$ as $k \to \infty$?

17. a) Find the average value of $f(x) = x^n$ over the interval $[0, 1]$ if n is any positive integer.

    b) What happens to the average value of f over $[0, 1]$ as $n \to \infty$?

18. If b is a positive constant, find the average value of the function f defined by $f(x) = x^2 + bx + 1$ on the interval $[-1, 1]$.

19 Let $f(x) = kx$ and $g(x) = x^3$ for $k > 0$. For what value of k will the average value of g equal the average value of f on the closed interval $[1, 3]$?

20. The level of air pollution a distance x miles from a tire factory is given by

$$L(x) = e^{-0.1x} + \frac{1}{x^2}.$$

What is the average level of pollution between 10 and 20 miles from the factory ?

21. Prove the Mean Value for Integrals by applying the Mean Value Theorem for derivatives to the function defined by $F(x) = \int_a^x f(t)\, dt$.

8.5   More Applications of the Definite Integral

In this section we consider several miscellaneous applications to illustrate the versatility of the definite integral. It should be clear from the examples and exercises in this chapter that if a quantity can be approximated by a sum of many terms, then it is a candidate for representation as a definite integral. The key requirement is that as the number of terms increases, the sum approaches a limit.

In applying the definite integral in any situation you start by identifying an interval $[a, b]$ and a functional quantity $f(x)$ defined on that interval. Then

1) divide the interval $[a, b]$ into subintervals of length Δx;

2) choose a single point x in one representative subinterval, calculate the desired function value for this subinterval at x and compute the product $f(x) \cdot \Delta x$;

3) create a Riemann sum by adding up the terms $f(x) \cdot \Delta x$ corresponding to each

     subinterval and take the limit of the sum; that is, evaluate $\displaystyle\int_a^b f(x)\, dx$.

For example, suppose a particle is moving with velocity given by the continuous function v at time t. We find the net distance traveled from $t = a$ to $t = b$ by

1) dividing the time interval into subintervals of length Δt;

2) choosing a point t_i in a representative subinterval, say the midpoint to be specific. If Δt is small then, since v is continuous, $v(t)$ is approximately equal to the constant value $v(t_i)$ on the subinterval. Thus, the term $v(t_i) \cdot \Delta t$ (velocity · time) is the approximate signed distance traveled in the time interval of length Δt.

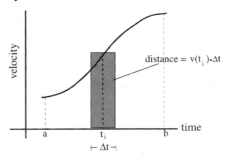

3) Adding up these small distances, $v(t_1) \cdot \Delta t + v(t_2) \cdot \Delta t + \ \ldots \ + v(t_n) \cdot \Delta t$, we have a

     Riemann sum approximation whose limiting value is the definite integral $\displaystyle\int_a^b v(t)\, dt$

     which is the net distance traveled by the particle from $t = a$ to $t = b$.

Density

Suppose an industrial plant emits a certain amount of pollutants and we wish to determine the amount of pollutants falling to the ground along some straight line from the plant. Suppose the density function $y = d(x)$ gives the number of particles of pollutant per foot (per day).

To find the total number of particles of pollutant on the interval $[a, b]$ we

1) divide the interval $[a, b]$ into n subintervals, each of width Δx feet;

2) choose a point x_i in a representative subinterval, say the right-hand endpoint to be specific. If Δx is small then we can safely assume that the density function d is approximately constant on the subinterval. Thus, the term $d(x_i) \cdot \Delta x$ (particles of pollutant per foot times the number of feet) is the approximate number of particles of pollutant in the subinterval Δx.

3) An approximation of the total number of particles of pollutant is given by the Riemann sum $d(x_1) \cdot \Delta x + d(x_2) \cdot \Delta x + \ldots + d(x_n) \cdot \Delta x$. By definition the limiting value of this sum is the definite integral $\int_a^b d(x)\, dx$.

Thus, if d is a continuous function that gives the density in number of particles of pollutant per unit length, then the total number of particles of pollutant in the interval $[a, b]$ is the definite integral $\int_a^b d(x)\, dx$.

Example 1: *Measuring Pollution Given Its Density*

A tire factory is located at the center of a town, which extends three miles either side of the factory. Let x be the distance from the factory. A straight highway goes through town and the density of particles of pollutant along the highway is given by

$$d(x) = 500(12 - x^2)$$

where $d(x)$ is the number of particles of pollutant per mile per day at the point x. Find the total number of particles of pollutant per day deposited in town along the highway.

Solution:

Using the results above, we have

$$\text{total number} = \int_a^b d(x)\, dx$$

$$= \int_{-3}^{3} 500(12 - x^2)\, dx$$

$$= \left[500\left(12x - \frac{x^3}{3}\right) \right]_{-3}^{3}$$

$$= 500\left[\left(36 - \frac{27}{3}\right) - \left(-36 + \frac{27}{3}\right) \right]$$

$$= 27{,}000 \text{ particles.}$$

Our first example was based on the idea of a density function measuring the number of particles of pollutant per mile deposited along a highway. Similarly, knowing the density of people per square mile we could determine the population of a given geographical region.

Example 2: *Determining the Number of Tents in a Campground*

A campground is in the shape of a rectangle measuring 1000 yards by 500 yards, with a river running along one of the 1000 yard sides. At a distance of x yards from the river, the density function $d(x) = -0.00015x + 0.05$ gives the number of tents per square yard on a recent Fourth of July weekend. Find the number of tents in the campground.

Solution:

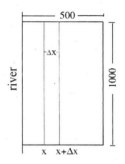

Divide the interval from $x = 0$ to $x = 500$ into n equal subintervals,

each of length Δx ($\Delta x = \dfrac{500 - 0}{n}$). The strip of campground land

(parallel to the river) determined by a representative interval from x to $x + \Delta x$ will have area $1000\,\Delta x$ square yards. Thus in that strip there will be, approximately,

$$(\text{density})(\text{area}) = (d(x))(1000\,\Delta x)$$

$$= (-0.00015x + 0.05)(1000\Delta x) \text{ tents.}$$

Adding the number of tents in each such strip produces the Riemann sum

$$d(x_1)(1000\Delta x) + d(x_2)(1000\Delta x) + \ ... \ + d(x_n)(1000\Delta x).$$

The limit of this sum as n increases without bound is a definite integral, giving the number of tents in the entire campground:

$$\int_0^{500} d(x) \cdot 1000 \ dx \ = 1000 \int_0^{500}(-0.00015x + 0.05) \ dx \ = 6250 \text{ tents.}$$

In science the notion of density is connected to the ideas of mass and unit measure. The density of a substance is defined as the mass of the substance divided by one or more of its dimensions. Thus, units like gm/cm³ (mass per unit of volume) are appropriate for density, but it is also conventional to use units like kg/m² (mass per unit area) or kg/m (mass per unit of length) depending on the application.

Thus, a rectangular brick with dimensions 10 cm, 5 cm, and 4 cm would have a volume of V = 10 x 5 x 4 = 200 cm³, and if it is made of a material having a constant density $d = 3$ g/cm³ its mass would be

$$\text{mass} = (\text{density})(\text{volume}) = (3 \ \frac{g}{cm^3})(200 \text{ cm}^3) = 600 \text{ gms.}$$

On the other hand, if a thin rod 8 meters long has a constant density of 5 kilograms per meter (kg/m), then the mass of the rod is

$$\text{mass} = (\text{density})(\text{length}) = (5 \ \frac{kg}{m})(8 \text{ m}) = 40 \text{ kg.}$$

If the density of a thin rod is not constant but varies from point to point we can set up a coordinate axis along the length of the rod so that the rod lies over the interval $[a, b]$. Then at each point in the interval it has density $d(x)$, $a \le x \le b$. If we divide the rod into small segments of length Δx, then each segment has approximately constant density and a mass approximately equal to $d(x) \cdot \Delta x$. An approximation of the total mass of the rod is given by the Riemann sum $d(x_1) \cdot \Delta x + d(x_2) \cdot \Delta x + \ldots + d(x_n) \cdot \Delta x$. If the density function d is continuous over the interval

[a, b] then by definition the limiting value of this Riemann sum is a definite integral $\int_a^b d(x)\, dx$.

Example 3: *Finding the Mass of a Rod Given Its Density*

A rod has length 8 meters. At a distance x meters from its left end, the density of the rod is

$$d(x) = 3 + 2x \quad \text{g/m.}$$

a) Write a Riemann sum with 4 terms that approximates the total mass of the rod.

b) Write and evaluate a definite integral that gives the exact total mass of the rod.

Solution:

a) We can set up a coordinate axis along the length of the rod so that the rod lies over the interval [0, 8]. To create a Riemann sum with 4 terms we divide the interval [0, 8] into 4 subintervals each with length $\Delta x = 2$ m. To approximate the mass of each of the 4 segments, we assume that the density is constant on each subinterval and is obtained by evaluating the density function $d(x) = 3 + 2x$ at the left-hand endpoint. A Riemann sum for the mass of the rod is the sum of the masses of the four segments:

$$
\begin{aligned}
R_4 &= d(x_1) \cdot \Delta x + d(x_2) \cdot \Delta x + d(x_3) \cdot \Delta x + d(x_4) \cdot \Delta x \\
&= (3 + 2x_1) \cdot \Delta x + (3 + 2x_2) \cdot \Delta x + (3 + 2x_3) \cdot \Delta x + (3 + 2x_4) \cdot \Delta x \\
&= (3 + 2 \cdot 0) \cdot 2 + (3 + 2 \cdot 2) \cdot 2 + (3 + 2 \cdot 4) \cdot 2 + (3 + 2 \cdot 6) \cdot 2 \\
&= 6 + 14 + 22 + 30 = 72 \text{ gms}
\end{aligned}
$$

b) As the number of subdivisions increases the width of each subinterval decreases and the limiting value of the Riemann sums is the definite integral

$$\int_0^8 (3 + 2x)\, dx$$

Thus, total mass $= \int_0^8 (3 + 2x)\, dx = \left[3x + x^2 \right]_0^8 = 88$ gms.

Work

In science, the term **work** is defined in a formal way that involves the application of a force to a body and the body's subsequent displacement. For example, if an object moves a distance d while subjected to a constant force F in the direction of motion, then the work W done by the force is defined to be

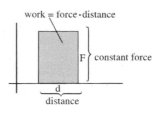

$$\text{work} = (\text{force}) \cdot (\text{distance}) \quad \text{or} \quad W = F \cdot d$$

If the force is measured in pounds and the distance in feet, then the unit of work is foot-pounds. For example, lifting a 50 lb box 3 feet requires $(50 \text{ lbs})(3\text{ft}) = 150$ foot-pounds. Calculus is needed when calculating the work done by a variable force because the amount of force changes as the object changes position.

Example 4: *Work Done by a Variable Force*

As a container moves along an assembly line, it is being filled with merchandise. When the container is at point x, the force needed to keep it moving is $f(x) = 4 + \dfrac{1}{2}x^2$ pounds. How much work is done in moving the container from $x = 0$ to $x = 6$ feet?

a) Divide the interval $[0, 6]$ into 3 equal subintervals. Assuming the force is constant on each subinterval and is obtained by using the value of the force at the left-hand endpoint, write a Riemann sum with 3 terms that approximates the work done in moving the object from $x = 0$ to $x = 6$.

b) Write and evaluate a definite integral that gives the exact amount of work done in moving the container from $x = 0$ to $x = 6$.

Solution:

a)   Work $\approx (4 + \dfrac{1}{2} \cdot 0^2)(2) + (4 + \dfrac{1}{2} \cdot 2^2)(2) + (4 + \dfrac{1}{2} \cdot 4^2)(2)$

$$= (4)(2) + (6)(2) + (12)(2) = 44 \text{ ft-lbs.}$$

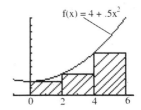

b)   If we divide the interval $[0, 6]$ into n subintervals, each of length Δx ($\Delta x = \dfrac{6 - 0}{n}$), a Riemann sum for the approximate work done is

$$(4 + \dfrac{1}{2}x_1^2)\Delta x + (4 + \dfrac{1}{2}x_2^2)\Delta x + \dots + (4 + \dfrac{1}{2}x_n^2)\Delta x.$$

As n grows large, this leads to the conclusion that

$$\text{Work} = \int_0^6 (4 + \dfrac{1}{2}x^2)\, dx = \left[4x + \dfrac{1}{6}x^3 \right]_0^6 = 60 \text{ ft-lbs.}$$

8.5   Exercises

1.   The acceleration (in ft/sec^2) of a car over 8 seconds is given by $a(t) = 3t - 12$.

   a) Write a Riemann sum with 8 terms that approximates the total change in velocity from $t = 0$ to $t = 8$.

   b) Find the exact increase in velocity from $t = 0$ to $t = 8$ by evaluating a definite integral.

2.   A rod is 10 centimeters long. At a distance of x centimeters from one end, the linear density of the rod is $3x^2$ grams per centimeter.

   a) Write a Riemann sum with 5 terms that approximates the total mass of the rod.

   b) Find the exact mass of the rod by evaluating a definite integral.

3.   Suppose silver is being extracted from a mine at the rate given by $A'(t) = 100e^{-0.2t}$, where $A(t)$ is measured in tons of silver and t in years from the opening of the mine.

   a) On a graph of A' draw a narrow rectangle of width Δt and show an input value t_i within the rectangle.

   b) Write an equation for ΔA, the approximate amount of silver extracted in the time interval from t to $t + \Delta t$.

   c) Use a definite integral to write an expression for A, the approximate total number of tons of silver mined from $t = 0$ to $t = b$.

   d) How much silver is mined in the first five years?

4.   The density function in people per mile for the population of the small coastal town of Westport, WA is given by $p(x) = 2000(2 - x)$ where x is the distance from the ocean. The town extends for 2 miles from the ocean.

   a) On a graph of the density function p draw a narrow rectangle of width Δx and show an input value x_i within the rectangle.

   b) Write an equation for ΔP, the approximate number of people within a distance of between x and $x + \Delta x$ miles.

   c) Write a Riemann sum that approximates the population of Westport.

   d) Find the exact population of Westport by evaluating a definite integral.

5.   During a two hour trip, a ferryboat uses oil at the rate of $r(t) = t\sqrt{4 - t^2}$ gal/hr. If the engine is started at $t = 0$, how much oil is used in the two hours?

6.   A city in the shape of a circle of radius 10 miles is growing with its population density a function of the distance from the center of the city. At a distance of r miles from the city center its population density is $d(r) = \dfrac{5000}{1+r}$ people/square mile.

In the figure, thin rings have been drawn around the center of the city. The area of the shaded ring can be approximated by the product $(2\pi r) \cdot \Delta r$ where r is the radius of the outer ring.

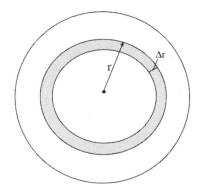

a) Write an equation for ΔP, the approximate number of people in the shaded ring.

b) Write a Riemann sum that approximates the population of the city.

c) Find the exact population of the city by evaluating a definite integral.

7.   Greater Seattle can be approximated by a semicircle of radius of 5 miles with its center on the shoreline of Puget Sound. Moving away from the center along a radius, the population density can be approximated by $d(x) = 20000e^{-.13x}$ people per square mile.

a) Write a Riemann sum that approximates the population of the city.

b) Find the exact population of the city by evaluating a definite integral.

8.   The density of cars (in cars per mile) down a 10 mile stretch of Route 3, a highway that connects Boston and Cape Cod, can be approximated by $d(x) = 83 + 34\cos(.5x)$ where x represents the distance in miles from the start of the 10-mile stretch.

a) Write a Riemann sum that approximates the total number of cars on this 10-mile stretch.

b) Find the total numbers of cars on the 10-mile stretch by evaluating a definite integral.

9.   A flat metal plate is shaped like the first quadrant region under the graph of $f(x) = \dfrac{1}{1+x}$ between $x = 0$ and $x = 4$. The density of the plate x units from the y-axis is given by x grams/cm^2.

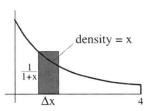

a) Assume the approximate mass ΔM of a strip of width Δx is:

$$\Delta M = (\text{area})(\text{density}) = \frac{x \cdot \Delta x}{1+x}.$$

Write a Riemann sum with 4 terms that approximates the total mass of the metal plate.

b) Write a definite integral that gives the exact value of the total mass of plate.

c) Evaluate the integral in part b).

10. A state forest is in the shape of a rectangle measuring 8 miles by 20 miles, with a highway running along one of the 20 mile sides. Deer in the forest are more numerous away from the highway. In fact, the density of their population is expressed by the function $D(x) = 0.25x + 1$ deer per square mile at a distance x miles from the highway.

a) Approximately how many deer are there in a strip of forest parallel to the highway and extending from 2 to 2.5 miles from the highway?

b) Write an expression for the approximate number of deer in a strip of forest parallel to the highway and extending from x to $x + \Delta x$ miles from the highway.

c) Subdividing the interval from $x = 0$ to $x = 8$ into 4 subintervals, each of width 2 miles, write a Riemann sum for the approximate number of deer in the forest.

d) Write a Riemann sum based on n subintervals of width $\Delta x = \dfrac{8}{n}$ for the total number of deer in the forest.

e) Find the number of deer in the forest by evaluating a definite integral.

11. An hour after the start of a Walk for Hunger, the number of walkers per mile a distance x miles from the start of the walk was given (approximately) by the model function $W(x) = 0.17x^3 - 3.1x^2 + 14.8x$.

a) Approximately how many walkers are located between 3 and 3.5 miles from the starting location?

b) Write an expression for the approximate number of walkers between x to $x + \Delta x$ miles from the starting location.

c) Subdividing the interval from $x = 0$ to $x = 6$ into n subintervals, each of width $\Delta x = \dfrac{6}{n}$ miles, write a Riemann sum for the approximate number of walkers at a distance of at most 6 miles from the starting location.

d) Find the number of walkers at a distance of at most 6 miles from the starting location by evaluating a definite integral.

12. Find the work done when

a) a constant force of 30 pounds along the x-axis moves an object from $x = 2$ to $x = 5$ feet;

b) a variable force of $f(x) = \dfrac{120}{x^2}$ pounds along the x-axis moves an object from $x = 2$ to $x = 3$ feet.

13. When a particle is located a distance x feet from the origin a force of $f(x) = \dfrac{x}{1 + x^2}$ pounds acts on it. How much work is done in moving the particle along the x-axis from $x = 1$ to $x = 3$?

14.   The table below was obtained experimentally for a variable force F, in pounds, required to move an object from $x = 3$ feet to $x = 5$ feet along the x-axis. Use the Trapezoid Rule with $n = 4$ to approximate the work done by force F.

x (ft)	3.0	3.5	4.0	4.5	5.0
F(x) (lb)	5.9	6.8	7.0	8.0	9.2

15.   The graph in the figure shows the force, F, in pounds, required to move an object 8 feet along the x-axis.

Using the Midpoint Rule with $n = 4$ to estimate the work done by the force F.

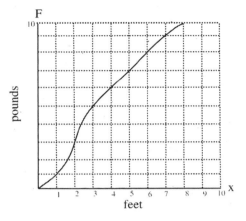

16.   The force, F, in lbs, required to stretch a spring x units beyond its natural length is proportional to x . Let $F(x) = cx$ where c is the constant of proportionality.

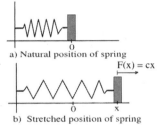

a) Natural position of spring

b) Stretched position of spring

a)   Suppose that 10 lbs of force are required to stretch a large spring 4 inches. Why does this mean that the spring constant $c = 30$ lbs/ft?

b)   To approximate the work done in stretching the spring from $x = 0$ to $x = 2$ feet, divide the interval [0, 2] into 4 equal subintervals. Assume the force is constant on each subinterval and is obtained by using the value of the force at the left-hand endpoint. Write a sum with 4 terms that approximates the work.

c)   Suppose the interval [0, 2] is divided into n equal subdivisions, each of width Δx ($\Delta x = \dfrac{2 - 0}{n}$), and assume the force is constant on each subinterval and is obtained by using the force at the smallest value of x on each subinterval. Write a Riemann sum that approximates the work done in stretching the spring from $x = 0$ to $x = 2$ feet.

d)   Write and evaluate a definite integral to show that the work done in stretching the spring from $x = 0$ to $x = 2$ feet equals 60 ft-lbs.

17.   Suppose a force of 12 lb is required to hold a spring stretched 4 inches beyond its natural length. How much work (in ft-lbs) is done in stretching the spring from its natural length to 6 inches beyond its natural length? First approximate the required work by a Riemann sum. Then express the work as a definite integral and evaluate it.

Chapter 8   Supplementary Problems

1.   Let R be the region enclosed by the graphs of $y = x^2$ and $y = 2x$.

a) Find the exact area of R.

b) Find the volume of the solid obtained by revolving R about the x-axis.

c) Find the volume of the solid obtained by revolving R about the y-axis.

d) Set up but do not evaluate the definite integral whose value is the volume of the solid obtained by revolving R about the horizontal line $y = -2$.

2.   Let R be the region in the first quadrant that is enclosed by the graphs of $y = \tan x$, the x-axis and the line $x = \dfrac{\pi}{4}$.

a) Find the exact area of R.

b) Find the volume of the solid obtained by revolving R about the x-axis.

c) Find the volume of the solid obtained by revolving R about the y-axis.

d) Set up but do not evaluate the definite integral whose value is the volume of the solid obtained by revolving R about the line $y = -2$.

3.   Let R be the region in the first quadrant that is enclosed by the graphs of $y = 4 - x^2$, $y = 3x$ and the y-axis.

a) Find the exact area of R.

b) Find the volume of the solid obtained by revolving R about the x-axis.

c) Find the volume of the solid obtained by revolving R about the y-axis.

d) Set up but do not evaluate the definite integral whose value is the volume of the solid obtained by revolving R about the line $y = -2$.

4.   A region S is bounded by the graphs of the following equations:

$y = \sqrt{x - 1}, \ y = \ln x, \ x = 2, \ x = 5.$

Set-up definite integrals that give each of the following quantities. You do not need to evaluate the integrals.

a) The area of S.

b) The volume of the solid when S is rotated about the y-axis.

c) The volume of the solid when S is rotated about the x-axis.

5. Let R be the region in the first quadrant bounded by the graph of $y = x + 1$, $y = 2x^3$, and $x = 0$. Set up, but do not evaluate, definite integrals that give

   a) the area of R;

   b) the volume when R is rotated about x-axis;

   c) the volume when R is rotated about y-axis.

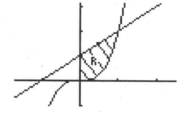

6. The following table lists the known values of a continuous function f.

x	1	1.2	1.4	1.6	1.8	2
$f(x)$	3.3	2.8	1.9	2.4	3.0	2.8

   a) Use the Trapezoid Rule with $n = 5$ to approximate the area of the region S bounded above by the graph of f and below by the x-axis on the interval $[1, 2]$.

   b) Use the Trapezoid Rule with $n = 5$ to approximate the volume of the solid generated when the region S (of part a) is rotated about the y-axis.

7. Let R be the region enclosed by the graph of $y = \ln x$, the line $x = e$ and the x-axis.

   a) Sketch region R.

   b) Find the area of region R by setting up and evaluating a definite integral.

   c) Find the volume of the solid generated by revolving R about the y-axis.

8. A particle moves along a coordinate line with velocity v at time t defined by $v(t) = t^2 - t$. If distance is measured in feet and time in seconds, determine the total distance traveled by the particle from $t = 0$ to $t = 2$.

9. The region shaded in the figure is rotated about the x-axis. Using the Trapezoid Rule with 5 equal subdivisions, approximate the volume of the resulting solid.

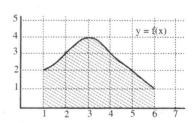

10. The first quadrant region enclosed by the y-axis and the graphs of $y = e^x$ and $y = k$, $k > 1$, is rotated about the x-axis. If the volume of the solid formed is 2π, find the value of k.

11. Let the base of a solid be the first quadrant region enclosed by the x-axis, the y-axis and the graph of $y = 1 - \dfrac{x^2}{4}$. If the cross sections perpendicular to the x-axis are squares, find the volume of the solid.

12.   Let the base of a solid be the first quadrant region enclosed by the x-axis and one arch of
       the graph of $y = \sin x$. If the cross sections perpendicular to the x-axis are squares, find
       the volume of the solid.

13.   A particle moves along the x-axis so that its velocity at any time $t \geq 0$ is given by
       $v(t) = 3t^2 - 6t$. The position at $t = 2$ is $x(2) = 4$.

       a) Write an expression for the position $x(t)$ of the particle at any time $t \geq 0$.

       b) For what values of t, $1 \leq t \leq 4$, is the particle's instantaneous velocity the same as its
          average velocity on the closed interval $[1, 4]$?

       c) Find the total distance traveled by the particle from time $t = 1$ to $t = 4$.

14.   Let R be the first quadrant region enclosed by the graph of $y = 2e^{-x}$ and the line $x = k$.

       a) Find the area of R in terms of k.

       b) Find the volume of the solid generated when R is rotated about the x-axis in terms of k.

       c) What is the limit of the volume in part b) as $k \to \infty$?

15.   Let R be the region in the first quadrant bounded by the graph of $f(x) = \sqrt{1 + e^{-x}}$, the
       line $x = 2$, and the x-axis.

       a) Write an integral that gives the area of R.

       b) Use the Trapezoidal Rule with $n = 4$ to approximate the area. You must show the
          numbers that lead to your answer.

       c) Determine, exactly, the volume of the solid produced when R is revolved about the
          x-axis.

16.   A rectangular swimming pool with vertical sides is 20 ft wide and 40 ft long. The depth of
       the water x ft from the shallow end of the pool is $2 + \dfrac{x^2}{200}$ ft. Find the volume of
       water in the pool.

17.   The following table represents the diameter of the cross-section of a tree at various heights
       above the ground. Both are measured in feet. Assume the cross-sections are all circular.

Height (ft.)	2	6	10	14	18	22	26	30
Diameter (ft.)	2.0	2.0	2.0	1.8	1.6	1.5	1.3	1.2

       a) How fast is the diameter of the tree changing 22 ft. above the ground? Indicate units of
          measure.

       b) Use the Trapezoidal Rule to approximate the volume of the tree from 14 ft. to 30 ft.
          above the ground. Indicate units of measure.

18.   The density of cars (in cars per mile) down a 10-mile stretch of the Massachusetts Turnpike starting at a toll plaza is given by $d(x) = 400 + 100\sin(\pi x)$ where x is the distance in miles from the toll plaza and $0 \le x \le 10$.

   a) Write a Riemann sum that approximates the total number of cars down the 10-mile stretch.

   b) Find the total number of cars on the 10-mile stretch by evaluating a definite integral.

19.   Oil has spilled into a straight river that is 100 meters wide.

   The density of the oil slick is $d(x) = \dfrac{50x}{1 + x^2}$ kilograms per

   square meter, where $x \ge 0$ is the number of meters downstream from the spill.

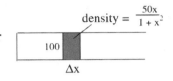

   a) Assume the density of the slick does not vary from one shore to another and the approximate amount of oil in one strip of the river of width Δx is:

   $$\text{density x area} = d(x) \cdot 100\Delta x.$$

   Write a Riemann sum with 8 terms that approximates how much oil is within 80 meters of the source of the spill.

   b) Write a definite integral that gives the exact amount oil that is within 80 meters of the source of the spill.

   c) Evaluate the integral in part b).

20.   Let the function F be defined by $F(x) = \displaystyle\int_0^x \dfrac{t^2 - 2t}{e^t}\, dt$.

   a) Find $F'(x)$ and $F''(x)$.

   b) If they exist, determine critical numbers for F.

   c) Discuss the concavity of the graph of F.

21.   Approximate the average rate of change of the function $f(x) = \displaystyle\int_0^x \sqrt{1 + \sin^2 t}\, dt$ over the interval $[1, 4]$.

22.   Use the Fundamental Theorem of Calculus to evaluate the following definite integrals.

   a) $\displaystyle\int_0^1 x\sqrt{4 - x^2}\, dx$ b) $\displaystyle\int_0^1 xe^{2x}\, dx$

23.   Sketch and label graphs of both the derivative and an antiderivative of the function in the
      figure. Make the antiderivative go through the point $(-1, 0)$.

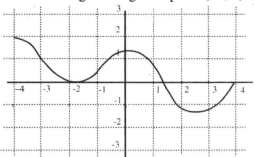

24.   Let the function f be defined by $f(x) = \begin{cases} x^2 + k, & \text{if } x \le 1 \\ 7 - kx, & \text{if } x > 1 \end{cases}$.

      a) For what values of k will f be continuous at $x = 1$? Justify your answer.

      b) Using the value of k found in part a), determine whether f is differentiable at $x = 1$.

      c) Let $k = 4$. Determine whether f is differentiable at $x = 1$. Justify your answer.

25.   Oil from a ruptured tanker spreads in a circular
      pattern. The function A whose graph is sketched
      in the figure gives the area of the oil slick, $A(t)$,
      measured in square miles, after t days. Find the
      rate at which the radius is changing after 4 days.

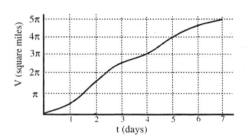

26.   The graph of the velocity $v(t)$, in ft/sec, of a car
      traveling on a straight road, for $0 \le t \le 35$, is shown in
      the figure.

      a) Find the average acceleration of the car, in ft/sec^2,
         over the interval $0 \le t \le 35$.

      b) Find one approximation for the acceleration of the
         car, in ft/sec^2, at $t = 20$. Show your computations.

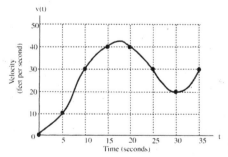

      c) Approximate $\int_5^{35} v(t)\,dt$ with a Riemann sum, using the midpoints of three subintervals
         of equal length. Explain the meaning of this integral.

27. Let f be the function given by $f(x) = \dfrac{x^2 + 7x + 3}{x^2}$.

a) In the viewing rectangle provided below, sketch a graph of f.

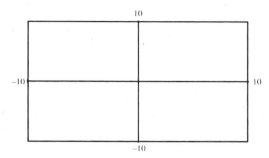

b) Write an equation for each asymptote of the graph of f.

c) Find the range of f exactly. Use $f'(x)$ to justify your answer.

d) Find the exact x-coordinates of all inflection points. Use $f''(x)$ to justify your answer.

28. The graph of g is shown at the right.

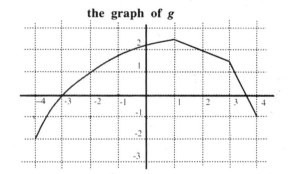

the graph of g

a) Find $\lim\limits_{x \to 1} g(x)$.

b) Find $\lim\limits_{h \to 0} \dfrac{g(2+h) - g(2)}{h}$.

c) Find $\lim\limits_{x \to ^-2} g'(x)$.

d) Find $g'(g(-3))$.

e) Find $\dfrac{dF}{dx}$ at $x = 2$,

if $F(x) = g(x^2 - g(x))$.

29. A snail starts moving along the x-axis at time $t = 0$. Her goal is to travel 2 feet. She moves more and more slowly, so that her speed is given in ft/min by

$$v(t) = \frac{1}{1 + t^2}, \quad t \geq 0.$$

a) What distance does the snail travel between $t = 0$ and $t = 1$?

b) How long does it take the snail to travel a distance of $\pi/3$ feet?

c) Will the snail ever reach her goal?

30. If $F(x) = \displaystyle\int_{-5}^{x} t(t-2)^3 \, dt$, on what intervals is F decreasing?

31.   A rectangular storage container with no top must be constructed using the following
       guidelines:

       • The volume is 36 cubic meters.
       • The length of the base is twice the width.
       • The cost of material for the sides and base is $5 per square meter.

       Find the dimensions and the cost of the least expensive container. Show your work
       and give an exact answer.

32.   Suppose that $f(0) = 3$ and f' is the function shown below. Let $g(x) = (x^2 + 1) \cdot f(x)$.

       a) Evaluate $g'(0)$.

       b) Is g increasing at $x = 1$? Justify briefly.

       c) Estimate $g''(0)$.

       d) Is g concave up at $x = 1$? Justify briefly.

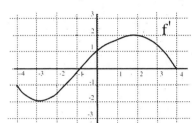

33.   Oil from an offshore rig located 3 miles from the shore is to be pumped to a refinery
       location on the edge of the shore that is 8 miles east of the rig. The cost of constructing a
       pipe in the ocean from the rig to the shore is 1.5 times as expensive as the cost of
       constructing on land. How should the pipe be laid to minimize cost?

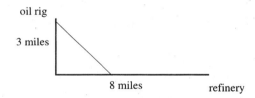

34.   Water flows into a tank at a rate of ($5t + 4$) gallons/min. Water flows out of the tank at a
       rate of $0.5t^2$ gallons/min. At $t = 0$ minutes the tank contains 100 gallons.

       (a) Write an expression for the amount of water in the tank at a given time t.

       (b) When will the quantity of water in the tank be a maximum and how much is the
           quantity of water at that time?

       (c) What was the average number of gallons of water in the tank in the first 5 minutes?

       (d) At what time during the first 5 minutes is the average number of gallons actually
           obtained?

CHAPTER 9

Differential Equations

9.1 Introduction

Solving a problem often means inventing a function which "fits" given data; in many problems the data concern the function's rate of change. For example, a function to predict the position of a moving object might be created from information about the object's velocity or acceleration; or, a radioactive substance might be disintegrating at a known rate and from this information a function which relates the amount present to time might be discovered. In such examples a model function is being determined from equations involving the derivatives of the model function. The equations that relate derivatives to other functions are called **differential equations**.

Each equation below is a differential equation:

$$y' = 0; \qquad y' = y; \qquad y' = 2xy; \qquad \frac{dy}{dx} = \frac{1}{1+x^2}; \qquad y'' = -y \ .$$

The **order** of a differential equation refers to its highest-order derivative. Thus, $y' = y$ is a **first-order differential equation**; $y'' = -y$ is a **second-order differential equation**.

A **solution** of a differential equation is any function that satisfies the differential equation. As the following example illustrates, a differential equation normally has many solution functions. To solve a differential equation means to find at least one – but preferably all – solution functions.

Example 1: *Solving a Differential Equation*

Find the solution(s) of the differential equation $y' = y$.

Solution:

A solution of the differential equation $y' = y$ is a function $y = f(x)$ that is its own derivative. The exponential function $y = e^x$ is famous for this reason. If $f(x) = e^x$, then $f'(x) = e^x = f(x)$; thus $f(x) = e^x$ is a solution of the differential equation.

In fact, every function of the form $f(x) = Ce^x$, where C is a constant, is a solution of the same differential equation, because if $f(x) = Ce^x$, then $f'(x) = Ce^x = f(x)$.

The solution $f(x) = Ce^x$ in Example 1 is called the **general solution** of the differential equation because it can be shown that any solution of the differential equation can be derived from the rule for $f(x)$ by choosing the constant C appropriately.

In other words, the differential equation $y' = y$ has an infinite family of solutions. Graphically, the general solution corresponds to a whole family of curves. Several of the solution curves are displayed in the figure; each is labeled with the corresponding value of the C. Each graph represents one solution of the differential equation $y' = y$. Not all solutions are shown.

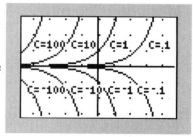

By choosing C correctly, we can find a solution curve that passes through a given point. For instance, if the solution $y = Ce^x$ to the differential equation $y' = y$ is to pass through the point $(0, 3)$, then $3 = Ce^0$, $C = 3$, and so $y = 3e^x$ is the desired function.

A differential equation typically has infinitely many solutions. However, for most practical applications it is a single function that is required for the solution. To obtain such a particular solution it is usually necessary first to find the general solution and then to determine specific values for the constants from the data given in the problem. The data are usually called **initial conditions** and the combination of an initial condition and a differential equation is called an **initial value problem**. The next example illustrates the idea.

Example 2:   *Finding a Particular Solution*

Find a function that satisfies both the differential equation $y' = \dfrac{2x}{x^2 + 1}$ and the initial condition $y(0) = 1$.

Solution:

In this instance, integrating both sides of the equation produces the general solution

$$y = \ln(x^2 + 1) + C.$$

For each constant C a solution is determined. Several of the solutions are displayed below:

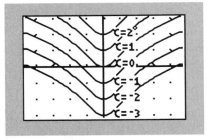

Solutions to $y' = \dfrac{2x}{x^2 + 1}$, $y = \ln(x^2 + 1) + C$ for various values of C.

If the initial condition $y(0) = 1$ is to be satisfied, then necessarily $1 = \ln(0 + 1) + C$. Thus $C = 1$ and the desired particular solution is $y = \ln(x^2 + 1) + 1$. The graph of $y = \ln(x^2 + 1) + 1$ is among those shown above. Note that this curve passes through the point $(0, 1)$, as it should, because $y(0) = 1$.

The connection between solving a first-order differential equation and finding antiderivatives is illustrated above in Example 2. Every time you antidifferentiate to find $\int f(x)\, dx$, you are actually solving the first order differential equation $y' = f(x)$.

In general, solving a differential equation is not an easy task. There is no systematic technique that enables us to solve all differential equations. However, in Section 9.2 we see how to draw rough graphs of the solution function when we have no explicit formula. And the methods in Section 9.3 will enable us to find numerical approximations to solutions.

9.1   Exercises

1.   Match the following graphs with the descriptions below.

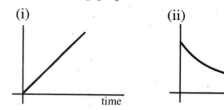

   (i)   (ii)   (iii)

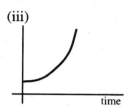

   a) The temperature of a hot cup of tea left on a table.

   b) The balance in an interest-bearing bank account into which a single deposit of $100 was made at time zero.

   c) The speed of a falling body.

2.   Determine whether or not the given functions are solutions of the second order differential equation $y'' - y = 0$.

   a) $y = \sin x$ b) $y = e^{2x}$ c) $y = 4e^{-x}$ d) $y = Ce^x$

3.   Show that for any constant C, the function $y = Ce^{-x^2}$ is a solution of the differential equation $y' + 2xy = 0$.

4.   Show that for any constant C, $y = \dfrac{x^2}{3} + \dfrac{C}{x}$ is a solution of the first order differential equation $xy' + y = x^2$.

5.   Find a function that satisfies both the differential equation $y' = x^2 - 4$ and the initial condition $y(3) = 2$.

6.   a) Show that $y = \dfrac{1}{1 + e^{-t}}$ satisfies the logistic differential equation $\dfrac{dy}{dt} = y(1 - y)$.

   b) What is the limiting value of y as time t increases without limit?

7.   a) Show that $y = \dfrac{C}{x}$ is a solution of the equation $xy' + y = 0$.

   b) Sketch solution curves $y = \dfrac{C}{x}$ for C = –2, –1, 1, 2 in a [–3, 3] x [–3, 3] viewing rectangle.

8.   a) Show that $y = Ce^x - x - 1$ is a solution of the equation $y' = x + y$.

   b) Sketch solution curves $y = Ce^x - x - 1$ for C = –3, –2, –1, 0, 1, 2, 3 in a [–5, 5] x [–5, 5] viewing rectangle.

   c) One solution curve is a straight line. Label this curve with the appropriate value of C.

9. Find the particular solution to the differential equation $y' = y$ whose graph passes through the specified point.

   a) $(0, -2)$ b) $(1, 5)$ c) $(-3, 8)$

10. Find a function that satisfies the given differential equation and the initial condition.

   a) $y' = 2\sin x$, $y(0) = 7$

   b) $y' = \dfrac{x}{\sqrt{9 + x^2}}$, $y(0) = 1$

11. In Exercise 4, you verified that the function $y = \dfrac{x^2}{3} + \dfrac{C}{x}$ is a solution to the differential equation $xy' + y = x^2$. Determine the constant C if the graph of the solution passes through the point $(1, -\dfrac{2}{3})$.

12. Consider the differential equation $\dfrac{dy}{dx} = \dfrac{3y}{x}$.

   a) Show that $y = Ax^3$ is a solution of the differential equation.

   b) Find a solution of the differential equation that satisfies the initial condition $y(-3) = 2$.

13. Consider the differential equation $\dfrac{dy}{dx} = 2y\cos x$

   a) Show that $y = Ae^{2\sin x}$ is a solution of the differential equation.

   b) Find a solution of the differential equation that satisfies the initial condition $y(\pi) = 5$.

14. Consider the differential equation $\dfrac{dy}{dx} = -2xy$

   a) Show that $y = e^{C-x^2}$ is a solution of the differential equation.

   b) Find a solution of the differential equation that satisfies the initial condition $y(0) = e$.

9.2   Solving Differential Equations Graphically: Slope Fields

Derivatives can always be thought of in terms of slope: for any input x, $f'(x)$ is the slope of the graph of f at the point (x, y). From this geometric point of view, a first order differential equation describes the slopes of solution curves at a given point.

Consider the differential equation $y' = y$ from Example 1 in Section 9.1. Any solution of this differential equation has the property that at a given point (x, y) in the plane, the slope of the solution curve is equal to its y coordinate. This means that if we choose the point $(1, 2)$, the curve's slope is 2; where it goes through a point with $y = 3$, its slope is 3. In the figure on the left below, a small line segment is drawn at some of the marked points showing the slope of the solution curve.

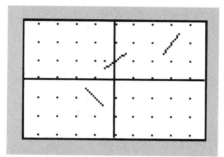
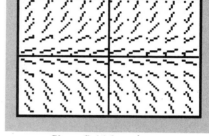

Tangent segments at (-1, -1), (0, 1) and (3, 3)   Slope field for $y' = y$

In the figure on the right, many of the tangent segments have been drawn creating the **slope field** for the differential equation $y' = y$. You can almost see the family of solutions and you can sketch a solution corresponding to a specific initial condition; start at any point in the plane and look at the slope field at that point and start to move in that direction. After a small step, look at the slope again and alter your direction if necessary. Continue to move across the plane in the direction of the slope field – "go with the flow" – and you will trace an approximation to a solution curve. The picture below illustrates the solution curve satisfying the initial condition $y(0) = 1$ obtained by sketching the solution curve passing through the point $(0, 1)$ and following the direction of the slope field.

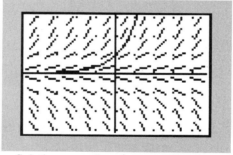

Solution curve for $y' = y$, $y(0) = 0$

The slope field for any first-order differential equation offers a simple and natural method of approximating solution curves. However, drawing a complete field of tangent segments by hand is a tedious task and is best left to a computer or graphing calculator. The following slope field program takes as input a differential equation $y' = f(x, y)$ where f could involve both x and y. The program computes the value of the derivative at several predetermined points (x, y) and then plots a short line segment centered at each point having the value of the derivative as slope. An appropriate viewing rectangle should be defined before executing the program. The program asks for the number of marks across and down; twelve across and eight down are good choices. Note that the function is stored in Y1.

```
 Program:SLOPEFLD
:Func:FnOff
:ClrDraw:ClrHome
:Pxl–Off(0, 0)
:Disp "HOW MANY MARKS?"
:Input "ACROSS",A
:Input "DOWN",B
:(Xmax–Xmin)/A→H
:(Ymax–Ymin)/B→K
:Xmin+.5H→Z
:Ymax–.5K→Y
```
```
:For(I,1,B)
:Z→X
:For(J,1,A)
:Y1ΔX/ΔY→M
:H/√(1+M²)→U
:KM/√(1+M²)→V
:X–.4U→P:Y–.4V→Q
:X+.4U→R:Y+.4V→S
:Line(P,Q,R,S)
:X+H→X:End
:Y–K→Y:End
```

Example 1: *Graphing a Slope Field*

Graph the slope field for the differential equation $y' = x + y$.

Solution:

Enter Y1 = X+Y. Press WINDOW and enter a [–2, 2] x [–1, 1] viewing rectangle. Execute the program SLOPEFLD to produce the following slope field.

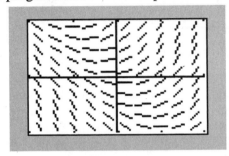

    The next example illustrates that a solution curve for a differential equation is not necessarily the graph of a function.

Example 2: *Graphing a Slope Field*

Graph the slope field for the differential equation $y' = -\dfrac{x}{y}$ and guess the form of the solution curves.

Solution:

Enter Y1 = –X/Y. Press WINDOW and enter a [–4.5, 4.5] x [–3, 3] viewing rectangle. Execute the program SLOPEFLD to produce the following slope field.

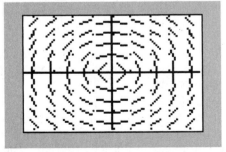

The slope field suggests that the solution curves look like circles centered at origin. In a later section we will obtain the solution analytically but, even without this, it can be verified that the circle is a solution. Letting r represent the radius, we have $x^2 + y^2 = r^2$.

Differentiating implicitly, we get

$$2x + 2y \cdot y' = 0 \quad \text{and}$$

$$y' = -\frac{x}{y}.$$

Summarizing we see that the slope field for a differential equation shows by means of short line segments the slope of the solution curve at a number of random points. Thus the slope field is useful for determining the general appearance of the solution curves and sketching the graph of a particular solution curve. Sometimes it is even possible to use such curves to give the formulas for the solution curves. The next example demonstrates that even without formulas, solution curves can be helpful in practice.

Example 3:   *Newton's Law of cooling*

Newton's Law of Cooling says that hot coffee in a 70°F room cools at a rate proportional to the difference between the coffee temperature and the room temperature. If $y(t)$ represents the coffee temperature at time t, then

$$y'(t) = k(y - 70).$$

If at a given time t, the coffee temperature is 190°F and its temperature is dropping at a rate of 12 degrees per minute, how much later did the coffee temperature reach 130°F?

Solution:

Letting $t = 0$ denote the starting time t, we have initial conditions $y'(0) = -12$ and $y(0) = 190$. Now to find the constant k we substitute in the differential equation.

$$-12 = k(190 - 70)$$

$$-12 = k(120)$$

$$k = -0.1.$$

Hence, the initial value problem is $y'(t) = -0.1(y - 70)$ with $y(0) = 190$. Using the slope field on a [0, 20] x [60, 220] viewing rectangle it is possible to sketch a solution curve starting at the point (0, 190) and conclude that the temperature reaches 130 degrees in approximately $t = 7$ minutes.

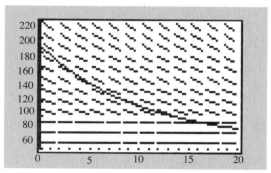

Technology Tip: *Drawing a Solution Curve*

To superimpose a solution curve $y = f(x)$ on a slope field, press [2nd] [Draw] and select the [6:DrawF] command and enter the formula for $f(x)$. We will see later by analytical methods that the solution curve here has equation $y = 70 + 120e^{-0.1t}$.

9.2   Exercises

1.   The calculator drawn slope field for the differential equation $y' = -2xy$ is shown in a [–3, 3] x [–3, 5] viewing rectangle in the figure below. The solution curve passing through the point (0, 2) is also displayed.

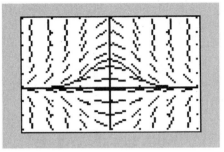

     a)   Sketch the solution curve through the point (0, 1).

     b)   Sketch the solution curve though the point (0, –2).

2.   The figure below shows the slope field for the differential equation $y' = x + y$ in a [–4.7, 4.7] x [–3.1, 3.1] viewing rectangle.

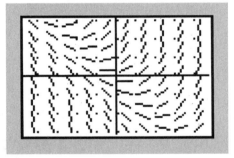

     a)   Sketch the solution curve through the point (0, 1).

     b)   Sketch the solution curve through the point (–2, 0).

3.   Consider the differential equation $y' = x + 1$.

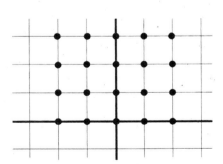

     a)   On a grid, like that shown in the figure, sketch a slope field for the given differential equation at the 20 points indicated.

     b)   Sketch the solution curve that passes through the point (–1, 0).

     c)   Use the SLOPEFLD program to verify the results in parts a) and b).

4. Consider the differential equation $y' = 1 - y$.

    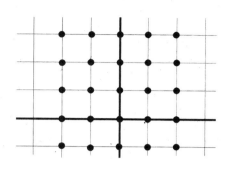

    a) On a grid, like that shown in the figure, sketch a slope field for the given differential equation at the 25 points indicated.

    b) Sketch the solution curve that passes through the point $(-1, 3)$.

    c) Use the SLOPEFLD program to verify the results in parts a) and b).

5. Consider the differential equation $y' = \dfrac{2x}{y}$.

    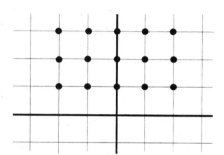

    a) On a grid, like that shown in the figure, sketch a slope field for the given differential equation at the 15 points indicated.

    b) Sketch the solution curve that passes through the point $(1, 2)$.

    c) Use the SLOPEFLD program to verify the results in parts a) and b).

6. Consider the differential equation $y' = x - y$.

    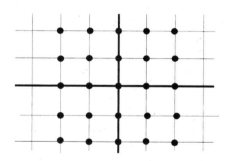

    a) On a grid, like that shown in the figure, sketch a slope field for the given differential equation at the 25 points indicated.

    b) Find an equation for the straight line solution that passes through the point $(1, 0)$.

    c) Use the SLOPEFLD program to verify the results in parts a) and b).

7. a) Use the SLOPEFLD program to sketch the slope field for the equation $\dfrac{dy}{dx} = 0.25y(4 - y)$.

    b) What information, if any, can you determine about concavity for the solution curves?

    c) What information, if any, can you determine about the solutions as $x \to +\infty$?

8. Given the implicitly defined function $x^2 - y^2 = C$.

    a) Show that $\dfrac{dy}{dx} = \dfrac{x}{y}$.

    b) Use the SLOPEFLD program to show that the solution curves are hyperbolas.

9.   The function defined by

$$F(x) = \frac{1}{\sqrt{2\pi}} \int_a^x e^{-.5t^2} \, dt$$

is called the **Normal Probability Density Function** and is very important in statistics.

a) Find $F'(x)$.

b) Using the SLOPEFLD program, sketch the slope field for the equation $Y_0 = F'(x)$
   in a $[-3, 3]$ x $[-.5, 1.1]$ viewing rectangle.

c) Sketch the solution curve through the point $(0, 0.5)$.

10.   Match the slope fields with their differential equations.

i) $y' = x$ ii) $y' = 2 - y$ iii) $y' = \sin x$ iv) $y' = x - y$

a)
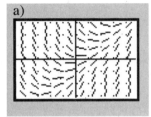
b)
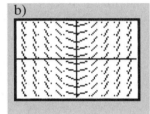
c)
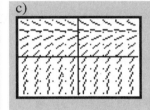
d)
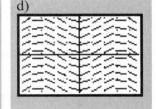

11.   Match the slope fields with their differential equations.

i) $y' = x + y$ ii) $y' = .5y$ iii) $y' = .5x - 1$ iv) $y' = -\dfrac{x}{y}$

a)
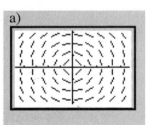
b)
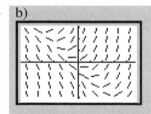
c)
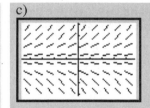
d)
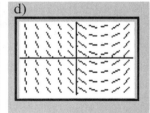

9.3   Solving Differential Equations Numerically: Euler's Method

In the preceding section we sketched solution curves for a differential equation using its slope field. In this section we consider a numerical version of the same idea: **Euler's method.** It describes how to move, step by step, through a slope field and is based on the tangent approximation

$$y(x_0 + \Delta x) \approx y(x_0) + \Delta x \cdot y'(x_0)$$

to approximate y near a point x_0 where the value $y(x_0)$ is known.

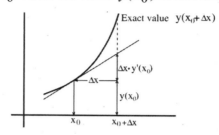

Here is how Euler's method works. Choose a starting point corresponding to an initial value and calculate the slope at that point using the differential equation. The slope tells you the direction to take. Move a short distance in that direction to a new point. Recalculate the slope from the differential equation, using the coordinates of the new point. Change direction to correspond to the new slope and move another short distance, and so on.

Example 1: *Using Euler's Method*

Use Euler's method to approximate the solution of $y' = y$ with initial conditions $y(0) = 1$ and a step-size of $\Delta x = 1$.

Solution:

The slope at the initial point $P_0 = (0, 1)$ is $y'(0) = 1$. We seek a nearby point, P_1, which is approximately on the graph of our solution curve. As we move to the next point P_1, y will increase by Δy where

$$\Delta y = (\text{slope at } P_0) \, \Delta x = 1 \cdot 1 = 1$$

so the y value at P_1 is

$$(y \text{ value at } P_0) + \Delta y = 1 + 1 = 2.$$

Thus, $P_1 = (1, 2)$.

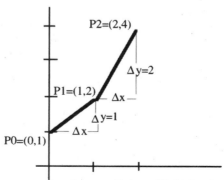

Euler approximation to solution of initial value problem
y' = y and y(0) = 1.

Now using the differential equation again, we have the slope at $P_1 = 2$, and in the second step

$$\Delta y = (\text{slope at } P_1) \ \Delta x = 2 \cdot 1 = 2$$

so the y value at P_2

$$(y \text{ value at } P_1) + \Delta y = 2 + 2 = 4.$$

Thus, $P_2 = (2, 4)$.

The following table summarizes our numerical work for two steps.

n	$P_n = (x_n, y_n)$	$\Delta y = \text{slope} \cdot \Delta x$	$y_{n+1} = y_n + \Delta y$
0	(0, 1)	$1 \cdot 1 = 1$	$1 + 1 = 2$
1	(1, 2)	$2 \cdot 1 = 2$	$2 + 2 = 4$
2	(2, 4)		

Using analytical methods (introduced in the next section), the exact solution to the given initial value problem is $y = e^x$. So our approximation $y(2) = 4$ is far from the exact value $y(2) = e^2 = 7.39$; however, by reducing the step size, Δx, we may obtain a better approximation.

In the following figure we have three Euler approximations and the exact solution curve for the initial value problem $y' = y$ and $y(0) = 1$.

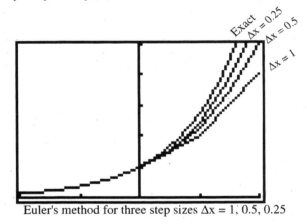

Euler's method for three step sizes $\Delta x = 1, 0.5, 0.25$

Below is an Euler program that takes as input a differential equation, the initial point, the step size and the number of steps. It returns the sketch of a step-wise approximation curve and stores the x values in list L1 and the corresponding approximating y values in list L2.

```
PROGRAM:EULER                    :N+1→dim (L1):N+1→dim (L2)
:Func:FnOff                      :For(I,1,N)
:Disp "DY/DX MUST"               :X+H→U: Y1→M
:Disp "BE IN Y1"                 :Y+MH→V
:Pause                           :X→ L1(I):Y →L2(I)
:Input "X START",A               :Line(U,V,X,Y)
:Input "Y START",B               :U→X
:Input "STEP SIZE",H             :V→Y
:Input "NO. STEPS",N             :End
:A→X :B→Y                        :U→L1(N + 1) : V→L2(N + 1)
```

Example 2: *Using a Calculator Program*

Given the initial value problem $y' = x + y$ with $y(0) = 1$, use the EULER program to determine an approximation to $y(1)$. Use a step size of $\Delta x = 0.25$.

Solution:

With $\Delta x = 0.25$ we require 4 steps to get from $x = 0$ to $x = 1$. Enter Y1 = X + Y. Press WINDOW and enter a [−4.7, 4.7] x [−.5, 5.2] viewing rectangle. Execute the program EULER entering X START = 0, Y START = 1, STEP SIZE = 0.25 and NO. STEPS = 4, to produce the following picture and approximating values.

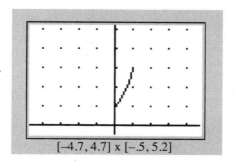 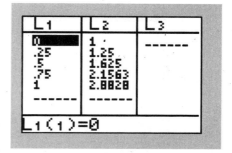

The function $y = 2e^x - x - 1$ is the exact solution to the initial value problem $y' = x + y$ with initial condition $y(0) = 1$. In the figure below, the Euler approximation is displayed in list L2 and the exact value for each input value is displayed in list L3.

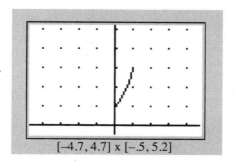

Notice that the values found by the Euler method (L2) get farther and farther away from the exact values (L3). This is a characteristic of Euler's method, and for this reason the method should be employed with caution. Euler's method may be modified to give more accurate approximations.

Evaluating Integrals

The value of the definite integral $\int_0^1 \sqrt{1 + x^3}\ dx$ is the value of the accumulation function

$F(x) = \int_0^x \sqrt{1 + t^3}\ dt$ at $x = 1$. The function F is, in turn, a solution of the initial value problem

$$F'(x) = \sqrt{1 + x^3}\ ,\quad F(0) = 0.$$

Thus, by solving the initial value problem numerically on the interval $0 \le x \le 1$, we can approximate the value of the definite integral as the value of the accumulation function at $x = 1$.

Example 3: *Approximating a Definite Integral*

Approximate the value of $\int_0^1 \sqrt{1+x^3}\, dx$ by numerically solving the corresponding initial

value problem: $y' = \sqrt{1+x^3}$, $y(0) = 0$.

Solution:

The results of executing the EULER program ($\Delta x = 0.1$) to estimate $y(1)$ if $y' = \sqrt{1+x^3}$, $y(0) = 0$ are displayed below.

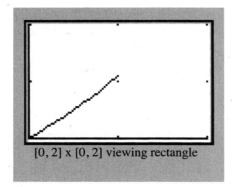

[0, 2] x [0, 2] viewing rectangle

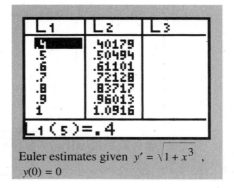

Euler estimates given $y' = \sqrt{1+x^3}$, $y(0) = 0$

$y(x) = \int_0^x \sqrt{1+t^3}\, dt$ is the solution of the initial value problem $y' = \sqrt{1+x^3}$, where

$y(0) = 0$ and $y(1) = \int_0^1 \sqrt{1+x^3}\, dx$. Thus we can use the Euler approximation of $y(1)$ to
approximate the integral

$$\int_0^1 \sqrt{1+x^3}\, dx \approx 1.0916.$$

In summary, we can say that Euler's method uses and produces numerical information. We can use Euler's method to produce a graph or table of values, but not a formula for a solution function. When using Euler's method, as for any other numerical approximation strategy, it is important to make the step size small enough. And, in general, the smaller the step size, the better the approximation.

9.3   Exercises

1.   Given the initial value problem $y' = \frac{1}{x}$ and $y(1) = 0$.

   a) Use 4 steps of Euler's method to determine an approximation of $y(2)$. Use a step size $\Delta x = 0.25$ and round to 4 decimal places.

   b) Using integration, solve the differential equation to find the exact value at $y(2)$.

   c) Is your approximation value of $y(2)$ bigger or smaller than the exact value? Use a slope field to explain your answer.

2.   Given the initial value problem $y' = \frac{1}{x^2 + 1}$ and $y(0) = 0$.

   a) Use 10 steps of Euler's method with $\Delta x = 0.1$ to determine an approximation of $y(1)$.

   b) Using integration, solve the differential equation to find the exact value at $y(1)$.

   c) Is your approximation value of $y(1)$ bigger or smaller than the exact value? Use a slope field to explain your answer.

3.   a) Show that $y = \tan x$ is a solution of the equation $y' = 1 + y^2$.

   b) Explain why Euler's method may not be used to approximate $y(2)$ if you start at $(0, 0)$.

4.   Use Euler's method with $\Delta x = 0.20$ to estimate $y(1)$ if $y' = y$ and $y(0) = 1$. What is the exact value of $y(1)$?

In Exercises 5–8, use the Euler program to find graphical solutions(graphs) and numerical solutions (tables) of the initial value problem. Each exercise gives the step size and the number of steps.

5. $y' = \frac{-y^2}{x}$, $y(2) = 2$; $\Delta x = 0.20$ and $n = 10$.

6. $y' = \frac{x}{y}$, $y(1) = 1$; $\Delta x = 0.10$ and $n = 20$.

7. $y' = x^2 + y$ $y(1) = 1$; $\Delta x = 0.20$ and $n = 10$.

8. $y' = 1 - y$, $y(0) = 0$; $\Delta x = 0.10$ and $n = 20$.

9.   The curve passing through $(1, 0)$ satisfies the differential equation $\frac{dy}{dx} = 4x + y$.

   Approximate the value of $y(2)$ using Euler's Method with two equal steps.

10.   Suppose a continuous function f and its derivative f' have values as given in the following table. Given that $f(1) = 2$, use Euler's method to approximate the value of $f(2)$.

x	1.0	1.5	2.0
$f'(x)$	0.4	0.6	0.8
$f(x)$	2.0		

11.   Consider the differential equation $y' = \dfrac{x}{1 + x^2}$.

   a) Starting at the point (0, 2), use Euler's method with two steps of size 1 to estimate $y(2)$.

   b) Using integration, solve the differential equation and find the exact value of $y(2)$. How does the exact value compare with the approximation in part a)?

12.   Consider the differential equation $y' = x \cdot \ln x$.

   a) Starting at the point $(1, -\frac{1}{4})$, use Euler's method with 6 steps of size .5 to estimate $y(4)$.

   b) Using integration, solve the differential equation and find the exact value of $y(4)$. How does the exact value compare with the approximation in part a)?

13.   Consider the differential equation $y' = x \cdot y^2$.

   a) Starting at the point (0, 1), use Euler's method with 4 steps of size 0.2 to approximate $y(0.8)$.

   b) Repeat part a) with 8 steps of size 0.1.

   c) Verify that $y = \dfrac{2}{2 - x^2}$ is the exact solution of the differential equation.

   d) Compute the exact value of $y(0.8)$ and compare it with the approximations in parts a) and b).

9.4   Solving Differential Equations Symbolically: Separation of Variables

We have seen how to sketch solutions of a differential equation using a slope field and how to calculate approximate numerical solutions with Euler's method. In this section we will find the equation of a solution curve for certain differential equations by a process known as **separating variables**.

First we consider a familiar problem, the differential equation $\frac{dy}{dx} = -\frac{x}{y}$ whose solution curves are the circles

$$x^2 + y^2 = C.$$

To obtain these solutions analytically we separate the variables, putting the x's on one side and all of the y's and $\frac{dy}{dx}$ on the other, giving

$$y \cdot \frac{dy}{dx} = -x.$$

Antidifferentiating each side separately gives

$$\int y \cdot \frac{dy}{dx} \, dx = \int -x \, dx$$

$$\frac{y^2}{2} + C_1 = \frac{-x^2}{2} + C_2.$$

Because both C_1 and C_2 are arbitrary constants, it is convenient to combine them into one, giving the circle equation

$$x^2 + y^2 = C, \text{ where } C = 2(C_2 - C_1).$$

If y is the number of individuals in a population at time t, the differential equation $y' = ky$, where k is a positive constant, states that the rate of increase, y', for the population at a certain time is proportional to the number of individuals y. This is known as **exponential growth**.

Example 1:   *Using Separation of Variables*

Find the general solution of $y' = ky$ where k is a constant.

Solution:

Separating variables, we get

$$\frac{1}{y} y' = k.$$

Integrating both sides yields

$$\ln|y| = kt + C.$$

Solving for $|y|$ leads to

$$|y| = e^{kt+C} = e^{kt} \cdot e^C = Ae^{kt}, \text{ where } A = e^C.$$

If we assume that $y > 0$, then the general solution to $y' = ky$ is

$$y = Ae^{kt} \text{ where } A \text{ and } k \text{ are constants.}$$

There are many important situations in which the rate of growth (or decay) of some quantity is proportional to the amount $y(t)$ present at time t. Examples are growth of a population, savings with interest compounded continously and decay of a radioactive substance. Because the differential equation $y' = ky$ is the appropriate model in all of these cases we summarize the results of Example 1 for future use.

> The solution of the initial value problem $y' = ky$, $y(0) = y_0$ is
>
> $$y(t) = y_0 e^{kt}.$$

The next example illustrates how the method of separating variables gives an exact solution to the differential equation in Example 3 on page 415.

Example 2: *Newton's Law of Cooling*

Newton's law of cooling states that the rate at which an object cools is proportional to the difference in temperature between the object and the surrounding medium. Let y denote the temperature (degrees Fahrenheit) of a cup of coffee at time t (minutes). Room temperature is 70 degrees and coffee temperature starts at 190 degrees. The coffee temperature at time t is described by the differential equation

$$\frac{dy}{dt} = -0.1(y - 70) \text{ with initial condition } y(0) = 190.$$

How hot is the coffee after 10 minutes?

Solution:

Separate the variables to obtain $\dfrac{1}{y - 70} \cdot \dfrac{dy}{dt} = -0.1.$

Antidifferentiating both sides with respect to t gives

$$\int \frac{1}{y - 70} \cdot \frac{dy}{dt}\ dt = \int -0.1\ dt,$$

$$\ln|y - 70| = -0.1t + C.$$

To solve for y, exponentiate both sides

$$|y - 70| = e^{-0.1t + C}.$$

Assuming $y > 70$ gives $y = e^{-0.1t} e^C + 70.$

Letting $e^C = A$, we have $y = Ae^{-0.1t} + 70.$

To find A use the initial condition $y(0) = 190$, and substituting we have

$$190 = Ae^0 + 70 \text{ and } A = 120.$$

Thus our solution is

$$y = 120e^{-0.1t} + 70.$$

At $t = 10$ we have $y(10) = 120e^{-0.1(0)} + 70 = 120e^{-1} + 70 \approx 114.1$ degrees.

9.4   Exercises

In Exercises 1–8, find the solution of the differential equation.

1. $y' = 2\sqrt{y} \cdot x$

2. $y' = -2xy$

3. $y' = 2x \sec y$

4. $y' = \dfrac{x-1}{y}$

5. $y' = \dfrac{\cos^2 y}{\sin^2 x}$

6. $y' = e^{x-y}$

7. $y' = y \ln x$

8. $y' = \dfrac{2x}{\sqrt{y-1}}$

In Exercises 9–16, find the solution of the differential equation that satisfies the given initial condition.

9. $y' = x^3 y,\quad y(0) = 3$

10. $y' = y^2 + 1,\quad y(1) = 0$

11. $y' = y \cos x,\quad y(0) = 1$

12. $y' = x\sqrt{y},\quad y(2) = 9$

13. $y' = \dfrac{x-1}{y},\quad y(2) = 5$

14. $y' = \dfrac{y}{x},\quad y(1) = 1$

15. $y' = \dfrac{y}{x+1},\quad y(0) = 1$

16. $y' = e^{x-y},\quad y(0) = 0$

17. Given the differential equation $y' = \dfrac{x}{y}$.

   a) Sketch the slope field for the equation on a [–4.7, 4.7] x [–3.1, 3.1] viewing rectangle.

   b) Sketch the solution curve passing through the point (0, 1).

   c) Find the general solution for the differential equation.

18. A glass of orange juice is taken out of a 40°F refrigerator and placed in a 65°F room. If the orange juice warms to a temperature of 50°F in an hour, then how long will take to warm to 60°F?

19. A murder victim is found at 9 PM. The temperature of the body is measured at 90°F. One hour later, the temperature of the body is 89°F. The temperature of the room has been maintained at a constant 70°F and the victim's normal temperature was 98.6°F.

   a) Assuming that the temperature, *T*, of the body obeys Newton's Law of Cooling, write a differential equation for *T*.

   b) Solve the differential equation to estimate the time the murder occurred.

20. The rate of change of the number of wolves $N(t)$ in a population is directly proportional to $350 - N(t)$, where time is in years. When $t = 0$, the population size is 200 and when $t = 3$, the population size has increased to 300.

   a) Write a differential equation that models the size of the wolf population.

   b) Solve the differential equation in part a) and use the result to predict the size of the wolf population when $t = 5$.

21. The population of Phoenix, Arizona was about 100,000 in 1950 and about 500,000 in 1970. Assuming an exponential growth rate, what predicted population would have been expected in 1980?

22. The population of the United States was about 92,000,000 in 1910 and about 123,000,000 in 1930. Assuming an exponential growth rate, what predicted population would have been expected in 1950? The actual population in 1950 was slightly more than 151,325,000.

23. The number of bacteria in a certain culture increases at a rate proportional to the population. The initial number is 20,000 and grows to 48,000 in three hours.

   a) Find an equation that models the growth rate of the bacteria.

   b) What is the population after seven hours?

   c) How long will it take for the population to double?

Chapter 9   Supplementary Problems

1.   Find an equation for the curve $y = f(x)$ that contains the point (0, 2) if $\frac{dy}{dx} = \frac{e^{2x}}{y}$ and

$f(x) > 0$ for all x.

2.   If $\frac{dy}{dt} = -3y$ and if $y = 1$ when $t = 0$, what is the value of t for which $y = \frac{1}{3}$?

3.   At each point (x, y) on a certain curve the slope of the cuve is $4x^3y$. Find an equation of
the curve if it contains the point (0, 5)

4.   If $\frac{dx}{dt} = kt$ and if $x = 2$ when $t = 0$ and $x = 6$ when $t = 1$, find the value of k.

5.   Given the differential equation $\frac{dy}{dx} = y^2 \sin x$.

a) Show that $y = \sec x$ is a solution of the differential equation.

b) Find the solution of the differential equation that satisfies the initial condition $y(0) = \frac{1}{2}$.

6.   The general solution of the differential equation $\frac{dy}{dx} = \frac{1 - 2x}{y}$ is a family of

A) straight lines   B) circles   C) hyperbolas   D) parabolas   E) ellipses

7.   A slope field for the differential equation $y' = y - e^{-x}$,
−2.5 ≤ x ≤ 2.5 and −2.5 ≤ y ≤ 2.5 is shown in the figure.

a) Use the program SLOPEFLD to plot the slope
field. Plot the particular solution that contains the
point (0, 0).

b) Sketch the solution curve that contains the point $(0, \frac{1}{2})$.

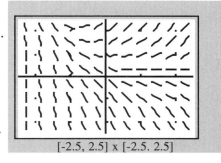

[-2.5, 2.5] x [-2.5. 2.5]

8.   Consider the differential equation given by $y' = x - 2$.

a) Sketch a slope field for this differential equation for $-4 \le x \le 4$ and $-3 \le y \le 3$.

b) Sketch the solution curve that satisfies the initial condition $y(0) = 1$ on the slope field in
part a).

9.   Given the differential equation $y' = y - 2$ with initial condition $y(0) = 1$.

a) Use Euler's method with 4 steps of size 0.2 to estimate $y(0.8)$.

b) Show that the exact solution of the differential equation is $y = 2 - e^x$.

c) Calculate the exact value of y when $x = 0.8$. How does this value compare with the
approximation in part a)?

10.   A slope field for the differential equation $y' = \dfrac{1}{2}xy$,

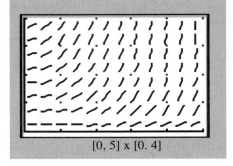

[0, 5] x [0. 4]

$0 \le x \le 5$ and $0 \le y \le 4$, is shown in the figure.

a) Use the program SLOPEFLD to plot the slope field. Sketch the particular solution that satisfies the initial condition $y(0) = 2$.

b) Solve the differential equation and find the particular solution that contains the point (0, 2).

c) Use the function in part b) to calculate the exact value of y when $x = 2$.

d) Starting at the point (0, 2), use Euler's method with two steps of size 1 to estimate $y(2)$. How does this value compare with the exact value in part c)?

11.   One thousand bacteria are started in a certain culture and the number of bacteria increase at a rate proportional to the number present.

a) If the number triples in 8 hours, how many are there in 12 hours?

b) In how many hours will the original number quadruple?

12.   The population $P(t)$ is increasing at a rate directly proportional to $300 - P(t)$, where the constant of proportionality is k.

a) If $P(0) = 100$, find $P(t)$ in terms of t and k.

b) If $P(2) = 200$, find k.

c) Find $\lim\limits_{t \to \infty} P(t)$.

13.   The thickness, $y(t)$, (in inches), of ice forming on a lake satisfies the differential equation $y'(t) = \dfrac{3}{y(t)}$, where t is measured in hours.

a) If $y(0) = 1$, find $y(t)$.

b) When is the thickness two inches?

14.   A roast is put in a $300°$F oven and heats according to the differential equation

$$\frac{dH}{dt} = k(300 - H)$$

where k is a positive constant and $H(t)$ is the temperature of the roast after t minutes.

a) If the roast is at $50°$F when put in the oven, $H(0) = 50$, find $H(t)$ in terms of k and t.

b) If $H(30) = 200°$F, find k.

15.   A certain population increases at a rate proportional to the square root of the population. If the population goes from 2500 to 3600 in five years, what is it at the end of t years? Assume that at $t = 0$ years the population is 2500.

16. The rate of change of volume V of a melting snowball is proportional to the surface area of the snowball, that is, $\frac{dV}{dt} = -kS$, where k is a positive constant. If the radius of the ball at $t = 0$ is $r = 2$ and at $t = 10$ is $r = 0.5$, show that $r = -\frac{3}{20}t + 2$.

17. A cup of hot water is placed in a refrigerator and the temperature of the water is recorded at various time intervals and listed in the following table.

Time (minutes)	0	5	12	16	22	30	35
Temp (° Fahrenheit)	159.6	143.3	121.3	111.8	102	88.4	78.5

a) Find one approximation for the rate of change of the temperature, in degrees/min, at $t = 16$. Show the computation you used to arrive at your answer.

b) Use the regression capabilites of your calculator to find an exponential function $y = E(t)$ that fits the data in the table. Plot the data and graph the model.

c) Calculate $E'(16)$ and compare the result the that obtained in part a).

18. a) Write an expression whose value is the trapezoid approximation for $\int_1^7 \frac{1}{\sqrt{x}}\, dx$ for

   $n = 2$ equal subintervals.

   b) Evaluate exactly the expression in part a).

19. Suppose a function f is defined so that its derivative is $f'(x) = \dfrac{1 - x^2}{(1 + x^2)^2}$.

   a) Determine the critical numbers of f.

   b) On which interval(s) is f increasing?

   c) Determine the x-coordinates of each inflection point.

20. To the right is the graph of the function f.

   a) Estimate $\int_0^4 f(x)\, dx$.

   b) Estimate $\int_0^6 f(x)\, dx$.

   c) Estimate the average value of f over on the interval $[0, 4]$.

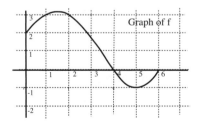
Graph of f

21. The function F is defined for all x by $F(x) = \int_0^{x^2} \sqrt{t^2 + 8}\, dt$.

   a) Determine $F'(x)$.

   b) Determine $F''(1)$.

22.   Find the points on the curve $y = x^4 - 6x^2 + 4$ where the tangent line is horizontal.

23.   If $f(x) = e^x - x^2 \arctan x$, estimate the interval(s) on which f is both increasing and concave down.

24.   If g is a differentiable function, where $g(4) = 2$ and $g'(4) = 5$, find $f'(4)$ if

    a) $f(x) = \sqrt{x} \cdot g(x)$;   b) $f(x) = \dfrac{x^2}{g(x)}$.

25.   If a tank holds 50 gallons of water, which drains from the bottom of the tank in 20 minutes, then Torricelli's Law gives the volume V of the water in the tank after t minutes as

$$V = 50(1 - \frac{t}{40})^2, \quad 0 \le t \le 20.$$

    a) Find the rate at which the water is draining from the tank after 5 minutes.

    b) At what time is the water flowing out of the tank the fastest?

26.   Find the points on the ellipse $x^2 + 2y^2 = 3$ where the tangent line has slope 1.

27.   If f and g are positive, increasing, concave upward functions on the interval (a, b), show that the function h defined by $h(x) = f(x) \cdot g(x)$ is concave upward on (a, b).

28.   Find the point on the hyperbola $xy = 8$ that is closest to the point $(3, 0)$.

29.   A basketball team plays in an arena with a seating capacity of 8000 spectators. With a ticket price of \$12, the average attendance per game is 5500. A market survey indicates that for each dollar that the ticket price is lowered, the average attendance will increase by 1000. What price will maximize revenue from ticket sales?

30.   Let R be the region bounded by the curves $y = 1 - x^2$ and $y = e^{2x}$. Estimate the following.

    a) The x-coordinates of the points of intersection.

    b) The area of region R.

    c) The volume of the solid generated when region R is rotated about the x-axis.

31.   A function f is defined on the interval $[0, 9]$ and the graph of its derivative f' is shown in the figure.

    a) On what interval(s) is the graph of f concave down?

    b) Using the Trapezoid Rule with $n = 9$, estimate $\displaystyle\int_0^9 f'(x)\, dx$.

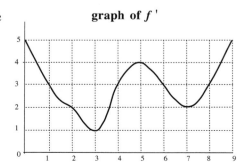

graph of f'

32.   The region bounded by the x-axis and the part of the graph of $y = \sin x$ between $x = 0$ and $x = \pi$ is separated into two regions by the line $x = p$. If the area of the region for $0 \le x \le p$ exceeds the area of the region for $p \le x \le \pi$ by one square unit, find the value of p.

33.   Let f be a function with $f(2) = 4$ and derivative $f'(x) = \sqrt{x^3 + 1}$. Using a tangent line approximation to the graph of f at $x = 2$, estimate $f(2.2)$.

34.   A particle moves along the x-axis in such a way that at time t, $1 \le t \le 8$, its position is given by $s(t) = \int_1^t \left[1 - x(\cos x) - (\ln x)(\sin x)\right] dx$.

   a) Write a formula for the velocity of the particle at time t.

   b) At what instant does the particle reach its maximum speed?

   c) When is the particle moving to the left?

   d) Find the total distance traveled by the particle from $t = 1$ to $t = 8$.

35.   Consider the following table of values for the differentiable function f.

x	1.0	1.2	1.4	1.6	1.8
$f(x)$	5.0	3.5	2.6	2.0	1.5

   a) Estimate $f'(1.4)$ and show your work.

   b) Give an equation for the tangent line to the graph of f at $x = 1.4$.

   c) What is the sign of $f''(1.4)$? Explain your answer.

   d) Using the data in the table, find a midpoint approximation with 2 equal subdivisions for
   $$\int_{1.0}^{1.8} f(x)\, dx.$$

36.   In the town of Sunset, Maine, the number of hours of daylight per day can be approximated by the function $D(x) = 3.27 \sin(0.018x - 1.5) + 12.22$, where x is the number of days since January first.

   a) By subdividing the month of February into four weeks and using the function values $D(x)$ at $x = 31, 38, 45,$ and 52, find an approximation to the number of hours of daylight during February.

   b) Write a Riemann sum to approximate the total number of hours of daylight during one year.

   c) Write and evaluate a definite integral to determine the total number of hours of daylight during one year.

37.   The region R is enclosed by the graphs of $y = \dfrac{x}{x^2 - 2}$, $y = 0$, $x = 2$, and $x = k$, where $k > 2$. If the area of R is 1 square unit, determine the value of k.

38.   The rate at which water is leaking from the town water tank is $r(t)$ gallons/hour, with t in hours.

   a) Write an expression approximating the amount of water that leaked out of the tank during the interval from time t to time $t + \Delta t$, where Δt is small.

   b) Write a Riemann sum approximating the total amount of water leaking from the tank between $t = 0$ and $t = 5$. Write an exact expression for this amount.

   c) By how much has the amount of water in the tank changed between $t = 0$ and $t = 5$ if $r(t) = 20e^{-.02t}$.

Exercises 39– 43 are Multiple Choice.

39.   The approximate value of $y = \sqrt{4 + \sin(2x)}$ at $x = 0.16$, obtained from the tangent line to the graph of $y = f(x)$ at $x = 0$ is

   A) 2.02   B) 2.04   C) 2.06   D) 2.08

40.   The maximum acceleration obtained on the interval $0 \le t \le 3$ by a particle whose velocity is given by $v(t) = \frac{2}{3}t^3 - 2t^2 + 3t + 10$ is

   A) 1   B) 3   C) 6   D) 9

41.   If $f(x) = \frac{100x}{x+1}$ for $0 \le x \le 8$, at what input x is the instantaneous rate of change equal to the average rate of change ?

   A) 1   B) 2   C) 3   D) 4

42.   What is the y-intercept of the line tangent to the graph of $f(x) = 3 + \int_0^x \frac{5}{1+t}\,dt$ at $x = 0$?

   A) 1   B) 3   C) 5   D) 7

43.   The graph of the function f, consisting of three line segments, is shown in the figure. Let $g(x) = \int_{-2}^{x} f(t)\,dt$

   Then $g''(0) =$

   (A) $\frac{3}{2}$ (B) $\frac{4}{3}$ (C) 0   (D) –3

graph of f

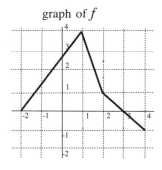

CHAPTER 10
L'Hopital's Rule, Improper Integrals and Partial Fractions

In this chapter we are interested in three topics. First we introduce a useful method for evaluating limits called **L'Hopital's Rule** . Then we extend the scope of integration by defining **improper integrals** in terms of limits. Finally we develop a strategy for integrating rational functions by breaking them into simpler parts using a method called **partial fractions**.

10.1   L'Hopital's Rule

The first important limit that we considered in Chapter 2 was the derivative formula

$$f'(a) = \lim_{h \to 0} \frac{f(a+h) - f(a)}{h}.$$

If f is continuous at $x = a$, then taking the limit of the numerator and denominator separately gives us an undefined expression

$$f'(a) = \lim_{h \to 0} \frac{f(a+h) - f(a)}{h} = \frac{0}{0}.$$

However, we know that derivatives are not always undefined. You may recall that to arrive at each rule for finding derivatives we used algebraic and trigonometric simplification, which was sometimes accompanied by clever manipulation or geometric argument. In this section we introduce a more direct strategy when considering similar problems about limits.

Here are two familiar limit problems: $\displaystyle\lim_{x \to 2} \frac{x^2 - 4}{x - 2}$ and $\displaystyle\lim_{x \to 0} \frac{\sin x}{x}$.

The two limits have a common feature. An attempt to find the limits by direct substitution leads to the undefined result $\frac{0}{0}$. It is customary to describe such limits as **indeterminate forms of type** $\frac{0}{0}$. As we shall see, a limit of this type can have any real number whatsoever as its value or can diverge to $+\infty$ or $-\infty$. The value of such a limit, if it converges, is not generally evident by inspection, so the term "indeterminate" is used to convey the idea that the limit can not be determined without some additional work.

In fact, we have already found values for the two limits above. Factoring and canceling the common factor $x - 2$ from the numerator and denominator allowed us to obtain

$$\lim_{x \to 2} \frac{x^2 - 4}{x - 2} = \lim_{x \to 2} \frac{(x + 2)(x - 2)}{x - 2} = \lim_{x \to 2}(x + 2) = 4 .$$

Graphical and numerical evidence led us to the conclusion that $\displaystyle\lim_{x \to 0} \frac{\sin x}{x} = 1$. Because graphical and numerical investigations and the technique of canceling factors apply only to a limited range of problems, it is desirable to have a general method for handling indeterminate forms. This is provided by **L'Hopital's Rule**.

L'Hopital's Rule

If $\displaystyle\lim_{x \to a} f(x) = 0$ and $\displaystyle\lim_{x \to a} g(x) = 0$ and $\displaystyle\lim_{x \to a} \frac{f'(x)}{g'(x)}$ exists, then

$$\lim_{x \to a} \frac{f(x)}{g(x)} = \lim_{x \to a} \frac{f'(x)}{g'(x)}.$$

L'Hopital's Rule, named for the French mathematician Guillaume Francois de L'Hopital (1661–1704), says that under appropriate conditions an indeterminate form can be evaluated by differentiating the numerator and denominator separately. Essentially, L'Hopital's Rule enables us to transform one limit problem into another limit problem that may be simpler.

The following examples are designed to show how L'Hopital's Rule works, and in what circumstances.

Example 1:   *Using L'Hopital's Rule*

Use L'Hopital's Rule to show that

a) $\lim\limits_{x \to 0} \dfrac{\sin x}{x} = 1,$ b) $\lim\limits_{x \to 0} \dfrac{1 - \cos x}{x} = 0$

Solution:

Both limits have the indeterminate $\dfrac{0}{0}$ form. By L'Hopital's Rule

a) $\lim\limits_{x \to 0} \dfrac{\sin x}{x} = \lim\limits_{x \to 0} \dfrac{\frac{d}{dx}\sin x}{\frac{d}{dx}x} = \lim\limits_{x \to 0} \dfrac{\cos x}{1} = 1.$

b) $\lim\limits_{x \to 0} \dfrac{1 - \cos x}{x} = \lim\limits_{x \to 0} \dfrac{\frac{d}{dx}(1 - \cos x)}{\frac{d}{dx}x} = \lim\limits_{x \to 0} \dfrac{\sin x}{1} = 0.$

Occasionally it is necessary to apply L'Hopital's Rule more than once to remove an indeterminate form, as illustrated in the next example.

Example 2:   *L'Hopital's Rule Applied More Than Once*

Evaluate $\lim\limits_{x \to 1} \dfrac{x^3 - 3x + 2}{1 - x + \ln x}.$

Solution:

This is an indeterminate $\dfrac{0}{0}$ form, and we find that

$$\lim\limits_{x \to 1} \frac{x^3 - 3x + 2}{1 - x + \ln x} = \lim\limits_{x \to 1} \frac{3x^2 - 3}{-1 + \frac{1}{x}}$$

This is still the indeterminate form $\dfrac{0}{0}$, so L'Hopital's Rule can be applied again:

$$\lim\limits_{x \to 1} \frac{3x^2 - 3}{-1 + \frac{1}{x}} = \lim\limits_{x \to 1} \frac{6x}{-\frac{1}{x^2}} = \frac{6}{-1} = -6.$$

In the figure a graph of the function f defined by $f(x) = \dfrac{x^3 - 3x + 2}{1 - x + \ln x}$ has been plotted in a [–2, 20] x [–1.5, 1.5] viewing rectangle. The graph has a hole at the point (1, –6), which supports our answer.

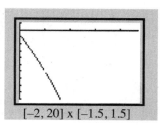

[–2, 20] x [–1.5, 1.5]

Extending L'Hopital's Rule

Another form of L'Hopital's Rule states that if the limit of $\dfrac{f(x)}{g(x)}$ as x approaches ∞ (or $-\infty$) produces the indeterminate form $\dfrac{0}{0}$ or $\dfrac{\infty}{\infty}$, then

$$\lim_{x \to \infty} \frac{f(x)}{g(x)} = \lim_{x \to \infty} \frac{f'(x)}{g'(x)}$$

provided the limit on the right exists.

Example 3:   *L'Hopital's Rule with an $\dfrac{\infty}{\infty}$ Form*

Evaluate $\displaystyle\lim_{x \to \infty} \frac{2x^2 - 3x + 1}{3x^2 + 5x - 2}$.

Solution:

We could divide numerator and denominator by x^2 and compute the limit as was suggested in Example 2 of Section 5.3. Instead, we note that this is of the form $\dfrac{\infty}{\infty}$ and apply L'Hopital's Rule.

$$\lim_{x \to \infty} \frac{2x^2 - 3x + 1}{3x^2 + 5x - 2} = \lim_{x \to \infty} \frac{4x - 3}{6x + 5}$$

$$= \lim_{x \to \infty} \frac{4}{6} = \frac{2}{3}$$

The graphs of the line $y = \dfrac{2}{3}$ and the function f defined by

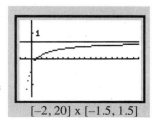

$[-2, 20] \times [-1.5, 1.5]$

$f(x) = \dfrac{2x^2 - 3x + 1}{3x^2 + 5x - 2}$ have been plotted in a $[-2, 20] \times [-1.5, 1.5]$ viewing rectangle. We have graphically supported our answer because the line appears to be a horizontal asymptote of the graph of f.

In addition to the forms $\dfrac{0}{0}$ and $\dfrac{\infty}{\infty}$, there are other indeterminate forms such as $\infty \cdot 0$, 1^∞, ∞^0, 0^0 and $\infty - \infty$. The following examples illustrate methods for evaluating these forms. Basically, the strategy is to convert each of these forms to $\dfrac{0}{0}$ or $\dfrac{\infty}{\infty}$ so that L'Hopital's Rule may be applied.

Example 4:   *Indeterminate Form $\infty \cdot 0$*

Evaluate $\lim\limits_{x \to \infty} x^2 e^{-x}$.

Solution:

As x goes to infinity, x^2 gets infinitely large and e^{-x} goes to 0.
A graph of $y = x^2 e^{-x}$ suggests that the expression goes to zero
as x becomes infinite. To show this, we first transform the
expression into a fraction.

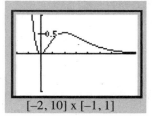

$[-2, 10] \times [-1, 1]$

$$\lim\limits_{x \to \infty} x^2 e^{-x} \to \infty \cdot 0 \qquad \text{L' Hopital' s rule does not apply yet.}$$

$$= \lim\limits_{x \to \infty} \frac{x^2}{e^x} \to \frac{\infty}{\infty} \qquad \text{L' Hopital' s rule does apply now.}$$

$$= \lim\limits_{x \to \infty} \frac{2x}{e^x} \to \frac{\infty}{\infty} \qquad \text{L' Hopital' s rule applies again.}$$

$$= \lim\limits_{x \to \infty} \frac{2}{e^x} \to \frac{2}{\infty} \qquad \text{L' Hopital' s rule is no longer needed.}$$

$$= 0$$

Indeterminate Exponential Forms

Limits that take the indeterminate form 1^∞, 0^0 and ∞^0 can sometimes be handled by taking
logarithms first. Then you can transform the result to a fraction and apply L'Hopital's Rule.

In Section 1.5 we used a table to investigate the values of $f(x) = \left(1 + \frac{1}{x}\right)^x$ as $x \to \infty$.
Now we find this limit with L'Hopital's Rule.

Example 5:   *Indeterminate Form 1^∞*

Evaluate $\lim\limits_{x \to \infty} \left(1 + \frac{1}{x}\right)^x$.

Solution:

Because substitution yields the indeterminate form 1^∞, we begin by assuming the limit exists
and is equal to y.

$$y = \lim\limits_{x \to \infty} \left(1 + \frac{1}{x}\right)^x.$$

Taking the natural logarithm of both sides produces

$$\ln y = \ln \left[\lim\limits_{x \to \infty} \left(1 + \frac{1}{x}\right)^x \right].$$

Because the natural logarithmic function is continuous, $\ln \left[\lim\limits_{x \to \infty} f(x) \right] = \lim\limits_{x \to \infty} [\ln f(x)]$,
and substituting on the right-side of this equation we obtain the following

$$\ln y = \lim_{x \to \infty}\left[x \cdot \ln\left(1 + \frac{1}{x}\right)\right] \qquad \text{(of type } \infty \cdot 0)$$

$$= \lim_{x \to \infty} \frac{\ln\left(1 + \frac{1}{x}\right)}{\frac{1}{x}} \qquad \text{(of type } \frac{0}{0})$$

$$= \lim_{x \to \infty} \frac{\frac{1}{1 + \frac{1}{x}}\left(-\frac{1}{x^2}\right)}{-\frac{1}{x^2}} \qquad (\text{Differentiate numerator and denominator.})$$

$$= \lim_{x \to \infty} \frac{1}{1 + \frac{1}{x}} = 1.$$

Now, because we have shown that $\ln y = 1$, we can conclude that $y = e$ and obtain

$$\lim_{x \to \infty}\left(1 + \frac{1}{x}\right)^x = e.$$

In the figure graphs of the line $y = e$ and the function f, defined

by $y = \left(1 + \frac{1}{x}\right)^x$, have been plotted in a $[-4, 6] \times [-2, 5]$

viewing rectangle. We have graphically supported our answer because the line appears to be a horizontal asymptote of the graph of f.

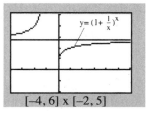

10.1   Exercises

In Exercises 1–4, estimate the limit graphically. Use L'Hopital's Rule to confirm your estimate.

1. $\lim\limits_{x \to 2} \dfrac{x^3 - 8}{x^2 - 4}$

2. $\lim\limits_{x \to 0} \dfrac{\sin(3x)}{x}$

3. $\lim\limits_{x \to 3} \dfrac{\sqrt{x+1} - 2}{x^2 - 9}$

4. $\lim\limits_{x \to \infty} \dfrac{x^3 - 4x^2}{6 - 6x^2}$

In Exercises 5–26, use L'Hopital's Rule where appropriate. If there is a more elementary method, use it. If L'Hopital's Rule does not apply, explain why.

5. $\lim\limits_{x \to 0} \dfrac{x}{\tan x}$

6. $\lim\limits_{x \to 1} \dfrac{x^2 - x}{\ln x}$

7. $\lim\limits_{x \to 2} \dfrac{\sin \pi x}{2 - x}$

8. $\lim\limits_{x \to 0} \dfrac{\cos x}{1 - \sin x}$

9. $\lim\limits_{x \to 0} \dfrac{2^x - 3^x}{x}$

10. $\lim\limits_{x \to 1} \dfrac{x^3 - 1}{x^3 + 3x - 4}$

11. $\lim\limits_{x \to 1} \dfrac{\ln x}{x - 1}$

12. $\lim\limits_{x \to 0} \dfrac{e^x - \cos x}{x \sin x}$

13. $\lim\limits_{x \to 0} \dfrac{\arcsin x}{x}$

14. $\lim\limits_{x \to 0} \dfrac{\sin(2x)}{1 - e^{2x}}$

15. $\lim\limits_{x \to 0} \dfrac{\tan x - x}{x - \sin x}$

16. $\lim\limits_{x \to 1} \dfrac{\ln x - x + 1}{x^2 - 2x + 1}$

17. $\lim\limits_{x \to 5} \dfrac{\sqrt{x-1} - 2}{x^2 - 25}$

18. $\lim\limits_{x \to 4} \dfrac{x - 4}{\sqrt[3]{x + 4} - 2}$

19. $\lim\limits_{x \to \infty} \dfrac{2 - 7x}{3 + 5x}$

20. $\lim\limits_{x \to \infty} \dfrac{2x^2 - 1}{x^2 - 2}$

21. $\lim\limits_{x \to \infty} \dfrac{\ln x}{x}$

22. $\lim\limits_{x \to \infty} \dfrac{e^x}{x}$

23. $\lim\limits_{x \to \infty} \dfrac{x^2}{e^{-x}}$

24. $\lim\limits_{x \to \infty} \dfrac{\sin x}{x}$

25. $\lim\limits_{x \to \infty} \dfrac{4x^2 + 15}{x + e^x}$

26. $\lim\limits_{x \to \infty} \dfrac{x - 2}{6x^2 - 10x - 4}$

In Exercises 27 – 30, use Example 5 as a model and evaluate the limit. Graph the function to confirm your answer.

27. $\lim\limits_{x\to\infty}\left(1+\dfrac{2}{x}\right)^{x}$

28. $\lim\limits_{x\to\infty}\left(1-\dfrac{1}{x}\right)^{x}$

29. $\lim\limits_{x\to\infty}\left(1-\dfrac{1}{x}\right)^{2x}$

30. $\lim\limits_{x\to\infty}\left(1+\dfrac{1}{x^2}\right)^{x}$

In Exercises 31–34, use the graph of f to estimate the limits.

31. $\lim\limits_{x\to2}\dfrac{f(x)}{x^2-4}$

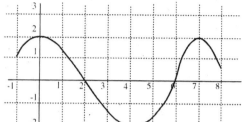

graph of f

32. $\lim\limits_{x\to7}\dfrac{f(x)}{(x-7)^2}$

33. $\lim\limits_{x\to0}\dfrac{f(x)}{x}$

34. $\lim\limits_{x\to4}\dfrac{f(x+2)}{f(x-2)}$

35.   Show that for any integer n

$$\lim\limits_{x\to\infty}\dfrac{x^n}{e^x}=0.$$

This shows that the exponential function grow faster than any positive power of x.

36.   Show that for any number $p>0$

$$\lim\limits_{x\to\infty}\dfrac{\ln x}{x^p}=0$$

This shows that the logarithmic function increases more slowly than any positive power of x.

In Exercises 37–39, give examples of polynomials $p(x)$ and $q(x)$ such that $\lim\limits_{x\to\infty}p(x)=\infty$ and $\lim\limits_{x\to\infty}q(x)=\infty$ and

37. $\lim\limits_{x\to\infty}\dfrac{p(x)}{q(x)}=3$

38. $\lim\limits_{x\to\infty}\dfrac{p(x)}{q(x)}=0$

39. $\lim\limits_{x\to\infty}\dfrac{p(x)}{q(x)}=\infty$

10.2   Improper Integrals

In the definition of the definite integral $\int_a^b f(x)\,dx$, it is assumed that the function f is
continuous on the finite interval $[a, b]$. It follows from the Extreme Value Theorem that f is
bounded on the interval $[a, b]$ in the sense that for some number M, $|f(x)| \le M$ for all x in
$[a, b]$. However, there are many applications in physics, economics and probability where it is
necessary to extend the concept of integral to the case where the interval of integration is infinite
and also to the case where f is unbounded at a finite number of points on the interval of integration
Collectively, these are called **improper integrals**.

Improper Integrals with Infinite Limits of Integration

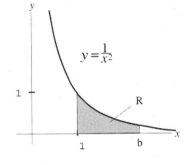

Consider the infinite region R that lies under the curve $y = \dfrac{1}{x^2}$,

above the x-axis and to the right of the line $x = 1$. You might
think that this region has infinite area. However, closer
examination reveals that the area of the region is finite. The part
of R that lies to the left of the line $x = b$ is shaded in the figure
and its area is

$$\int_1^b \frac{1}{x^2}\,dx = -\frac{1}{x}\Big]_1^b = 1 - \frac{1}{b}$$

Now $\lim\limits_{b\to\infty}\left(1 - \dfrac{1}{b}\right) = 1$, so it seems natural to define the area of region R to be 1 and we write

$$\int_1^\infty \frac{1}{x^2}\,dx = \lim_{b\to\infty}\int_1^b \frac{1}{x^2}\,dx = 1.$$

Here is a general definition.

Improper Integrals with Infinite Limits of Integration

1. If f is continuous on $[a,\infty)$, then

$$\int_a^\infty f(x)\,dx = \lim_{b\to\infty}\int_a^b f(x)\,dx.$$

2. If f is continuous on $[-\infty,b)$, then

$$\int_{-\infty}^b f(x)\,dx = \lim_{a\to-\infty}\int_a^b f(x)\,dx.$$

The improper integrals $\int_a^\infty f(x)\,dx$ and $\int_{-\infty}^b f(x)\,dx$ are called **convergent** if the
corresponding limit exists and **divergent** if the limit does not exist.

Example 1: *A Divergent Improper Integral*

Determine whether the improper integral $\int_1^\infty \frac{1}{x}\,dx$ is convergent or divergent.

Solution:

$$\int_1^\infty \frac{1}{x}\,dx = \lim_{b\to\infty}\int_1^b \frac{1}{x}\,dx$$

$$= \lim_{b\to\infty}\left[\ln x\right]_1^b$$

$$= \lim_{b\to\infty}(\ln b - 0)$$

$$= \infty$$

The limit does not exist as a finite number and so the improper integral $\int_1^\infty \frac{1}{x}\,dx$ diverges.

Comparing the result of Example 1 with the example at the beginning of this section we see that

$$\int_1^\infty \frac{1}{x^2}\,dx \quad \text{converges} \qquad \text{and} \qquad \int_1^\infty \frac{1}{x}\,dx \quad \text{diverges.}$$

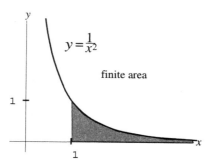

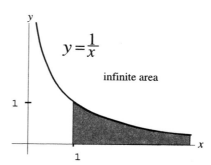

Geometrically, the curves $y = \frac{1}{x^2}$ and $y = \frac{1}{x}$ are similar for $x > 0$; however, the shaded region under $y = \frac{1}{x^2}$ has finite area whereas the corresponding region under $y = \frac{1}{x}$ has infinite area. Why is it that one improper integral converges and the other diverges? The explanation is that, although both $\frac{1}{x^2}$ and $\frac{1}{x}$ approach 0 as $x \to \infty$, $\frac{1}{x^2}$ approaches 0 faster than $\frac{1}{x}$.

If both limits of integration are infinite, then we consider the integral to be a sum of two improper integrals of the type we have just introduced.

$$\int_{-\infty}^\infty f(x)\,dx = \int_{-\infty}^c f(x)\,dx + \int_c^\infty f(x)\,dx$$

where c is any real number. If either or both of the two improper integrals diverge, then we say that $\int_{-\infty}^\infty f(x)\,dx$ diverges.

Example 2:   *Infinite Upper and Lower Limits of Integration*

Determine whether the improper integral $\displaystyle\int_{-\infty}^{\infty}\frac{1}{1+x^2}dx$ is convergent or divergent.

Solution:

The graph of $y = \dfrac{1}{1+x^2}$ is a bell-shaped curve having

symmetry about the line $x = 0$. In other words,

$f(x) = \dfrac{1}{1+x^2}$ is an even function and we can write

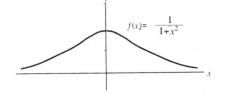

$$\int_{-\infty}^{\infty}\frac{1}{1+x^2}dx = \int_{-\infty}^{c}\frac{1}{1+x^2}dx + \int_{c}^{\infty}\frac{1}{1+x^2}dx = 2\int_{0}^{\infty}\frac{1}{1+x^2}dx.$$

The improper integral $\displaystyle\int_{-\infty}^{\infty}\frac{1}{1+x^2}dx$ can be evaluated as

$$\int_{0}^{\infty}\frac{1}{1+x^2}dx = \lim_{b\to\infty}\left[\arctan x\right]_{0}^{b} = \lim_{b\to\infty}\arctan(b) = \frac{\pi}{2}$$

Hence, the improper integral converges, and

$$\int_{-\infty}^{\infty}\frac{1}{1+x^2}dx = 2\left(\frac{\pi}{2}\right) = \pi.$$

Improper Integrals with Unbounded Integrands

There is another way for an integral $\displaystyle\int_{a}^{b}f(x)\,dx$ to be improper. The interval $[a, b]$ may be finite but the function may be discontinuous and unbounded at some points in the interval. A function f is unbounded at $x = c$ if it has arbitrarily large values near $x = c$. Geometrically, this occurs when the line $x = c$ is a vertical asymptote to the graph of f at $x = c$, as shown in the following figures.

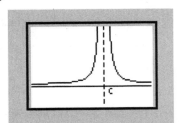 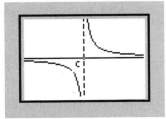

Two functions that are unbounded at $x = c$

If f is unbounded at $x = c$ and $a \le c \le b$, the integral $\displaystyle\int_{a}^{b}f(x)\,dx$ is not even defined (only bounded functions are integrable). However, it is possible to define $\displaystyle\int_{a}^{b}f(x)\,dx$ as an improper integral in certain cases.

For example, suppose

$$f(x) = \frac{1}{\sqrt{x}} \quad \text{for } 0 < x \le 1.$$

Then f is unbounded at $x = 0$ and $\int_0 f(x)\,dx$ is not defined.

However, $f(x) = \frac{1}{\sqrt{x}}$ is continuous on every interval $[b, 1]$ for

$b > 0$, as shown in the figure. For any such interval $[b, 1]$, we have

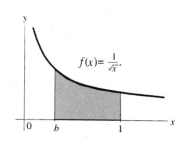

$$\int_b^1 \frac{1}{\sqrt{x}}\,dx = \int_b^1 x^{-1/2}\,dx = 2\sqrt{x}\Big]_b^1 = 2 - 2\sqrt{b}$$

If we let $b \to 0$ through positive values, we see that

$$\lim_{b \to 0^+} \int_b^1 \frac{1}{\sqrt{x}}\,dx = \lim_{b \to 0^+} (2 - 2\sqrt{b}) = 2.$$

This is called a convergent improper integral with value 2, and it seems natural to say

$$\int_0^1 \frac{1}{\sqrt{x}}\,dx = \lim_{b \to 0^+} \int_b^1 \frac{1}{\sqrt{x}}\,dx = 2$$

In this example, f is unbounded at the left endpoint of the interval of integration, but similar reasoning would apply if it were unbounded at the right endpoint. In addition, an integral can be improper because the integrand is unbounded inside the interval of convergence rather than at an endpoint. In this case, we break the given integral into two (or more) improper integrals so that the integrand tends to infinity only at endpoints.

Improper Integrals with Unbounded Integrands

1. If f is unbounded at $x = a$ and $\int_c^b f(x)\,dx$ exists for all c such that $a < c \le b$, then

$$\int_a^b f(x)\,dx = \lim_{c \to a^+} \int_c^b f(x)\,dx.$$

   If the limit exists, we say that the improper integral **converges**; otherwise, the improper integral **diverges**.

2. If f is unbounded at $x = b$ and $\int_a^c f(x)\,dx$ exists for all c such that $a \le c < b$, then

$$\int_a^b f(x)\,dx = \lim_{c \to b^-} \int_a^c f(x)\,dx.$$

3. If f is unbounded at $x = c$ where $a < c < b$ and improper integrals $\int_a^c f(x)\,dx$

   and $\int_c^b f(x)\,dx$ both converge, then

$$\int_a^b f(x)\,dx = \int_a^c f(x)\,dx + \int_c^b f(x)\,dx$$

   We say the integral on the left diverges if either of the integrals on the right diverges.

Example 3: *Improper Integral at a Right Endpoint*

Determine whether $\displaystyle\int_{1}^{2}(x-2)^{-2/3}\,dx$ is convergent or divergent.

Solution:

Let $f(x) = (x-2)^{-2/3}$. This function is unbounded at the right endpoint of the interval of integration and is continuous on $[1, t]$ for any t with $1 \le t < 2$. We find that

$$\int_{1}^{2}(x-2)^{-2/3}\,dx = \lim_{t\to 2^-}\int_{1}^{t}(x-2)^{-2/3}\,dx = \lim_{t\to 2^-} 3(x-2)^{1/3}\Big]_{1}^{t}$$

$$= \lim_{t\to 2^-}\left[3(t-2)^{1/3} - 3(1-2)^{1/3}\right] = \lim_{t\to 2^-}\left[3(t-2)^{1/3} + 3\right] = 3$$

The improper integral converges to 3.

If a vertical asymptote appears between the endpoints than we split the interval into two pieces at the vertical asymptotes. Again, if either or both of these improper integrals diverges, we say the entire original improper integral diverges.

Example 4: *Improper Integral at an Interior Point*

Determine whether $\displaystyle\int_{1}^{4}(x-2)^{-2/3}\,dx$ is convergent or divergent.

Solution:

This is similar to the previous example, except the vertical asymptote $x = 2$ occurs in the interior of the interval shown in the figure. Thus, we can write the following.

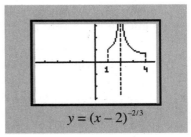

$y = (x-2)^{-2/3}$

$$\int_{1}^{4}(x-2)^{-2/3}\,dx = \lim_{b\to 2^-}\int_{1}^{b}(x-2)^{-2/3}\,dx + \lim_{a\to 2^+}\int_{a}^{4}(x-2)^{-2/3}\,dx$$

We have already computed the first of these improper integrals and found it to converge to 3. As for the second improper integral,

$$\lim_{a\to 2^+}\int_{a}^{4}(x-2)^{-2/3}\,dx = \lim_{a\to 2^+} 3(x-2)^{1/3}\Big]_{a}^{4} = 3\sqrt[3]{2}$$

Since both integrals converge, we can conclude that the entire improper integral converges to $3 + 3\sqrt[3]{2}$.

10.2   Exercises

In Exercises 1–16, evaluate the integral or state that it diverges.

1. $\displaystyle\int_{1}^{\infty} \frac{1}{x^3}\, dx$ 2. $\displaystyle\int_{0}^{1} \frac{1}{\sqrt[3]{x}}\, dx$

3. $\displaystyle\int_{2}^{\infty} \frac{1}{(x-1)^2}\, dx$ 4. $\displaystyle\int_{1}^{\infty} \frac{2}{x^2}\, dx$

5. $\displaystyle\int_{0}^{\infty} e^{-x}\, dx$ 6. $\displaystyle\int_{0}^{\infty} \frac{2}{1+x^2}\, dx$

7. $\displaystyle\int_{-\infty}^{0} e^{2x}\, dx$ 8. $\displaystyle\int_{-1}^{0} \frac{1}{\sqrt{x+1}}\, dx$

9. $\displaystyle\int_{0}^{\infty} \frac{x}{1+x^2}\, dx$ 10. $\displaystyle\int_{1}^{2} \frac{1}{x-1}\, dx$

11. $\displaystyle\int_{0}^{2} \frac{1}{\sqrt{2-x}}\, dx$ 12. $\displaystyle\int_{0}^{\infty} \left(\frac{1}{x} - \frac{1}{x+1} \right) dx$

13. $\displaystyle\int_{0}^{\infty} xe^{-x}\, dx$ 14. $\displaystyle\int_{0}^{\pi/2} \frac{\sin x}{\sqrt{1 - \cos x}}\, dx$

15    $\displaystyle\int_{-\infty}^{\infty} xe^{-x^2}\, dx.$ 16. $\displaystyle\int_{-1}^{1} \frac{1}{x}\, dx$

17.   Find the area, if it exists, of the region between the curve $y = \dfrac{4}{1+x^2}$ and its asymptote.

18.   Find the volume, if its exists, of the solid generated when the region bounded on the left by $x = 1$, below by $y = 0$, above by $y = \dfrac{1}{x}$, is rotated about the x-axis.

19.   Find the area, if it exists, of the region below the curve $y = \dfrac{4}{\sqrt{1-x^2}}$, between its asymptotes, and above the x-axis.

20.   Find the area of the region bounded by the graph of $y = e^{-|x|}$ and the x-axis.

21.   Find the volume, if its exists, of the solid generated when the region in the first quadrant under $y = e^{-x^2}$ is rotated about the y-axis.

22.   Let f be the function defined by $f(x) = x^{-2/3}$ for $0 < x \le 1$ and let R be the first quadrant region between the graph of f and the x-axis.

     a)   Determine whether region R has finite area. Justify your answer.

     b)   Determine whether the solid generated by revolving R about the x-axis has finite volume. Justify your answer.

10.3   Partial Fractions

Adding fractions is a skill practiced in algebra. For example,

$$\frac{1}{x-2} + \frac{1}{x+2} = \frac{2x}{x^2-4}.$$

The reverse process, in which a quotient is expressed as the sum of two or more simpler fractions, called **partial fractions**, is a useful technique for antidifferentiating rational functions. The basic idea is to "divide and conquer": We rewrite the given rational function as a sum of simpler rational functions and then antidifferentiate the terms separately.

For example, the evaluation of the integral $\int \frac{3x}{x^2+x-2} dx$ can be accomplished if the integrand is expressed as the sum of simpler fractions. We begin by factoring the denominator

$$\frac{3x}{x^2+x-2} = \frac{3x}{(x+2)(x-1)}$$

and assuming that there are constants A and B such that

$$\frac{3x}{x^2+x-2} = \frac{A}{(x+2)} + \frac{B}{(x-1)}.$$

If we multiply both sides of this equation by $(x+2)(x-1)$ the result is the equation

$$3x = A(x-1) + B(x+2)$$

What we need to do is find constants A and B so that this equation above is true for all x. For instance, if we let $x = -2$, then we have

$$-6 = A(-3) + B \cdot 0, \text{ so } A = 2.$$

Similarly, if we let $x = 1$, we find that $B = 1$. Thus,

$$\frac{3x}{x^2+x-2} = \frac{2}{(x+2)} + \frac{1}{(x-1)}.$$

An alternative procedure for finding A and B uses the fact that two polynomials of the same degree are equal if and only if their corresponding coefficients are equal. Thus, from the equation

$$3x = A(x-1) + B(x+2) = (A+B)x + (-A+2B)$$

we equate coefficients and obtain

$$A + B = 3 \text{ and } 2B - A = 0.$$

Solving simultaneously, we again find that $A = 2$ and $B = 1$.

Having succeeded in expressing the integrand as the sum of partial fractions, we have

$$\int \frac{3x}{x^2+x-2} dx = \int \frac{2}{(x+2)} + \frac{1}{(x-1)} dx = 2\ln|x+2| + \ln|x-1| + C$$

This result can be simplified so that

$$\int \frac{3x}{x^2+x-2} dx = \ln|x+2|^2|x-1| + C.$$

Example 1: *Distinct Linear Factors*

Find $\int \dfrac{x^2 - x + 4}{x^3 - 3x^2 + 2x} dx$

Solution:

We factor the denominator and set

$$\frac{x^2 - x + 4}{x(x-1)(x-2)} = \frac{A}{x} + \frac{B}{x-1} + \frac{C}{x-2}$$

where the constants A, B and C are to be determined. It follows that

$$x^2 - x + 4 = A(x-1)(x-2) + Bx(x-2) + Cx(x-1).$$

Since this equation must hold for all real numbers, we see that

if $x = 0$,   then $4 = 2A$ and $A = 2$

if $x = 1$,   then $4 = -B$ and $B = -4$

if $x = 2$,   then $6 = 2C$ and $C = 3$

Finally, the original integral equals

$$\int \left(\frac{A}{x} + \frac{B}{x-1} + \frac{C}{x-2} \right) dx \;=\; 2\ln|x| - 4\ln|x-1| + 3\ln|x-2| + C$$

$$= \ln \frac{x^2|x-2|^3}{(x-1)^4} + C.$$

In the examples above we have integrated a rational function in which the degree of the polynomial in the numerator has been less than the degree of the polynomial in the denominator (called proper rational functions). If in a rational function the degree of the numerator is greater than the degree of the denominator then by a simple process of long division the improper rational function can be expressed as the sum of a polynomial and a proper rational function.

Example 2: *An Improper Rational Function*

Find $\int \dfrac{x^3}{1+x^2} dx$

Solution:

In this case the denominator cannot be factored. But the numerator has degree three and the denominator only two, and so we can long divide the numerator by the denominator and obtain

$$\frac{x^3}{1+x^2} = x - \frac{x}{1+x^2}$$

Again, we have written the integrand as the sum of simpler terms, each of which is relatively easy to antidifferentiate.

$$\int x \, dx - \int \frac{x}{1+x^2} dx = \frac{x^2}{2} - \frac{1}{2}\ln|1+x^2| + C$$

10.3   Exercises

1.   Consider the partial fraction decomposition equation

$$\frac{2}{(1-x)(1+x)} = \frac{A}{1-x} + \frac{B}{1+x}$$

   a) Multiply both sides of this equation by $1-x$ and then substitute $x=1$ to show that $A=1$.

   b) Multiply both sides of this equation by $1+x$ and then substitute $x=-1$ to show that $B=1$.

   c) Find $\displaystyle\int \frac{2}{1-x^2}\,dx$.

2.   Consider the partial fraction decomposition equation

$$\frac{x-1}{(x-3)(x-2)} = \frac{A}{x-3} + \frac{B}{x-2}$$

   a) Multiply both sides of this equation by $x-3$ and then substitute $x=3$ to show that $A=2$.

   b) Multiply both sides of this equation by $x-2$ and then substitute $x=2$ to show that $B=-1$.

   c) Find $\displaystyle\int \frac{x-1}{(x-3)(x-2)}\,dx$.

In Exercises 3–12, find the antiderivative.

3. $\displaystyle\int \frac{1}{x^2 - x}\,dx$

4. $\displaystyle\int \frac{2}{x^2 + x}\,dx$

5. $\displaystyle\int \frac{x-1}{x^2 - 7x + 12}\,dx$

6. $\displaystyle\int \frac{4}{x^2 + 2x - 3}\,dx$

7. $\displaystyle\int \frac{1}{x^2 - 9}\,dx$

8. $\displaystyle\int \frac{x^2}{x-4}\,dx$

9. $\displaystyle\int \frac{x^2}{x^2 + x - 2}\,dx$

10. $\displaystyle\int \frac{x^3 + 2}{x^2 - x}\,dx$

11. $\displaystyle\int \frac{10 - 2x}{2x^2 + x - 1}\,dx$

12. $\displaystyle\int \frac{6}{x^3 - x}\,dx$

In Exercises 13–16, evaluate the definite integrals. Use your calculator to verify your result.

13. $\displaystyle\int_{3}^{4} \frac{2}{x^2 - 2x}\,dx$

14. $\displaystyle\int_{4}^{6} \frac{12}{x^2 - x - 2}\,dx$

15. $\displaystyle\int_{2}^{3}\frac{x^{3}}{x^{2}-1}dx$ 16. $\displaystyle\int_{1}^{2}\frac{6}{x^{2}+x}$

17. A particle is moving along a line so that if v feet per second is the velocity of the particle at t seconds, then

$$v=\frac{t+3}{t^{2}+3t+2}.$$

Find the distance traveled by the particle from the time $t=0$ to the time $t=2$.

18. Find the average value of $f(x)=\dfrac{x^{2}}{x^{2}+x-2}$ over the interval $[3,5]$.

19. The function f is defined by $f(x)=\dfrac{x}{x^{2}-2x-3}$ for $0\le x\le 2$. Find the area of the region bounded above by the x-axis and below by the graph of f.

20. Let R be the region in the first quadrant that is bounded above the graph of

$$y=\frac{3}{(x-1)(4-x)}$$ and the lines $x=2$ and $x=3$.

a) Find the exact area of R.

b) Find the volume of the solid obtained by revolving R about the y-axis.

10.4   The Logistic Equation

In this section we discuss a model for population growth, the logistic model, which is more sophisticated than the exponential growth model that was introduced in Section 9.4. We will make use of all available tools – slope fields from Section 9.2, Euler's method from Section 9.3 and the explicit solution of separable differential equations from Section 9.4.

The Logistic Model

In Section 9.4 we considered the exponential growth equation

$$y' = ky$$

where $k > 0$ is the growth constant, to model populations. This is an accurate model for the growth of populations during the initial growth phase. Since the solution of this differential equation is

$$y = y_0 e^{kt} ,$$

the growth rate k is not usually sustainable. For as $t \to +\infty$, so does e^{kt} (since $k > 0$) and hence $y_0 e^{kt} \to +\infty$. No population can grow without bound, due to the constraints of finite living space and a finite food supply. Thus, in practice, there often exists some upper limit L past which growth cannot occur. In other words, there is a maximum population L, the **carrying capacity**, that the environment is capable of sustaining in the long run. In such cases, we assume the rate of growth to be not only proportional to the existing quantity, but also to the difference between the existing quantity y and the upper limit L. A model that incorporates these assumptions is the differential equation

$$y' = ky\left(1 - \frac{y}{L}\right)$$

which is called the **logistic differential equation** since it models growth that is limited by the supply of necessary resources. The solution to this logistic differential equation is called the **logistic growth model**. Notice that if y is small compared to L, then $\frac{y}{L}$ is close to 0 and so $y' \approx ky$. But if $y \to L$ (the population approaches the carrying capacity), then $\frac{y}{L} \to 1$, so $y' \to 0$. Also notice that if the population lies between 0 and L the right side of the equation is positive, so $y' > 0$ and the population is increasing. On the other hand, if the population exceeds the carrying capacity ($y > L$), then $1 - \frac{y}{L}$ is negative, so $y' < 0$ and the population is decreasing.

Slope Fields

In the following example a slope field for a logistic differential equation is displayed.

Example 1: *Slope Field*

Draw a slope field for the logistic equation $y' = ky\left(1 - \frac{y}{L}\right)$ with $k = 0.08$ and carrying capacity $L = 1000$. Draw possible solutions to the differential equation if $y(0) = 100$ and $y(0) = 1300$.

Solution:

The figure below shows the slope field for the logistic model. It also includes a characteristic S-shaped solution curve that corresponds to the initial condition $y(0) = 100$. Notice that for each fixed value of y, that is, along each horizontal line, the slopes are all the same because y' depends on y and not on t. As expected, the slopes are small near $y = 0$ and near $y = L$; they are steepest around $y = L/2$. For $y > L$, the slopes are negative, meaning that if the population is above the carrying capacity, the population will decrease as illustrated by the solution curve that corresponds to the initial condition $y(0) = 1300$.

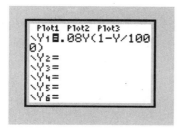 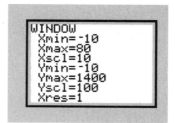 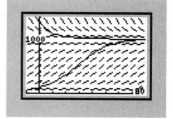

Euler's Method

We now use Euler's method to obtain numerical estimates for solutions of the logistic differential equation.

Example 2: *Using Euler's Method*

Use Euler's method with step size 10 to estimate the population sizes $y(40)$ and $y(80)$, where y is the solution of the initial-value problem

$$y' = 0.08y\left(1 - \frac{y}{1000}\right) \text{ and } y(0) = 100.$$

Solution:

With $\Delta t = 10$ we require 8 steps to get from $t = 0$ to $t = 80$. Enter $Y_1 = 0.08Y(1 - Y/1000)$. Press **WINDOW** and enter a $[-10, 80] \times [-10, 1400]$ viewing rectangle. Execute the program EULER entering X START = 0, Y START = 100, STEP SIZE = 10 and NO. STEPS = 8, to produce the following picture and approximating values.

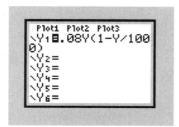 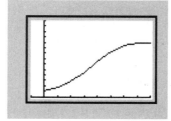 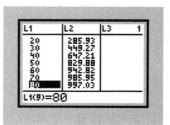

Thus, our estimates for the population sizes at times $t = 40$ and $t = 80$ are

$$y(40) \approx 647.21 \text{ and } y(80) \approx 997.03$$

The Analytic Solution to the Logistic Equation

The logistic differential equation can be solved exactly using separation of variables and a partial fraction decomposition.

To prepare for integration we rewrite the equation $y' = ky\left(1 - \frac{y}{L}\right)$ in the form

$$y' = \frac{k}{L}y(L-y).$$

Separating variables we obtain

$$\frac{L}{y(L-y)}y' = k$$

We can easily evaluate the integral on the right-hand side to get

$$\int k\,dt = kt + C,$$

but we need a partial fraction expansion on the left-hand side. Because we have two distinct linear factors, we write

$$\frac{L}{y(L-y)} = \frac{A}{y} + \frac{B}{L-y}$$

$$L = A(L-y) + By$$

Letting $y = 0$ in the second equation we get $A = 1$, and letting $y = L$ in the same equation we get $B = 1$. We can now integrate to get

$$\int \frac{L}{y(L-y)}\,dy = \int \frac{1}{y}\,dy \ + \int \frac{1}{L-y}\,dy$$

$$= \ln|y| - \ln|L-y| + C$$

$$= \ln\left|\frac{y}{L-y}\right| + C$$

Equating the two integrals and multiplying through by (-1), we have

$$\ln\left|\frac{L-y}{y}\right| = -kt - C.$$

Exponentiating both sides gives

$$\left|\frac{L-y}{y}\right| = e^{-kt-C} = e^{-C}e^{-kt},$$

so

$$\frac{L-y}{y} = Ae^{-kt} \quad \text{where} \quad A = \pm e^{-C}.$$

We find A by substituting $y = y_0$ when $t = 0$, which gives

$$\frac{L-y_0}{y_0} = Ae^0 = A$$

Thus

$$\frac{L-y}{y} = Ae^{-kt} \quad \text{where} \quad A = \frac{L-y_0}{y_0}.$$

Since $\dfrac{L-y}{y} = \dfrac{L}{y} - 1$ we have

$$\frac{L}{y} = 1 + Ae^{-kt}.$$

Solving for y we obtain a solution to the logistic differential equation.

$$y = \frac{L}{1 + Ae^{-kt}} \quad \text{where} \quad A = \frac{L - y_0}{y_0}$$

Example 3: *Solving the Logistics Equation*

Write a solution to the initial-value problem

$$y' = 0.08y\left(1 - \frac{y}{1000}\right) \quad \text{and} \quad y(0) = 100$$

and use it to find population sizes $y(40)$ and $y(80)$.

Solution:

The differential equation is the logistic equation with $k = 0.08$, carrying capacity $L = 1000$ and initial population $y_0 = 100$. So the general solution gives the population at time t as

$$y(t) = \frac{1000}{1 + Ae^{-0.08t}} \quad \text{where} \quad A = \frac{1000 - 100}{100} = 9$$

Thus,

$$y(t) = \frac{1000}{1 + 9e^{-0.08t}}.$$

So the population sizes when $t = 40$ and $t = 80$ are

$$y(40) = \frac{1000}{1 + 9e^{-3.2}} \approx 731.6 \quad \text{and} \quad y(80) = \frac{1000}{1 + 9e^{-6.4}} \approx 985.3.$$

Example 4: *Rate of Growth*

Consider the logistic differential equation $\frac{dy}{dt} = ky\left(1 - \frac{y}{L}\right)$ where $y(t)$ is the population at time t, k is a growth constant and L is the limiting value of the population. Show that the solution $y(t)$ of the logistic equation, having $0 < y(0) < L$, is increasing most rapidly when its value is $\frac{L}{2}$

Solution:

$\frac{d}{dt}\left(\frac{dy}{dt}\right) = k - \frac{2ky}{L} = 0$ when $y = \frac{L}{2}$. By the first derivative test, $\frac{dy}{dt}$ is maximum when $y = \frac{L}{2}$.

10.4   Exercises

In Exercises 1–7, the questions refer to the logistic curve

$$P(t) = \frac{600}{1 + 4e^{-.02t}}$$

which models a population of bacteria that are growing in a culture at time t, where t is measured in hours.

1.   Find a) $\lim\limits_{t \to +\infty} P(t)$ and b) $\lim\limits_{t \to -\infty} P(t)$

2.   Find the horizontal asymptotes of the graph of P.

3.   Draw a graph of P.

4.   Draw the graph of P' and find the intervals on which P is a) increasing, b) decreasing.

5.   Draw the graph of P'' and find the intervals on which P is a) concave up, b) concave down.

6.   Find the points of inflection of the graph of P.

7.   After how many hours does the population size reach 300?

8.   Consider the logistic differential equation and initial conditions

$$y' = 0.7y\left(1 - \frac{y}{100}\right), \quad y(0) = 10$$

The equation has the form $y' = ky\left(1 - \frac{y}{L}\right)$

  a) List the values of k and L.

     b) Compute $A = \dfrac{L - y_0}{y_0}$.

  c) Obtain a solution to the differential equation by substituting in the general solution:

$$y = \frac{L}{1 + Ae^{-kt}} \quad \text{where} \quad A = \frac{L - y_0}{y_0}$$

9.   Consider the logistic differential equation and initial conditions

$$y' = 0.0002y(150 - y), \quad y(0) = 20$$

  a) Transform the equation into the form $y' = ky\left(1 - \frac{y}{L}\right)$

  b) List the values of k and L.

  c) Compute $A = \dfrac{L - y_0}{y_0}$.

  d) Obtain a solution to the differential equation by substituting in the general solution:

$$y = \frac{L}{1 + Ae^{-kt}} \quad \text{where} \quad A = \frac{L - y_0}{y_0}$$

Exercises 10 –13, solve the given initial-value problem.

10. $y' = 0.5y\left(1 - \dfrac{y}{500}\right),$ $y(0) = 100$ 11. $y' = 0.0015y\left(1 - \dfrac{y}{6000}\right),$ $y(0) = 1000$

12. $y' = 0.0004y(100 - y),$ $y(0) = 40$ 13. $y' = 0.0015y(150 - y),$ $y(0) = 30$

14.   Suppose that a population of bacteria grows according to the logistic equation

$\dfrac{dP}{dt} = 0.01P - 0.0002P^2$, where P is the population measured in thousands and t is time measured in days.

a) What is the i) carrying capacity and, ii) the value of k?

b) A slope field for this equation is given below.

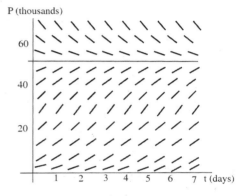

i)   Where are the slopes close to 0?

ii) Where are the slope values the largest?

iii) Where are the solutions increasing?

iv) Where are the solutions decreasing?

c) Use the slope field to sketch solutions for initial populations of 10, 30 and 70.

i)   What do these solution curves have in common and how do they differ?

ii) Which of the solutions have inflection points?

15.   Suppose that a population P grows according to a logistic model.

a) Write the differential equation for this situation with $k = 0.01$ and carrying capacity of 1000.

b) Solve the differential equation in part a) with initial condition $t = 0$ (days) and population $P = 50$.

c) Find the population for $t = 10$ days and $t = 100$ (days).

d) Sketch a graph of the solution of the differential equation.

16.   Match the slope fields with their differential equations.

i) $y' = 0.06y\left(1 - \dfrac{y}{70}\right)$ ii) $y' = -0.5\sqrt{y}$ iii) $y' = 0.06y\left(1 - \dfrac{y}{140}\right)$ iv) $y' = 0.065y$

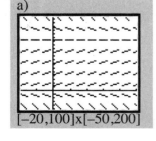

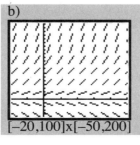

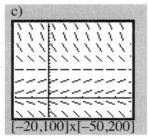

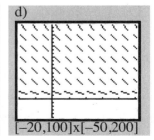

17. A population of bacteria in a culture is found to be growing according to a logistic model.

   a) Write the differential equation for this situation with $k = 0.07$ and carrying capacity of 1000.

   b) Solve the differential equation in part a) with initial condition $t = 0$ (hours) and population $P = 40$.

   c) Find the population for $t = 25$ (hours) and $t = 100$ (hours).

   d) Sketch a graph of the solution of the differential equation.

18. A confined and protected wildlife area is stocked with two timber wolves that grow according to the logistic model. Five years later 10 wolves were counted in this area. The area is estimated to able to support 30 wolves. How many wolves will there be in this area 10 years after their introduction?

19. In a model of epidemics, the number of infected individuals in a population at a time t is a solution of the logistic differential equation $\frac{dy}{dt} = 0.0002y(300 - y)$, where y is the number of infected individuals in the community and t is the time in days.

   a) Assume that 10 people were infected at the initial time $t = 0$. Find a solution for the differential equation.

   b) How many days will it take for half of the population to be infected?

20. Termites have infested a old boat house. The table below gives the size of the termite population after t days.

Time (Days)	0	1	2	3	4	5	6	7	8	9	10
Number of Termites	18	35	69	130	231	377	549	710	832	908	952

   a) Plot the data and use the plot to estimate the carrying capacity of the termite population.

   b) Find both an exponential and logistic model for these data.

   c) Compare the predicted values with the observed values, both in a table and with graphs.

   d) Use the logistic model to estimate the number of termites after 7 days.

21. Andover High has 1000 students. On day 0 four students start a rumor that spreads logistically. A day later 25 students know it.

   a) How many students have heard the rumor after 3 days?

   b) When is the rumor spreading the fastest?

22. Consider the logistic equation $\frac{dP}{dt} = kP\left(1 - \frac{P}{L}\right)$ where $P(t)$ is the size of the population at t ime t, and k and L are positive constants.

   a) If $P < L$ show that the graph of the solution curve $y = P(t)$ is always increasing.

   b) Using implicit differentiation find $\frac{d^2P}{dt^2}$

   c) Show that the solution curve $y = \frac{L}{1 + Ae^{-kt}}$ where $A = \frac{L - y_0}{y_0}$, has an inflection point at the $\left(\frac{\ln A}{k}, \frac{L}{2}\right)$.

Chapter 10 Supplementary Problems

1.   Find   a) $\lim\limits_{x\to 0}\dfrac{\cos x + 2x - 1}{3x}$ b) $\lim\limits_{x\to\infty}\dfrac{\ln x}{\sqrt{x}}$

2.   Find   a) $\lim\limits_{x\to\infty}\dfrac{\ln x}{\sqrt[3]{x}}$ b) $\lim\limits_{x\to 0}\dfrac{\tan x - x}{x^3}$

3.   Find the area, if it exists, of the region in the first quadrant under the graph of $y = e^{-2x}$.

4.   If f' is continuous, use L'Hopital's Rule to show that

$$\lim_{h\to 0}\frac{f(x+h) - f(x-h)}{2h} = f'(x).$$

Explain the meaning of this equation with the aid of a diagram.

5.   If an initial amount P of money is invested at an interest rate r compounded n times a year, the value of the investment after t years is

$$A = P\left(1 + \frac{r}{n}\right)^{nt}.$$

If we let $n \to \infty$, we refer to the continuous compounding of interest. Use L'Hopital's Rule to show that if interest is compounded continuously, then the amount after n years is

$$A = Pe^{rt}$$

6.   Find the volume, if its exists, of the solid generated when the region in the first quadrant under $y = e^{-x}$ is rotated about the x-axis.

7.   Find the volume, if its exists, of the solid that is generated when the region in the first quadrant under $y = e^{-x}$ is rotated about the y-axis.

8.   Let f be the function defined by $f(x) = \dfrac{1}{x}$ for $x \ge 1$ and let R be the region between the graph of f and the x-axis.

a) Determine whether region R has finite area. Justify your answer.

b) Determine whether the solid generated by revolving R about the x-axis has finite volume. Justify your answer.

9.   Let R be the region in the first quadrant that is bounded above by the graph of

$$y = \frac{4}{(x+1)(3-x)}$$ and the lines $x = 0$ and $x = 2$.

a) Find the exact area of R.

b) Find the volume of the solid obtained by revolving R about the y-axis.

10.   Find the average value of $f(x) = \dfrac{2x}{(x+1)(x+3)}$ over the interval [0, 2].

11.   Let R be the region in the first quadrant that is bounded above the graph of $y = \dfrac{x+2}{x^2+x}$ and
the lines $x = 1$ and $x = 3$.

   a) Find the exact area of R.

   b) Find the volume of the solid obtained by revolving R about the y-axis.

12.·   Suppose that a certain population grows according to an exponential model.

   a) Write a differential equation for this situation with relative growth rate of $k = 0.01$.

   b) Solve the differential equation in part a) with initial condition $t = 0$ (days) and
       population $P = 1000$.

   c) Find the population when $t = 10$ days and $t = 100$ days.

   d) After how many days does the population reach 2000?

13.   A population of bacteria in a culture is found to be growing according to a logistic model.

   a) Write the differential equation for this situation with $k = 0.04$ and carrying
       capacity of 800.

   b) Solve the differential equation in part a) with initial condition $t = 0$ (hours) and
       population $P = 40$.

   c) Find the population for $t = 25$ hours and $t = 100$ hours.

   d) Sketch a graph of the solution of the differential equation.

14.   Consider the logistic growth model defined by

$$\frac{dP}{dt} = 0.2P\left(1 - \frac{P}{10}\right) \text{ lbs / hr} \quad P(0) = 0.8 \text{ lbs}$$

   a) Obtain the formula for the solution to this initial-value problem.

   b) How large will P be after i) 3 hours, ii) after 10 hours?

   c) When will P reach one-half the carrying capacity?

15.   Match the slope fields with their differential equations.

i) $y' = y\left(1 - \frac{y}{3}\right)$ ii) $y' = 3\cos(x)$ iii) $y' = y$ iv) $y' = x + y$

a)
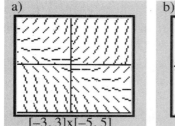
$[-3, 3] \times [-5, 5]$

b)
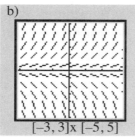
$[-3, 3] \times [-5, 5]$

c)
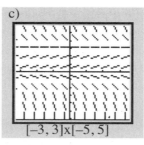
$[-3, 3] \times [-5, 5]$

d)
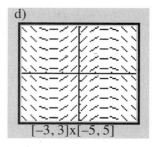
$[-3, 3] \times [-5, 5]$

16.   A small pond can support up to 1000 fish so that the rate of growth of the fish population is jointly proportional to the number of fish present and the difference between 1000 and the number present. The pond initially contained 100 fish.

a) If there are approximately 141 fish after 10 weeks, find a mathematical model expressing the number of fish present as a function of the number of weeks the population has been growing.

b) Use the model in part a) to predict the size of the fish population in 1 year.

c) Estimate in how many weeks the growth of the fish population will be the greatest.

17.   In a small town of population 1200 the rate of growth of a flu epidemic is jointly proportional to the number of people who have the flu and the number of people who do not have it.

a) If five days ago 100 people in the town had the flu and today 300 people have it, find a mathematical model describing the epidemic.

b) Use the model in part a) to predict how many people will be infected in 20 days from now.

18.   Solve the following initial-value problems.

a) $\dfrac{dy}{dx} = \dfrac{1}{xy}$, $y(1) = 1$ b) $y' = x^2 y$, $y(0) = 1$

19.   Suppose the position, $p(t)$, of a particle at time t in seconds along a line marked in meters is given by $p(t) = 2t^2 - 3$.

a) Find the average velocity of the particle between $t = 2$ and $t = 5$.

b) Find the average velocity between $t = 2$ and $t = 2 + h$, for some constant h.

c) Find the instantaneous velocity of the particle at $t = 2$.

20.   Consider the curve $x + xy + 2y^2 = 6$. Find the slope of the line tangent to the curve at the point $(2, 1)$.

21.   If the function f is defined by $f(x) = 4x^3 - 21x^2 + 36x - 4$ then on what intervals is the graph of f is decreasing and concave up?

22. Find the slope of the tangent line to the graph of $y = \int_1^x (t - 2\sqrt{t})\, dt$ at its point of inflection.

23. It is estimated that t years from now the population of a certain country will be $f(t) = \dfrac{160}{1 + 8e^{-.15t}}$ million. In approximately how many years will the population be growing most rapidly?

24. A function f is defined for all real numbers so that its derivative, f', is given by
$$f'(x) = \frac{x^2 - x^3}{x^2 + 1}.$$

a) Determine the critical numbers of f .

b) Determine the x-coordinate of each local minimum point of f.

c) Determine the x-coordinate of each inflection point of the f-graph.

d) Determine the intervals on which f is decreasing.

25. Use the Trapezoid rule with $n = 4$ to approximate the integral $\int_1^5 f(x)\, dx$ for the function whose graph is shown in the figure.

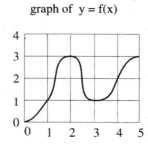

graph of $y = f(x)$

26. A boater finds herself 2 miles from the nearest point to a straight shoreline which is 5 miles from a shore side motel. She plans to row to the shore and then walk to the motel. If she can walk 4 mph, but only row at 3 mph, toward what point on shore should she row in order to reach the motel in the shortest time?

27. Let R be the first quadrant region enclosed by the graph of $y = \sqrt{x}$, the lines $x = 1$ and $x = k$, where $k > 1$. If the area of region R is $\dfrac{14}{3}$ unit^2 , find the value of k .

28. The region in the first quadrant enclosed by the graph of $y = x^{1/3}$ and the lines $x = 1$ and $x = 8$ is revolved about the x-axis. Find the volume of the solid generated.

29. Let R be the region bounded above by the graph of $y = 2k - x^2$ and below by $y = 2x^2 - k$, where $k > 0$. Find in terms of k

a) the area of region R ;

b) the perimeter of the largest rectangle (with sides parallel to the coordinate axes) that can be inscribed in region R.

30.　At noon, an experimenter has 50 grams of a radioactive isotope. Nine days later only 45 grams remain. To the nearest day, how many days after the experiment started will there be only 20 grams?

31.　Suppose the function f is defined by $f(x) = 2\cos x + \sin 4x$. How many times on the interval $[0, \pi]$ does the average value of f equal the instantaneous value of f?

32.　Explain briefly why each of the following statements is false. A counterexample or appropriate graph is sufficient.

a) The function f defined by $f(x) = e^{2x} - ae^x$ is increasing for every a .

b) If g is continuous at $x = a$, then for all functions f the composite function $h(x) = f[g(x)]$ is continuous at $x = a$.

c) The integral $\displaystyle\int_{-\infty}^{0} e^x \, dx$ diverges.

d) $\displaystyle\lim_{x \to 0} \frac{\arcsin x}{x}$ does not exist.

e) If $\displaystyle\int_{-1}^{1} f(t) \, dt = \int_{-1}^{1} g(t) \, dt$, then $f = g$.

33.　Determine m so that the region above the line $y = mx$ and below the parabola $y = 2x - x^2$ has an area of 36 square units.

34.　A stone is thrown vertically upward from the ground with an initial velocity of 25 ft/sec.

a) How long will the stone be going up?

b) How high will the stone go?

c) How long does it take the stone to reach the ground?

d) With what speed will the stone strike the ground?

35.　Assume that the function f is continuous and decreasing on the interval $[0, 4]$ and that the following is a table showing some function values.

x	0	0.5	1	1.5	2	2.5
$f(x)$	5	4	3	2.5	2	1

a) From the data given, find two estimates of $\displaystyle\int_{0}^{4} f(x) \, dx$.

b) Obtain a different estimate for the integral by taking an average value of f over each subinterval.

c) Determine whether your estimates are less than or greater than the value of the definite integral.

36.   The function g is defined on the interval $[0, 6]$ by

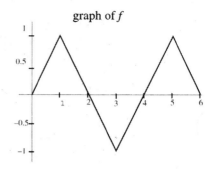

$g(x) = \int_0^x f(t)\,dt$ where f is the function graphed in
the figure.

a)   For what values of x, $0 \le x \le 6$, does g have a
relative maximum? Justify your answer.

b)   For what values of x is the graph of g concave
down? Justify your answer.

c)   Sketch a graph of the function g . List the
coordinates of all critical points and inflection
points.

37.   The slope field for the differential equation

$$\frac{dy}{dx} = 0.5xy , \quad 0 \le x \le 5 \text{ and } 0 \le y \le 4,$$

is shown in the figure at the right.

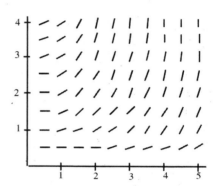

a)   Sketch the solution curve that satisfies the initial
condition $y(0) = 2$ on the slope field.

b)   Solve the differential equation and find the
particular solution that contains the point $(0, 2)$.

c)   Use the function in part (b) to calculate the exact
value of y when $x = 2$.

38.   Find the average rate of change of y with respect to x on the closed interval $[0, 3]$ if

$$\frac{dy}{dx} = \frac{1}{(x+1)(x+2)}.$$

39.   The infinite first quadrant region bounded by the curve $y = e^{-x^2}$ is rotated about the
y-axis to generate a solid. Find the volume of the solid.

40.   A particle is moving along a line with position equation $s(t) = 2t^2 - \ln(1+t)$, where
time $t \ge 0$ is measured in seconds and its position in feet.

a) What is the speed of the particle at $t = 1$?

b) At what time t does the particle's velocity equal its acceleration?

c) How many times does the particle change direction? Justify your answer.

d) What is the minimum velocity of the particle? Show the analysis that leads to
your conclusion.

CHAPTER 11
Taylor Polynomials and Series

11.1   Polynomial Approximations of Functions

Determining values for a particular function is usually an important part of any problem solution. However, the evaluation of most functions is at best difficult if not impossible. Polynomial functions are an exception since computation of their values is essentially an arithmetic problem.

If a function can be approximated by a polynomial and the difference between the function values and the polynomial values is small, then we can use the values of the polynomial as approximations of the function values. Consider a function f that has n derivatives at $x = c$. Our goal is to find a polynomial function that approximates f at $x = c$. We begin by considering the important special case where $c = 0$.

For example, suppose we are interested in evaluating the function $f(x) = e^x$ near the point $x = 0$ by evaluating an approximating polynomial. Note that at $x = 0$, the exponential function and its derivative $f'(x) = e^x$ both have the value of 1. The first degree polynomial p_1 defined by

$$p_1(x) = 1 + x \qquad \textbf{1st-degree approximation}$$

also has $p_1(0) = 1$ and $p_1'(0) = 1$, so the polynomial p_1 agrees with the exponential function and its derivative at $x = 0$.

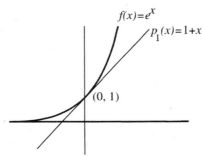

In the figure above we see that the graph of p_1 is tangent to f at $x = 0$ and thus the values of p_1 should be close to the values of f when x is near 0.

If we approximate $f(x) = e^x$ by a 2nd degree polynomial which agrees with f and its first two derivatives at $x = 0$, we might expect a better approximation of f. For the polynomial p_2 with

$$p_2(x) = 1 + x + \frac{x^2}{2} \qquad \textbf{2nd-degree approximation}$$

we have $p_2'(x) = 1 + x$ and $p_2''(x) = 1$ so that $p_2(0) = 1$, $p_2'(0) = 1$ and $p_2''(0) = 1$. Thus, p_2 agrees with f and its first two derivatives at $x = 0$. In the coordinate plane to the right, we see that the graph of p_2 approximates the graph of f more closely than does the line $p_1(x) = 1 + x$ near the point (0, 1).

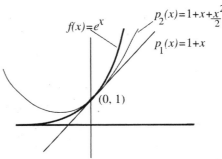

465

If you continue this pattern, requiring that the values of the approximating polynomial p_n and its first n derivatives match those of $f(x) = e^x$ at $x = 0$, you obtain the following:

$$p_n(x) = 1 + x + \frac{x^2}{2!} + \frac{x^3}{3!} + \frac{x^4}{4!} + \dots + \frac{x^n}{n!} \qquad \textbf{nth-degree approximation}$$

The symbol $n!$, read "n factorial," represents $n \cdot (n-1) \cdot \dots \cdot 3 \cdot 2 \cdot 1$, the product of the positive integers less than or equal to n.. As a matter of convenience, for $n = 0$, we define $0! = 1$. Notice that $8! = 8 \cdot 7!$ and $10! = 10 \cdot 9 \cdot 8 \cdot 7!$ so that , for all positive integers n ,

$$n! = n \cdot (n-1)!$$

Note also that $n!$ increases rapidly; $1! = 1$, $2! = 2$, $3! = 6$, $4! = 24$, $5! = 120$ and $6! = 720$.

Taylor Polynomials

A simple scheme exists for creating a polynomial which can be used to approximate a function f near a given point $x = c$. Suppose we want to approximate a function $f(x)$ for x near 0 by a polynomial of degree n.

$$f(x) \approx p_n(x) = a_0 + a_1 x + a_2 x^2 + \dots + a_n x^n$$

We want the value of the function f and the values of the first n derivatives of f to agree with those of the polynomial $p_n(x)$ and of the corresponding derivatives of p_n at the point $x = 0$.

For the value of the polynomial $p_n(x) = a_0 + a_1 x + a_2 x^2 + \dots + a_n x^n$ to agree with the value of f at $x = 0$, we must set $a_0 = f(0)$.

The choice of the coefficient a_1 is based on our desire that $f'(0) = p'_n(0)$. Differentiating $p_n(x)$ yields

$$p'_n(x) = a_1 + 2a_2 x + 3a_3 x^2 + \dots + na_n x^{n-1} ,$$

and substituting $x = 0$ shows that $p'_n(0) = a_1$, so we choose $a_1 = f'(0)$.

Differentiating again, we have

$$p''_n(x) = 2a_2 + 3 \cdot 2a_3 x + \dots + n(n-1)a_n x^{n-2}$$

so that $p''_n(0) = 2a_2$. Since we want $p''_n(0) = f''(0)$, we set $a_2 = \dfrac{f''(0)}{2}$.

Differentiating and substituting a third time, $p'''_n(0) = 3 \cdot 2 \cdot 1 \cdot a_3$ so we choose $a_3 = \dfrac{f'''(0)}{3!}$.
After differentiating n times, we choose

$$a_n = \frac{f^{(n)}(0)}{n!} ,$$

where $f^{(n)}$ designates the nth derivative of f. Therefore, if we make the following choice of coefficients,

$$a_0 = f(0), \quad a_1 = f'(0), \quad a_2 = \frac{f''(0)}{2!}, \quad a_3 = \frac{f'''(0)}{3!}, \quad ..., \quad a_n = \frac{f''(0)}{n!},$$

our approximating polynomial p_n is defined by

$$p_n(x) = f(0) + f'(0)x + \frac{f''(0)}{2!}x^2 + \frac{f'''(0)}{3!}x^3 + \cdots + \frac{f^{(n)}(0)}{n!}x^n.$$

The polynomial p_n which agrees with f and its n derivatives at $x = 0$ is called a **Taylor polynomial** about $x = 0$, or a Taylor Polynomial at $x = 0$. It is named in honor of the 18th century English mathematician Brook Taylor.

> If the function f has n derivatives at $x = 0$, then the polynomial
>
> $$p_n(x) = f(0) + f'(0)x + \frac{f''(0)}{2!}x^2 + \frac{f'''(0)}{3!}x^3 + \cdots + \frac{f^{(n)}(0)}{n!}x^n$$
>
> is called a **Taylor polynomial for f at $x = 0$**.

The calculations are easy because the derivatives are evaluated at $x = 0$. Taylor polynomials at $x = 0$ are called **Maclaurin polynomials**. Later in this section we will consider Taylor polynomial approximations centered at $x = c$ where $c \neq 0$.

Example 1: *Finding a Taylor Polynomial for $f(x) = e^x$*

Find Taylor (or Maclaurin) polynomials p_1, p_2, p_3 and p_n for $f(x) = e^x$ at $x = 0$.
Solution:

If $f(x) = e^x$, then $f'(x) = f''(x) = f'''(x) = ... = f^{(n)}(x) = e^x$ and it follows that

$f(0) = f'(0) = f''(0) = f'''(0) = ... = f^{(n)}(0) = 1$. Thus, by the definition of a Taylor polynomial at $x = 0$ we have

$$p_1(x) = f(0) + f'(0) \cdot x = 1 + x$$

$$p_2(x) = f(0) + f'(0) \cdot x + \frac{f''(0)}{2!} \cdot x^2 = 1 + x + \frac{x^2}{2}$$

$$p_3(x) = f(0) + f'(0) \cdot x + \frac{f''(0)}{2!} \cdot x^2 + \frac{f'''(0)}{3!} \cdot x^3 = 1 + x + \frac{x^2}{2} + \frac{x^3}{6}$$

$$p_n(x) = f(0) + f'(0) \cdot x + \frac{f''(0)}{2!} \cdot x^2 + ... + \frac{f^{(n)}(0)}{n!} \cdot x^n = 1 + x + \frac{x^2}{2} + ... + \frac{x^n}{n!}$$

The following figures show the graph of $f(x) = e^x$ and the approximating Taylor polynomials of degree 1, 2, 3. Notice that each successive approximation remains close to the exponential curve over a larger interval of x-values.

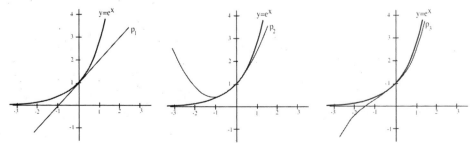

To check the accuracy of these approximating polynomials, compare, for instance, $f(1)$ and $p_5(1)$. Numerical calculations show that

$$p_5(1) = 1 + 1 + \frac{1}{2} + \frac{1}{6} + \frac{1}{24} + \frac{1}{120} = \frac{163}{60} \approx 2.717$$

$$f(1) = e \approx 2.718$$

Finding a Taylor polynomial for the sine function is only slightly more difficult than for the exponential function.

Example 2: *Finding a Maclaurin Polynomial for* $\sin x$

Find Taylor (or Maclaurin) polynomials p_1, p_3, p_5 and p_7 for $f(x) = \sin x$ at $x = 0$.

Solution:

If $f(x) = \sin x$, then $f(0) = \sin 0 = 0$, and

$$f'(x) = \cos x \implies f'(0) = \cos 0 = 1$$
$$f''(x) = -\sin x \implies f''(0) = -\sin 0 = 0$$
$$f'''(x) = -\cos x \implies f'''(0) = -\cos 0 = -1$$
$$f^{(4)}(x) = \sin x \implies f^{(4)}(0) = \sin 0 = 0$$

Through repeated differentiation, you can see that the pattern 1, 0, –1, 0 continues, and the Taylor polynomial approximation of degree 7 is

$$p_7(x) = 0 + x + 0 \cdot \frac{x^2}{2!} - \frac{x^3}{3!} + 0 \cdot \frac{x^4}{4!} + \frac{x^5}{5!} + 0 \cdot \frac{x^6}{6!} - \frac{x^7}{7!}.$$

Therefore, we obtain

$$p_1(x) = x; \qquad\qquad p_3(x) = x - \frac{x^3}{3!};$$

$$p_5(x) = x - \frac{x^3}{3!} + \frac{x^5}{5!}; \qquad p_7(x) = x - \frac{x^3}{3!} + \frac{x^5}{5!} - \frac{x^7}{7!}.$$

Here are the graphs of $f(x) = \sin x$ and the first four Taylor polynomials that approximate f.

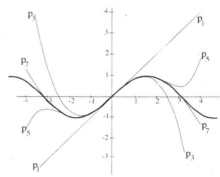

All four Taylor polynomials fit the sine curve well near the $x = 0$. Differences appear as we examine values of x further away from 0 in either direction. The fit improves as the degree of the polynomial increases.

Notice that only the terms with odd powers appear in a Taylor polynomial for $\sin x$. Therefore, a graph of the Taylor polynomial for $\sin x$ is symmetric with respect to the origin, as is the graph of $f(x) = \sin x$. Recall that functions whose graphs are symmetric with respect to the origin are known as odd functions.

Example 3:   *Finding a Maclaurin Polynomial for* $\cos x$

Find the Maclaurin polynomials p_0, p_2, p_4 and p_6 for $f(x) = \cos x$ at $x = 0$. Use p_6 to approximate the value of $\cos(0.1)$.

Solution:

If $f(x) = \cos x$, then $f(0) = \cos 0 = 1$, and

$$f'(x) = -\sin x \implies \quad f'(0) = -\sin 0 = 0$$

$$f''(x) = -\cos x \implies \quad f''(0) = -\cos 0 = -1$$

$$f'''(x) = \sin x \implies \quad f'''(0) = \sin 0 = 0$$

$$f^{(4)}(x) = \cos x \implies \quad f^{(4)}(0) = \cos 0 = 1$$

Through repeated differentiation, you can see that the pattern $0, -1, 0, 1$ continues, and the Taylor polynomial approximation of degree 6 is

$$p_n(x) = 1 + 0 \cdot x - \frac{x^2}{2!} + 0 \cdot \frac{x^3}{3!} + \frac{x^4}{4!} + 0 \cdot \frac{x^5}{5!} - \frac{x^6}{6!}.$$

Therefore, we obtain

$$p_0(x) = 1 \qquad\qquad\qquad\qquad p_2(x) = 1 - \frac{x^2}{2!};$$

$$p_4(x) = 1 - \frac{x^2}{2!} + \frac{x^4}{4!}; \qquad\qquad p_6(x) = 1 - \frac{x^2}{2!} + \frac{x^4}{4!} - \frac{x^6}{6!}.$$

Using p_6, we obtain the approximation $\cos(0.1) \approx 1 - \dfrac{(0.1)^2}{2!} + \dfrac{(0.1)^4}{4!} - \dfrac{(0.1)^6}{6!} = 0.995004$, which coincides with the calculator value to six decimal places.

Here are the graphs of $f(x) = \cos x$ and the first four Taylor polynomials that approximate f.

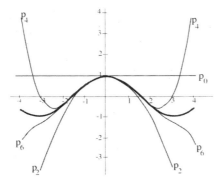

All four Taylor polynomials fit the cosine curve well near $x = 0$ and successive approximations remain close to the cosine curve for larger intervals of x-values.

Notice that only the terms with even powers appear in a Maclaurin polynomial for $\cos x$. Therefore, a graph of the Taylor polynomial for $\cos x$ is symmetric with respect to the y-axis as is the graph of $f(x) = \cos x$. Recall that functions whose graphs are symmetric with respect to the y-axis are known as even functions.

Taylor Polynomials Around $x = c$

We may wish to use a polynomial to approximate a function about a point other than $x = 0$. Suppose we want a polynomial p that approximates the function f at $x = c$. In this case, we modify the definition of a Taylor approximation of f at $x = 0$ in two ways. First the graph of p must be shifted horizontally so that what originally occurred at $x = 0$ now occurs at $x = c$. This is accomplished by replacing x with $(x - c)$. Second, the function value and the derivative values must be evaluated at $x = c$ rather than at $x = 0$. This means that $f(c)$ must be substituted for $f(0)$, $f'(c)$ must be substituted for $f'(0)$, $f''(c)$ must be substituted for $f''(0)$, and so on. Our modified definition for a Taylor polynomial approximation around $x = c$ is

---

If the function f has n derivatives at $x = c$, then the polynomial

$$p_n(x) = f(c) + f'(c)(x - c) + \frac{f''(c)(x - c)^2}{2!} + \frac{f'''(c)(x - c)^3}{3!} + \cdots + \frac{f^{(n)}(c)(x - c)^n}{n!}$$

is called the **Taylor polynomial for f at $x = c$.**

Example 4:   *Finding a Taylor Polynomial for $f(x) = \ln x$ at $x = c$.*

Find the Taylor polynomials p_1, p_2, p_3 and p_4 for $f(x) = \ln x$ at $c = 1$.

Solution:

Since $f(x) = \ln x$, we have $f(1) = 0$ and

$$f'(x) = \frac{1}{x} \quad \Rightarrow \quad f'(1) = 1$$

$$f''(x) = -\frac{1}{x^2} \quad \Rightarrow \quad f''(1) = -1$$

$$f'''(x) = \frac{2!}{x^3} \quad \Rightarrow \quad f'''(1) = 2$$

$$f^{(4)}(x) = -\frac{3!}{x^4} \quad \Rightarrow \quad f^{(4)}(1) = -6$$

Therefore, the Taylor polynomials are

$$p_1(x) = f(1) + f'(1) \cdot (x - 1) = (x - 1)$$

$$p_2(x) = f(1) + f'(1) \cdot (x - 1) + \frac{f''(1)}{2!}(x - 1)^2 = (x - 1) - \frac{1}{2}(x - 1)^2$$

$$p_3(x) = f(1) + f'(1) \cdot (x - 1) + \frac{f''(1)}{2!}(x - 1)^2 + \frac{f'''(1)}{3!}(x - 1)^3$$

$$= (x - 1) - \frac{1}{2}(x - 1)^2 + \frac{1}{3}(x - 1)^3$$

$$p_4(x) = f(1) + f'(1) \cdot (x - 1) + \frac{f''(1)}{2!}(x - 1)^2 + \frac{f'''(1)}{3!}(x - 1)^3 + \frac{f^{(4)}(1)}{4!}(x - 1)^4$$

$$= (x - 1) - \frac{1}{2}(x - 1)^2 + \frac{1}{3}(x - 1)^3 - \frac{1}{4}(x - 1)^4.$$

The following figure compares the graphs of p_1, p_2, p_3 and p_4 with the graph of $f(x) = \ln x$. Notice that near $x = 1$ the graphs are indistinguishable and as n increases, p_n becomes a better and better approximation to $f(x) = \ln x$.

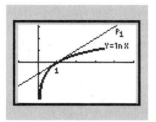 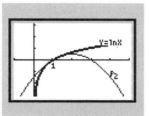 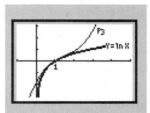 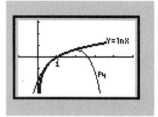

Program for Graphing Taylor Polynomials

Here is a program that plots a function $f(x) = \sin(x)$ and its Taylor polynomials of degree $n = 1, 2$ and 3 at the point $x = a$. First enter the program TDERIV to store $f(x)$ and its first three derivatives. Then execute the TPOLY to have the formulas for the linear, quadratic and cubic polynomials at $x = a$ inserted into the Y= menu and then graphed. Notice that when the approximating polynomials are plotted, Trace and Zoom may be used.

```
PROGRAM:TDERIV
:"sin(X)" →Y1
:"cos(X)" →Y2
:"−sin(X)"→Y3
: "−cos(X)"→Y4
```

```
PROGRAM:TPOLY
:prgm TDERIV
:Input " A=",A
:Y1(A) →F
:Y2(A) →G
:Y3(A)/2→H
:Y4(A)/6→I
:"G(X−A)+F"→Y2
:"H(X−A)²+Y2"→Y3
:"I(X−A)³+Y3"→Y4
:DispGraph
```

If $f(x) = \sin(x)$ and $A = 1$, then the output of the program TPOLY is

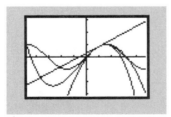

Note that successive polynomials stay close to the *sin* function for longer and longer intervals.

The function and derivatives defined in the subprogram TDERIV can be changed so that Taylor polynomials for any function may be explored.

11.1   Exercises

In Exercises 1–6, find the Taylor (Maclaurin) polynomial of degree n approximating the function f at $x = 0$. Then plot f and p_n on the same axes. Choose a viewing rectangle to show clearly the relationship between f and p_n.

1. $f(x) = \sqrt{x+1}$, $n = 2$ 2. $f(x) = \ln(1+x)$, $n = 2$

3. $f(x) = e^{2x}$, $n = 3$ 4. $f(x) = \sin(x) + \cos(x)$, $n = 4$

5. $f(x) = \dfrac{1}{1-x}$, $n = 2$ 6. $f(x) = xe^x$, $n = 3$

In Exercises 7–10, find a formula for the third-degree Maclaurin polynomial $p_3(x)$ that has the given values for the function f and its derivatives at $x = 0$ and use it to approximate $f(0.2)$.

7. $f(0) = 3$, $f'(0) = -4$, $f''(0) = 8$ and $f'''(0) = 4$

8. $f(0) = 0$, $f'(0) = 1$, $f''(0) = 2$ and $f'''(0) = 3$

9. $f(0) = 1$, $f'(0) = 2$, $f''(0) = 8$ and $f'''(0) = 48$

10. $f(0) = 1$, $f'(0) = -2$, $f''(0) = 8$ and $f'''(0) = -24$

In Exercises 11–14, find the Taylor polynomial of degree n approximating the function f at $x = c$. Then plot f and p_n on the same axes. Choose a viewing rectangle to show clearly the relationship between f and p_n.

11. $f(x) = x^{1/3}$, $n = 2$, $c = 8$ 12. $f(x) = \dfrac{1}{x}$, $n = 3$, $c = 1$

13. $f(x) = \sqrt{x}$, $n = 3$, $c = 4$ 14. $f(x) = \cos x$, $n = 2$, $c = \dfrac{\pi}{3}$

15.   Calculate the values of $\sin(.4)$ and $\sin(\dfrac{\pi}{12})$ using the seventh degree Maclaurin polynomial

$$\sin(x) \approx x - \frac{x^3}{3!} + \frac{x^5}{5!} - \frac{x^7}{7!} \; .$$

Compare your answers with what the calculator gives you.

16.   Let $f(t) = \dfrac{t}{t+1}$ and $G(x) = \displaystyle\int_0^x f(t)\, dt$.

a) Find a formula for the fourth-degree Maclaurin polynomial for f.

b) Find a formula for the fourth-degree Maclaurin polynomial for G.

17. The first four derivatives of $f(x) = \cos^2 x$ are $f'(x) = -2\sin x \cos x$, $f''(x) = 2 - 4\cos^2 x$, $f'''(x) = 8\sin x \cos x$ and $f^{(4)}(x) = 16\cos^2 x - 8$.

    a) Find the fourth-degree Maclaurin approximation $p_4(x)$ to f.

    b) Find the fourth-degree Taylor polynomial approximation $p_4(x)$ associated with f at $x = \pi$.

18. The first three derivatives of $f(x) = (x+4)^{3/2}$ are $f'(x) = \dfrac{3(x+4)^{1/2}}{2}$, $f''(x) = \dfrac{3}{4(x+4)^{1/2}}$,

    and $f'''(x) = \dfrac{-3}{8(x+4)^{3/2}}$.

    a) Find the third-degree Maclaurin approximation $p_3(x)$ to f.

    b) Find the third-degree Taylor polynomial approximation $p_3(x)$ associated with f at $x = -3$.

19. Graphically estimate the interval over which $P_3(x) = x - \dfrac{x^3}{6}$ approximates $\sin(x)$ with an accuracy of 0.1 or less. That is, determine the interval over which $|P_3(x) - \sin(x)| < 0.1$.   [Hint: Graph $Y1 = |P_3(x) - \sin(x)|$ and $Y2 = 0.1$]

20. If $\cos(x)$ is replaced by $P_2(x) = 1 - \dfrac{x^2}{2}$ and $-.5 < x < .5$, estimate the maximum error by graphing $P_2(x) - \cos(x)$.

21. Consider the function that satisfies the differential equation $f'(x) = \dfrac{1}{1+x}$ and contains the point (0, 1). The solution of this differential equation can be approximated using Taylor polynomials.

    a) Write the fifth-degree Taylor polynomial for f at $x = 0$.

    b) Estimate the value of $f(0.25)$ using the fifth-degree Taylor polynomial found in part a).

    c) Verify that the function defined by $f(x) = \ln(x+1) + 1$ satisfies the differential equation. Use this function to find $f(0.25)$ and compare this value with the approximation $p_5(0.25)$ made in part b).

22. Consider the function that satisfies the differential equation $f'(x) = \ln x$ and contains the point (1, 1). The solution of this differential equation can be approximated using Taylor polynomials.

    a) Write the fifth-degree Taylor polynomial for f at $x = 1$.

    b) Estimate the value of $f(1.75)$ using the fifth-degree Taylor polynomial found in part a).

    c) Verify that the function defined by $f(x) = x \ln x - x + 2$ satisfies the differential equation. Use this function to find $f(1.75)$ and compare this value with the approximation $p_4(1.75)$ made in b).

11.2   Sequences and Series

One of the underlying ideas in calculus is that we often solve a complicated problem by generating a sequence of approximations that tend to get closer and closer to the exact solution. This has been the strategy with derivatives, integrals, Newton's root-finding method, and Taylor polynomials. This section is an introduction to sequences – what they are, what it means for them to converge or diverge and how to find their limits. Sequences will support much of our previous work as well as prepare the ground for the next topic – series.

What is a Sequence?

Any ordered list of numbers is called a **sequence.** For example, the following list of reciprocals

$$1, \frac{1}{2}, \frac{1}{3}, \frac{1}{4}, \frac{1}{5}, \cdots$$

is a sequence called the **harmonic** sequence.

Formally speaking, a sequence is a function whose domain is the set of positive integers. The values of the sequence are called **terms** of the sequence. Instead of using $f(n)$ to denote the value corresponding to the input n, it is customary to use a subscripted letter such as a_n; that is, $a_n = f(n)$. Thus, for the harmonic sequence we have

$$a_1 = 1, \quad a_2 = \frac{1}{2}, \quad a_3 = \frac{1}{3}, \quad a_4 = \frac{1}{4}, \quad a_5 = \frac{1}{5}, \cdots$$

Notice that as the index gets larger in the harmonic sequence, the terms of the sequence get closer and closer to 0. In this case we say the sequence has a limit 0, or the sequence **converges** to 0, and write

$$\lim_{n \to \infty} \frac{1}{n} = 0.$$

Our main interest in sequences is their limits. We need to know when sequences do and do not have limits and how to find the limits when they exist. For the simplest sequences, limits are not difficult to find.

Example 1: *Determining Convergence and Divergence*

Determine whether the following sequences converge or diverge.

$$\text{a)} \ \ a_n = \frac{n}{n+1} \qquad \text{b)} \ \ b_n = \frac{(-1)^n}{n} \qquad \text{c)} \ \ c_n = (-1)^n$$

Solution:

Write out the first few terms of each sequence:

a) $\dfrac{1}{2}, \dfrac{2}{3}, \dfrac{3}{4}, \dfrac{4}{5}, \cdots$ suggest that a_n converges to 1. In symbols $\lim\limits_{n \to \infty} \dfrac{n}{n+1} = 1$.

b) $-1, \dfrac{1}{2}, -\dfrac{1}{3}, \dfrac{1}{4}, -\dfrac{1}{5}, \cdots$ suggest that b_n oscillates in sign, but converges to 0.

Thus, $\lim\limits_{n \to \infty} \dfrac{(-1)^n}{n} = 0.$

c) $-1, \ 1, \ -1, \ 1, \ -1, \cdots$, suggest that the terms oscillate but never settle on a single limit. This sequence diverges.

Evaluating and Graphing Sequences

Just as with functions, we can graphically and numerically investigate the behavior of a sequence. The TI-83 has a built-in sequence mode that makes defining and graphing sequences easier. We will use the harmonic sequence to illustrate these features.

From the Mode menu select **Seq** mode. Press $\boxed{\text{Y=}}$ and enter $u(n) = 1/n$. Note that sequence functions must be defined in terms of the input variable n (use $\boxed{\text{X,T,}\theta,n}$). Set the window variables as shown, graph and trace. The horizontal axis is for the input variable n, while the vertical axis is for the terms of the sequence $u(n)$.

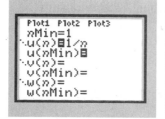  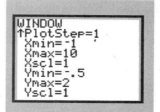 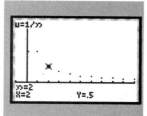

Using the Table feature or from the Home screen, values for the harmonic sequence can be displayed

 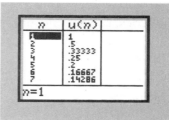 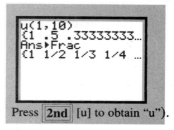

Press $\boxed{\text{2nd}}$ [u] to obtain "u").

Example 2: *Determining Convergence and Divergencex*

Show graphically that the alternating harmonic sequence $a_n = \dfrac{(-1)^n}{n}$ converges to 0.

Solution:

Define $u(n)$ as shown below and set nMin = 1 (so the first value of n used is 1). Press $\boxed{\text{WINDOW}}$ and set nMax = Xmax = 20 (to compute and plot terms up to a_{20}), Xmin = –1 (to see the vertical axis), PlotStart = 1 (the first term plotted will be $u(n)$, where $n = 1$), PlotStep = 1 (n values will increase by 1), Ymin = –1 and Ymax = 1 (since $-1 \leq a_n \leq .5$ for $1 \leq n \leq 20$).

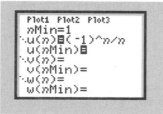 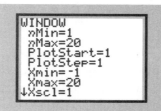 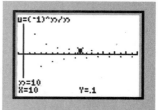

Technology Tip: *Using Sequences*

Sequences may also be calculated using the **sequence** command found in the List menu. Press $\boxed{\text{2nd}}$ [List] and in the OPS menu, select [5:seq( ]. Enter seq(1/N, N, 1, 10) to display the first 10 terms of the harmonic sequence.

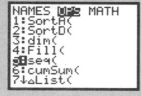 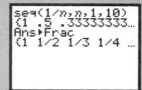

Introduction to Infinite Series

The point of the preliminary work in Taylor polynomials and sequences has been to prepare the way for the study of series.

Informally, if $\{a_n\}$ is an infinite sequence, then the infinite sum

$$a_1 + a_2 + a_3 + \cdots + a_n + \cdots$$

is called an **infinite series** or simply a **series**. Thus, a series results from adding the terms of a sequence $a_1, a_2, a_3, \cdots$. Sigma notation is often convenient to use when discussing sums and a series is commonly denoted by the symbol $\displaystyle\sum_{n=1}^{\infty} a_n$ or $\displaystyle\sum a_n$. So we have

$$\sum_{n=1}^{\infty} a_n = a_1 + a_2 + a_3 + \cdots + a_n + \cdots$$

Whenever you find sigma notation confusing, write out enough terms explicitly to be sure you understand what the notation represents.

To find the sum of a series, consider the following **sequence of partial sums.**

$$S_1 = a_1$$
$$S_2 = a_1 + a_2$$
$$S_3 = a_1 + a_2 + a_3$$
$$\vdots$$
$$S_n = a_1 + a_2 + a_3 + \cdots + a_n$$

These partial sums form a new sequence $\{S_n\}$, which may or may not converge. If the sequence of partial sums converges to a limit S, the series is said to converge and we call S its sum.

Convergent and Divergent Series

For the infinite series $\displaystyle\sum a_n$, the *n*th partial sum is given by

$$S_n = a_1 + a_2 + a_3 + \cdots + a_n.$$

If the sequence of partial sums $\{S_n\}$ converges to S, then the series $\displaystyle\sum a_n$ **converges**. The limit S is called the **sum of the series.**

If $\{S_n\}$ diverges, then the series **diverges.**

It is clear that in a series, there are two different sequences to be considered. The first is the original sequence whose terms $a_1, a_2, a_3, \cdots$ are being added up. The second is the sequence of partial sums $S_1, S_2, S_3, \cdots$. The sum of the series, when it exists, is not the limit of terms $a_1, a_2, a_3, \cdots$, but the limit of the sequence of partial sums $S_1, S_2, S_3, \cdots$. Keeping these related but different sequences separate is essential.

Example 3:   *Convergent and Divergent Series of Constants*

a) The series $\displaystyle\sum_{n=1}^{\infty}\frac{1}{2^n}=\frac{1}{2}+\frac{1}{4}+\frac{1}{8}+\frac{1}{16}+\cdots$ has the following partial sums.

$$S_1=\frac{1}{2}$$

$$S_2=\frac{1}{2}+\frac{1}{4}=\frac{3}{4}$$

$$S_3=\frac{1}{2}+\frac{1}{4}+\frac{1}{8}=\frac{7}{8}$$

$$S_4=\frac{1}{2}+\frac{1}{4}+\frac{1}{8}+\frac{1}{16}=\frac{15}{16}$$

We note that $S_1=1-\dfrac{1}{2}$, $\;S_2=1-\dfrac{1}{4}$, $\;S_3=1-\dfrac{1}{8}$ and $S_4=1-\dfrac{1}{16}$, and, in general, we have

$$S_n=1-\frac{1}{2^n}.$$

Because the sequence of partial sums converges to $\displaystyle\lim_{n\to\infty}\left(1-\frac{1}{2^n}\right)=1$, it follows that the series converges and its sum is 1.

b) The series

$$\sum_{n=1}^{\infty}1=1+1+1+\cdots$$

diverges because $S_n=n$ and the sequence of partial sums diverges.

Example 4:   *Finding the Sum of a Series of Constants*

Find the sum of the convergent series

$$\sum_{n=1}^{\infty}\frac{1}{n(n+1)}=\frac{1}{1\cdot2}+\frac{1}{2\cdot3}+\frac{1}{3\cdot4}+\cdots$$

Solution:

Displaying the first few partial sums we discover a pattern.

$$S_1=\frac{1}{1\cdot2}=\frac{1}{2}$$

$$S_2=S_1+\frac{1}{2\cdot3}=\frac{1}{2}+\frac{1}{2\cdot3}=\frac{2}{3}$$

$$S_3=S_2+\frac{1}{3\cdot4}=\frac{2}{3}+\frac{1}{3\cdot4}=\frac{3}{4}$$

$$S_4=S_3+\frac{1}{4\cdot5}=\frac{3}{4}+\frac{1}{4\cdot5}=\frac{4}{5}$$

A reasonable guess for the general term is $S_n = \dfrac{n}{n+1}$ and the sum of the series is

$\lim\limits_{n\to\infty} \dfrac{n}{n+1} = 1$. The series in this example is a special type where it is possible to find an explicit formula for nth partial sum. We will learn that discovering a pattern in the sequence of partial sums is a rare event.

The *n*th Test for Divergence

Notice how the partial sums were computed in Example 4. Each sum is the preceding sum plus the next term in the series; that is

$$S_n = S_{n-1} + a_n.$$

This observation leads to a simple but important fact about convergence. A series $\sum a_n$ converges if and only if its partial sums converge to a limit. For this to occur the difference $S_n - S_{n-1}$ between successive partial sums must tend to zero. This difference is simply the nth term. Since the nth term of the series is $a_n = S_n - S_{n-1}$, the limit of a_n, as $n \to \infty$, must be zero if the series converges:

$$\lim_{n\to\infty} a_n = \lim_{n\to\infty}(S_n - S_{n-1}) = \lim_{n\to\infty} S_n - \lim_{n\to\infty} S_{n-1} = S - S = 0,$$

where S is the sum of the series.

> If the series $\sum a_n$ converges, then $\lim\limits_{n\to\infty} a_n = 0$.

Thus, for a series to converge, its terms must tend to zero. This statement provides a convenient and useful test for the divergence of a series $\sum a_n$, since it is equivalent to its contrapositive statement: If the nth term of a sequence does not tend to zero, then the series formed from that sequence must diverge.

> ### The *n*th Term Test for Divergence
>
> If $\lim\limits_{n\to\infty} a_n \neq 0$, then the series $\sum a_n$ diverges.

Example 5: *Using the nth Term Test for Divergence*

Show that the series $\displaystyle\sum_{n=1}^{\infty} n \sin\frac{1}{n} = \sin 1 + 2\sin\frac{1}{2} + 3\sin\frac{1}{3} + \cdots$ diverges.

Solution:

The series $\displaystyle\sum_{n=1}^{\infty} n \sin\frac{1}{n}$ diverges because its general term does not approach zero.

$$\lim_{n\to\infty} n\sin\frac{1}{n} = \lim_{n\to\infty} \frac{\sin(1/n)}{1/n} = \lim_{t\to 0} \frac{\sin t}{t} \quad (t = 1/n)$$

$$= 1 \neq 0.$$

A word of caution: If we find $\lim_{n\to\infty} a_n \neq 0$, we know that $\sum a_n$ diverges. However, if we know only that $\lim_{n\to\infty} a_n = 0$, we cannot decide about the convergence or divergence of $\sum a_n$. The harmonic series $\sum \frac{1}{n}$ is a classic example of a series whose nth term goes to zero but which diverges. We will investigate the behavior of the harmonic series in Examples 9 and 10 later in this section.

Geometric Series

Geometric series are the simplest and most important class of infinite series. In general, a geometric series has the form

$$\sum_{n=0}^{\infty} ar^n = a + ar + ar^2 + ar^3 + ar^4 + \cdots, \quad a \neq 0;$$

a is called the leading term and r is the ratio. Note that each term is r times the previous term. The series of Example 3, $\sum_{n=1}^{\infty} \frac{1}{2^n}$, is geometric with $a = \frac{1}{2}$ and $r = \frac{1}{2}$.

Geometric series have one great advantage: It's easy to decide whether they converge or diverge and, if they converge, to find their limits. We now show how the limit of a geometric series depends on the factor r. It is easy to see that the series diverges if $r = 1$ or $r = -1$.

If $r \neq \pm 1$, the nth partial sum is

$$S_n = a + ar + ar^2 + ar^3 + \cdots + ar^n.$$

Multiplying both sides by r, we obtain

$$rS_n = ar + ar^2 + ar^3 + \cdots + ar^{n+1};$$

subtracting this second equation from the original equation for S_n, we have

$$S_n - rS_n = a - ar^{n+1}.$$

We can solve for S_n, provided $r \neq 1$: $\quad S_n = \dfrac{a(1 - r^{n+1})}{1 - r}$

With this formula for partial sums, we can determine the limit of the geometric series

$$\sum_{n=0}^{\infty} ar^n = \lim_{n\to\infty} \frac{a(1 - r^{n+1})}{1 - r}.$$

If $r > 1$ or $r < -1$, then r^{n+1} will become larger and larger in magnitude as $n \to \infty$ and the limit will not exist.

If $-1 < r < 1$ then $r^{n+1} \to 0$ as $n \to \infty$ and

$$\sum_{n=0}^{\infty} ar^n = \frac{a(1 - 0)}{1 - r} = \frac{a}{1 - r}.$$

If $|r| < 1$, the geometric series

$$\sum_{n=0}^{\infty} ar^n = a + ar + ar^2 + ar^3 + \cdots$$

converges to $\dfrac{a}{1-r}$. If $|r| \geq 1$, the series diverges.

This completely settles the issue for geometric series. We know which ones converge and which ones diverge, and for the convergent ones we know what the sums must be.

Example 6: *Convergent and Divergent Geometric Series*

a) The series $\displaystyle\sum_{k=0}^{\infty} \frac{2}{(-3)^k} = 2 - \frac{2}{3} + \frac{2}{9} - \frac{2}{27} + \cdots$

   is geometric with ratio $r = -\dfrac{1}{3}$ and $a = 2$. Because $|r| < 1$ the series converges and its sum is

$$S = \frac{a}{1-r} = \frac{2}{1+(1/3)} = \frac{3}{2}.$$

b) The series $\displaystyle\sum_{n=0}^{\infty} \left(\frac{3}{2}\right)^n = 1 + \frac{3}{2} + \frac{9}{4} + \frac{27}{8} + \cdots$

   is geometric with ratio $r = \dfrac{3}{2}$ and $a = 1$. Because $|r| \geq 1$ the series diverges.

Sometimes the terms in a series are constants and sometimes the terms are functions. For example, consider the series

$$1 + x + x^2 + x^3 + \cdots + x^n + \cdots$$

The terms in this series are functions of the form

$$x^0, \; x, \; x^2, \; x^3, \cdots, \; x^n, \cdots$$

In summation notation we have

$$\sum_{n=0}^{\infty} x^n = 1 + x + x^2 + x^3 + x^4 + \cdots$$

Note that this series starts with $n = 0$ and so the first term is $x^0 = 1$.

Example 7: *Summing a Series of Functions*

   Find the sum of the series $\displaystyle\sum_{n=0}^{\infty} x^n$, where $|x| < 1$.

Solution:

   The series $\displaystyle\sum_{n=0}^{\infty} x^n = 1 + x + x^2 + x^3 + x^4 + \cdots$ is geometric with $a = 1$ and $r = x$. Since

   $|x| < 1$, $\displaystyle\sum_{n=0}^{\infty} x^n$ converges and its sum is

$$\sum_{n=0}^{\infty} x^n = \frac{a}{1-r} = \frac{1}{1-x}.$$

Thus, over the interval $(-1, 1)$ the function $f(x) = \dfrac{1}{1-x}$ can be represented by an infinite series of functions.

Convergence of Series Graphically and Numerically

Unhappily, the accumulated roundoff errors connected with summing a large number of small terms means that the calculator is not always a useful tool in the study of series. However, the following examples do illustrate ways in which a calculator is helpful.

Example 8:   *Using Parametric Mode to Display a Sequence of Partial Sums*

Use Parametric mode to display a graph and table of the sequence of partial sums for the convergent series $\displaystyle\sum_{n=1}^{\infty} \frac{1}{2^n}$.

Solution:

In Parametric mode using the Dot graph style, enter

$$X_1T = T$$

$$Y_1T = \text{sum (seq}(1/2\text{^}T,T,\,1,T)).$$

Press $\boxed{\textbf{WINDOW}}$ and set Tmin = 1, Tmax = 10, Tstep = 1,   Xmin = 1, Xmax = 10, Ymin = –1, Ymax = 2. Note that Tstep must equal 1 when parametric mode is used to plot sequences or partial sums. Then set TblStart = 1 and ΔTbl = 1.

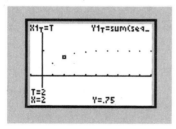 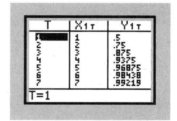

Example 9:   *Partial Sums for the Harmonic Series*

Display in a table the sequence of partial sums for the divergent harmonic series.

Solution:

Place the calculator in Seq mode. Define the sequence of partial sums by entering the values for nMin, u(n) and u(nMin). In Table Setup, set TblStart = 1 and ΔTbl = 1.

We continue our investigation of the harmonic series $\sum \frac{1}{n}$ by using the statistical features of the TI-83 to make a list of its partial sums.

Example 10:   *Fitting a Curve to Data*

Given the harmonic series $\sum \frac{1}{n}$. Plot the sequence of partial sums and fit a log regression curve to the data.

Solution:

Assuming that u(*n*) is defined as in Example 9, place the calculator in Func mode. Enter the first 25 positive integers in list L1 using L1 = seq(N,N,1,25) and the first 25 values of u(*n*) in list L2 using L2 = u(L1). (Press   2nd   [u].) Define the scatter plot using L1 and L2, and plot using ZoomStat.

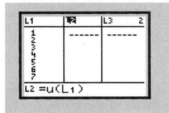   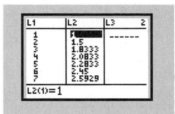   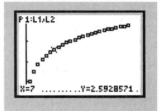

Fit the partial sum sequence with a logarithmic regression curve.

      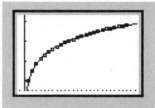

Does this suggest that the harmonic series diverges?

The sum and seq(uence) commands may also be employed to study series. These commands are found in the List menus OPS [5:seq(] and MATH [5:sum(].

11.2   Exercises

In Exercises 1– 6, find the first five terms of the sequence of partial sums. Graph the sequence of partial sums. Does it appear that the series converges or diverges? If the series converges, find its sum.

1. $\displaystyle\sum_{n=0}^{\infty}\left(-\frac{1}{2}\right)^{n}$
                  2. $1+\dfrac{1}{3}+\dfrac{1}{5}+\dfrac{1}{7}+\dfrac{1}{9}+\cdots$

3. $2+\dfrac{1}{2}+\dfrac{1}{8}+\dfrac{1}{32}+\dfrac{1}{128}+\cdots$
       4. $1-\dfrac{4}{3}+\dfrac{16}{9}-\dfrac{64}{27}+\dfrac{256}{81}-\cdots$

5. $\displaystyle\sum_{n=1}^{\infty}\frac{3}{10^{n}}$
                    6. $\displaystyle\sum_{n=0}^{\infty}\cos(n\pi)$

In Exercises 7–10 a sequence $\{a_{n}\}$ is defined. Determine:

a) whether $\{a_{n}\}$ is convergent;   b) whether $\displaystyle\sum_{n=1}^{\infty}a_{n}$ is convergent.

7. $a_{n}=\dfrac{n}{n+1}$
                  8. $a_{n}=\dfrac{3}{2^{n}}$

9. $a_{n}=\dfrac{3n-1}{2n+1}$
               10. $a_{n}=\left(-\dfrac{3}{5}\right)^{n}$

11.   Find the first four terms a_{1},a_{2},a_{3} and a_{4} of the series having as a sequence of partial sums

     a) $\{S_{n}\}=\dfrac{1}{2},\dfrac{2}{3},\dfrac{3}{4},\dfrac{4}{5},\cdots$
            b) $\{S_{n}\}=\dfrac{1}{2},\dfrac{1}{3},\dfrac{1}{4},\dfrac{1}{5},\cdots$

12.   Find the sum of each of the following convergent geometric series.

     a) $\displaystyle\sum_{n=1}^{\infty}\left(\dfrac{3}{4}\right)^{n}$
     b) $1-\dfrac{2}{3}+\dfrac{4}{9}-\dfrac{8}{27}+\cdots$
     c) $\dfrac{1}{16}+\dfrac{1}{32}+\dfrac{1}{64}+\cdots$

13.   Verify that the following series diverge.

     a) $\displaystyle\sum_{n=1}^{\infty}\left(\dfrac{3}{2}\right)^{n}$
     b) $\displaystyle\sum_{k=1}^{\infty}\left(2+\dfrac{1}{2^{k}}\right)$
     c) $\displaystyle\sum_{n=0}^{\infty}\sin(\dfrac{n\pi}{2})$

In Exercises 14 – 23, determine whether the given series converges or diverges. If the series converges, find the sum.

14. $\displaystyle\sum_{n=1}^{\infty} (-1/3)^n$ 15. $\displaystyle\sum_{n=0}^{\infty} (1/3)^n$

16. $\displaystyle\sum_{n=1}^{\infty} e^{-n}$ 17. $\displaystyle\sum_{n=1}^{\infty} (-1)^n$

18. $\dfrac{1}{11} - \dfrac{10}{11^2} + \dfrac{100}{11^3} - \dfrac{1000}{11^4} + \cdots$ 19. $\displaystyle\sum_{n=0}^{\infty} \sin(n\pi)$

20. $\displaystyle\sum_{n=1}^{\infty} \dfrac{2^n}{3^{n-1}}$ 21. $.01 + .02 + .04 + .08 + \cdots$

22. $7 + \dfrac{7}{5} + \dfrac{7}{25} + \dfrac{7}{125} + \cdots$ 23. $\displaystyle\sum_{n=0}^{\infty} (-3)^{1-n}$

24.   Find two different geometric series, each of which converges to $\dfrac{5}{4}$.

25.   Use geometric series to show :

     a) $\displaystyle\sum_{k=0}^{\infty} (-x)^k = \dfrac{1}{1+x}$ if $-1 < x < 1$

     b) $\displaystyle\sum_{k=0}^{\infty} (-1)^k (2x)^k = \dfrac{1}{1+2x}$ if $-\dfrac{1}{2} < x < \dfrac{1}{2}$

     c) $\displaystyle\sum_{k=0}^{\infty} (x-3)^k = \dfrac{1}{4-x}$ if $2 < x < 4$

In Exercises 26–29, a) find the common ratio of the geometric series; b) write the function that gives the sum of the series; and c) in a single viewing rectangle graph the partial sum S_3.

26. $x - x^3 + x^5 - x^7 + x^9 - \cdots$ 27. $1 + 2x + 4x^2 + 8x^3 + \cdots$

28. $1 - \dfrac{x}{2} + \dfrac{x^2}{4} - \dfrac{x^3}{8} + \cdots$ 29. $x + 2x^2 + 4x^3 + 8x^4 + \cdots$

30.   Consider the series $\displaystyle\sum_{n=1}^{\infty} \left(\dfrac{1}{n} - \dfrac{1}{n+1} \right)$.

     a)   Compute the first four partial sums of the series.

     b)   Find a formula for the nth partial sum S_n of the series.

     c)   Show that the series converges, and find the sum of the series.

31.   Three people Adam, Bill and Charley, decide to divide an apple as follows. First they divide it into fourths, each taking a quarter. Then they divide the leftover quarter in fourths, each taking a quarter, and so on. Show that each gets a third of the apple.

32.   A rubber ball is dropped from a height of six feet. Each time it hits the ground it rebounds to $\frac{5}{8}$ its previous height. Find the total distance traveled by the ball.

33.   Suppose the government pumps an extra $1 billion into the economy. Assume that each business and individual saves 25% of its income and spends the rest, so that of the initial $1 billion, 75% is spent by individuals and business. Of that amount, 75% is spent and so on. What is the total increase in spending due to the government action? (This is called the *multiplier effect* in economics.)

34.   Determine whether the following statements are true or false.

   a) If $\lim\limits_{n \to \infty} a_n = 0$, then $\sum a_n$ converges.

   b) If $\lim\limits_{n \to \infty} a_n \neq 0$, then $\sum a_n$ diverges.

   c) If $\sum a_n$ diverges, then $\lim\limits_{n \to \infty} a_n \neq 0$.

   d) If $\sum a_n$ converges, then $\lim\limits_{n \to \infty} a_n = 0$.

11.3   Power Series

In the first section of this chapter we considered Taylor polynomial approximations for functions. For example, given the exponential function $f(x) = e^x$, it was shown that for inputs near 0,

$$e^x \approx 1 + x + \frac{x^2}{2!} + \frac{x^3}{3!} + \frac{x^4}{4!} + \ldots + \frac{x^n}{n!} \ .$$

We generally found that the polynomial approximation grew more and more accurate as we took more and more terms. Hence, if we could add up all the terms, we might expect actual equality,

$$e^x = 1 + x + \frac{x^2}{2!} + \frac{x^3}{3!} + \frac{x^4}{4!} + \cdots \ = \sum_{n=0}^{\infty} \frac{x^n}{n!}.$$

Series obtained in this way, by taking all the terms in the Taylor expansion are called **Taylor Series**. In general, given a function f, its Taylor (Maclaurin) series at $x = 0$ will be

$$f(0) + f'(0)x + \frac{f''(0)}{2!}x^2 + \frac{f'''(0)}{3!}x^3 + \cdots = \sum_{n=0}^{\infty} \frac{f^{(n)}(0)}{n!} x^n \ .$$

In addition, just as we have Taylor polynomials centered at points other than 0, we can also have a Taylor series centered at $x = c$. Assuming all derivatives of a given function f exist at $x = c$, we have the following Taylor series

$$f(c) + f'(c)(x-c) + \frac{f''(c)}{2!}(x-c)^2 + \frac{f'''(c)}{3!}(x-c)^3 + \cdots = \sum_{n=0}^{\infty} \frac{f^{(n)}(c)}{n!}(x-c)^n \ .$$

We can define Taylor (Maclaurin) series for $\sin x$ and $\cos x$ in a similar way to the Taylor series for e^x. We will verify shortly that for all x, we can write the following.

$$e^x = 1 + x + \frac{x^2}{2!} + \frac{x^3}{3!} + \frac{x^4}{4!} + \frac{x^5}{5!} + \cdots \ = \sum_{n=0}^{\infty} \frac{x^n}{n!}$$

$$\sin x = x - \frac{x^3}{3!} + \frac{x^5}{5!} - \frac{x^7}{7!} + \frac{x^9}{9!} - \cdots \ = \sum_{n=0}^{\infty} (-1)^n \frac{x^{2n+1}}{(2n+1)!}$$

$$\cos x = 1 - \frac{x^2}{2!} + \frac{x^4}{4!} - \frac{x^6}{6!} + \frac{x^8}{8!} - \cdots \ = \sum_{n=0}^{\infty} (-1)^n \frac{x^{2n}}{(2n)!}$$

In this section we will see that several important functions including the exponential, sine and cosine functions can be represented exactly by an infinite series called a **power series**. Taylor series and Maclaurin series are special cases of a power series.

We begin with the following definition.

An infinite series of the form

$$\sum_{n=0}^{\infty} a_n x^n = a_0 + a_1 x + a_2 x^2 + a_3 x^3 + \cdots + a_n x^n + \cdots$$

is called a **power series centered at $x = 0$**. More generally, a series of the form

$$\sum_{n=0}^{\infty} a_n (x - c)^n = a_0 + a_1 (x - c) + a_2 (x - c)^2 + \cdots + a_n (x - c)^n + \cdots$$

is called a **power series centered at c**, where c is a constant.

Roughly speaking, a power series is a polynomial with infinitely many terms. More precisely, it is the limit of a sequence of polynomials:

$$p_0(x) = a_0$$
$$p_1(x) = a_0 + a_1 x$$
$$p_2(x) = a_0 + a_1 x + a_2 x^2$$
$$p_3(x) = a_0 + a_1 x + a_2 x^2 + a_3 x^3$$

where $p_n(x) = a_0 + a_1 x + a_2 x^2 + a_3 x^3 + \cdots + a_n x^n$ is a polynomial of degree n. These polynomials play the role of partial sums, and

$$\sum_{n=0}^{\infty} a_n x^n = \lim_{n \to \infty} p_n(x)$$

To evaluate a power series $\sum_{n=0}^{\infty} a_n x^n = \lim_{n \to \infty} p_n(x)$ at a specific input value $x = x_0$, we substitute the

value of x_0 and obtain a series of constants: $\sum_{n=0}^{\infty} a_n x_0^n$ that may or may not converge.

This raises two important questions about a power series:

1) For what x's does the series converge?

2) To what function does it converge; that is, what is the sum $S(x)$ of the series?

We can quickly answer our two questions for one such power series.

Example 1:   *Convergence of a Power Series*
For what x's does the power series

$$\sum_{n=0}^{\infty} x^n = 1 + x + x^2 + x^3 + \cdots$$

converge, and what is its sum?

Solution:

We studied this series in Example 7 of the previous section. It is a geometric series centered at $x = 0$ with common ratio $r = x$. We know that an infinite geometric series converges if the common ratio has absolute value less than 1. Thus, our power series $\sum x^n$ converges on the interval $-1 < x < 1$, also centered at $x = 0$. And, it has a sum $S(x)$ given by

$$S(x) = \frac{1}{1-x}, \quad -1 < x < 1.$$

This is typical behavior, as we shall see shortly. A power series either converges for all x, converges on a finite interval with the same center as the series, or converges only at the center.

Radius and Interval of Convergence

The set of x-values for which a power series converges is called the **interval of convergence** of the series. The interval of convergence of the Maclaurin series for $f(x) = \frac{1}{1-x}$ is $-1 < x < 1$. This interval of convergence is the domain over which the series is equal to $f(x) = \frac{1}{1-x}$, so we can write

$$\frac{1}{1-x} = 1 + x + x^2 + x^3 + \cdots \quad \text{if } -1 < x < 1.$$

There are several ways to determine the interval of convergence of a power series. One of the simplest and most useful is by means of a ratio test.

The Ratio Test

An important characteristic of a series of numbers is the behavior of the ratio of consecutive terms. For an infinite geometric series

$$a + ar + ar^2 + ar^3 + \cdots + ar^n + \cdots,$$

the ratio of consecutive terms is r. We know that an infinite geometric series converges for r values between -1 and 1, and diverges for all other r values.

Recall that if there is any hope that $\sum a_n$ converges, we must have $\lim_{n\to\infty} a_n = 0$. The ratio test studies the rate at which terms a_n decrease as n increases and is based on a comparison of the series $\sum a_n$ with a suitable geometric series. The decrease from term a_n to the next term a_{n+1} can be measured by the ratio $\frac{a_{n+1}}{a_n}$.

The Ratio Test

Let $\sum a_n$ be a series of nonzero terms, and suppose $\lim_{n\to\infty} \frac{|a_{n+1}|}{|a_n|} = L$.

Then,

a) the series converges if $L < 1$,

b) the series diverges if $L > 1$,

c) the test is inconclusive if $L = 1$.

The Ratio Test can be proved by comparing the given series to a geometric series. For large n, $\dfrac{|a_{n+1}|}{|a_n|} \approx L$. Then, if j is such a large n, $|a_{j+1}| \approx L|a_j|$. Then $|a_{j+2}| \approx L|a_{j+1}| \approx L^2|a_j|$ and $|a_{j+3}| \approx L|a_{j+2}| \approx L^3|a_j|$. In general, we then see that for large n

$$|a_{j+n}| \approx L^n|a_n|.$$

In other words, the infinite series for large n is approximately a geometric series with ratio L. Thus, we expect the series $\sum a_n$ to converge if $L < 1$ and diverge if $L > 1$.

The following three examples illustrate how the Ratio Test may be used to determine values of x for which a power series is convergent.

Example 2: *Finding the Interval of Convergence*

For what values of x is the series $\displaystyle\sum_{n=0}^{\infty} n!\, x^n$ convergent?

Solution:

Using the Ratio Test and letting $a_n = n!\, x^n$ and $a_{n+1} = (n+1)!\, x^{n+1}$ we have

$$\lim_{n\to\infty} \frac{|a_{n+1}|}{|a_n|} = \lim_{n\to\infty} \left| \frac{(n+1)!\, x^{n+1}}{n!\, x^n} \right| \qquad \text{Note: } \left|\frac{a}{b}\right| = \left|\frac{a}{b}\right| \text{ and } (n+1)! = (n+1)\cdot n!$$

$$= \lim_{n\to\infty} |(n+1)x|$$

So by the Ratio Test, this series diverges when $x \neq 0$. Thus, the series converges only when $x = 0$.

Example 3: *Finding the Interval of Convergence*

Find the interval of convergence for the power series

$$\sum_{n=0}^{\infty} \frac{x^n}{n!} = 1 + x + \frac{x^2}{2!} + \frac{x^3}{3!} + \frac{x^4}{4!} + \dots$$

which is the Maclaurin series for $f(x) = e^x$.

Solution:

Using the Ratio Test and letting $a_n = \dfrac{x^n}{n!}$ and $a_{n+1} = \dfrac{x^{n+1}}{(n+1)!}$ we have

$$\lim_{n\to\infty} \frac{|a_{n+1}|}{|a_n|} = \lim_{n\to\infty} \left| \frac{\dfrac{x^{n+1}}{(n+1)!}}{\dfrac{x^n}{n!}} \right|$$

$$= \lim_{n\to\infty} \left| \frac{x^{n+1}}{x^n} \right| \left| \frac{n!}{(n+1)!} \right|$$

$$= \lim_{n\to\infty} \frac{|x|}{n+1} = 0 < 1 \text{ for all } x.$$

Thus, by the Ratio Test the given series converges for all values of x. This means that the Maclaurin series for $f(x) = e^x$ equals $f(x)$ for all x.

Example 4: *Finding the Interval of Convergence*

Find the interval of convergence for

$$\sum_{n=0}^{\infty} \frac{x^n}{(n+1)2^n} = 1 + \frac{1}{2} \cdot \frac{x}{2} + \frac{1}{3} \cdot \frac{x^2}{2^2} + \frac{1}{4} \cdot \frac{x^3}{2^3} + \cdots$$

Graph the partial sum polynomials $p_n(x) = \sum_{n=0}^{\infty} \frac{x^n}{(n+1)2^n}$ for $n = 1, 2, 3, 4, 5$ on the same axes.

Solution:

Using the Ratio Test and letting $a_n = \frac{x^n}{(n+1)2^n}$ and $a_{n+1} = \frac{x^{n+1}}{(n+2)2^{n+1}}$ we have

$$\lim_{n \to \infty} \frac{|a_{n+1}|}{|a_n|} = \lim_{n \to \infty} \left| \frac{\frac{x^{n+1}}{(n+2)2^{n+1}}}{\frac{x^n}{(n+1)2^n}} \right| = \lim_{n \to \infty} \left| \frac{x^{n+1}}{x^n} \right| \cdot \left| \frac{(n+1)2^n}{(n+2)2^{n+1}} \right|$$

$$= \frac{|x|}{2} \cdot \lim_{n \to \infty} \frac{n+1}{n+2} = \frac{|x|}{2}$$

This series converges when $L = \frac{|x|}{2} < 1$ and diverges when $\frac{|x|}{2} > 1$. Consequently, the series converge when $|x| < 2$ and diverge when $|x| > 2$. We say the interval of convergence (without endpoints) of the series is $-2 < x < 2$. At the endpoints of the interval, in this case $x = -2$ and $x = 2$, the series may or may not converge. We consider this question in Section 11.5.

In the following figure, the polynomial graphs represent the partial sums p_n for $n = 1, 2, 3, 4, 5$.

$$p_1(x) = 1 + \frac{x}{4}$$

$$p_2(x) = 1 + \frac{x}{4} + \frac{x^2}{12}$$

$$p_3(x) = 1 + \frac{x}{4} + \frac{x^2}{12} + \frac{x^3}{32}$$

$$p_4(x) = 1 + \frac{x}{4} + \frac{x^2}{12} + \frac{x^3}{32} + \frac{x^4}{80}$$

$$p_5(x) = 1 + \frac{x}{4} + \frac{x^2}{12} + \frac{x^3}{32} + \frac{x^4}{80} + \frac{x^5}{192}$$

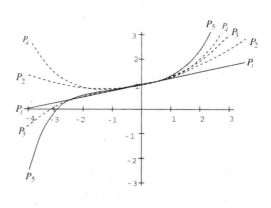

On the interval $-2 < x < 2$ the graphs are converging, but outside that interval the graphs are rapidly going off to infinity or negative infinity. Thus, one can estimate the interval of convergence by viewing the first few partial sums. Notice it is difficult to determine convergence at the endpoints $x = -2$ and $x = 2$ from the graphs.

For the power series we have looked at so far, the set of x-values for which the series is convergent has turned out to be a single real number for the series $\sum_{n=0}^{\infty} n! x^n$ as we saw in Example 2, all the real numbers for the series $\displaystyle\sum_{n=0}^{\infty}\frac{x^n}{n!}$ in Example 3, and a finite interval for the series

$\displaystyle\sum_{n=0}^{\infty}\frac{x^n}{(n+1)2^n}$ in Example 4. The following is true in general.

For a given power series $\displaystyle\sum_{n=0}^{\infty}a_n(x-c)^n$ there are only three possibilities:

(i) The series converges for all x.

(ii) The series converges only when $x = c$.

(iii) There is a positive number R such that the series converges if $|x-c| < R$ and diverges if $|x-c| > R$. The series may or may not converge at either of the endpoints $x = c - R$ and $x = c + R$.

The number R in case (iii) is called the **radius of convergence** of the power series. By convention, the radius of convergence is $R = \infty$ in case (i) and $R = 0$ in case (ii).

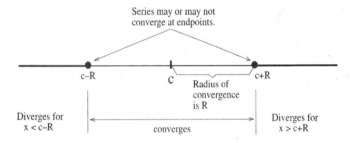

The Ratio Test can be used to determine the radius of convergence R in most cases.

Example 5: *Finding the Radius of Convergence*

Find the radius of convergence of the series $\displaystyle\sum_{n=0}^{\infty}2(x-3)^n$

Solution:

Using the Ratio Test and letting $a_n = 2(x-3)^n$ and $a_{n+1} = 2(x-3)^{n+1}$ we have, for $x \neq 3$

$$\lim_{n\to\infty}\frac{|a_{n+1}|}{|a_n|} = \lim_{n\to\infty}\left|\frac{2(x-3)^{n+1}}{2(x-3)^n}\right|$$

$$= \lim_{n\to\infty}|x-3|.$$

By the Ratio Test, the series converges for $|x-3| < 1$ and diverges for $|x-3| > 1$

Therefore, the radius of convergence of the series is $R = 1$. Note that the radius of convergence is the distance from the midpoint of the interval of convergence, $x = 3$ in this case, to one of the endpoints.

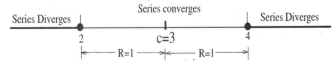

In summary, it is possible for a power series $\sum a_n(x-c)^n$ to converge for all x; if that happens, we take R to be ∞. At the other extreme, the series may converge only at $x = c$; in that case we take R to be 0. If $R > 0$, then at the endpoints of the interval, $x = c - R$ and $x = c + R$, different things may happen to the series. The radius of convergence tells us where the switch from convergence to divergence takes place; however, it does not tell us what happens at the place where the switch occurs. Thus, an interval of convergence may include either, both or neither of its endpoints. However, in practice, what matters most is the radius of convergence. What happens at the endpoints can be interesting, but it is usually less important.

A Reminder: The Ratio Test never determines the behavior at the endpoints of the interval, for the limit of the ratio at each endpoint will be 1.

In this section we have seen that elementary functions that are infinitely differentiable can be written as infinite polynomials, or power series. We have also examined ways of determining the radius of convergence for a power series, so that we can determine the domain over which a power series equals the original function. So far, we have written a power series that equals a given function by constructing a Taylor series about some value $x = c$. In the next section, we will examine alternative methods for writing a power series that can be applied to functions for which it is difficult to construct the series directly from the nth-order derivatives.

11.3   Exercises

1. Find the first three non-zero terms in the Maclaurin series for f.

    a) $f(x) = \ln(1 + 2x)$ b) $f(x) = xe^{-x}$

In Exercises 2–5,

    a) Find the first four terms of the Maclaurin series for the given function.

    b) By graphing the function and several of its approximating Taylor polynomials, estimate the interval of convergence of the series.

2. $f(x) = \dfrac{1}{1 + x}$ 3. $f(x) = \sqrt{1 + x}$

4. $f(x) = e^{-x}$ 5. $f(x) = \displaystyle\int_0^x \dfrac{1}{1 + t}\, dt$

In Exercises 6–8, show that these familiar series for elementary functions converge for all real numbers.

6. $\sin x = x - \dfrac{x^3}{3!} + \dfrac{x^5}{5!} - \dfrac{x^7}{7!} + \cdots + (-1)^n \dfrac{x^{2n+1}}{(2n+1)!} + \cdots$

7. $\cos x = 1 - \dfrac{x^2}{2!} + \dfrac{x^4}{4!} - \dfrac{x^6}{6!} + \cdots + (-1)^n \dfrac{x^{2n}}{(2n)!} + \cdots$

8. $e^{-x} = 1 - x + \dfrac{x^2}{2!} - \dfrac{x^3}{3!} + \dfrac{x^4}{4!} + \cdots + (-1)^n \dfrac{x^n}{n!} + \cdots$

In Exercises 9–20, find the radius of convergence of the given power series.

9. $\displaystyle\sum_{n=0}^{\infty} \dfrac{x^n}{(n+2)}$ 10. $\displaystyle\sum_{n=0}^{\infty} \dfrac{x^n}{n^2}$

11. $\displaystyle\sum_{n=0}^{\infty} \dfrac{4^n}{n!}(x-2)^n$ 12. $\displaystyle\sum_{n=0}^{\infty} \dfrac{(x-2)^n}{n3^n}$

13. $\displaystyle\sum_{n=0}^{\infty} \dfrac{3^n(x+2)^n}{n!}$ 14. $\displaystyle\sum_{n=0}^{\infty} \dfrac{3^n(x+2)^n}{n+1}$

15. $\displaystyle\sum_{n=1}^{\infty} \dfrac{n!}{n^4}(x-3)^n$ 16. $\displaystyle\sum_{n=0}^{\infty} (-1)^n \dfrac{x^{2n+1}}{2n+1}$

17. $x - \dfrac{x^2}{4} + \dfrac{x^3}{9} - \dfrac{x^4}{16} + \dfrac{x^5}{25} - \cdots$ 18. $1 + 2x + \dfrac{4x^2}{2!} + \dfrac{8x^3}{3!} + \dfrac{16x^4}{4!} + \dfrac{32x^5}{5!} + \cdots$

19. $\dfrac{x}{4} + \dfrac{x^2}{4^2} + \dfrac{x^3}{4^3} + \dfrac{x^4}{4^4} + \dfrac{x^5}{4^5} + \cdots$ 20. $\dfrac{x}{3} + \dfrac{2x^2}{5} + \dfrac{3x^3}{7} + \dfrac{4x^4}{9} + \dfrac{5x^5}{11} + \cdots$

In Exercises 21–24, find the Maclaurin series for f and determine its radius of convergence.

21. $f(x) = \dfrac{1}{(1+x)^2}$ 22. $f(x) = \dfrac{1}{1-2x}$

23. $f(x) = \dfrac{1}{4+x^2}$ 24. $f(x) = xe^x$

25.   The Taylor series about $x = 1$ for a certain function f converges for all x in the interval of convergence. The nth derivative of f at $x = 1$ is given by $f^{(n)}(1) = 3^n$ and $f(1) = 1$.

     a)   Write the third-degree Taylor polynomial for f about $x = 1$.

     b)   Find the radius of convergence for the Taylor series for f about $x = 1$.

26.   The Taylor series about $x = 0$ for a certain function f converges for all x in the interval of convergence. The nth derivative of f at $x = 0$ is given by $f^{(n)}(0) = \dfrac{n \cdot n!}{3^n}$ and $f(0) = 0$.

     a)   Write the third degree Taylor polynomial for f about $x = 0$.

     b)   Find the radius of convergence for the Taylor series for f about $x = 0$.

27.   The Taylor series about $x = 6$ for a certain function f converges for all x in the interval of convergence. The nth derivative of f at $x = 6$ is given by $f^{(n)}(6) = (-1)^n \cdot 4^n \cdot \sqrt{n} \cdot n!$ and $f(6) = 1$.

     a)   Write the third degree Taylor polynomial for f about $x = 6$.

     b)   Find the radius of convergence for the Taylor series for f about $x = 6$.

28.   Let f be the function defined by $f(x) = \dfrac{1}{x}$.

     a)   Find $f(1) = 1$ and $f^{(n)}(1)$ for $n = 1$ to $n = 3$, where $f^{(n)}$ is the nth derivative of f.

     b)   Write the first four nonzero terms and the general term for the Taylor series expansion of $f(x)$ about $x = 1$.

     c)   Determine the radius of convergence for the series in part b). Show your reasoning.

11.4   Representing Functions with Power Series

Finding a Taylor series for a function means finding coefficients which in turn means computing derivatives. This is how we derived the Taylor series for e^x, $\sin x$ and $\cos x$. However, for many functions, computing Taylor series coefficients by differentiation can be difficult and tedious. The goal of this section is to employ several methods including substitution, differentiation and antidifferentiation (integration) to create new series from series we already know.

Geometric Power Series

The series

$$a + ar + ar^2 + ar^3 + \cdots + ar^n + \cdots$$

converges to $\dfrac{a}{1-r}$ if $|r| < 1$. Using this formula and a little algebraic ingenuity we can produce many other useful series.

Example 1: *Finding a Geometric Power Series*

Find a Maclaurin series for the function $f(x) = \dfrac{1}{1+x}$ and its radius of convergence.

Solution:

Writing $f(x)$ in the form $\dfrac{a}{1-r}$ produces $\dfrac{1}{1+x} = \dfrac{1}{1-(-x)}$.

We recognize this as the sum of a geometric series with first term $a = 1$ and ratio $r = -x$.

Thus, the power series for *f* at $x = 0$ is

$$1 - x + x^2 - x^3 + x^4 + \cdots + (-1)^n x^n + \cdots.$$

The series converges for $|-x| < 1$ or $|x| < 1$ which implies the radius of convergence is $R = 1$.

Example 2: *Finding a Geometric Power Series Centered at $x = 1$*

Find a power series centered at $x = 1$ for the function $f(x) = \dfrac{5}{3-x}$. What is its radius of convergence?

Solution:

As in Example 1, we put $\dfrac{5}{3-x}$ in the form $\dfrac{a}{1-r}$, but this time we want the ratio *r* in terms of $x - 1$ rather than *x*. We have

$$\frac{5}{3-x} = \frac{5}{3-[(x-1)+1]} = \frac{5}{2-(x-1)}$$

$$= \frac{1}{2} \cdot \frac{5}{1 - \dfrac{x-1}{2}} = \frac{\dfrac{5}{2}}{1 - \dfrac{x-1}{2}}.$$

This is the sum of a geometric series with first term $a = \dfrac{5}{2}$ and ratio $r = \dfrac{x-1}{2}$. The series is

$$\frac{5}{2} + \frac{5}{2^2}(x-1) + \frac{5}{2^3}(x-1)^2 + \cdots + \frac{5}{2^{n+1}}(x-1)^n + \cdots.$$

The series converges for $\left|\dfrac{x-1}{2}\right| < 1$ or $|x-1| < 2$. Thus, the radius of convergence is $R = 2$.

New Series by Substitution

Example 3: *Finding a Power Series by Substitution*

Find a Taylor series representation for the function $f(x) = e^{-x^2}$ by substitution.

Solution:

Starting with the series $e^x = 1 + x + \dfrac{x^2}{2!} + \dfrac{x^3}{3!} + \dfrac{x^4}{4!} + \dfrac{x^5}{5!} + \cdots$

and substituting $-x^2$ for x we obtain

$$e^{-x^2} = 1 + (-x^2) + \frac{(-x^2)^2}{2!} + \frac{(-x^2)^3}{3!} + \frac{(-x^2)^4}{4!} + \cdots, \quad \text{for all } x.$$

Simplifying shows that the Taylor series for e^{-x^2} is

$$e^{-x^2} = 1 - x^2 + \frac{x^4}{2!} - \frac{x^6}{3!} + \frac{x^8}{4!} - \cdots, \quad \text{for all } x.$$

A straightforward computation of coefficients by calculation of derivatives would give the same result, but with a great deal more work.

Example 4: *Finding a Power Series by Substitution*

Find a Taylor series that represents $f(x) = \cos\sqrt{x}$.

Solution:

Starting with the series $\cos x = 1 - \dfrac{x^2}{2!} + \dfrac{x^4}{4!} - \dfrac{x^6}{6!} + \dfrac{x^8}{8!} - \cdots$

and substituting $\sqrt{x}$ for x we obtain the (alternating) series

$$\cos\sqrt{x} = 1 - \frac{x}{2!} + \frac{x^2}{4!} - \frac{x^3}{6!} + \frac{x^4}{8!} + \cdots$$

This series converges for all x in the domain of $\cos\sqrt{x}$; that is, for $x \geq 0$.

New Series by Differentiation and Integration

If a power series is used to define a function f, so that

$$f(x) = \sum_{n=0}^{\infty} a_n (x - c)^n.$$

then the domain of f is the interval of convergence for the power series. A function defined by power series in this way is continuous on the interval of convergence. In fact, the function f can be differentiated and antidifferentiated term-by-term as any polynomial can be.

If the power series $\sum a_n(x-c)^n$ has radius of convergence $R > 0$, then the function f defined by

$$f(x) = \sum_{n=0}^{\infty} a_n(x-c)^n,$$

is differentiable on the interval $(c - R, c + R)$ with

(a) $f'(x) = a_1 + 2a_2(x-c) + 3a_3(x-c)^2 + \cdots = \sum_{n=1}^{\infty} na_n(x-c)^{n-1}$ and

(b) $\int f(x)\,dx = C + a_0(x-c) + a_1\dfrac{(x-c)^2}{2} + a_2\dfrac{(x-c)^3}{3} + \cdots \;\; = C + \sum_{n=0}^{\infty}\dfrac{a_n(x-c)^{n+1}}{n+1}.$

The series obtained by differentiating in (a) or integrating in (b) has the same radius of convergence as the original series $\sum a_n(x-c)^n$.

Example 5:   *Finding a Power Series by Differentiation*

Find a power series representation for $\dfrac{1}{(1-x)^2}$ by differentiation. What is its radius of convergence?

Solution:

We start with the geometric series

$$\frac{1}{1-x} = 1 + x + x^2 + x^3 + x^4 + \cdots = \sum_{n=0}^{\infty} x^n,$$

then differentiate both sides to obtain

$$\frac{1}{(1-x)^2} = 1 + 2x + 3x^2 + 4x^3 + \cdots = \sum_{n=1}^{\infty} nx^{n-1}.$$

The radius of convergence of the differentiated series is the same as the radius of convergence of the original series, namely $R = 1$.

Caution: Although the radius of convergence remains the same when a power series is differentiated or integrated, this does not mean that interval of convergence remains the same. It may happen that the original series converges at an endpoint, whereas the differentiated series diverges there.

Term-by-term integration of a power series is illustrated in the following examples.

Example 6:   *Finding a Power Series by Integration*

By integrating an appropriate geometric series find a power series representation for $f(x) = \ln(1 + x)$ and its radius of convergence.

Solution:

We notice that $f'(x) = \dfrac{1}{1+x}$ and find the required series by integrating the power series for $\dfrac{1}{1+x}$ found in Example 1.

$$\ln(1+x) = \int \frac{1}{1+x}\, dx = C + x - \frac{x^2}{2} + \frac{x^3}{3} - \frac{x^4}{4} + \cdots = \sum_{n=0}^{\infty} \frac{(-1)^n x^{n+1}}{n+1} + C$$

The constant of integration C is zero, since when $x = 0$ the equality becomes $0 = 0 + C$, so we have

$$\ln(1+x) = x - \frac{x^2}{2} + \frac{x^3}{3} - \frac{x^4}{4} + \cdots = \sum_{n=0}^{\infty} \frac{(-1)^n x^{n+1}}{n+1}.$$

The radius of convergence is the same as for the original series, $R = 1$.

Example 7:   *Finding a Power Series by Integration*

By integrating an appropriate series find a power series representation for $f(x) = \arctan x$

Solution:

Now $\dfrac{d}{dx}(\arctan x) = \dfrac{1}{1+x^2}$. We can represent $\dfrac{1}{1+x^2}$ as the sum of a geometric series with initial term $a = 1$ and ratio $r = -x^2$.

$$\frac{1}{1+x^2} = 1 - x^2 + x^4 - x^6 + \cdots + (-1)^n x^{2n} + \cdots \qquad \text{for } -1 < x < 1.$$

Integrating both sides we obtain

$$\arctan x = C + \int \frac{1}{1+x^2}\, dx$$

$$= C + x - \frac{x^3}{3} + \frac{x^5}{5} - \frac{x^7}{7} + \cdots + (-1)^n \frac{x^{2n+1}}{2n+1} + \cdots$$

for some constant C. Putting $x = 0$ we see that $C = \arctan 0 = 0$, so

$$\arctan x = x - \frac{x^3}{3} + \frac{x^5}{5} - \frac{x^7}{7} + \cdots + (-1)^n \frac{x^{2n+1}}{2n+1} + \cdots \qquad \text{for } -1 < x < 1$$

Since the radius of convergence of series for $\dfrac{1}{1+x^2}$ is 1, the radius of convergence of this series for $f(x) = \arctan x$ is also 1.

If we try to compute the series by differentiating repeatedly, we appreciate the quick way we obtained this series representation of the function $f(x) = \arctan x$.

Since a function f defined by a power series has a derivative f' that is also represented by a power series, we can continue to differentiate for higher order derivatives f'', f''', $f^{(4)}$, and so on indefinitely. In other words, f is infinitely differentiable, and all derivatives have the same radius of convergence.

Combining Power Series

On the intersection of their intervals of convergence power series can be added, subtracted, multiplied by constants and powers of x, and the results are once again power series. In other words, power series for functions f and g can be combined to produce power series for

$f \pm g$, $c \cdot f$, and $x^n \cdot f$

Example 8: *Adding Two Power Series*

Find a power series, centered at $x = 0$, for $f(x) = \dfrac{3 - 4x}{2x^2 - 3x + 1}$

Solution:

Using partial fractions, we can obtain the decomposition (without showing steps):

$$\frac{3 - 4x}{2x^2 - 3x + 1} = \frac{1}{1 - x} + \frac{2}{1 - 2x}$$

Both of the terms on the right side of equation above can be expressed as the sum of a geometric series.

$$\frac{1}{1 - x} = 1 + x + x^2 + x^3 + x^4 + \cdots = \sum_{n=0}^{\infty} x^n \qquad \text{for } -1 < x < 1$$

$$\frac{2}{1 - 2x} = 2 + 4x + 8x^2 + 16x^3 + 32x^4 + \cdots = 2\sum_{n=0}^{\infty} (2x)^n \qquad \text{for } -\frac{1}{2} < x < \frac{1}{2}$$

Because we are dealing with two power series, we need to be careful about finding the interval of convergence. It is the intersection of the individual intervals of convergence, so we note that

$$(-1,\ 1) \cap \left(-\frac{1}{2},\ \frac{1}{2} \right) = \left(-\frac{1}{2},\ \frac{1}{2} \right).$$

Thus,

$$\frac{3 - 4x}{2x^2 - 3x + 1} = \sum_{n=0}^{\infty} x^n + 2\sum_{n=0}^{\infty} (2x)^n = \sum_{n=0}^{\infty} \left[x^n + 2(2x)^n \right] = \sum_{n=0}^{\infty} \left(1 + 2^{n+1} \right) x^n$$

$$= 3 + 5x + 9x^2 + 17x^3 + 33x^4 + \cdots \quad \text{converges on } \left(-\frac{1}{2},\ \frac{1}{2} \right).$$

Example 9: *Multiplying a Power Series by x*

Find a power series, centered at $x = 0$, for $f(x) = xe^{-2x}$

Solution:

First we substitute $-2x$ in the power series for e^x.

$$e^{-2x} = 1 + (-2x) + \frac{(-2x)^2}{2!} - \frac{(-2x)^3}{3!} + \cdots + \frac{(-2x)^n}{n!} + \cdots,$$

or $e^{-2x} = 1 - 2x + 2^2 \dfrac{x^2}{2!} - 2^3 \dfrac{x^3}{3!} + \cdots + (-2)^n \dfrac{x^n}{n!} + \cdots$

which can be written as

$$e^{-2x} = \sum_{n=0}^{\infty} (-2)^n \frac{x^n}{n!}.$$

Multiplying both sides by x gives us

$$xe^{-2x} = x - 2x^2 + 2^2 \frac{x^3}{2!} + 2^3 \frac{x^4}{3!} + \cdots + (-2)^n \frac{x^{n+1}}{n!} + \cdots$$

$$xe^{-2x} = \sum_{n=0}^{\infty} (-2)^n \frac{x^{n+1}}{n!} \qquad \text{for all } x.$$

For easy reference we display a short list of power series for basic calculus functions.

Function	Series	Radius of Convergence
e^x	$1 + x + \dfrac{x^2}{2!} + \dfrac{x^3}{3!} + \dfrac{x^4}{4!} + \dfrac{x^5}{5!} + \cdots$	$R = \infty$
$\sin x$	$x - \dfrac{x^3}{3!} + \dfrac{x^5}{5!} - \dfrac{x^7}{7!} + \dfrac{x^9}{9!} - \cdots$	$R = \infty$
$\cos x$	$1 - \dfrac{x^2}{2!} + \dfrac{x^4}{4!} - \dfrac{x^6}{6!} + \dfrac{x^8}{8!} - \cdots$	$R = \infty$
$\dfrac{1}{1-x}$	$1 + x + x^2 + x^3 + x^4 + \cdots$	$R = 1$
$\arctan x$	$x - \dfrac{x^3}{3} + \dfrac{x^5}{5} - \dfrac{x^7}{7} + \dfrac{x^9}{9} - \cdots$	$R = 1$

11.4   Exercises

In Exercises 1–8, find the power series representation for f centered about $x = 0$ and specify its radius of convergence. Each is somehow related to a geometric series (see Examples 1 and 2).

1. $f(x) = \dfrac{2}{1+x}$ 2. $f(x) = \dfrac{1}{2+x}$

3. $f(x) = \dfrac{1}{1-x^2}$ 4. $f(x) = \dfrac{1}{3+2x}$

5. $f(x) = \dfrac{x^2}{1+x}$ 6. $f(x) = \dfrac{2}{5-x}$

7. $f(x) = \displaystyle\int_0^x \dfrac{t}{1-t}\, dt$ 8. $f(x) = \displaystyle\int_0^x \dfrac{2}{1-t^2}\, dt$

In Exercises 9–14, use the power series representation of e^x to produce a power series representation of the function f.

9. $f(x) = e^{2x}$ 10. $f(x) = e^{\frac{x}{2}}$

11. $f(x) = 1 - e^{-x}$ 12. $f(x) = \dfrac{1-e^{-x}}{x}$

13. $f(x) = \dfrac{e^{x^2} - 1}{x^2}$ 14. $f(x) = xe^{-x}$

In Exercises 15–22, find a power series representation of the function f centered at $x = c$. Specify the radius of convergence of the power series.

15. $f(x) = \sin x, \;\; c = \dfrac{\pi}{2}$ 16. $f(x) = \dfrac{3}{1-3x}, \;\; c = 0$

17. $f(x) = e^{-x}, \;\; c = -2$ 18. $f(x) = \ln x, \;\; c = 1$

19. $f(x) = \dfrac{1}{1-x} + \dfrac{1}{1+x}, \;\; c = 0$ 20. $f(x) = \dfrac{\sin x}{x}, \;\; c = 0$

21. $f(x) = \sqrt{1+x}, \;\; c = 1$ (first three nonzero terms)

22. $f(x) = \tan x, \;\; c = 0$ (first two nonzero terms)

23.   By differentiating the geometric series

$$\dfrac{1}{1-x} = 1 + x + x^2 + x^3 + x^4 + \cdots \qquad |x| < 1$$

find a power series that represents $\dfrac{1}{(1-x)^2}$. What is its radius of convergence?

24. Use the Maclaurin series for e^x and e^{-x} to find the Maclaurin series for hyperbolic cosine function

$$\cosh x = \frac{e^x + e^{-x}}{2}$$

25. Use the Maclaurin series for e^x and e^{-x} to find the Maclaurin series for hyperbolic sine function

$$\sinh x = \frac{e^x - e^{-x}}{2}.$$

26. Use an appropriate identity to find the Maclaurin series for $f(x) = \sin x \cos x$.

27. Find the Maclaurin series for $g(x) = e^x - 1$ and use this series to find

$$\lim_{x \to 0} \frac{e^x - 1}{x}.$$

28. Find a series expansion at $c = 1$ for the function $f(x) = \frac{1}{x}$ by each of the following methods.

a) Differentiating f repeatedly to compute coefficients.

b) Using the identity $\dfrac{1}{x} = \dfrac{1}{1 - [-(x-1)]}$ and expanding in a geometric series.

29. Use power series and term-by-term differentiation to confirm that

$$\frac{d}{dx} \sin(x^2) = \cos(x^2) \cdot 2x$$

30. Given the function f defined by $f(x) = \sqrt{4+x}$.

a) Find the first three nonzero terms in the Maclaurin series for the function f.

b) Use the results found in part a) to find the first three terms in Maclaurin series for the function g defined by $g(x) = \sqrt{4 + x^3}$.

c) Find the first four nonzero terms in the Maclaurin series for the function h. such that $h'(x) = \sqrt{4 + x^3}$ and $h(0) = 2$.

31. a) Find the first four terms in the Maclaurin series for $f(x) = \dfrac{1}{1 - 3x}$.

b) Find the interval of convergence for the series in part a).

c) Use partial fractions and the result from part a) to find the Maclaurin series for

$$g(x) = \frac{1}{(1 - 3x)(1 - x)}.$$

11.5   Testing Convergence at Endpoints

The Ratio Test establishes the radius of convergence for a power series; however, there remains the question of convergence at the endpoints of the convergence interval when the radius of convergence is a finite, nonzero number. The goal of this section is to develop tests for convergence of series of constants that can be used at the endpoints of the intervals of convergence of a power series.

Convergence Criterion for a Bounded Monotone Sequence

There is a major property of sequences that supports several methods of testing for convergence. Given a series $\sum a_n$ whose terms are positive, $a_n > 0$, it follows that the corresponding sequence of partial sums $\{S_n\}$ is increasing. If the increasing sequence of partial sums $\{S_n\}$ is also bounded above, $S_n \leq M$ for some real number M, then the sequence of partial sums is convergent. In general

> **Convergence Criterion for a Bounded Monotonic Sequence**
>
> If a sequence $\{a_n\}$ is bounded and monotonic then it converges to some finite limit L.

By "monotonic" we mean either $a_{n+1} > a_n$ for all n (monotone increasing) or $a_{n+1} < a_n$ for all n (monotone decreasing). In other words, a sequence is montonic if it is either increasing or decreasing. The following figure illustrates a bounded monotone increasing sequence $\{a_n\}$ and suggests there must be a limit.

Example 1:   *Bounded and Monotonic Sequences*

a)   The sequence $a_n = \dfrac{n}{n+1}$ is both bounded and increasing and thus must converge.

b)   The divergent sequence $b_n = \dfrac{n^2}{n+1}$ is increasing but not bounded above.

c)   The divergent sequence $c_n = (-1)^n$ is bounded but not increasing.

In view of this convergence criterion, deciding whether a series with nonnegative terms is convergent or divergent means we need to determine whether the increasing sequence of partial sums $\{S_n\}$ is bounded from above.

The Integral Test

A powerful tool in determining whether or not a series of nonnegative terms converges or diverges is the **Integral Test**. This test, which uses a definite integral as an upper bound for a monotone increasing sequence, relates the convergence of a series to that of an improper integral.

The Integral Test

Let f be a decreasing continuous function that has positive values $f(x)$ for all $x \geq 1$.

The infinite series $\displaystyle\sum_{n=1}^{\infty} f(n)$

a) converges if the improper integral $\displaystyle\int_{1}^{\infty} f(x)\,dx$ converges;

b) diverges if the improper integral $\displaystyle\int_{1}^{\infty} f(x)\,dx$ diverges.

Drawing rectangles to approximate the area under the curve provides a geometric way of seeing that the series $\displaystyle\sum_{n=1}^{\infty} f(n)$ and the improper integral $\displaystyle\int_{1}^{\infty} f(x)\,dx$ either both converge or both diverge.

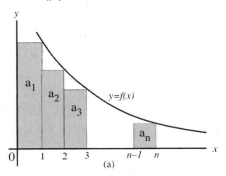
(a)

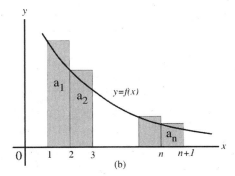
(b)

Because $f(x)$ is positive, the integral $\displaystyle\int_{1}^{n} f(x)\,dx$ is equal to the area of the region under the graph of f for $1 \leq x < n$. In Figure (a) above, the total area of the inscribed rectangles is less than the area under the curve from $x = 1$ to $x = n$, so

$$a_2 + a_3 + a_4 + \cdots + a_n < \int_{1}^{n} f(x)\,dx$$

$$a_1 + a_2 + a_3 + a_4 + \cdots + a_n < a_1 + \int_{1}^{n} f(x)\,dx \qquad \text{(Adding } a_1 \text{ to both sides)}$$

In Figure (b), the total area of the circumscribed rectangles is greater than the area under the curve from $x = 1$ to $x = n+1$, so

$$\int_{1}^{n+1} f(x)\,dx < a_1 + a_2 + a_3 + a_4 + \cdots + a_n$$

If we let $S_n = a_1 + a_2 + a_3 + a_4 + \cdots + a_n$ be the nth partial sum of the series $\displaystyle\sum_{k=1}^{\infty} a_k$, then combining inequalities we obtain

$$\int_{1}^{n+1} f(x)\,dx \;<\; S_n < a_1 + \int_{1}^{n} f(x)\,dx$$

If the improper integral $\int_1^{\infty} f(x)\, dx$ converges to a limit L, then from the right-hand inequality above

$$S_n < a_1 + \int_1^n f(x)\, dx < a_1 + \int_1^{\infty} f(x)\, dx = a_1 + L$$

Thus, each partial sum is less than $a_1 + L$, and by the convergence criterion for a bounded monotone sequence, the series converges. On the other hand, if the improper integral diverges, the left side of the double inequality above guarantees that the series diverges.

An important series is $\sum_{n=1}^{\infty} \frac{1}{n}$, which is called the **harmonic series**. In the next example, the Integral Test is used to show that this series diverges.

Example 2:   *Applying the Integral Test*

Use the Integral Test to show that the harmonic series $\sum_{n=1}^{\infty} \frac{1}{n}$ diverges.

Solution:

Because the function f defined by $f(x) = \frac{1}{x}$ is positive, continuous and decreasing for $x \geq 1$, the conditions of the Integral Test are satisfied.

$$\int_1^{\infty} \frac{1}{x}\, dx = \lim_{b \to \infty} \int_1^b \frac{1}{x}\, dx = \lim_{b \to \infty} \left[\ln b - \ln 1 \right] = \infty.$$

The integral diverges so the harmonic series diverges.

Convergent and Divergent p-series

The harmonic series $\sum \frac{1}{n}$ is a special case of a class of series called **p-series**. A *p*-series is an infinite series of the form

$$\sum_{n=1}^{\infty} \frac{1}{n^p} = 1 + \frac{1}{2^p} + \frac{1}{3^p} + \cdots + \frac{1}{n^p} + \cdots$$

where $p > 0$. If we define $f(x) = \frac{1}{x^p}$, then f meets the criteria of the Integral Test, and

$$\int_1^t \frac{1}{x^p} = \begin{cases} \ln t & \text{if } p = 1 \\ \dfrac{t^{(-p+1)} - 1}{-p + 1} & \text{if } p \neq 1 \end{cases}$$

Now let $t \to \infty$. If $p = 1$, $\ln t \to \infty$; if $p < 1$, then $-p + 1 > 0$ and $t^{(-p+1)} \to \infty$. In either case, the integral diverges. But, if $p > 1$, then $-p + 1 < 0$ and $t^{(-p+1)} \to 0$; in this case, the integral converges. Thus we have

---

p-series Test

The *p*-series $\sum_{n=1}^{\infty} \frac{1}{n^p}$ is convergent if $p > 1$ and divergent if $p \leq 1$.

Example 3:   *Applying the p-series Test*

Determine whether the following series converge or diverge.

a) $\displaystyle\sum_{n=1}^{\infty} \frac{1}{\sqrt{n}}$
    b) $\displaystyle\sum_{n=1}^{\infty} \frac{1}{\sqrt{n^3}}$

Solution:

a) Here $\sqrt{n} = n^{1/2}$, $p = \dfrac{1}{2} \le 1$ and the series diverges.

b) Here $\sqrt{n^3} = n^{3/2}$, $p = \dfrac{3}{2} > 1$ and the series converges.

Alternating Series

A series in which terms are alternately positive and negative is an **alternating series**. Here are two examples.

$$\sum_{n=1}^{\infty} (-1)^{n+1} \frac{1}{n} = 1 - \frac{1}{2} + \frac{1}{3} - \frac{1}{4} + \frac{1}{5} - \cdots \qquad \text{(First term is positive.)}$$

$$\sum_{n=1}^{\infty} (-1)^{n} \frac{1}{n!} = -1 + \frac{1}{2!} - \frac{1}{3!} + \frac{1}{4!} - \frac{1}{5!} + \cdots \qquad \text{(First term is negative.)}$$

In general, just knowing that $\lim\limits_{n \to \infty} a_n = 0$ tells us very little about the convergence of the series $\sum a_n$; however, it turns out that an alternating series must converge if its terms consistently shrink in size and approach zero.

Alternating Series Test

If $a_n > 0$, then an alternating series

$$\sum_{n=1}^{\infty} (-1)^{n} a_n \quad \text{or} \quad \sum_{n=1}^{\infty} (-1)^{n+1} a_n$$

converges if both of the following conditions are satisfied:

1) $\lim\limits_{n \to \infty} a_n = 0$;

2) $\{a_n\}$ is a decreasing sequence; that is, $a_{n+1} < a_n$ for all n.

Example 4:   *Using the Alternating Series Test*

Is the alternating harmonic series

$$\sum_{n=1}^{\infty} (-1)^{n+1} \frac{1}{n} = 1 - \frac{1}{2} + \frac{1}{3} - \frac{1}{4} + \frac{1}{5} - \cdots$$

convergent or divergent?

Solution:

Let $a_n = \dfrac{1}{n}$. The series satisfies

1) $\displaystyle\lim_{n\to\infty} a_n = \lim_{n\to\infty} \dfrac{1}{n} = 0$;

2) $a_{n+1} < a_n$ because $\dfrac{1}{n+1} < \dfrac{1}{n}$.

Thus, the series is convergent by the Alternating Series Test.

A formal proof of the Alternating Series Test is not very enlightening, but the following figure gives a picture of the idea behind the proof. Let $\{S_n\}$ be the sequence of partial sums for the alternating series. We first plot $S_1 = a_1$ on a number line. To find S_2 we subtract a_2, so S_2 is to the left of S_1. Then to find S_3 add a_3. so S_3 is to the right of S_2. But since $a_3 < a_2$, S_3 is to the left of S_1. Continuing in this fashion, we see that the partial sums oscillate back and forth. Since $a_n \to 0$, the successive steps are becoming smaller and smaller. The even partial sums $S_2, S_4, S_6, \cdots$ are increasing and the odd partial sums $S_1, S_3, S_5, \cdots$ are decreasing. Thus, it seems plausible that both are converging on some number S, which is the sum of the series.

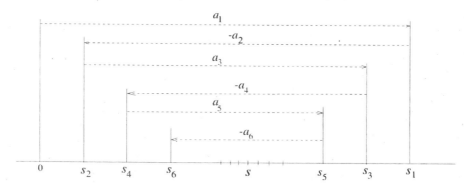

If a series converges, then the nth partial sum S_n can be used to approximate the sum S of the series. In many case it is difficult to determine the accuracy of the approximation. However, for an alternating series, there is a simple way of estimating the error involved.

Error Estimate for Alternating Series

Suppose an alternating series $\displaystyle\sum_{n=1}^{\infty} a_n$ satisfies the conditions of the Alternating Series Test; namely,

$$\lim_{n\to\infty} a_n = 0 \quad \text{and} \quad \{a_n\} \text{ is a decreasing sequence } (a_{n+1} < a_n).$$

If the series has sum S, then

$$|S - S_n| < a_{n+1}$$

where S_n is the nth partial sum of the series.

In other words, if an alternating series satisfies the conditions of the Alternating Series Test, you can approximate the sum of the series by using the nth partial sum, S_n, and your error will have an absolute value no greater than the first term left off (namely, a_{n+1}).

Example 5:   *Error Estimates for an Alternating Series*

Given the convergent alternating harmonic series

$$\sum_{n=1}^{\infty} (-1)^{n+1} \frac{1}{n} = 1 - \frac{1}{2} + \frac{1}{3} - \frac{1}{4} + \frac{1}{5} - \cdots \quad .$$

a) Estimate the sum of the series by taking the sum of the first four terms. How accurate is this estimate?

b) How many terms of this series would we need to take in order to get a partial sum S_n within 0.005 (two decimal places) of the sum S.

Solution:

a) Let $a_n = \frac{1}{n}$ and let S denote the actual sum of the series. The error estimate tells us that

$$\left| S - S_4 \right| \le a_5$$

where S_4 is the sum of the first four terms of the series. Using a calculator, we find

$$S_4 = 1 - \frac{1}{2} + \frac{1}{3} - \frac{1}{4} \approx 0.583333$$

and $a_5 = \frac{1}{5} = 0.2$. Thus if we estimate S by $S_4 \approx 0.583333$, we incur an error of 0.2, which means that

$$\left| S - S_4 \right| \le 0.2$$
$$0.583333 - 0.2 \le S \le 0.583333 + 0.2$$
$$0.383333 \le S \le 0.783333$$

b) We want $\left| S - S_n \right| \le 0.005$, and this will hold if $a_{n+1} \le 0.005$. Because $a_{n+1} = \frac{1}{n+1}$, we require $\frac{1}{n+1} \le 0.005$, which is satisfied if $n \ge 199$. Thus, we need 199 terms to get two decimal places of accuracy. This gives you an idea of how slowly the alternating harmonic series converges. The sum of the first 199 terms done on a calculator would be

$$\sum_{n=1}^{199} (-1)^{n+1} \frac{1}{n} = 0.695653$$

Absolute and Conditional Convergence

When a series has both positive and negative terms, as in the case with alternating series, the notions of **absolute convergence** and **conditional convergence** become useful.

The alternating harmonic series $\sum (-1)^{n+1} \frac{1}{n}$ and the series $\sum (-1)^{n+1} \frac{1}{n^2}$ both converge. But there is a difference.

$$\sum \left| (-1)^{n+1} \frac{1}{n} \right| = \sum \frac{1}{n} \text{ diverges} \quad \text{while} \quad \sum \left| (-1)^{n+1} \frac{1}{n^2} \right| = \sum \frac{1}{n^2} \text{ converges.}$$

Without absolute values, the alternating harmonic series converges because the negative terms subtract off enough to make the sum finite; this series is said to **converge conditionally**. The second series converges even if the minus signs are replaced by plus signs; this series is said to **converge absolutely**.

In general, if the series $\sum a_n$ converges, then $\sum |a_n|$ may converge or diverge. The two cases that can occur are given the following special names.

Absolutely Convergent Series

The series $\sum a_n$ is called **absolutely convergent** if the series of absolute values $\sum |a_n|$ converges.

Conditionally Convergent Series

The series $\sum a_n$ is called **conditionally convergent** if it converges but the series of absolute values $\sum |a_n|$ diverges

---

The Integral and the p-series convergence tests that we have developed in this section cannot be applied to a series that has mixed terms or that does not strictly alternate. In such cases, it is often useful to apply the following result.

Absolute Convergence Implies Convergence

If the series $\sum |a_n|$ converges, then $\sum a_n$ converges.

Any series of constants can be classified as (1) absolutely convergent, (2) conditionally convergent, or (3) divergent. To classify a given series $\sum a_n$, first determine if the series is absolutely convergent by testing $\sum |a_n|$. If this series diverges, then test $\sum a_n$ to see if it is conditionally convergent or divergent.

Example 6: *Absolute and Conditional Convergence*

Determine whether each of the following series of constants is convergent or divergent. Classify any convergent series as absolutely or conditionally convergent.

a) $\displaystyle\sum_{n=1}^{\infty} (-1)^n \frac{2n^2}{n^2 + 10n} = -\frac{2}{11} + \frac{1}{3} - \frac{6}{13} + \frac{4}{7} - \frac{2}{3} + \cdots$

b) $\displaystyle\sum_{n=1}^{\infty} (-1)^n \frac{1}{\sqrt{n+1}} = -\frac{1}{\sqrt{2}} + \frac{1}{\sqrt{3}} - \frac{1}{\sqrt{4}} + \frac{1}{\sqrt{5}} + \cdots$

c) $\displaystyle\sum_{n=1}^{\infty} \frac{(-1)^{n-1}}{n^4} = 1 - \frac{1}{2^4} + \frac{1}{3^4} - \frac{1}{4^4} + \cdots$

Solution:

a) The series alternates. However, $\displaystyle\lim_{n \to \infty} \frac{2n^2}{n^2 + 10n} = 2 \neq 0$. Since $\displaystyle\lim_{n \to \infty} a_n \neq 0$, the series diverges by the nth-Term Test for Divergence

b) The given series can be shown to be convergent by the Alternating Series Test. Moreover, it is easy to show that the corresponding series of positive terms $\sum_{n=1}^{\infty} \frac{1}{\sqrt{n+1}}$ diverges by the Integral Test. Thus, the original alternating series is conditionally convergent.

c) This series converges by the Alternating Series Test. In fact, it is absolutely convergent because $\left|\frac{(-1)^{n-1}}{n^4}\right| = \frac{1}{n^4}$, and $\sum_{n=1}^{\infty} \frac{1}{n^4}$ is a convergent p-series ($p = 4 > 1$).

Summary of Convergence and Divergence Test for a Series of Constants

We now have a several tests that can be used to investigate a series of constants for convergence or divergence. The following summary may be helpful in deciding which test to apply.

Test	Series	Convergence or Divergence	Comments
nth-term	$\sum a_n$	Diverges if $\lim_{n\to\infty} a_n \neq 0$	Inconclusive if $\lim_{n\to\infty} a_n = 0$
Geometric series	$\sum_{n=1}^{\infty} ar^{n-1}$	(i) Converges with sum $S = \frac{a}{1-r}$ if $|r| < 1$ (ii) Diverges if $|r| \geq 1$	
Integral	$\sum_{n=1}^{\infty} a_n$ $a_n = f(n)$	(i) Converges if $\int_1^{\infty} f(x)\,dx$ converges (ii) Diverges if $\int_1^{\infty} f(x)\,dx$ Diverges	The function f obtained from $a_n = f(n)$ must be continuous, positive, and decreasing.
p-series	$\sum_{n=1}^{\infty} \frac{1}{n^p}$	(i) Converges if $p > 1$ (ii) Diverges if $p \leq 1$	
Alternating series	$\sum (-1)^n a_n$ $a_n > 0$	Converges if $a_n > a_{n+1}$ for every n and $\lim_{n\to\infty} a_n = 0$	Applicable only to an alternating series
$\sum |a_n|$	$\sum a_n$	If $\sum |a_n|$ converges, then $\sum a_n$ converges.	Useful for series that contain both positive and negative terms.

Employing these tests we can now determine the convergence of power series at the endpoints of the interval of convergence, where the Ratio Test is inconclusive.

Example 7: *Finding the Complete Interval of Convergence*

The Taylor series expansion for the function $f(x) = \ln x$ about $x = 1$ is

$$\ln x = (x-1) - \frac{1}{2}(x-1)^2 + \frac{1}{3}(x-1)^3 - \frac{1}{4}(x-1)^4 + \cdots$$

and, by the Ratio Test, we know that the series converges in the interval (0, 2). Determine whether the series converges at the endpoints $x = 0$ and $x = 2$.

Solution:

At $x = 0$, the series becomes $-1 - \dfrac{1}{2} - \dfrac{1}{3} - \dfrac{1}{4} - \cdots$, which diverges because it is the opposite of

the divergent harmonic series. At $x = 2$, the series becomes $1 - \dfrac{1}{2} + \dfrac{1}{3} - \dfrac{1}{4} + \cdots$, which

converges because it meets the criteria of the Alternating Series Test. Therefore the complete interval of convergence is the interval $(0, \ 2]$.

Our objective in this section has been to develop tests for convergence that can be employed at the endpoints of the intervals of convergence for power series. There are three possibilities at each endpoint. The series could diverge, it could converge absolutely, or it could converge conditionally.

Testing a Power Series for Convergence

Here are the steps to be followed in determining the interval of convergence for a power series $\sum a_n (x - c)^n$.

1) Use the Ratio Test to find the x-values for which the series converges absolutely. This is usually an open interval $(c - R, c + R)$ where R is the radius of convergence. In some instances, the series converges for all values of x. In some rare cases, the series only converges at $x = c$.

2) If the interval of convergence is finite, test for convergence at each endpoint using one or more of the convergence tests we have developed in this section.

Example 8: *Finding the Interval of Convergence*

For what values of x do the following series converge?

a) $\displaystyle\sum_{n=1}^{\infty} nx^n$
b) $\displaystyle\sum_{n=1}^{\infty} \dfrac{x^n}{n^2}$

Solution:

a) Using the Ratio Test and letting $a_n = nx^n$ and $a_{n+1} = (n+1)x^{n+1}$ we have

$$\lim_{n \to \infty} \frac{|a_{n+1}|}{|a_n|} = \lim_{n \to \infty} \left| \frac{(n+1)x^{n+1}}{nx^n} \right| = \lim_{n \to \infty} \left| \frac{x^{n+1}}{x^n} \right| \left| \frac{n+1}{n} \right| = \lim_{n \to \infty} \left| 1 + \frac{1}{n} \right| |x| = |x|$$

The series converges absolutely for $|x| < 1$; that is, for all x, $-1 < x < 1$. When $x = 1$, we

obtain the series $\displaystyle\sum_{n=1}^{\infty} n(1)^n$, which clearly diverges by the nth-Term Divergence Test.

Similarly, if $x = -1$, the series $\displaystyle\sum_{n=1}^{\infty} n(-1)^n$ diverges by the nth-Term Divergence Test.

Thus, the interval of convergence for the given series is the open interval $(-1, 1)$.

b) Using the Ratio Test and letting $a_n = \dfrac{x^n}{n^2}$ and $a_{n+1} = \dfrac{x^{n+1}}{(n+1)^2}$, we have

$$\lim_{n \to \infty} \frac{|a_{n+1}|}{|a_n|} = \lim_{n \to \infty} \left| \frac{\dfrac{x^{n+1}}{(n+1)^2}}{\dfrac{x^n}{n^2}} \right| = \lim_{n \to \infty} \left[\left| \frac{x^{n+1}}{x^n} \right| \, \left| \frac{n^2}{(n+1)^2} \right| \right] = |x| \cdot \lim_{n \to \infty} \left| \frac{n^2}{(n+1)^2} \right| = |x|$$

The series converges absolutely for $|x| < 1$; that is, for all x, $-1 < x < 1$. When $x = 1$, we obtain the convergent p-series

$$\sum_{n=1}^{\infty} \frac{1}{n^2} = \frac{1}{1^2} + \frac{1}{2^2} + \frac{1}{3^2} + \frac{1}{4^2} + \cdots \qquad\qquad \text{Converges when } x = 1$$

When $x = -1$, we obtain the convergent alternating series

$$\sum_{n=1}^{\infty} \frac{(-1)^n}{n^2} = -\frac{1}{1^2} + \frac{1}{2^2} - \frac{1}{3^2} + \frac{1}{4^2} - \cdots \qquad\qquad \text{Converges when } x = -1$$

Therefore, the interval of convergence for the given series is the closed interval $[-1,\ 1]$.

11.5   Exercises

In Exercises 1–6, apply the Integral Test to determine whether the series of constants is convergent or divergent

1. $\displaystyle\sum_{n=1}^{\infty} \frac{1}{4n^2}.$

2. $\displaystyle\sum_{n=1}^{\infty} \frac{1}{n^2 + n}$

3. $\displaystyle\sum_{n=1}^{\infty} \frac{1}{2n + 1}$

4. $\displaystyle\sum_{n=1}^{\infty} \frac{1}{\sqrt{2n + 1}}$

5. $\displaystyle\sum_{n=1}^{\infty} \frac{n}{n^2 + 1}$

6. $\displaystyle\sum_{n=1}^{\infty} \frac{\arctan n}{n^2 + 1}$

7.   In each part, determine whether the given *p*-series converges or diverges.

   a) $\displaystyle\sum_{n=1}^{\infty} \frac{1}{n^3}$

   b) $\displaystyle\sum_{n=1}^{\infty} \frac{1}{n\sqrt{n}}$

   c) $\displaystyle\sum_{n=1}^{\infty} n^{-2/3}$

   d) $\displaystyle\sum_{n=1}^{\infty} n^{-4/3}$

   e) $\displaystyle\sum_{n=1}^{\infty} \frac{1}{\sqrt[4]{n}}$

   f) $\displaystyle\sum_{n=1}^{\infty} \frac{1}{\sqrt[3]{n^5}}$

In Exercises 8–13, apply the Alternating Series Test to determine whether the series of constants converges or diverges. Discuss the absolute or conditional convergence of each series.

8. $\displaystyle\sum_{n=2}^{\infty} (-1)^n \frac{1}{2n - 3}$

9. $\displaystyle\sum_{n=0}^{\infty} (-2)^n$

10. $\displaystyle\sum_{n=1}^{\infty} (-1)^n \frac{n}{e^n}$

11. $\displaystyle\sum_{n=3}^{\infty} (-1)^{n+1} \frac{\ln n}{n}$

12. $\displaystyle\sum_{n=1}^{\infty} (-1)^n \frac{1}{\sqrt{n}}$

13. $\displaystyle\sum_{n=1}^{\infty} (-1)^{n+1} \frac{n}{n + 1}$

In Exercises 14–19, the given series satisfies the hypotheses of the Alternating Series Test.

   a) Find an upper bound for the error if the sum of the first four terms is used as an approximation to the sum of the series.

   b) Find the smallest value of *n* for which the *n*th partial sum approximates the sum of the series within 0.005 (two decimal places).

14. $\displaystyle\sum_{n=1}^{\infty} \frac{(-1)^{n+1}}{2n}$

15. $\displaystyle\sum_{n=1}^{\infty} \frac{(-1)^{n+1}}{n!}$

16. $\displaystyle\sum_{n=1}^{\infty} \left(\frac{-1}{3}\right)^{n+1}$

17. $\displaystyle\sum_{n=1}^{\infty} \frac{(-1)^{n+1}}{n^2}$

18. $\displaystyle\sum_{n=1}^{\infty} \frac{(-1)^{n+1}}{n^3}$

19. $\displaystyle\sum_{n=1}^{\infty} \left(\frac{-1}{5}\right)^n$

In Exercises 20–33, find the interval of convergence for the given power series, including convergence or divergence at the endpoints of the interval.

20. $\displaystyle\sum_{n=1}^{\infty} \frac{x^n}{n}$

21. $\displaystyle\sum_{n=1}^{\infty} \frac{x^n}{2n^2}$

22. $\displaystyle\sum_{n=1}^{\infty} \frac{n}{4^n}(x+3)^n$

23. $\displaystyle\sum_{n=1}^{\infty} \frac{(x-2)^n}{n3^n}$

24. $\displaystyle\sum_{n=0}^{\infty} \frac{(-1)^n(x+1)^n}{2^n}$

25. $\displaystyle\sum_{n=0}^{\infty} \frac{x^{2n+1}}{(2n+1)!}$

26. $\displaystyle\sum_{n=0}^{\infty} (x-5)^n$

27. $\displaystyle\sum_{n=1}^{\infty} \frac{(x+1)^n}{3^n}$

28. $\displaystyle\sum_{n=1}^{\infty} \frac{2^n x^n}{n!}$

29. $\displaystyle\sum_{n=1}^{\infty} \frac{(x+2)^n}{\sqrt{n}}$

30. $1 - x^2 + \dfrac{x^4}{2!} - \dfrac{x^6}{3!} + \dfrac{x^8}{4!} - \cdots$

31. $x - \dfrac{x^2}{2} + \dfrac{x^3}{3} - \dfrac{x^4}{4} + \cdots$

32. $1 - \dfrac{1}{2}(x-3) + \dfrac{1}{3}(x-3)^2 - \dfrac{1}{4}(x-3)^3 + \cdots$

33. $1 - \dfrac{3x}{\sqrt{2}} + \dfrac{9x^2}{\sqrt{3}} - \dfrac{27x^3}{\sqrt{4}} + \dfrac{81x^4}{\sqrt{5}} - \cdots$

34. Determine whether the following statements are True or False.

a) Every alternating series is convergent.

b) $\sum a_n$ converges conditionally if $\sum |a_n|$ diverges.

c) A convergent series with positive terms is absolutely convergent.

d) The alternating harmonic series converges conditionally.

35. Let f be the function defined by $f(x) = \displaystyle\sum_{n=1}^{\infty} \frac{\ln(n+1)}{n+1}x^n$ for all values of x for which the series converges.

a) Find the interval of convergence of this series. Justify your answer.

b) Find $f(0)$ and $f'(0)$.

36. a) Write the Maclaurin series for $f(x) = \ln(1+x)$. Include an expression for the general term.

b) For what values of x does the series in part a) converge?

c) Estimate the error in evaluating $f(x) = \ln\left(\dfrac{3}{2}\right)$ by using the first four terms of the series in part a).

11.6   Taylor's Formula with Remainder

In Section 11.1 we saw that a Taylor polynomial $p_n(x)$ approximates a given function f closely if x is near the base point $x = c$ and that the approximation improves as the degree n increases. An approximation process is only valuable if you have some idea about the size of the possible error. To measure the accuracy of approximating a function value $f(x)$ by the Taylor polynomial $p_n(x)$, we need an estimate of the difference between $f(x)$ and each of its Taylor polynomials. We start with the concept of a **remainder** $R_n(x)$, defined as follows:

$$f(x) = p_n(x) + R_n(x)$$

Thus, $R_n(x) = f(x) - p_n(x)$. The absolute value of $R_n(x)$ is called the **error** associated with the approximation $p_n(x)$. That is:

$$\text{Error} = |R_n(x)| = |f(x) - p_n(x)|.$$

Taylor's Formula with Remainder

Let f be a function whose $(n+1)$st derivative, $f^{(n+1)}(x)$, exists for each x in an open interval containing c. Then for any x in that interval ,

$$f(x) = f(c) + f'(c)(x - c) + \frac{f''(c)(x - c)^2}{2!} + \cdots + \frac{f^{(n)}(c)(x - c)^n}{n!} + R_n(x)$$

where the remainder $R_n(x)$ (or error) is given by the formula

$$R_n(x) = \frac{f^{(n+1)}(z)}{(n + 1)!}(x - c)^{(n+1)} \qquad \textbf{The Lagrange Remainder}$$

where z is some number between x and c.

This formula expresses $f(x)$ as the sum of the nth Taylor polynomial about $x = c$ plus a remainder, or error term,

$$R_n(x) = \frac{f^{(n+1)}(z)}{(n + 1)!}(x - c)^{(n+1)},$$

in which z is some unspecified number between c and x. The value of z depends on c, x, and n. Observe that the Lagrange form of the remainder, $R_n(x)$, looks just like the next term in the series, except that c in $f^{(n+1)}(c)$ has been replaced by an unknown number z between c and x.

Historically, the remainder was not due to Taylor, but rather to a French – born in Italy but considered French – mathematician Joseph Louis Lagrange (1736-1813). For this reason, $R_n(x)$ is called the **Lagrange form** of the remainder. There is at least one other version of Taylor's Formula in which the remainder is expressed as an integral.

When applying Taylor's Formula, we would not expect to be able to find the exact value of z. Rather, we would attempt to find bounds for the derivative $f^{(n+1)}(z)$ from which we will be able to tell how large the remainder $R_n(x)$ is. Thus, for the purpose of approximating values of a function, we restate Taylor's Formula in the following way.

Taylor's Inequality

Suppose $p_n(x)$ is the nth-degree polynomial approximation for the function f about $x = c$ and M is the maximum value of $\left|f^{(n+1)}(x)\right|$ on the interval $[c, b]$ (or $[b, c]$ if $b < c$). Then the error in using the polynomial value $p_n(b)$ to estimate $f(b)$ is bounded by $\dfrac{M}{(n+1)!}|b - c|^{n+1}$ (know as the **Lagrange Error Bound**).

Thus, the remainder $R_n(x)$ in Taylor's Formula satisfies the inequality

$$\left|R_n(x)\right| \le \frac{M}{(n+1)!}|b - c|^{n+1}.$$

Example 1:   *Determining the Accuracy of an Approximation*

Let f be a function with 5 derivatives on the interval $[2, 3]$ and assume that $\left|f^{(5)}(x)\right| < 0.2$ for all x in the interval $[2, 3]$. If a fourth-degree Taylor polynomial for f at $c = 2$ is used to estimate $f(3)$, calculate the Lagrange error bound to determine the accuracy of the approximation?

Solution:

Using Taylor's Inequality with $n = 4$, $b = 3$ and $c = 2$ we obtain

$$\left|R_n(x)\right| \le \frac{M}{(n+1)!}|b - c|^{n+1} = \frac{M}{5!}|3 - 2|^5$$

where M is the maximum value of the fifth derivative of f on the closed interval $[2, 3]$. Because $\left|f^{(5)}(x)\right| < 0.2$, we take $M = 0.2$ and obtain

$$\left|R_n(x)\right| \le \frac{0.2}{120} \cdot 1^5 = \frac{0.2}{120} < 0.00167.$$

Thus, the error (the difference between approximation $p_4(3)$ and the actual value $f(3)$) is less than 0.00167.

Example 2:   *Determining the Accuracy of an Approximation*

Estimate the error that results when $\sin x$ is replaced by $x - \dfrac{1}{6}x^3$ for $|x| < 0.2$. Support the answer graphically.

Solution:

Since $f(x) = \sin x$, the successive derivatives are $\cos x$, $-\sin x$, and $-\cos x$. Because $|\cos x| \le 1$ and $|\sin x| \le 1$, this means we have a simple estimate for the nth derivative of f, namely $\left|f^{(n)}(x)\right| \le 1$ for all n and x. Now using Taylor's Inequality with $M = 1$, $b = \pm 0.2$, $n = 3$ and $c = 0$ we have

$$\left|R_n(x)\right| \le \frac{M}{(n+1)!}|b - c|^{n+1} \le \frac{1}{4!}|\pm 0.2|^4 = 0.000067.$$

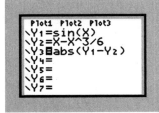 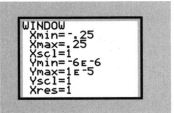 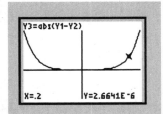

The maximum absolute error on the interval $[-0.2, 0.2]$ is $\left| \sin x - (x - \frac{1}{6}x^3) \right| = 2.664128 \times 10^{-6}$, which is indeed less than the bound, 6.7×10^{-5}.

Example 3: *Approximating a Value to a Desired Accuracy*

Find n so that the nth-order Taylor polynomial for $f(x) = e^x$ approximates the value of $f(1) = e$ correct to three decimal places.

Solution:

The strategy is to find a Taylor polynomial for $f(x) = e^x$ at an appropriate number c. We choose c to be a number close to $x = 1$ at which $f(c), f'(c), ..., f^{(n)}(c)$ can be computed because the Taylor approximation is best in the vicinity of $x = c$. In this case $c = 0$ is the logical choice. Since $f'(x) = e^x, f''(x) = e^x, \cdots, f^{(n)}(x) = e^x$ for every integer $n \geq 0$, we have $f^{(n)}(0) = 1$ and Taylor's Formula gives

$$e^x = 1 + x + \frac{x^2}{2!} + \frac{x^3}{3!} + \cdots + \frac{x^n}{n!} + R_n(x),$$

Since e^x is positive and increasing with x, we have $\left| f^{(n)}(x) \right| = e^x \leq e^1$ for all x in $[0, 1]$. Hence, we take $M = e$, or, for easier calculation, we can let $M = 3$. Now using Taylor's Inequality with $M = 3$, $b = 1$ and $c = 0$ we have

$$\left| R_n(x) \right| \leq \frac{M}{(n+1)!} |b - c|^{n+1} = \frac{3}{(n+1)!} |1 - 0|^5 = \frac{3}{(n+1)!}$$

To get an approximation for $e = e^1$ correct to three decimal places, we need to have

$$\frac{3}{(n+1)!} \leq 0.0005.$$

Employing trial and error, we see that $3 / 8! = 3 / 40320 \approx 0.000074$, but $3 / 7! = 3 / 5040 \approx 0.00059$, so we can be sure $n = 7$ will do, but not $n = 6$. Thus,

$$e^x = 1 + 1 + \frac{1}{2!} + \frac{1}{3!} + \frac{1}{4!} + \frac{1}{5!} + \frac{1}{6!} + \frac{1}{7!} \approx 2.7182 \approx 2.718$$

to three decimal places.

Example 4: *Determining the Accuracy of an Approximation Graphically*

Using a calculator, estimate the values of x for which the Taylor polynomial $p_3(x) = x - \frac{x^3}{3}$ approximates $f(x) = \arctan x$ to four decimal places.

Solution:

If $p_3(x)$ is to approximate $\arctan x$ accurate to four decimal places, then we need

$$\left| \arctan x - p_3(x) \right| < 0.00005.$$

The following figure shows the graph of

$$y = \left| \arctan x - \left(x - \frac{x^3}{3} \right) \right|$$

and the line $y = 0.00005$ plotted in a $[-0.25, 0.25] \times [-6 \cdot 10^{-5}, 6 \cdot 10^{-5}]$ viewing rectangle.

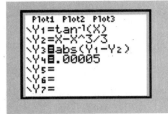 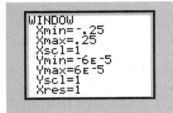 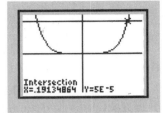

We determine that the graph of $y = |\arctan x - p_3(x)|$ and the line $y = 0.00005$ intersect at

$$x = -0.191489 \quad \text{and} \quad x = 0.191489$$

Thus, we conclude that when $-0.191489 < x < 0.191489$, $p_3(x)$ approximates $\arctan x$ to four decimal places

Example 5: *Approximating a Definite Integral to a Desired Accuracy*

Find an estimate for $\displaystyle\int_0^1 e^{-x^2}\, dx$ that is accurate to three decimal places.

Solution:

Find a Maclaurin series for e^{-x^2}.

$$e^x = 1 + x + \frac{x^2}{2!} + \frac{x^3}{3!} + \frac{x^4}{4!} + \frac{x^5}{5!} + \cdots$$

$$e^{(-x^2)} = 1 + (-x^2) + \frac{(-x^2)^2}{2!} + \frac{(-x^2)^3}{3!} + \frac{(-x^2)^4}{4!} + \cdots = 1 - x^2 + \frac{x^4}{2!} - \frac{x^6}{3!} + \frac{x^8}{4!} - \frac{x^{10}}{5!} + \cdots$$

$$\int_0^1 e^{-x^2}\, dx = \int_0^1 \left(1 - x^2 + \frac{x^4}{2!} - \frac{x^6}{3!} + \frac{x^8}{4!} - \frac{x^{10}}{5!} + \cdots \right) dx$$

$$= \left(x - \frac{x^3}{3} + \frac{x^5}{5 \cdot 2!} - \frac{x^7}{7 \cdot 3!} + \frac{x^9}{9 \cdot 4!} - \frac{x^{11}}{11 \cdot 5!} + \frac{x^{13}}{13 \cdot 6!} - \cdots \right) \Bigg]_0^1$$

$$= 1 - \frac{1}{3} + \frac{1}{5 \cdot 2!} - \frac{1}{7 \cdot 3!} + \frac{1}{9 \cdot 4!} - \frac{1}{11 \cdot 5!} + \frac{1}{13 \cdot 6!} - \cdots$$

Because this is an alternating series whose terms decrease in value, we find the first term less than 0.0005 in absolute value. By trial and error we obtain

$$\frac{1}{11(5!)} \approx 0.000758 \quad \text{and} \quad \frac{1}{13(6!)} \approx 0.000107 .$$

So, if we truncate after 6 terms,

$$\int_0^1 e^{-x^2}\, dx \approx 1 - \frac{1}{3} + \frac{1}{5(2!)} + \frac{1}{7(3!)} + \frac{1}{9(4!)} - \frac{1}{11(5!)} = 0.746729$$

and we have an approximation with the desired accuracy.

11.6　Exercises

1.　Let f be a function that has derivatives of all orders for all real numbers x. Assume $f(0) = 6$, $f'(0) = 8$, $f''(0) = 30$, $f'''(0) = 48$ and $\left|f^{(4)}(x)\right| \le 75$ for all x in the closed interval $[0, 1]$.

　　a) Find the third-degree Taylor polynomial about $x = 0$ for the function f.

　　b) Use your answer to part a) to estimate the value of $f(0.2)$.

　　c) What is the maximum possible error for the approximation made in part b)?

2.　Let f be a function that has derivatives of all orders on the interval $(-1, 1)$. Assume $f(0) = 1$, $f'(0) = \dfrac{1}{2}$, $f''(0) = -\dfrac{1}{4}$, $f'''(0) = \dfrac{3}{8}$, and $\left|f^{(4)}(x)\right| \le 6$ for all x in the interval $(0, 1)$.

　　a) Find the third-degree Taylor polynomial about $x = 0$ for the function f.

　　b) Use your answer to part a) to estimate the value of $f(0.5)$.

　　c) What is the maximum possible error for the approximation made in part b)?

3.　Estimate the error that results when $\sqrt{1 + x}$ is replaced by $1 + \dfrac{1}{2}x$ if $0 < x < 0.01$. Confirm your answer graphically. [Hint: See Example 4]

4.　Find the Lagrange error bound that results when $\arctan x$ is replaced by $x - \dfrac{x^3}{3}$ if $|x| < 0.2$ and $\left|f^{(4)}(x)\right| \le 4$.

5.　Find the Lagrange error bound that results when e^x is replaced by $1 + x + \dfrac{x^2}{2}$ if $|x| < 0.1$.

6.　Estimate the error that results when $\ln(x + 1)$ is replaced by $x - \dfrac{1}{2}x^2$ if $|x| < 0.1$

7.　Find the Lagrange error bound that results when using the Taylor polynomial $p_2(x)$ for $f(x) = \sqrt[3]{x}$ at $x = 8$ on the interval $[7, 9]$.

8.　Find an approximation of $\ln(1.1)$ that is accurate to three decimal places.

9.　Let f be the function defined by $f(x) = \sqrt{x}$

　　a) Find the second-degree Taylor polynomial about $x = 4$ for the function f.

　　b) Use your answer to part a) to estimate the value of $f(4.2)$.

　　c) Find a bound on the error for the approximation in part b).

11.   Let $f(x) = \sum\limits_{n=0}^{\infty} \dfrac{x^n}{2^n}$ for all x for which the series converges.

     a) Find the interval of convergence of this series.

     b) Use the first three terms of this series to approximate $f(-\tfrac{1}{2})$.

     c) Estimate the error involved in the approximation in part b). Show your reasoning.

In Exercises 12–14, use a Taylor polynomial to estimate the number with three decimal place accuracy.

12. $\cos(0.2)$

13. $e^{-0.1}$

14. $\displaystyle\int_0^1 \dfrac{\sin x}{x}\,dx$

15.   Suppose the function f is defined so that $f(1) = 1$, $f'(1) = 2$ and $f''(x) = (1 + x^3)^{-1}$ for $x > -1$.

     a) Write a second-degree Taylor polynomial for f about $x = 1$.

     b) Use the result in part a) to approximate $f(1.5)$.

     c) Find an upper bound for the approximation error in part b).

16.   Suppose the function f is defined so that $f(1) = \dfrac{1}{2}$, $f'(1) = -\dfrac{1}{2}$ and $f''(x) = \dfrac{2(3x^2 - 1)}{(x^2 + 1)^3}$.

     a) Write a second-degree Taylor polynomial for f about $x = 1$.

     b) Use the result in part a) to approximate $f(1.5)$.

     c) If $|f''(x)| \le \dfrac{1}{2}$ for all x in $[1, 1.5]$, find an upper bound for the approximation error in part b).

17.   Estimate the range of values of x for which the approximation $e^x \cos x = 1 + x - \dfrac{1}{3}x^3$ is accurate within 0.001.

18.   The first four derivatives of $f(x) = \dfrac{1}{\sqrt{1+x}}$ are

$$f'(x) = \frac{-1}{2(x+1)^{3/2}}$$

$$f''(x) = \frac{3}{4(x+1)^{5/2}}$$

$$f'''(x) = \frac{-15}{8(x+1)^{7/2}}$$

$$f^{(4)}(x) = \frac{105}{16(x+1)^{9/2}}$$

a) Find the third-degree Taylor approximation $p_3(x)$ to f at $x = 0$.

b) Use your answer to part a) to find an approximation of $f(0.5)$.

c) Estimate the error involved in the approximation in part b). Show your reasoning.

19.   The first three derivatives of $f(x) = \ln(1+x^2)$ are

$$f'(x) = \frac{2x}{1+x^2}$$

$$f''(x) = \frac{2(1-x^2)}{(1+x^2)^2}$$

$$f'''(x) = -\frac{4x(3-x^2)}{(1+x^2)^3}$$

a) Find the third-degree Taylor approximation $p_3(x)$ to f at $x=0$.

b) Use your computer or graphing calculator to find an interval over which the graph of $p_3(x)$ appears to fit the graph of f well.

c) Using an appropriate y-scale, graph $y = |f(x) - p_3(x)|$ on the interval you determined in part b), and determine the maximum error in the approximation in that interval.

Chapter 11 Supplementary Problems

1. Let f be the function defined by $f(x) = e^{-4x^2}$.

   a) Find the first four terms and general term of the power series for f about $x = 0$.

   b) Find the interval of convergence of the power series f about $x = 0$. Show the analysis that leads to your conclusion.

   c) Use term-by-term differentiation to show that $f'(x) = -8xe^{-4x^2}$.

2. Let f be the function defined by $f(x) = \sum_{n=1}^{\infty} \frac{x^n}{n3^n}$ for all x for which the series converges.

   a) Find the radius of convergence of this series.

   b) Use the first three terms of this series to find an approximation of $f(-1)$.

   c) Estimate the amount of error in the approximation in part b) Justify your answer.

3. Let f be the function defined by $f(t) = \frac{2}{1-t^2}$ and G be the function defined by $G(x) = \int_0^x f(t) \, dt$.

   a) Find the first four nonzero terms and the general term for the power series expansion of the function f about $t = 0$.

   b) Find the first four nonzero terms and the general term for the power series expansion of the function G about $t = 0$.

   c) Find the interval of convergence of the power series in part b). (Your solution must include an analysis that justifies your answer.)

4. Write out the first five terms of each of the following power series and determine the radius of convergence of each.

   a) $\displaystyle\sum_{n=0}^{\infty} nx^n$ b) $\displaystyle\sum_{n=0}^{\infty} n!x^n$ c) $\displaystyle\sum_{n=0}^{\infty} \frac{n^2 x^n}{2^n}$

5. Use power series to evaluate the following limits. Check your answers using l'Hopital's rule.

   a) $\displaystyle\lim_{x \to 0} \frac{1 - \cos x}{x^2}$ b) $\displaystyle\lim_{x \to 0} \frac{e^x - 1}{x}$ c) $\displaystyle\lim_{x \to 0} \frac{x - \arctan x}{x^3}$

6. Show that the series $\displaystyle\sum_{n=1}^{\infty} \frac{x^n}{n}$ converges on the interval $[-1, 1)$ but that term-by-term differentiation is not valid at $x = -1$.

7. Is $1 - x + \dfrac{x^2}{2} - \dfrac{x^4}{8} + \dfrac{x^6}{15} - \cdots$ the Maclaurin series representation of the function f shown in the graph? Justify your answer.

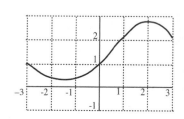

8.   The Maclaurin series for the function f is given by

$$\sum_{n=0}^{\infty} \frac{x^n}{2^n} = 1 + \frac{x}{2} + \frac{x^2}{4} + \frac{x^3}{8} + \frac{x^4}{16} + \cdots$$

on its interval of convergence.

a) Find the interval of convergence of the Maclaurin series for f. Justify your answer.

b) Find the first four terms and general term for the Maclaurin series for $f'(x)$.

c) Use the Maclaurin series you found in part b) to find the value of $f'\left(\frac{1}{2}\right)$

9.   The Taylor series about $x = 2$ for a certain function f converges for all x in the interval of convergence. The nth derivative of f at $x = 2$ is given by $f^{(n)}(2) = \frac{n!}{n \cdot 3^n}$ and $f(2) = 0$.

a) Write the third degree Taylor polynomial for f about $x = 2$.

b) Find the radius of convergence for the Taylor series for f about $x = 2$.

10.   The Taylor series about $x = 3$ for a certain function f converges for all x in the interval of convergence. The nth derivative of f at $x = 3$ is given by $f^{(n)}(3) = \frac{n+1}{2n+1} \cdot \frac{n!}{3^n}$ and $f(3) = 0$.

(a) Write the third degree Taylor polynomial for f about $x = 3$.

(b) Find the radius of convergence for the Taylor series for f about $x = 3$.

11.   Derive the formula.

$$\frac{1}{1+x^2} = 1 - x^2 + x^4 - x^6 + \cdots + (-1)^n x^{2n} + \cdots$$

in three ways, as follows.

a) Replace t by $-x^2$ in the geometric series for $1/(1-t)$.

b) Differentiate the series for $\arctan x$.

c) Divide 1 by $1 + x^2$.

12.   Find the elementary function represented by the power series $\sum_{n=0}^{\infty} \frac{(2x)^n}{n!}$ by manipulating a more familiar power series.

13.   a) What is the Taylor series for $\sin(x^2)$?

b) Use the first three terms of the series in part a) to approximate the integral $\int_0^1 \sin(x^2)\, dx$.

14. Let f be a function that has derivatives of all orders for all real number x. Assume $f(0) = 0$, $f'(0) = 2$, $f''(0) = 4$, $f'''(0) = 8$ and $|f^{(4)}(x)| \le 44$ for all x in the closed interval $[0, 1]$.

   a) Find the third-degree Taylor polynomial about $x = 0$ for the function f.

   b) Use your answer to part a) to estimate the value of $f(0.2)$. What is the maximum possible error in making this estimate?

15. Let f be a function that has derivatives of all orders for all x in the closed interval $[-0.5, 0.5]$. Assume $f(0) = 1$, $f'(0) = -\frac{1}{2}$, $f''(0) = -\frac{1}{4}$, and $|f^{(3)}(x)| \le 3$ for all x in the closed interval $[-0.5, 0.5]$.

   a) Find the second-degree Taylor polynomial about $x = 0$ for the function f.

   b) Use your answer to part a) to estimate the value of $f(0.2)$. What is the maximum possible error in making this estimate?

16. For each of the following functions, find its Taylor series about $x = 0$ without computing derivatives. Instead, modify a familiar Taylor series by integrating it, differentiating it, making a substitution in it, or some combination of these.

   a) $f(x) = e^{-2x}$ b) $f(x) = \dfrac{1}{(1-x)^2}$

   c) $f(x) = \arctan x$ d) $f(x) = \dfrac{\ln(1+x)}{x}$

In Exercises 17–20, find the interval of convergence for the given power series, including convergence or divergence at the endpoints of the interval.

17. $\displaystyle\sum_{n=1}^{\infty} \frac{x^n}{4^n \sqrt{n}}$ 18. $\displaystyle\sum_{n=0}^{\infty} \frac{x^n}{n+2}$

19. $\displaystyle\sum_{n=0}^{\infty} \frac{x^n}{n^2+1}$ 20. $\displaystyle\sum_{n=1}^{\infty} \frac{(x-3)^n}{2n}$

21. a) Show that if $0 \le x \le \frac{1}{2}$, then $\sin x = x - \frac{x^3}{3!} + R(x)$ where $|R(x)| < 0.00026$.

   b) Use the results of part a) to find an approximate value of $\displaystyle\int_0^{1/\sqrt{2}} \sin x^2 \, dx$.

In Exercises 22–24, use Taylor's Inequality to determine the accuracy of the approximation.

22. $\cos(0.3) \approx 1 - \dfrac{(0.3)^2}{2!} + \dfrac{(0.3)^4}{4!}$

23. $\sin(0.4) \approx 0.4 - \dfrac{(0.4)^3}{2 \cdot 3}$

24. $\arctan(0.5) \approx 0.5 - \dfrac{(0.5)^3}{3}$

25. Suppose the function f is defined so that $f(0) = -1$, $f'(0) = 2$ and $f''(x) = \dfrac{-4}{(x+1)^3}$.

a) Write a Maclaurin polynomial for f.

b) Use the result in part a) to approximate $f(0.5)$.

c) If $|f'''(x)| \le 12$ for all x in $[0, 0.5]$, find an upper bound for the approximation error in part b).

26. Suppose that the function f is defined so that $f(1) = 1$, $f'(1) = 2$ and $f''(x) = \dfrac{1}{1+x^3}$ for $x > -1$.

a) Approximate $f(1.5)$ using a quadratic polynomial.

b) Find an upper bound on the approximation error made in part a).

27. The first three derivatives of $f(x) = (x+4)^{3/2}$ are $f'(x) = \dfrac{3(x+4)^{1/2}}{2}$, $f''(x) = \dfrac{3}{4(x+4)^{1/2}}$ and

$f'''(x) = \dfrac{-3}{8(x+4)^{3/2}}$.

a) Give the first four terms of the Taylor series for f at $c = -3$.

b) Give the quadratic Taylor polynomial, $p_2(x)$, at $c = 0$.

c) Suppose that $x \ge 0$ and that $p_2(x)$ is used to approximate f. Show that the error in this

approximation does not exceed $\dfrac{x^3}{128}$.

28. Determine the interval of convergence of the power series $\displaystyle\sum_{n=1}^{\infty} \dfrac{(x-2)^n}{n3^n}$.

29. Evaluate the following limits.

a) $\displaystyle\lim_{x \to 1} \dfrac{\ln(2x-1)}{x-1}$ b) $\displaystyle\lim_{x \to 1} \dfrac{x^2 - 2x + 1}{1 - x + \ln x}$ c) $\displaystyle\lim_{x \to 0} \dfrac{a^x - b^x}{x}$

30. Determine whether the following statements are true or false.

a) The power series $x - \dfrac{x^2}{2} + \dfrac{x^3}{3} - \dfrac{x^4}{4} + \dots$ converges on the interval $(-1, 1]$.

b) $e^x = \displaystyle\sum_{n=0}^{\infty} \dfrac{x^2}{n!}$

c) The sum of the series $5 - \dfrac{10}{3} + \dfrac{20}{9} - \dfrac{40}{27} + \dots$ is 15.

d) $\dfrac{1}{1+x} = \displaystyle\sum_{n=0}^{\infty} nx^n$

e) If $\displaystyle\sum a_k$ converges then $\displaystyle\lim_{k \to \infty} a_k = 0$.

31. The table below shows how the speed of a braking car (in feet per second) changes second by second.

t (sec)	0	1	2	3	4	5	6
v (ft/sec)	110.0	99.8	90.9	83.2	76.4	70.4	65.1

   a) Estimate the distance traveled during this 6-second interval.

   b) Suppose the car's speed is modeled by the function $f(t) = \dfrac{44000}{(t+20)^2}$ during the time interval $0 \le t \le 6$. Use the function f to calculate the exact distance traveled by the car.

32. The infinite region beneath the curve $y = \dfrac{5}{x+1}$ in the first quadrant is revolved about the x-axis to generate a solid. Find the volume of the solid.

33. The base of a solid is the first quadrant region in the xy-plane beneath the curve $y = kx - x^2$. Each of the solid's cross-sections perpendicular to the x–axis has the shape of a rectangle with height kx. If the volume of the solid is $\dfrac{8}{3}$ cubic units, find the value of k.

34. A particle moves along a line in such a way that at time t, $1 \le t \le 8$, its position is given by

$$s(t) = \int_1^t [1 - x(\cos x) - (\ln x)(\sin x)]\, dx.$$

   a) Write a formula for the velocity of the particle at time t.

   b) At what instant does the particle reach it maximum speed?

   c) When is the particle moving to the left?

   d) Find the total distance traveled by the particle from $t = 1$ to $t = 8$.

35. Consider the following table of values for the differentiable function f.

x	1.0	1.2	1.4	1.6	1.8
f(x)	5.0	3.5	2.6	2.0	1.5

   a) Estimate $f'(1.4)$.

   b) Give an equation for the tangent line to the graph of f at $x = 1.4$.

   c) What is the sign of $f''(1.4)$? Explain your answer.

   d) Using the data in the table, find a midpoint approximation with 2 equal subdivisions for

$$\int_{1.0}^{1.8} f(x)\, dx.$$

CHAPTER 12

Parametric Equations, Vectors and Polar Coordinates

12.1 Parametric Equations

If a particle is moving in a coordinate plane, its path may not be easy to describe with the graph of a function. However, in this section we will show how such a path can be specified by expressing the x – and y –coordinates separately as functions of a third variable.

Suppose a particle is moving in a coordinate plane and functions f and g exist so that at time t the x – and y –coordinates of the particle's position are determined by two equations, called **parametric equations**.

$$\begin{cases} x = f(t) \\ y = g(t) \end{cases}$$

Thus, at time t_0 the particle is at the point $(f(t_0), g(t_0))$.

Example 1: *Modeling the Motion of a Particle with Parametric Equations*

Functions f and g are defined as follows: $f(t) = 2t$ and $g(t) = t^2$, where $t \geq 0$. A particle is moving in the coordinate plane with its position (x, y) given by the parametric equations $x = f(t)$ and $y = g(t)$. Find the position of the particle at times 0, 1, and 3, and determine the path of the particle.

Solution:

If (x, y) is a point on the path of the particle, then x is determined by a value of f and y is determined by a value of g so that

$$x = 2t \text{ and } y = t^2.$$

At time $t = 0$ the particle is at $(0, 0)$.
At time $t = 1$ the particle is at $(2, 1)$.
At time $t = 3$ the particle is at $(6, 9)$.

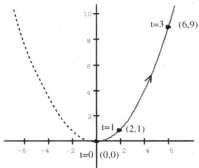

If we solve the equation $x = 2t$ for t and substitute for t in the equation $y = t^2$, we get $y = \dfrac{x^2}{4}$ with $x \geq 0$ as an equation of the path of the particle. The path is seen to be the right half of a parabola. The particle starts at $(0, 0)$ and as t increases the particle moves up along the graph of the parabola.

Parametric Equations:
$f(t) = 2t$ and $g(t) = t^2$

Rectangular Equation:
$$y = \frac{x^2}{4}$$

When the point (x, y) is described by two coordinate functions f and g so that x is the value f relates to t and y is the value g relates to t, then we say that we have a **parametrization** of the points (x, y) in terms of the **parameter t**. A collection of points (x, y) defined in this way is called a **plane curve**.

It is also convenient to designate the two coordinate functions by x and y, in spite of the fact that some confusion may arise over different meanings given to the letters.

In Example 1, it was given that the parameter t was any nonnegative real number, $t \geq 0$. Sometimes the values of the parameter t are restricted to a finite interval.

For instance, the parametric curve

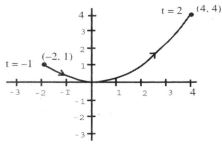

$$x = 2t \text{ and } y = t^2, \ -1 \le t \le 2,$$

shown in the figure, is the part of the parabola in Example 1 that starts at point $(-2, 1)$ and ends at $(4, 4)$. The arrows indicate the direction the curve is traced as t increases from -1 to 2.

Most calculators have a parametric graphing mode. As with rectangular graphs, the calculator generates an internal table of values first, then connects points with short line segments.

Example 2:   *Sketching a Parametric Curve with a Calculator*

Suppose that the position of a particle in the plane at time t is given by the parametric equations $x = 2t$ and $y = t^2$, $0 \le t \le 3$. Describe the motion of the particle.

Solution:

In Parametric mode, define the functions, set the window values and graph.

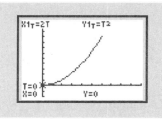  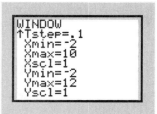

The particle is at $(0, 0)$ at $t = 0$ and at $(6, 9)$ when $t = 3$; these are the **initial** and **terminal** points of the particle's path.

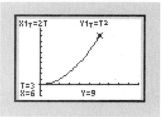

Note that in plotting points in increasing values of t, the path of the point is traced out in a specific direction. This is called the **orientation** of the curve. Also note that the points determined by parametric equations may only be a part of the graph of the equation obtained by eliminating the parameter.

Time is not the only possible parameter. For example, if (x, y) is a point on the circle $x^2 + y^2 = 1$ and θ is the measure of the angle *POA* in the figure, then

$$x = \cos\theta \text{ and } y = \sin\theta .$$

Since $x^2 + y^2 = \cos^2\theta + \sin^2\theta = 1$, the parametric equations

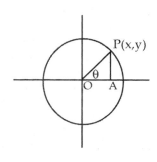

$$\begin{cases} x = \cos\theta \\ y = \sin\theta \end{cases}$$

define the unit circle in terms of the parameter θ.

Eliminating the Parameter

When a collection of points (x, y) is defined by means of parametric equations, it is sometimes possible to find an equation relating x and y but not involving the parameter. For instance, the most familiar parametrization of points is the one that describes the unit circle by the two equations

$$\begin{cases} x = \cos\theta \\ y = \sin\theta. \end{cases}$$

To eliminate the parameter θ, we square both sides of both equations and add, so that we get

$$x^2 + y^2 = \cos^2\theta + \sin^2\theta.$$

But since $\cos^2\theta + \sin^2\theta = 1$, we have

$$x^2 + y^2 = 1.$$

In this case the collection of points (x, y) defined by the parametric equations above is equivalent to the collection of pairs satisfying the Cartesian equation $x^2 + y^2 = 1$. However, the equation obtained by eliminating a parameter from a pair of parametric equations is not always equivalent to the original parametric equation. That is, there may be points (x, y) that satisfy the equation in x and y but that do not satisfy the parametric equations.

Example 3:   *Using Trigonometry to Eliminate a Parameter*

Sketch the curve defined by the parametric equations

$$\begin{cases} x = \cos\theta \\ y = \cos 2\theta \end{cases}$$

by eliminating the parameter and adjusting the domain of the resulting rectangular equation.

Solution:

Since $\cos 2\theta = 2\cos^2\theta - 1$, these equations can be written

$$\begin{cases} x = \cos\theta \\ y = 2\cos^2\theta - 1. \end{cases}$$

Thus, we can see that $y = 2x^2 - 1$. In this form it would appear that x can be any real number and y can be any number greater than or equal to -1. In fact, if we reconsider the original parametric equations, we find that both $x-$ and $y-$ coordinates must be between -1 and 1, so that we have only a piece of the parabola $y = 2x^2 - 1$, namely, the piece with $-1 \le x \le 1$. Then we find that as θ increases in the original parametric equations we go back and forth over this same restricted part of the parabola.

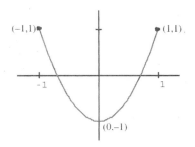

The next example also illustrates that the points determined by parametric equations may be only part of the graph of the equation obtained by eliminating the parameter.

Example 4:   *Eliminating a Parameter*

Sketch the curve defined by the parametric equations

$$\begin{cases} x = \cos^2\theta \\ y = \sin^2\theta. \end{cases}$$

Solution:

To eliminate the parameter, we add the two equations obtaining

$$x + y = 1.$$

The graph of $x + y = 1$ is a line; however, if we consider the original parametric equations, we note that both x– and y– coordinates must be between 0 and 1, so that the parametric equations describe a line segment which is the part of the line $x + y = 1$ in the first quadrant.

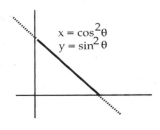

Finding Parametric Equations

There are some paths that can be described more easily by parametric equations than by a single equation in x and y. The next example is a classic.

Example 5:   *Parametric Equations for a Cycloid*

A circle of radius r rolls along the x-axis. Find parametric equations of the path traced out by the point of the circle that starts at the origin.

Solution:

We take the angle θ through which the circle has rolled at a given moment as the parameter. Then the coordinates of the point P are given by

$$\begin{cases} x = OR - QC \\ y = RC + PQ \end{cases}.$$

Since the circle is rolling, the length of the segment $\overline{OR}$ is the same as the length of the arc PR,.

$$PR = \frac{\theta}{2\pi} \cdot 2\pi r = \theta r, \text{ where } \theta \text{ is the measure of } \angle RCP.$$

Since $\angle QCP = \theta - \frac{\pi}{2}$, we have $QC = r\cos(\theta - \frac{\pi}{2})$ and $PQ = r\sin(\theta - \frac{\pi}{2})$.

Hence $\begin{cases} x = r\theta - r\cos(\theta - \frac{\pi}{2}) \\ y = r + r\sin(\theta - \frac{\pi}{2}). \end{cases}$

Simplifying, we obtain

$$\begin{cases} x = r(\theta - \sin\theta) \\ y = r(1 - \cos\theta) \end{cases}$$

The curve described by these equations is called a **cycloid**.

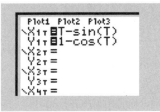
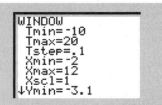
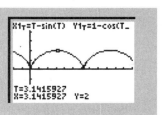

Slope

The slope of a graph described parametrically can be found directly from the parametric equations. Suppose a collection of points (x, y) has been defined parametrically by coordinate functions f and g so that

$$x = f(t) \quad \text{and} \quad y = g(t).$$

A secant line determined by the points $(f(t_0), g(t_0))$ and $(f(t_0 + h), g(t_0 + h))$ has a slope $m(h)$ where

$$m(h) = \frac{g(t_0 + h) - g(t_0)}{f(t_0 + h) - f(t_0)}$$

or, for $h \neq 0$,

$$m(h) = \frac{\dfrac{g(t_0 + h) - g(t_0)}{h}}{\dfrac{f(t_0 + h) - f(t_0)}{h}}$$

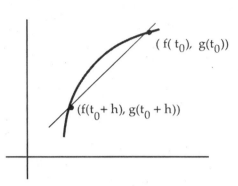

Now suppose that the coordinate functions f and g are differentiable and $f'(t_0) \neq 0$. Then

$$\lim_{h \to 0} m(h) = \frac{g'(t_0)}{f'(t_0)} .$$

Thus, we can define the slope of the tangent to our graph at a point in terms of the derivatives of the coordinate functions.

If the graph of the parametric equations is also the graph of a differentiable function F, and if F relates y to x, then $F(x) = y$. Replacing x with $f(t)$ and y with $g(t)$ we obtain

$$F[f(t)] = g(t).$$

By the Chain Rule

$$F'[f(t)] \cdot f'(t) = g'(t)$$

or

$$F'(x) = \frac{g'(t)}{f'(t)} .$$

With $F'(x) = \frac{dy}{dx}$, $f'(t) = \frac{dx}{dt}$ and $g'(t) = \frac{dy}{dt}$, the derivative for F' may be written

$$\frac{dy}{dx} = \frac{\dfrac{dy}{dt}}{\dfrac{dx}{dt}} \quad \text{if} \quad \frac{dx}{dt} \neq 0$$

Example 6 :   *Finding Slope*

Find the slope of the line tangent to the point on the circle

$$\begin{cases} x = \cos\theta \\ y = \sin\theta \end{cases}$$

where $\theta = \frac{\pi}{3}$.

Solution:

Since $\frac{dx}{d\theta} = -\sin\theta$ and $\frac{dy}{d\theta} = \cos\theta$, we have at $\theta = \frac{\pi}{3}$,

$$\frac{dy}{dx} = \frac{\frac{dy}{d\theta}}{\frac{dx}{d\theta}} = \frac{\cos\theta}{-\sin\theta} = \frac{1/2}{-\sqrt{3}/2} = -\frac{\sqrt{3}}{3}.$$

In addition to plotting parametric curves, the TI-83 can calculate the slope of a smooth parametric curve at a point (x_0, y_0) and then overdraw a plot of the tangent line to the curve at this point. To illustrate, we begin with a parametric plot of the curve in Example 6 using the following window settings.

 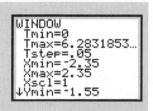 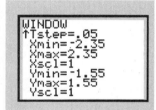 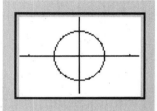

To plot the tangent line at the point $T = \frac{\pi}{3}$, press **2nd** [DRAW] and select item [5:Tangent] from the Draw menu. At the prompt, set $T = \frac{\pi}{3}$; press **ENTER** and the tangent line is overdraw and the slope of the tangent line is displayed.

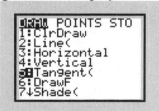 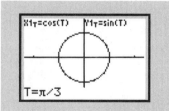 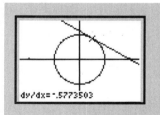

Example 7:   *Finding Where $\frac{dy}{dx}$ Does Not Exist for a Curve*

Find the points at which $\frac{dy}{dx}$ does not exist for the curve defined parametrically by

$x = \sec\theta$ and $y = \tan\theta$.

Solution:

$$\frac{dx}{d\theta} = \sec\theta \cdot \tan\theta \quad \text{and} \quad \frac{dy}{d\theta} = \sec^2\theta. \quad \text{Thus,} \quad \frac{dy}{dx} = \frac{\frac{dy}{d\theta}}{\frac{dx}{d\theta}} = \frac{\sec^2\theta}{\sec\theta \cdot \tan\theta} = \frac{\sec\theta}{\tan\theta}.$$

This derivative is undefined whenever $\tan\theta = 0$, which happens when $\theta = \frac{n\pi}{2}$.

Example 8:   *Finding the Slope of a Cycloid*

Find the slope of the tangent to the cycloid

$$\begin{cases} x = \theta - \sin\theta \\ y = 1 - \cos\theta \end{cases}$$

at the point $\theta = \dfrac{2}{3}\pi$.

Solution:

We have $\dfrac{dx}{d\theta} = 1 - \cos\theta$ and $\dfrac{dy}{d\theta} = \sin\theta$. Therefore, the

derivative $\dfrac{dy}{dx}$ is given by $\dfrac{dy}{dx} = \dfrac{\sin\theta}{1 - \cos\theta}$. When $\theta = \dfrac{2}{3}\pi$,

$$\frac{dy}{dx} = \frac{\sqrt{3}/2}{3/2} = \frac{\sqrt{3}}{3} \approx .57735027 \; .$$

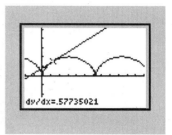

The Second Derivative

The formula $\;\; y' = \dfrac{dy}{dx} = \dfrac{\dfrac{dy}{dt}}{\dfrac{dx}{dt}}\;\;$ was obtained with the help of the Chain Rule. For information about

the concavity of a parametrically defined curve we need the second derivative, $\dfrac{d^2y}{dx^2}$. Letting

$y' = \dfrac{dy}{dx}$ and employing the Chain Rule again,

$$\frac{d^2y}{dx^2} = \frac{dy'}{dx} = \frac{\dfrac{dy'}{dt}}{\dfrac{dx}{dt}} = \frac{\dfrac{d\left(\dfrac{dy}{dx}\right)}{dt}}{\dfrac{dx}{dt}}$$

Thus, for the second derivative $\dfrac{d^2y}{dx^2}$ in terms of the parameter t we have

$$\frac{d^2y}{dx^2} = \frac{\dfrac{d\left(\dfrac{dy}{dx}\right)}{dt}}{\dfrac{dx}{dt}}$$

Example 9:   *Finding Concavity*

For the cycloid in Example 8, determine the concavity on the interval $(0, 2\pi)$.

Solution:

Since $\dfrac{dx}{d\theta} = 1 - \cos\theta$ and $\dfrac{dy}{d\theta} = \sin\theta$, we have $\dfrac{dy}{dx} = \dfrac{\sin\theta}{1 - \cos\theta}$

Thus, $\dfrac{d\left(\dfrac{dy}{dx}\right)}{d\theta} = \dfrac{(1 - \cos\theta)(\cos\theta) - (\sin\theta)(\sin\theta)}{(1 - \cos\theta)^2} = \dfrac{-1}{(1 - \cos\theta)}$

and

$$\frac{d^2 y}{dx^2} = \frac{\dfrac{d\left(\dfrac{dy}{dx}\right)}{d\theta}}{\dfrac{dx}{d\theta}} = \frac{-\dfrac{1}{1 - \cos\theta}}{1 - \cos\theta} = -\frac{1}{(1 - \cos\theta)^2}$$

Since the second derivative is always negative on the interval $(0, 2\pi)$, the graph is concave down.

Example 10:   *Points of Inflection*

Show that the curve defined parametrically by $x = t^3 - 4t$ and $y = 3t^2 + 4t$ has no point of inflection.

Solution:

$\dfrac{dx}{dt} = 3t^2 - 4$ and $\dfrac{dy}{dt} = 6t + 4$. Thus

$$\frac{dy}{dx} = \frac{\dfrac{dy}{dt}}{\dfrac{dx}{dt}} = \frac{6t + 4}{3t^2 - 4} \quad \text{and} \quad \frac{d^2 y}{dx^2} = \frac{\dfrac{d\left(\dfrac{dy}{dx}\right)}{dt}}{\dfrac{dx}{dt}} = \frac{-6(3t^2 + 4t + 4)}{(3t^2 - 4)^3}$$

But the discriminate for the polynomial $3t^2 + 4t + 4$ is $b^2 - 4ac = 16 - 4(12) = -32$. Therefore, the second derivative is never zero – a necessary condition for an inflection point.

12.1   Exercises

In Exercises 1–6, show that the graph of each pair of parametric equations is part (or all) of the parabola $y = x^2$. In each case, describe the graph, being precise about the specific portion of the parabola that is indicated.

1. $\begin{cases} x = t \\ y = t^2 \end{cases}$

2. $\begin{cases} x = \sin t \\ y = 1 - \cos^2 t \end{cases}$

3. $\begin{cases} x = e^t \\ y = e^{2t} \end{cases}$

4. $\begin{cases} x = 1 - \dfrac{1}{t^2} \\ y = 1 - \dfrac{2}{t^2} + \dfrac{1}{t^4} \end{cases}$

5. $\begin{cases} x = \sqrt{t-1} \\ y = t - 1 \end{cases}$

6. $\begin{cases} x = \sec t \\ y = 1 + \tan^2 t \end{cases}$

In Exercises 7–14, sketch the curve represented by the parametric equations (indicate the direction of the curve), and write the corresponding rectangular equation by eliminating the parameter.

7. $\begin{cases} x = 1 - t^2 \\ y = t \end{cases}$

8. $\begin{cases} x = 2t + 1 \\ y = t - 2 \end{cases}$

9. $\begin{cases} x = \cos\theta \\ y = |\sin\theta| \end{cases}$

10. $\begin{cases} x = \arccos t \\ y = 2t^2 - 1 \end{cases}$

11. $\begin{cases} x = \sec t \\ y = \tan t \end{cases}$

12. $\begin{cases} x = 2t + 3 \\ y = 4t^2 - 9 \end{cases}$

13. $\begin{cases} x = \cos^2\theta - \sin^2\theta \\ y = 2\sin\theta \cos\theta \end{cases}$

14. $\begin{cases} x = 9t - 12 \\ y = -6t + 10 \end{cases}$

15.   The equation $x^2 + y^2 = 1$ can be parametrized by letting $x = \dfrac{t^2 - 1}{t^2 + 1}$. Show that the

y-coordinate is determined by $y = \dfrac{2t}{t^2 + 1}$.

In Exercises 16–21, find an equation of the line tangent to the graph of the curve for the given value of the parameter.

16. $\begin{cases} x = \dfrac{1}{t} \\ y = t \end{cases}$ for $t = -1$

17. $\begin{cases} x = e^t \\ y = e^{-t} \end{cases}$ for $t = \ln 2$

18. $\begin{cases} x = \ln t \\ y = \arcsin t \end{cases}$ for $t = \dfrac{1}{2}$

19. $\begin{cases} x = \sqrt{t} \\ y = \dfrac{1}{4}(t^2 - 4) \end{cases}$ for $t = 4$

20. $\begin{cases} x = t^2 - t + 2 \\ y = t^3 - 3t \end{cases}$ for $t = -1$ 21. $\begin{cases} x = 3\cos t \\ y = 2\sin t \end{cases}$ for $t = \dfrac{\pi}{6}$

In Exercise 22-25, express $\dfrac{dy}{dx}$ and $\dfrac{d^2y}{dx^2}$ in terms of parameter t.

22. $\begin{cases} x = 4 - 2t \\ y = \ln 4t \end{cases}$ 23. $\begin{cases} x = \ln t \\ y = t^2 - t \end{cases}$

24. $\begin{cases} x = \cos t \\ y = \cos 2t \end{cases}$ 25. $\begin{cases} x = \ln t \\ y = t\ln t \end{cases}$

26. Show that the cycloid of Example 5 does not have a derivative $\dfrac{dy}{dx}$ at every point.

27. During the time period from $t = 0$ to $t = 5$ seconds, a particle moves along the path given by $x(t) = \sin(\pi t)$ and $y(t) = .5t^2$.

   a) Find the position of the particle when $t = 4$.

   b) Using your calculator, sketch a graph of the path of the particle from $t = 0$ to $t = 5$. Determine the direction of the particle along its path.

   c) Find $\dfrac{dy}{dx}$ in terms of t..

   d) Find an equation of the tangent line to the path of the particle at $t = 4$.

28. During the time period from $t = 0$ to $t = 6$ seconds, a particle moves along the path given by $x(t) = \sqrt{t}$ and $y(t) = \dfrac{1}{4}(t^2 - 4)$.

   a) Find the position of the particle at $t = 4$.

   b) Using your calculator, sketch a graph of the path of the particle from $t = 0$ to $t = 6$. Determine the direction of the particle along its path.

   c) Find $\dfrac{dy}{dx}$ in terms of t.

   d) Find an equation of the tangent line to the path of the particle at $t = 4$.

   e) Find $\dfrac{d^2y}{dx^2}$ at $t = 4$.

29. During the time period from $t = 0$ to $t = 5$ seconds, a particle moves along the path given by $x(t) = \cos(\pi t)$ and $y(t) = \sin(\pi t)$.

   a) Find the position of the particle when $t = 2.5$.

   b) Using your calculator, sketch a graph of the path of the particle from $t = 0$ to $t = 5$. Indicate the direction of the particle along its path.

   c) Find $\dfrac{dy}{dx}$ in terms of t.

   d) Find an equation of the tangent line to the path of the particle at $t = 2.5$.

   e) How many times is the particle at the point found in part a)?

30.   A curve C is defined by the parametric equations $x(t) = 2t + 1$ and $y(t) = t^2 - 3t + 5$.

     a) Find $\dfrac{dy}{dx}$ in terms of t.

     b) Find an equation of the tangent line to C at $t = 4$.

     c) Find the points on the curve and the corresponding values of t where the tangent line is horizontal.

     d) Find the values of t for which x is increasing and y is decreasing. How do the values of $\dfrac{dx}{dt}$ and $\dfrac{dy}{dt}$ support your conclusion?

31.   A curve C is defined by the parametric equations $x = \dfrac{1}{\sqrt{t+1}}$ and $y = \dfrac{t}{t+1}$ for $t \geq 0$.

     a) Find $\dfrac{dy}{dx}$ in terms of t.

     b) Find an equation of the tangent line to C at $t = 3$.

     c) Find an equation for the curve C in terms of x and y.

32.   Graph the curve defined by the parametric equations

$$x = \cos^2(2t) \quad \text{and} \quad y = \sin(2t), \quad -\frac{\pi}{2} \leq t \leq \frac{\pi}{2}.$$

     a) How would you change one of the equations so that the parabola opens to the right instead of to the left?

     b) How would you change the equations so that the parabola opened upward?

33.   Let $x = \tan t$ and $y = \sec t$, $0 \leq t \leq 2\pi$. Graph the curve defined by the parametric equations. How would you change the equations to obtain a hyperbola that opens to the left and right instead of up and down?

34.   Study the curves with parametric equations

$$x = \cos t + a \sin t$$
$$y = \sin t + b \cos t$$

     Here, a and b are constants. Start with the circle $(a = b = 0)$.

     a) What other values of a and b give circles?

     b) What other curves are possible?

     c) What effect does changing a and b have on the curve?

Exploratory Worksheet

Purpose: Using parametric equations to create curves.

Consider the functions listed here:
$$\sin(t), \quad \cos(t), \quad \sin(2t), \quad \cos(2t), \quad \sin(3t), \quad \cos(3t).$$

Using one function for $f(t)$ and one (perhaps the same one) for $g(t)$, there are 36 pairs of parametric equations $\begin{cases} x = f(t) \\ y = g(t) \end{cases}$ that can be formed. Of the 36 curves formed this way, find one that:

a) is a straight line segment;

b) is part of a parabola;

c) takes the shape of a figure eight;

d) takes the shape of the letter S.

12.2   Length of an Arc Described by Parametric Equations

In this section we consider the length of an arc which has been described by parametric equations. Suppose a set of points (x, y) is parameterized by coordinate functions f and g so that

$$x = f(t) \quad \text{and} \quad y = g(t).$$

Also suppose that f and g are continuous functions on the interval $[a, b]$. It can be shown that the continuity of the coordinate functions f and g insures that the collection of points $(f(t), g(t))$ for $a \le t \le b$ is a smooth curve with no holes or gaps. We can begin creating an approximating sum which will lead to a definition of arc length by considering a subdivision of the interval $[a, b]$ into n equal subdivisions given by $a = t_0 < t_1 < t_2 < ... < t_n = b$. For each endpoint t_k, there corresponds a point $(f(t_k), g(t_k))$ on our arc. Joining these points in order with segments and summing the lengths of the segments we obtain the number

$$\sqrt{[f(t_1) - f(t_0)]^2 + [g(t_1) - g(t_0)]^2} + \sqrt{[f(t_2) - f(t_1)]^2 + [g(t_2) - g(t_1)]^2} + ... + \sqrt{[f(t_n) - f(t_{n-1})]^2 + [g(t_n) - g(t_{n-1})]^2}$$

as an approximation of the length of the arc from the point $((f(t_0), g(t_0))$ to the point $(f(t_n), g(t_n))$.

If we add more intervals to the subdivision of the interval $[a, b]$ by considering additional points t_i between a and b, the collection of line segments more closely fits the curve and the sum given above more nearly approximates the length of the curve.

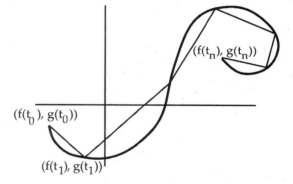

Also, if both f and g are differentiable functions so that the Mean Value Theorem holds, in each interval $[t_{k-1}, t_k]$ there are inputs $\overline{t}_k$ and t_k^* such that

$$f'(\overline{t}_k) = \frac{f(t_k) - f(t_{k-1})}{t_k - t_{k-1}} \quad \Rightarrow \quad f(t_k) - f(t_{k-1}) = f'(\overline{t}_k) \cdot \Delta t$$

and

$$g'(t_k^*) = \frac{g(t_k) - g(t_{k-1})}{t_k - t_{k-1}} \quad \Rightarrow \quad g(t_k) - g(t_{k-1}) = g'(t_k^*) \cdot \Delta t$$

where $\Delta t = t_k - t_{k-1}$.

Substituting in the approximating sum and removing the common factor $(\Delta t)^2$ from the radicands we obtain

$$\sqrt{[f'(\overline{t}_1)]^2 + [g'(t_1^*)]^2} \, \Delta t + \sqrt{[f'(\overline{t}_2)]^2 + [g'(t_2^*)]^2} \, \Delta t + ... + \sqrt{[f'(\overline{t}_n)]^2 + [g'(t_n^*)]^2} \, \Delta t$$

This is **not** a Riemann Sum because of the two intermediate inputs $\overline{t}_k$ and t_k^* in each interval.

However, it can be proved that as each subinterval approaches zero length, the difference between the sum above and the Riemann Sum

$$\sqrt{[f'(\overline{t}_1)]^2 + [g'(\overline{t}_1)]^2} \cdot \Delta t + \sqrt{[f'(\overline{t}_2)]^2 + [g'(\overline{t}_2)]^2} \cdot \Delta t + \ldots + \sqrt{[f'(\overline{t}_n)]^2 + [g'(\overline{t}_n)]^2} \cdot \Delta t$$

can be made arbitrarily small. The consequence of this is that the two summations have the same limit. But the limit of the Riemann Sum is a definite integral, so that

$$\lim_{n \to \infty} \left[\sqrt{[f'(\overline{t}_1)]^2 + [g'(\overline{t}_1)]^2} \cdot \Delta t + \ldots + \sqrt{[f'(\overline{t}_n)]^2 + [g'(\overline{t}_n)]^2} \cdot \Delta t \right] = \int_a^b \sqrt{[f'(t)]^2 + [g'(t)]^2} \, dt$$

Thus, when it exists, we call the length L of an arc defined by parametric equations $x = f(t)$, $y = g(t)$ on the interval $[a, b]$

$$L = \int_a^b \sqrt{[f'(t)]^2 + [g'(t)]^2} \, dt$$

Example 1. *Finding the Length of a Curve*

Find the length of the curve C defined for $0 \le t \le \pi$ by

$$\begin{cases} x = e^t \cos t \\ y = e^t \sin t \end{cases}$$

Solution:

We have $\dfrac{dx}{dt} = e^t \cos t - e^t \sin t = e^t (\cos t - \sin t)$

$\dfrac{dy}{dt} = e^t (\sin t) + e^t \cos t = e^t (\cos t + \sin t).$

Then $\left(\dfrac{dx}{dt}\right)^2 + \left(\dfrac{dy}{dt}\right)^2 = e^{2t}(\cos t - \sin t)^2 + e^{2t}(\cos t + \sin t)^2$

$= e^{2t}(\cos^2 t - 2\cos t \sin t + \sin^2 t + \cos^2 t + 2\cos t \sin t + \sin^2 t)$

$= 2e^{2t}.$

Thus, we obtain

$$\int_0^\pi \sqrt{[f'(t)]^2 + [g'(t)]^2} \, dt = \int_0^\pi \sqrt{2e^{2t}} \, dt = \int_0^\pi \sqrt{2}\, e^t \, dt = \sqrt{2}\left[e^t\right]_0^\pi = \sqrt{2}(e^\pi - 1).$$

12.2   Exercises

In Exercises 1–7, find the length of the arc described.

1. $\begin{cases} x = t^2 \\ y = t^3 \end{cases}$ $0 \le t \le 1$

2. $\begin{cases} x = 2\cos t \\ y = 2\sin t \end{cases}$ $0 \le t \le 2\pi$

3. $\begin{cases} x = t^2 \\ y = \dfrac{t^3}{3} - t \end{cases}$ $1 \le t \le 2$

4. $\begin{cases} x = \theta - \sin\theta \\ y = 1 - \cos\theta \end{cases}$ $0 \le \theta \le 2\pi$

5. $\begin{cases} x = e^{-t}\cos t \\ y = e^{-t}\sin t \end{cases}$ $0 \le t \le \pi$

6. $\begin{cases} x = \cos^3\theta \\ y = \sin^3\theta \end{cases}$ $0 \le \theta \le \dfrac{\pi}{2}$

7. $\begin{cases} x = 1 + \arctan t \\ y = 1 - \ln\sqrt{1 + t^2} \end{cases}$ $0 \le t \le 1$

8. The position of a particle at any time $t \ge 0$ is given by $x(t) = 1 - t^2$ and $y(t) = \dfrac{2}{3}t^3$.

    a) Find $\dfrac{dy}{dx}$ in terms of t.

    b) Find an equation of the tangent line to the path of the particle at $t = 3$

    c) Approximate the distance traveled by the particle from $t = 0$ to $t = 3$.

9. During the time period from $t = 0$ to $t = \pi$ the position of a particle is given by $x(t) = 2\sin t$ and $y(t) = 2\cos t$.

    a) Find $\dfrac{dy}{dx}$ in terms of t.

    b) Find an equation of the tangent line to the path of the particle at $t = \dfrac{\pi}{4}$.

    c) Write and evaluate an integral expression, in terms of sine and cosine, that gives the distance the particle travels from time $t = 0$ to $t = \pi$.

10. A curve is given parametrically by the equations $x(t) = t^2 - 1$ and $y(t) = 2e^t$.

    a) Find $\dfrac{dy}{dx}$ in terms of t.

    b) Find an equation of the line tangent to the curve at $t = 1$.

    c) Write an integral expression representing the length of the curve over the interval $0 \le t \le 2$ and evaluate it numerically with your calculator.

12.3   Motion in a Plane

The motion of a particle moving along a line has been discussed. Given a position function s which relates a particle's line coordinate to time t, the velocity v of the particle was defined as the first derivative of the position function $v = \dfrac{ds}{dt}$, and the acceleration a was defined as the second derivative of the position function: $a = \dfrac{d^2 s}{dt^2}$.

Example 1:   *Motion Along a Line*

A particle moves on a coordinate line so that its position at time t is given by $s(t) = t^3 - 3t^2$. Describe the particle's motion for $t \geq 0$.

t	$s(t)$	$v(t)$	$a(t)$
0	0	0	−6
1	−2	−3	0
2	−4	0	6
3	0	9	12

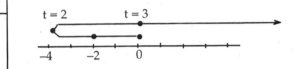

Now we are going to consider motion in a plane. A mathematical model which is useful in analyzing the motion of a particle in a plane is an algebraic structure called a **vector space.** We shall not study a vector space in its complete generality, but we will consider a particular instance of a vector space.

Ordered pairs of real numbers are familiar objects and they are the basic elements in the particular vector space we are going to consider. We will use the notation $\langle a, b \rangle$ for the ordered pair that refers to a vector, so as to not confuse it with the ordered pair (a, b) that refers to a point in the plane. It is common to denote vectors with bold type, as in $\boldsymbol{v}$, or with an arrow on top, as in $\overrightarrow{A}$. We begin by defining the sum $\oplus$ of two vectors.

Vector Addition

If $\boldsymbol{u} = \langle a, b \rangle$ and $\boldsymbol{v} = \langle c, d \rangle$, then the vector $\boldsymbol{u} \oplus \boldsymbol{v}$ is defined by

$$\boldsymbol{u} \oplus \boldsymbol{v} = \langle a, b \rangle \oplus \langle c, d \rangle = \langle a + c, b + d \rangle.$$

Since vector addition has been defined in terms of real number addition, it is no surprise that the $\oplus$ operation inherits the following properties:

a) There is a $\oplus$ identity, $\boldsymbol{0} = \langle 0, 0 \rangle$;

b) Every vector $\boldsymbol{u} = \langle a, b \rangle$ has an inverse $\langle -a, -b \rangle$, and we write $-\boldsymbol{u} = \langle -a, -b \rangle$;

c) The operation $\oplus$ is both associative and commutative.

As a matter of fact, if S denotes the set of all ordered pairs of real numbers, the structure $\{S, \oplus \}$ is a commutative group.

A peculiar multiplication operation combines the field of reals with the group $\{S, \oplus\}$.

Multiplication of a Vector by a Real number

If r is a real number and $\boldsymbol{u} = \langle a, b \rangle$, then the vector $r \cdot \boldsymbol{u}$ is defined by

$$r \cdot \boldsymbol{u} = \langle ra, rb \rangle.$$

Note that the product of a real number r and a vector $\langle a, b \rangle$ is another vector.

Example 2:   *Vector Operations*

If $u = \langle 2, 3 \rangle$, $v = \langle -3, 4 \rangle$, and $r = 6$, compute $u \oplus v$, $[r \cdot u\;] \oplus [r \cdot v]$, and $r \cdot \langle u \oplus v \rangle$.

Solution:

$$u \oplus v = \langle 2, 3 \rangle \oplus \langle -3, 4 \rangle = \langle -1, 7 \rangle$$

$$[r \cdot u\;] \oplus [r \cdot v] = [6 \cdot \langle 2, 3 \rangle] \oplus [6 \cdot \langle -3, 4 \rangle] = \langle 12, 18 \rangle \oplus \langle -18, 24 \rangle = \langle -6, 42 \rangle$$

$$r \cdot \langle u \oplus v \rangle = 6 \cdot [\langle 2, 3 \rangle \oplus \langle -3, 4 \rangle]\; = \; 6 \cdot \langle -1, 7 \rangle = \langle -6, 42 \rangle$$

The new operation $\cdot$ satisfies a number of familiar properties; for vectors u and v and real numbers r and s :

a) $r \cdot \langle u \oplus v \rangle = r \cdot u \oplus r \cdot v,$

b) $\langle r+s \rangle \cdot u = r \cdot u \oplus s \cdot u$,

c) $\langle r \cdot s \rangle \cdot u = r \cdot \langle s \cdot u \rangle$.

The set S of vectors with the operations of $\oplus$ and $\cdot$ is called a vector space and is denoted by V_2. The ordered pairs are called **vectors** and the real numbers are called **scalars**. The numbers a and b in the vector $\langle a, b \rangle$ are called **components**.

As with subtraction of real numbers, subtraction of vectors is defined by the addition of the additive inverse.

Subtraction of Vectors

For vectors u and v ,

$$u \ominus v = u \oplus -v,$$

Finally, vectors are equal if their corresponding components are equal.

Equal Vectors:

Vectors $u = \langle a, b \rangle$ and $v = \langle c, d \rangle$ are equal if and only if $a = c$ and $b = d$.

Geometric Representation of Vectors

Vectors can be represented by " arrows " (line segments which have a head and a tail). An arrow representing the vector $\langle 2, -3 \rangle$ is constructed by starting at any point T in the coordinate plane, moving right from T parallel to the x-axis a distance 2 to the point Q , moving down from Q parallel to the y-axis a distance 3 to a point H, then connecting T to H and placing the head of the arrow at H.

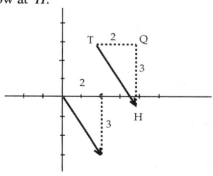

Another arrow representing the vector $\langle 2, -3 \rangle$ can be created by choosing T at the origin with the head at $H = (2, -3)$. This particular arrow is called the **standard representation** of the vector $\langle 2, -3 \rangle$.

Note that given an arrow from T to H, the components of the vector which it represents can be determined by constructing a right triangle with legs parallel to the axes and having the arrow as its hypotenuse.

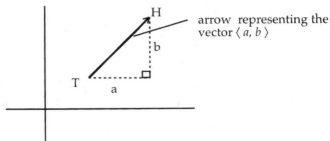

There are many arrows representing the same vector; however, the arrows have all the same length and point in the same direction. There is, however, a unique vector for each arrow.

For vectors $u = \langle 2, 3 \rangle$ and $v = \langle -4, 4 \rangle$, the vector sum $u \oplus v$ can be determined geometrically. Graph the standard representative of u ; then from the head of the arrow representing u construct the arrow which represents v; the arrow from $(0, 0)$ to $(-2, 7)$ is the representative of the vector $u \oplus v = \langle -2, 7 \rangle$.

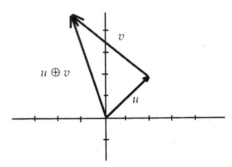

Next, the geometric effect of multiplying the vector $u = \langle 2, 3 \rangle$ by different scalars is illustrated.

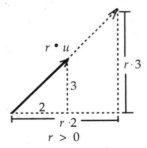
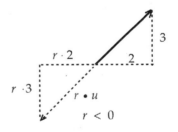

The Standard Unit Vectors i and j

Two vectors that play an important role in V_2 are given special names:

$$i = \langle 1, 0 \rangle \quad \text{and} \quad j = \langle 0, 1 \rangle.$$

The importance of i and j is based on the fact that any vector in V_2 can be expressed as a scalar times i plus a scalar times j (a **linear combination** of the **standard unit vectors i and j**). For example:

$$\langle 2, 3 \rangle = \langle 2, 0 \rangle \oplus \langle 0, 3 \rangle = 2 \langle 1, 0 \rangle \oplus 3 \langle 0, 1 \rangle = 2i + 3j.$$

Length of a Vector

The length of an arrow which represents a vector $\langle a, b \rangle$ is $\sqrt{a^2 + b^2}$.

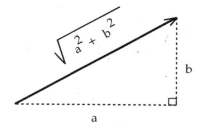

Thus we have

> **Length of a Vector**
> The length of a vector $\boldsymbol{u} = \langle a, b \rangle$, denoted by $|\boldsymbol{u}|$, is defined by,
> $$|\boldsymbol{u}| = \sqrt{a^2 + b^2}$$

A vector of length 1 is called a **unit vector**.

Example 3: *The Length of a Vector*

     If $\boldsymbol{u} = \langle -3, 4 \rangle$, find the length of the vector $\boldsymbol{u}$.

Solution:

$$|\boldsymbol{u}| = \sqrt{(-3)^2 + 4^2} = \sqrt{25} = 5.$$

Vector–Valued Functions

Having defined the vector space V_2, we now employ vector-valued functions to describe the motion of a particle in a plane. For example, suppose a particle moves in the plane so that at time t its position is determined by the values of the function $\boldsymbol{R}$ with

$$\boldsymbol{R}(t) = \cos t \cdot \boldsymbol{i} \oplus \sin t \cdot \boldsymbol{j}$$

t	$\boldsymbol{R}(t)$
0	$\langle 1, 0 \rangle$
$\dfrac{\pi}{6}$	$\langle \dfrac{\sqrt{3}}{2}, \dfrac{1}{2} \rangle$
$\dfrac{\pi}{2}$	$\langle 0, 1 \rangle$
$-\dfrac{\pi}{2}$	$\langle 0, -1 \rangle$

The table to the right illustrates that $\boldsymbol{R}$ has real number inputs and the corresponding outputs are vectors.

Although the vector-valued function $\boldsymbol{R}$ is not graphable, the path of the particle can be traced as follows: representing vectors in the range of $\boldsymbol{R}$ by arrows starting at the origin, the head of the arrow gives the position of the particle at time t.

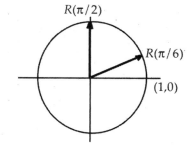

From the graph above it appears that the function $\boldsymbol{R}(t) = \cos t \cdot \boldsymbol{i} \oplus \sin t \cdot \boldsymbol{j}$ gives the location of a particle which is traveling on the unit circle.

If, as above, a vector-valued function R is defined in terms of vectors i and j so that for real-valued functions f and g, $R(t) = f(t) \cdot i \oplus g(t) \cdot j$, the functions f and g are called **component functions**. For example, if $R(t) = \cos t \cdot i \oplus \sin t \cdot j$, then $f(t) = \cos t$ and $g(t) = \sin t$ are the component functions. Unless stated otherwise, the domain of a vector-valued function R is considered to be the intersection of the domains of the component functions f and g, For instance, the domain of $R(t) = (\ln t) \cdot i \oplus \sqrt{1-t} \cdot j$ is the interval $(0, 1]$.

Example 4: *The Path of a Particle*

A particle moves in the plane so that at time $t \geq 0$ its position is determined by the values of the vector-valued function R with $R(t) = 2t \cdot i \oplus (3-t) \cdot j$. Graph the path of the particle.

Solution:

Some of the vector values that R relates to t are listed in the table. Plotting the values in a coordinate plane, we see that the terminal points for each arrow lie on a straight line.

t	R (t)
0	$\langle 0, 3 \rangle$
1	$\langle 2, 2 \rangle$
2	$\langle 4, 1 \rangle$
3	$\langle 6, 0 \rangle$
4	$\langle 8, -1 \rangle$

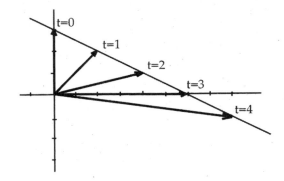

If the two component functions for R are defined by $x = 2t$ and $y = 3 - t$ we can solve the first for t, and substitute in the second to obtain

$$y = -\frac{1}{2}x + 3$$

This is clearly an equation whose graph is a straight line.

12.3   Exercises

1.   If $u = \langle -1,\ 2 \rangle$, sketch the arrows that represent

      a) $u \oplus i$ b) $u \oplus j$ c) $u \oplus i \oplus j$

      d) $-j$ e) $2u \ominus 2j$

2.   If $u = 2i \oplus 5j$ and $v = i \oplus j$, find the length of

      a) u b) v c) $u \oplus v$

3.   If $u = \left\langle \dfrac{3}{2}, 2 \right\rangle$, what is the length of the vector $\dfrac{1}{|u|} \cdot u$?

4.   If $|u| = |v|$, does $u = v$?

5.   Find a pair of vectors so that :

      a) $2u \oplus v = \langle 1,\ 7 \rangle$ b) $u \ominus 3v = \langle -12,\ -8 \rangle$

6.   If $u = \langle 1, 1 \rangle$ and $v = \langle -1, 0 \rangle$, determine scalars r and s so that

      a) $r \cdot u \oplus s \cdot v = \langle 2,\ -6 \rangle$ b) $r \cdot u \oplus s \cdot v = \langle -3, 7 \rangle$

7.   A vector-valued function R is defined by $R(t) = (2t - 1) \cdot i \oplus (4 - t) \cdot j$. Determine

      a) $R(1) \oplus R(3)$ b) $|R(2)|$ c) $|R(4) \ominus R(5)|$

In Exercises 8–12, the position vector for a particle moving in the plane is determined by values of the function R. Describe the path of the point.

8. $R(t) = t \cdot i \oplus \sqrt{4 - t^2} \cdot j, \quad -2 \le t \le 2$

9. $R(t) = (t + 1) \cdot i \oplus (t - 1) \cdot j$

10. $R(t) = (\cos t) \cdot i \oplus (1 - \sin^2 t) \cdot j, \quad 0 \le t \le \pi$

11. $R(t) = \dfrac{1}{t - 1} \cdot i \oplus \dfrac{2}{t} \cdot j$

12. $R(t) = \sec(\pi t) \cdot i \oplus \tan(\pi t) \cdot j, \quad 0 \le t \le \dfrac{1}{2}$

12.4   Limits and Differentiability of Vector-Valued Functions

Many definitions used in the calculus of real-valued functions can be applied to vector-valued functions. For example, you can add vector-valued functions, take the limit of a vector-valued function and differentiate a vector-valued function. The basic strategy is to extend the definitions on a component by component basis. The component by component extension of operations with real-valued functions to vector-valued functions is illustrated in the following definition of the limit of a vector-valued function.

> **Limit of a Vector-valued Function**
>
> If R is a vector-valued function such that $R(t) = f(t) \cdot i \oplus g(t) \cdot j$, then
>
> $$\lim_{t \to a} R(t) = [\lim_{t \to a} f(t)] \cdot i \oplus [\lim_{t \to a} g(t)] \cdot j$$
>
> provided that f and g have limits at $t = a$.

Thus, we take limits of vector-valued functions by taking the limits of their component functions.

Example 1:   *Limit of a Vector-Valued Function*

If R is defined by $R(t) = \cos t \cdot i \oplus \sin t \cdot j$, find $\lim_{t \to \pi/2} R(t)$.

Solution:

Since the component functions cosine and sine are continuous at $\frac{\pi}{2}$, $\lim_{t \to \pi/2} \cos t = 0$ and $\lim_{t \to \pi/2} \sin t = 1$. Thus

$$\lim_{t \to \pi/2} R(t) = 0 \cdot i \oplus 1 \cdot j = j.$$

We say that a vector-valued function $R(t) = f(t) \cdot i \oplus g(t) \cdot j$ is continuous at the input a provided $\lim_{t \to a} R(t) = R(a)$. This amounts to saying that R is continuous at $t = a$ if its component functions f and g are continuous at $t = a$.

The Derivative of a Vector Function

The derivative of a vector-valued function parallels that given for real-valued functions.

> **Derivative of a Vector-valued Function**
>
> If R is a vector-valued function, its derivative R' at the input a is defined by
>
> $$R'(a) = \lim_{h \to 0} \frac{R(a+h) \ominus R(a)}{h}$$

Note that $R(a+h) \ominus R(a)$ is the difference between two vectors and h is a scalar.

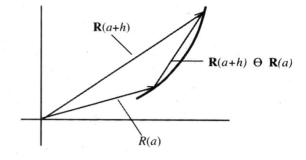

If the vector-valued function R is defined by $R(t) = f(t) \cdot i \oplus g(t) \cdot j$, and we assume the component functions f and g are differentiable at $t = a$, then $R'(a)$ exists and the derivative vector can be calculated by differentiating separately the component functions of R. To see why this is true we apply the definition of the derivative.

$$R'(a) = \lim_{h \to 0} \frac{R(a+h) \ominus R(a)}{h}$$

$$= \lim_{h \to 0} \frac{f(a+h) \cdot i \oplus g(a+h) \cdot j \ominus f(a) \cdot i \ominus g(a) \cdot j}{h}$$

$$= \lim_{h \to 0} \left\{ \left[\frac{f(a+h) - f(a)}{h} \right] i \oplus \left[\frac{g(a+h) - g(a)}{h} \right] j \right\}$$

$$= \left\{ \lim_{h \to 0} \left[\frac{f(a+h) - f(a)}{h} \right] \right\} i \oplus \left\{ \lim_{h \to 0} \left[\frac{g(a+h) - g(a)}{h} \right] \right\} j$$

$$= f'(a) \cdot i \oplus g'(a) \cdot j$$

Example 2: *Derivative of a Vector-Valued Function*

     If $R(t) = \cos t \cdot i \oplus \sin t \cdot j$, find $R'(t)$.

Solution:

     $R'(t) = -\sin t \cdot i \oplus \cos t \cdot j$

When a vector-valued function is used to predict the position of a particle moving in a plane, the derivatives R' and R'' have special significance.

Velocity Vector and Acceleration Vector

If R is a vector-valued function, R' is called the **velocity vector** and R'' is called the **acceleration vector**.

Example 3:   *Velocity and Acceleration Vectors*

If the position vector for a moving particle is determined by $R(t) = \frac{1}{2}t^2 \cdot i \oplus t \cdot j$, find the velocity and acceleration vectors at $t = 2$. Sketch arrows that represent $R(2)$, $R'(2)$, and $R''(2)$.

Solution:

$$R'(t) = t \cdot i \oplus j \;\;\Rightarrow\; R'(2) = 2 \cdot i \oplus j$$
$$R''(t) = i \;\Rightarrow\;\; R''(2) = i$$

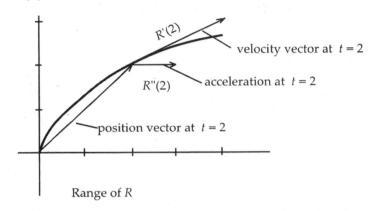

From the above example it appears that the arrow representing the velocity vector indicates the direction of motion. The length of the velocity vector also has meaning. Since the distance traveled along the graph of the vectors $R(t) = f(t) \cdot i \oplus g(t) \cdot j$ from $t = a$ to $t = x$ is determined by the function

$$s = \int_a^x \sqrt{[f'(t)]^2 + [g'(t)]^2} \; dt,$$

the rate of change of distance along the graph is

$$\frac{ds}{dt} = \sqrt{[f'(t)]^2 + [g'(t)]^2} \;.$$

Thus $\frac{ds}{dt}$ is called the **speed of a particle** moving along the graph of $R(t) = f(t) \cdot i \oplus g(t) \cdot j$, and we note that the speed at time t is the length of the velocity vector: Speed = $|R'(t)|$.

Example 4:   *Speed of a Particle*

A particle travels in the plane so that its position at time t is given by

$$R(t) = t \cdot i \oplus \ln(\sec t) \cdot j, \; 0 \le t \le \frac{\pi}{2}.$$ Determine the speed of the particle at $t = \frac{\pi}{4}$.

Solution:

$$R'(t) = i \oplus \tan t \cdot j$$

$$R'\left(\frac{\pi}{4}\right) = i \oplus \tan\left(\frac{\pi}{4}\right) \cdot j = i \oplus j$$

$$\left| R'\left(\frac{\pi}{4}\right) \right| = \sqrt{1+1} = \sqrt{2}$$

In the examples above we have found the velocity and acceleration vectors by differentiating the position function. Calculus applications often require solving the reverse problem – finding the position function for a given velocity or acceleration function.

Example 5: *Integrating a Vector-Valued Function*

The acceleration vector of a particle moving in a plane is given by the vector-valued function $R''(t) = 12t \cdot i \oplus 2 \cdot j$. When $t = 0$, the particle is at the point $(0, -1)$ and its velocity is $-i$. Find its position at time $t = 2$.

Solution:

To find the position function, we integrate twice, each time using initial conditions to solve for the constant of integration. The velocity vector is

$$R'(t) = \int R''(t)\,dt = \int [12t \cdot i \oplus 2 \cdot j]\,dt = 6t^2 \cdot i \oplus 2t \cdot j \oplus C$$

where $C = C_1 \cdot i \oplus C_2 \cdot j$. Letting $t = 0$ and applying the initial condition

$R'(0) = -1 \cdot i \oplus 0 \cdot j$, we obtain

$$R'(0) = -1 \cdot i \oplus 0 \cdot j = C_1 \cdot i \oplus C_2 \cdot j \Rightarrow C_1 = -1 \text{ and } C_2 = 0$$

Thus, the velocity vector is

$$R'(t) = (6t^2 - 1) \cdot i \oplus 2t \cdot j .$$

Integrating once more we obtain

$$R(t) = \int R'(t)\,dt = \int [(6t^2 - 1) \cdot i \oplus 2t \cdot j]\,dt = (2t^3 - t) \cdot i \oplus t^2 \cdot j \oplus C$$

where $C = C_1 \cdot i \oplus C_2 \cdot j$. Letting $t = 0$ and applying the initial condition

$R(0) = 0 \cdot i \oplus -1 \cdot j$, we obtain

$$R(0) = 0 \cdot i \oplus -1 \cdot j = C_1 \cdot i \oplus C_2 \cdot j \Rightarrow C_1 = 0 \text{ and } C_2 = -1.$$

Thus the position vector is

$$R(t) = (2t^3 - t) \cdot i \oplus (t^2 - 1) \cdot j , \text{ and}$$

at $t = 2$ the position is $R(2) = (2 \cdot 2^3 - 2) \cdot i \oplus (4 - 1) \cdot j = 14 \cdot i \oplus 3 \cdot j$.

12.4   Exercises

1.   If $R(t) = \dfrac{\sin t}{t} \cdot i \oplus \ln(1-t) \cdot j$, find $\lim\limits_{t \to 0} R(t)$.

2.   If $R(t) = \dfrac{\tan 2t}{t} \cdot i \oplus e^{-t^2} \cdot j$, find $\lim\limits_{t \to 0} R(t)$.

In Exercises 3–10, the motion of a particle in the plane is described by a vector equation. Sketch the path of the particle. At the given time *t*, represent the velocity and acceleration vectors as arrows starting from the position of the particle at the given time. Determine the speed of the particle at time *t*.

3. $R(t) = t \cdot i \oplus 2t^2 \cdot j$, $t = 1$

4. $R(t) = 2\sin t \cdot i \oplus 2\cos t \cdot j$, $t = \dfrac{\pi}{4}$

5. $R(t) = 2t \cdot i \oplus 2e^{-t} \cdot j$, $t = 0$

6. $R(t) = \sin t \cdot i \oplus \cos^2 t \cdot j$, $t = \dfrac{\pi}{2}$

7. $R(t) = t^2 \cdot i \oplus e^t \cdot j$, $t = 1$

8. $R(t) = (2-t) \cdot i \oplus \sqrt{t} \cdot j$, $t = 1$

9. $R(t) = \dfrac{2}{3}(1+t)^{3/2} \cdot i \oplus t\ln t \cdot j$, $t = 1$

10. $R(t) = \dfrac{1}{t} \cdot i \oplus \ln t \cdot j$, $t = 1$

11.   A particle moves in the plane so that its position vector and velocity are perpendicular at all times *t*. Show that the path of the particle is a circle centered at the origin.

12.   The acceleration vector of a particle moving in a plane is given by the vector-valued function $R''(t) = e^{-t} \cdot i \oplus e^t \cdot j$. When $t = 0$, the particle is at the point $(1, 2)$ and its velocity is $2 \cdot i$. Find its position at time $t = 1$.

13.   The velocity vector of a particle moving in a plane is given by the vector-valued function $R'(t) = \cos 2t \cdot i \oplus -\sin t \cdot j$. When $t = 0$, the particle is at the point $(3, -2)$. Find its position at time $t = \pi$.

14.   The acceleration vector of a particle moving in a plane is given by the vector-valued function $R''(t) = 6t \cdot i \; \oplus 2e^t \cdot j$. When $t = 1$, the particle is at the point $(0, 2e)$ and its velocity is $2 \cdot i \; \oplus 2e \cdot j$. Find its position at time $t = 0$.

15.   The velocity vector of a particle moving in a plane is given by the vector-valued function $R'(t) = \dfrac{1}{1+t} \cdot i \; \oplus 2t \cdot j \quad$ for $t \geq 0$.

     a) Find the particle's position as a vector-valued function of t if $R(1) = \ln 2 \cdot i$

     b) Approximate the distance the particle travels from $t = 0$ to $t = 2$.

16.   The acceleration vector of a particle moving in a plane is given by the vector-valued function $R''(t) = -2\sin t \cdot i \; \oplus \cos t \cdot j$. When $t = \dfrac{\pi}{2}$, the particle is at the point $(2,1)$ and its velocity is j. Find its position at time $t = \pi$.

17.   A moving particle has position $\langle x(t), y(t) \rangle$ at time t. The position of the particle at time $t = 2$ is $\langle -3, 5 \rangle$ and the velocity vector at time $t > 0$ is given by $\langle -2t, 2t^2 \rangle$.

     a) Find the acceleration vector at time $t = 2$.

     b) Find the position of the particle at time $t = 3$.

     c) For what time $t > 0$ does the tangent line to the path of the particle at $(x(t), y(t))$ have slope -3?

     d) Approximate the distance traveled by the particle from $t = 0$ to $t = 3$.

12.5　Polar Coordinates

In all of our work so far we have considered functions whose inputs and outputs were related to coordinate systems on two perpendicular lines. We now consider a way of assigning coordinates to points of the plane in which reference is made to a directed angle and the distance along a ray. This new system is called the **polar coordinate** system.

Let $\vec{OA}$ be any ray of the plane (called the **polar axis**, with point O called the **pole**), and let P be any point of the plane. If $\angle AOP = \theta$ (measured in radians and using the convention that $\theta > 0$ if $\angle AOP$ is measured counterclockwise while $\theta < 0$ if $\angle AOP$ is measured clockwise) and if $OP = r$, then one way of giving polar coordinates to P is to call them (r, θ).

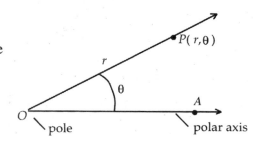

Since $\angle AOP$ can be measured in innumerable ways, differing from one another by multiples of 2π, other ways of designating the same point P are to use $(r, \theta + 2n\pi)$ where n is any integer.

We also adopt the convention that the coordinates of P can be given by $(-r, \theta + \pi)$. This corresponds to finding the ray $\vec{OQ}$ forming an angle $\theta + \pi$ with $\vec{OA}$ (note that $\vec{OQ}$ is opposite to $\vec{OP}$) and then, since $-r$ is negative, going out r units on the ray opposite to $\vec{OQ}$, namely $\vec{OP}$.

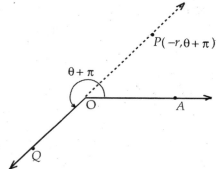

Thus, we find that the coordinates of a point P can be given in many ways by changing the angle specified. On the other hand, for any one given pair of polar coordinates (r, θ), there is a unique point P that those polar coordinates designate.

Example 1:　*Locating Points in a Polar Coordinate Plane*

Find the points in the plane with polar coordinates $P(2, \frac{\pi}{6})$, $Q(1, \frac{2\pi}{3})$, $R(-1, \frac{\pi}{2})$, and $S(-\frac{3}{2}, \frac{3\pi}{4})$.

Solution:

The points are plotted in the graph.

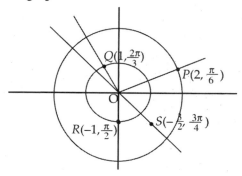

If we imagine two coordinate systems for the plane, a Cartesian system as well as a polar system, one superimposed upon the other, we can determine the relationships between the coordinate systems to be set up so that the polar axis of the one coincides with the positive part of the x-axis of the other and so that the ray $\theta = \frac{\pi}{2}$ coincides with the positive part of the y-axis. Then if a point P has polar coordinates (r, θ), the Cartesian coordinates of P will be given by

$$(x, y) = (r\cos\theta, r\sin\theta) \text{ or } \begin{cases} x = r\cos\theta \\ y = r\sin\theta \end{cases}$$

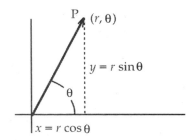

A conversion from Cartesian to polar coordinates is somewhat more difficult. From the fact that $x = r\cos\theta$ and $y = r\sin\theta$, we find that $r^2 = x^2 + y^2$ and $\tan\theta = \frac{y}{x}$.

Example 2:   *Converting From Polar to Cartesian Coordinates*

Find polar coordinates of the point with Cartesian coordinates $(-2, 2\sqrt{3})$.

Solution:

From the equations above, $r^2 = 4 + 12 = 16$ and $\tan\theta = -\sqrt{3}$. Since the point $(-2, 2\sqrt{3})$ is in the second quadrant, $\tan\theta = -\sqrt{3} \Rightarrow \theta = \frac{2\pi}{3}$. With r = 4, one set of polar coordinates is $(r, \theta) = (4, \frac{2\pi}{3})$. We can also use the coordinates $(r, \theta) = (-4, -\frac{\pi}{3})$, making use of a ray in the fourth quadrant along with a negative r.

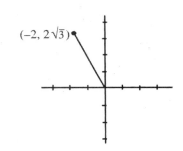

If we have an equation involving r and θ, its graph consists of all points in the plane having at least one set of polar coordinates satisfying the equation. If the equation happens to specify r as a function of θ, as many do, then various inputs θ are taken and the corresponding outputs r computed in order to find the points. Note that this is a violation of our convention to interpret the ordered pair (a, b) as having the output b for the input a. However, polar coordinates have historically been given as (r, θ), and we will prolong the tradition.

Example 3:   *Graphing Polar Equations*

Sketch a graph of the set of points with polar coordinates equation $r = 2\cos\theta$.

Solution:

We make a table of inputs θ and outputs $r = 2\cos\theta$ and use the pairs to plot points.

θ	r
0	2
$\dfrac{\pi}{6}$	$\sqrt{3}$
$\dfrac{\pi}{3}$	1
$\dfrac{\pi}{2}$	0
$\dfrac{2\pi}{3}$	-1
$\dfrac{5\pi}{6}$	$-\sqrt{3}$
π	-2

θ	r
$\dfrac{7\pi}{6}$	$-\sqrt{3}$
$\dfrac{4\pi}{3}$	-1
$\dfrac{3\pi}{2}$	0
$\dfrac{5\pi}{3}$	1
$\dfrac{11\pi}{6}$	$\sqrt{3}$
2π	2

$r = 2\cos\ \theta$

The polar graph of $r = 2\cos\theta$ appears to be a circle and we shall verify that as follows: Multiply both sides of the equation by r to obtain

$$r^2 = 2r\cos\theta.$$

Since $r^2 = x^2 + y^2$ and $x = r\cos\theta$, we substitute and obtain an equation for the graph in rectangular coordinates.

$$x^2 + y^2 = 2x$$

or

$$(x - 1)^2 + y^2 = 1$$

This is the equation for a circle with center at $(1, 0)$ and radius 1 and is in agreement with the polar graph in Example 3.

Symmetry

Three tests for symmetry in polar graphs are useful. For example, a polar curve is

a) symmetric about the origin if substitution
   of $-r$ for r produces an equivalent equation;

b) symmetric about the *x*-axis if substitution of $-\theta$
   for θ produces an equivalent equation;

c) symmetric about the *y*-axis if substitution of
   $\pi - \theta$ for θ produces an equivalent equation.

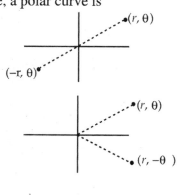

It will become clear in the following examples that if two of the three symmetries hold, the remaining one holds automatically. However, it is possible for a graph to have symmetry properties which the above rules fail to exhibit.

Example 4: *Sketching a Polar Graph*

Sketch a graph of the polar curve $r = 1 - \cos \theta$.

Solution:

Since $\cos(-\theta) = \cos \theta$, the equation is unchanged when $-\theta$ is substituted for θ and the graph is symmetric about the *x*-axis. Therefore we focus on the portion of the graph in the first and second quadrants to determine the shape of the curve for $0 \le \theta \le \pi$. Plotting a few points and using the symmetry about the *x*-axis we have

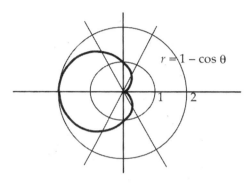

θ	r
0	0
$\dfrac{\pi}{6}$	$1 - \dfrac{\sqrt{3}}{2} \approx .13$
$\dfrac{\pi}{3}$	$\dfrac{1}{2}$
$\dfrac{\pi}{2}$	1
$\dfrac{2\pi}{3}$	$\dfrac{3}{2}$
$\dfrac{5\pi}{6}$	$1 + \dfrac{\sqrt{3}}{2} \approx 1.87$
π	2

This heart-shaped curve is called a **cardioid**, from the Greek (kordia), meaning heart.

In sketching graphs of polar equations it is often helpful to have a Cartesian graph of the same equation, for it can indicate the intervals in which r is positive or negative, increasing or decreasing.

Example 5. *Using a Cartesian Graph as a Aid in Polar Graphing*

Sketch a graph in polar coordinates of the set of points (r, θ) satisfying $r = 1 + 2\cos \theta$.

Solution:

We start with a Cartesian graph of $r = 1 + 2\cos \theta$

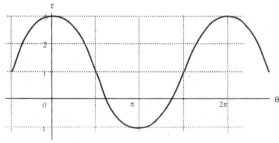

We then see that r decreases on the interval $[0, \pi]$, being positive when $0 \le \theta < \dfrac{2\pi}{3}$,

0 when $\theta = \dfrac{2\pi}{3}$ and negative when $\dfrac{2\pi}{3} < \theta \le \pi$. The interval $[\pi, 2\pi]$, on which r is increasing, can be analyzed similarly. If we translate this information into a polar graph, we see that when $|r|$ is decreasing, the points of the graph are getting closer to the pole; when r is positive, then the corresponding point is on the ray making an angle θ with the polar axis;

and when r is negative, the point is on the opposite ray. Thus we obtain the following polar graph, called a **limaçon** (from the Latin word limax, meaning snail).

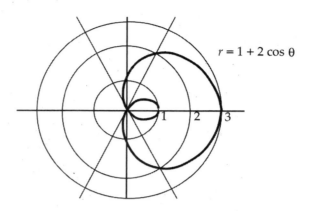

$r = 1 + 2 \cos \theta$

Example 6: *Graphing a Rose Curve*

Sketch the curve $r = 2 \sin 3\theta$.

Solution:

Since $\sin 3(\pi - \theta) = \sin 3\theta$, the graph is symmetric about the y -axis.

Now from the Cartesian graph of $r = 2 \sin 3\theta$, we see that r increases from 0 to 2 as θ increases from 0 to $\frac{\pi}{6}$. Then, as

θ increases from $\frac{\pi}{6}$ to $\frac{\pi}{3}$, r decreases from 2 to 0. This completes the right loop of the curve.

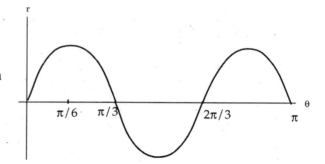

As θ increases from $\frac{\pi}{3}$ to $\frac{2\pi}{3}$, r is negative, first decreasing to –2 and then increasing to 0 , thus tracing the lower loop. For $\frac{2\pi}{3} < \theta < \pi$, r is positive and increasing to 2 and then decreasing to 0 to form the left loop of the curve. The curve is called a **three-leaved rose**.

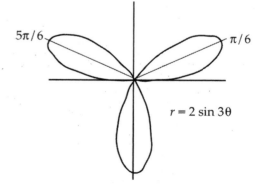

$r = 2 \sin 3\theta$

Example 7. *Graphing a Lemniscate*

Sketch the curve $r^2 = 4 \sin 2\theta$.

Solution:

Since $(-r)^2 = r^2$, the graph is symmetric about the pole. More importantly, since $r^2 \geq 0$, the function is only defined for values of θ such that $\sin 2\theta \geq 0$. From the Cartesian graph of $r = 2\sqrt{\sin 2\theta}$ it is clear that $\sin 2\theta \geq 0$ for $0 \leq \theta \leq \frac{\pi}{2}$ or $\pi \leq \theta \leq \frac{3\pi}{2}$. Using the Cartesian graph and the symmetry about the pole, we obtain the following polar graph. This curve is called a **lemniscate**.

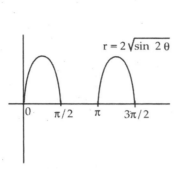

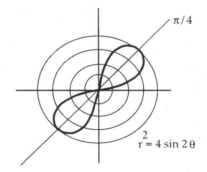

Technology and Graphs of Polar Equations

Graphs of polar equations are drawn on the TI-83 by changing to Polar mode (Press $\boxed{\text{MODE}}$ and select [Pol]), entering the polar equations in the Y= list, and setting the appropriate Window values. Notice that, in addition to designating a viewing rectangle, it is necessary to specify the range of θ-values, θmin and θmax, and the corresponding θ-step size.

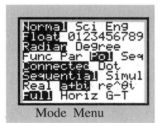

Mode Menu

Y= list

Window menu

Window menu

Graphing calculators make graphs of equations given in polar coordinates by translating each point of the polar graph, (r, θ), into its equivalent point in (x, y) coordinates.

Example 8: *Using a Calculator to Graph a Polar Curve*

Sketch the cardioid in Example 4 given by the polar equation $r = 1 - \cos \theta$.

Solution:

In polar mode, define the polar curve, set the window values and graph. For functions that are periodic on the interval $0 \leq \theta \leq 2\pi$ (such as $\sin (n\theta)$ or $\cos (n\theta)$), the interval $0 \leq \theta \leq 2\pi$ is usually sufficient to generate a complete graph.

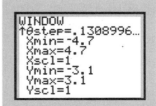

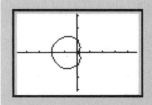

12.5   Exercises

1. Plot the points with the following polar coordinates.

a) $(2, \frac{\pi}{3})$

b) $(4, -\frac{\pi}{6})$

c) $(1, \frac{3\pi}{4})$

d) $(-2, \pi)$

e) $(\frac{3}{2}, \frac{5\pi}{6})$

f) $(-\frac{3}{2}, \frac{5\pi}{6})$

g) $(-1, 2\pi)$

h) $(3, \frac{7\pi}{3})$

i) $(3, -\frac{5\pi}{6})$

2. Find the Cartesian coordinates of each of the points in Exercise 1.

3. For each of the points given by the following Cartesian coordinates, find the corresponding polar coordinates.

a) $(2, 2\sqrt{3})$

b) $(\sqrt{3}, -1)$

c) $(-4, -4)$

d) $(\sqrt{3}, 1)$

e) $(-3, 3)$

f) $(-2, 2\sqrt{3})$

4. Transform the following equations of graphs from Cartesian coordinates to polar coordinates and sketch the graph.

a) $x = 4$

b) $x^2 + y^2 = 4$

c) $x^2 + y^2 = 2x$

d) $2x + y = 4$

e) $x^2 + (y-1)^2 = 1$

f) $(x^2 + y^2)^2 = 4(x^2 - y^2)$

5. Transform the following equations of graphs from polar coordinates to Cartesian coordinates and sketch the graph.

a) $r = 2\sin\theta$

b) $r = 2\cos\theta + 3\sin\theta$

c) $r = 2\tan\theta$

d) $r = \cos(2\theta)$

e) $r(1 - \cos\theta) = 2$

f) $r^2\cos(2\theta) = 4$

In Exercises 6–23, use a graphing calculator to sketch the graph of the polar curve and indicate any symmetry about either axis or the origin.

6. $r = \sin\theta$

7. $r = 2\cos\theta$

8. $r = 1 + \cos\theta$

9. $r = 1 - \sin\theta$

10. $r = 1 - 2\sin\theta$

11. $r = 4\cos 3\theta$

12. $r = \cos 4\theta$

13. $r = \sin 5\theta$

14. $r = 2 - \cos\theta$

15. $r = 1 + 2\sin\theta$

16. $r = 1 + \sin^2\theta$

17. $r^2 = 4\cos(2\theta)$

18. $r = 2\sin(\theta - \dfrac{\pi}{3})$

19. $r^2 = 2\cos\theta$

20. $r = \dfrac{1}{1 - \cos\theta}$

21. $r = \dfrac{2}{2 - \cos\theta}$

22. $r = 5$

23. $\theta = \dfrac{\pi}{3}$

24. Find the points of intersection of the circle $r = \cos\theta$ and the cardioid $r = 1 - \cos\theta$.

25. What is the minimum length of the interval for θ that will plot

   a) the complete circle $r = \sin\theta$?

   b) the complete circle $r = 2\cos\theta$?

26. Describe the general features of the graph of the function $r = f(\theta)$ in polar coordinates if

   a) f is even;

   b) f is odd.

27. a) Sketch the polar graph of $r = \sin(n\theta)$ for $n = 1, 2, 3, 4$.

   b) What can you conclude about the graph of $r = \sin(n\theta)$ for n positive integer?

28. a) Sketch the polar graph of $r = \cos(n\theta)$ for $n = 1, 2, 3, 4$.

   b) What can you conclude about the graph of $r = \cos(n\theta)$ for n positive integer?

29. Formulate and test a conjecture about the graph of $r = a + b\sin\theta$. What roles do the constants a and b play?

12.6   Area and Arc Length in Polar Coordinates

To find the area of a region bounded by the graph of a polar equation, we use an approximation based on the area of a sector of a circle to form a Riemann Sum.

If we have a sector of a circle of radius r, the sector having a central angle θ (measured in radians), then the area of the sector is

$$A = \frac{\theta}{2\pi} \cdot \pi \, r^2 = \frac{1}{2}\,\theta r^2.$$

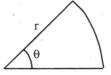

The area of the sector of a circle is $A = \frac{1}{2}\,\theta r^2$

If we have a region of the plane bounded by the polar curve $r = f(\theta)$ and the rays $\theta = \alpha$ and $\theta = \beta$ then a subdivision of the interval $[\alpha, \beta]$ into n equal subdivisions, namely

$$\alpha = \theta_0 < \theta_1 < \theta_2 < ... < \theta_{n-1} < \theta_n = \beta \,,$$

can be used to divide the region into n wedges, the kth wedge being bounded by the curve and the two rays $\theta = \theta_{k-1}$ and $\theta = \theta_k$.

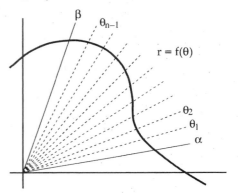

If we choose $\overline{\theta}_k$ in the interval $[\theta_{k-1}, \theta_k]$, then the area of the kth wedge is approximated by the area of the circular sector of radius $f(\overline{\theta}_k)$ and the central angle $\theta_k - \theta_{k-1}$. Hence the area of the kth wedge is approximately

$$\frac{1}{2}\,[f(\overline{\theta}_k)]^2\Delta\theta, \qquad \text{where } \Delta\theta = \theta_k - \theta_{k-1}.$$

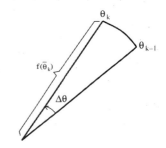

and the entire region has an area that is approximated by the Riemann Sum

$$\frac{1}{2}\,[f(\overline{\theta}_1)]^2 \cdot\Delta\theta + \frac{1}{2}\,[f(\overline{\theta}_2)]^2 \cdot \Delta\theta + ... + \frac{1}{2}\,[f(\overline{\theta}_n)]^2 \cdot \Delta\theta.$$

If f is a continuous function on $[a, b]$, then $\frac{1}{2}f^2$ is continuous; as n gets larger each central angle decreases in size and the limit of the Riemann Sum is given by a definite integral. Hence we have the area of the region in polar coordinates given by

$$\text{Area} = \int_{\alpha}^{\beta} \frac{1}{2}[f(\theta)]^2 \, d\theta$$

Example 1 *Area of a Polar Region*

Find the area of the region bounded by the cardioid $r = 2 + 2\cos\theta$.

Solution:

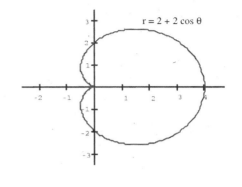

The graph is given on the right. The region is bounded by $r = 2 + 2\cos\theta$, $\theta = 0$ and $\theta = 2\pi$.
Making use of symmetry, we find the area of the upper half of the region and double the result. Hence the area is

$$A = 2\int_0^\pi \frac{1}{2}(2 + 2\cos\theta)^2\, d\theta$$

$$= \int_0^\pi \left(4 + 8\cos\theta + 4\cos^2\theta\right) d\theta$$

Using the fact that $\cos^2\theta = \frac{1}{2}(1 + \cos 2\theta)$,

$$A = \int_0^\pi (6 + 8\cos\theta + 2\cos 2\theta)\, d\theta$$

$$= [6\theta + 8\sin\theta + \sin 2\theta]_0^\pi = 6\pi \text{ units}^2.$$

The area of the region could also have been found by integrating from $\theta = 0$ to $\theta = 2\pi$.

One place where care must be taken involves the overlap of regions. For example, to find the area of the limaçon defined by $r = 1 + 2\cos\theta$ (see Example 5 of Section 12.5), the entire region is bounded by the rays $\theta = -\frac{2\pi}{3}$ and $\theta = \frac{2\pi}{3}$. If an integral were computed on the interval $[-\pi, \pi]$ or $[0, 2\pi]$, the area of the loop inside the limaçon would be counted twice.

Example 2: *Find the Area of a Region Between Two Polar Curves*

Find the area of the region in polar coordinates that is inside the limaçon $r = 2 + \cos\theta$ and outside the circle $r = 5\cos\theta$.

Solution:

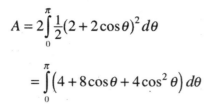

The graph at the right indicates the region of interest. If we find the area of the region inside the limaçon for $\frac{\pi}{3} \le \theta \le \frac{5\pi}{3}$ and subtract the area of the region inside the circle for $\frac{\pi}{3} \le \theta \le \frac{2\pi}{3}$ we will have the desired area. Hence

$$A = \int_{\pi/3}^{5\pi/3} \frac{1}{2}(4 + 4\cos\theta + \cos^2\theta)\, d\theta - \int_{\pi/3}^{2\pi/3} \frac{1}{2}(25\cos^2\theta)\, d\theta$$

$$= [2\theta + 2\sin\theta]_{\pi/3}^{5\pi/3} + \frac{1}{4}\int_{\pi/3}^{5\pi/3} \frac{1}{2}(1 + \cos 2\theta)\, d\theta - \frac{25}{4}\int_{\pi/3}^{2\pi/3} \frac{1}{2}(1 + \cos 2\theta)\, d\theta$$

$$= [2\theta + 2\sin\theta]_{\pi/3}^{5\pi/3} + \frac{1}{4}\left[\theta + \frac{1}{2}\sin 2\theta\right]_{\pi/3}^{5\pi/3} - \frac{25}{4}\left[\theta + \frac{1}{2}\sin 2\theta\right]_{\pi/3}^{2\pi/3}$$

$$= \left(\frac{10\pi}{3} - \sqrt{3} - \frac{2\pi}{3} - \sqrt{3}\right) + \frac{1}{4}\left(\frac{5\pi}{3} - \frac{\sqrt{3}}{4} - \frac{\pi}{3} - \frac{\sqrt{3}}{4}\right) - \frac{25}{4}\left(\frac{2\pi}{3} - \frac{\sqrt{3}}{4} - \frac{\pi}{3} - \frac{\sqrt{3}}{4}\right)$$

$$= \frac{8\pi}{3} - 2\sqrt{3} + \frac{\pi}{3} - \frac{\sqrt{3}}{8} - \frac{25\pi}{3} + \frac{25\sqrt{3}}{8} = \frac{11\pi}{12} + \sqrt{3} \approx 4.612\,\text{units}^2$$

You should check this result numerically to verify that it is correct.

Arc Length in Polar Coordinates

In order to find the length of an arc determined by the equation $r = f(\theta)$ in polar coordinates, we note that parametric equations in Cartesian coordinates for the same arc may be obtained from the fact that

$$\begin{cases} x = r\cos\theta \\ y = r\sin\theta. \end{cases}$$

Thus, we have the parametric equations

$$\begin{cases} x = f(\theta)\cos\theta \\ y = f(\theta)\sin\theta \end{cases}$$

and the length of the arc is

$$L = \int_{\theta=\alpha}^{\theta=\beta} \sqrt{\left(\frac{dx}{d\theta}\right)^2 + \left(\frac{dy}{d\theta}\right)^2}\, d\theta$$

$$= \int_{\alpha}^{\beta} \sqrt{(f'(\theta)\cos\theta - f(\theta)\sin\theta)^2 + (f'(\theta)\sin\theta + f(\theta)\cos\theta)^2}\, d\theta.$$

Simplifying this, we obtain

$$\boxed{L = \int_{\alpha}^{\beta} \sqrt{[(f'(\theta)]^2 + [(f(\theta)]^2}\, d\theta}$$

Example 3:   *Find the Length of a Polar Curve*

Find the length of the logarithmic spiral $r = e^{\theta}$ on the interval $[0, 2\pi]$.

Solution:

With $r = e^{\theta}$, $f(\theta) = e^{\theta}$, so $f'(\theta) = e^{\theta}$ and $[(f'(\theta)]^2 + [(f(\theta)]^2 = 2e^{2\theta}$. Hence the arc length is

$$\int_{0}^{2\pi} \sqrt{2e^{2\theta}}\, d\theta = \sqrt{2}\int_{0}^{2\pi} \sqrt{e^{\theta}}\, d\theta = \sqrt{2}\left[e^{\theta}\right]_{0}^{2\pi} = \sqrt{2}\,(e^{2\pi} - 1).$$

12.6   Exercises

In Exercises 1–10, sketch the curve and find the area of the region bounded by it.

1. $r = \cos\theta$ 2. $r = 2\sin(2\theta)$

3. $r = 1 - \sin\theta$ 4. $r = 2 + \cos\theta$

5. $r = \cos(3\theta)$ 6. $r = 2(1 + \sin\theta)$

7. $r^2 = \sin\theta$ 8. $r^2 = 2\sin(2\theta)$

9. $r = 4 + 2\sin\theta$ 10. $r = 1 - 2\sin\theta$

In Exercises 11-16, use your calculator to graph the polar equation. Set up an integral for finding the area of the indicated region and use the integration capabilities of the calculator to approximate the definite integral accurate to two decimal places.

11. Bounded by $r = \theta$, $\theta = 0$ and $\theta = \pi$.

12. Bounded by $r = \theta$, $\theta = 0$ and $\theta = 4\pi$.

13. Bounded by $r = 2\cos(2\theta)$, $\theta = -\dfrac{\pi}{4}$ and $\theta = \dfrac{\pi}{4}$.

14. Inside the circle $r = 4\cos\theta$ and outside $r = 2$.

15. Inside both $r = 2\sin\theta$ and $r = 2\cos\theta$.

16. Inside the cardioid $r = 1 + \cos\theta$ and outside the circle $r = \cos\theta$.

In Exercises 17-21, find the length of the indicated arc.

17. On $r = e^{2\theta}$ from $\theta = 0$ to $\theta = \dfrac{\pi}{2}$.

18. On $r = \theta^2$ from $\theta = \dfrac{\pi}{4}$ to $\theta = 1$.

19. On $r = 2 - 2\cos\theta$ from $\theta = 0$ to $\theta = 2\pi$.

20. On $r = \sec\theta$ from $\theta = 0$ to $\theta = \dfrac{\pi}{4}$.

21. On $r = a(1 + \cos\theta)$ from $\theta = 0$ to $\theta = 2\pi$.

Chapter 12   Supplementary Problems

1.   A curve C is defined by the parametric equations $x(t) = 4 - t$ and $y(t) = \dfrac{t}{t-4}$.

   a) Find $\dfrac{dy}{dx}$ in terms of t.

   b) Find an equation of the tangent line to C at $t = 2$.

   c) Find an equation for the curve C in terms of x and y.

2.   A particle has position $\langle x(t), y(t) \rangle$ at time t. The position of the particle at time $t = 1$ is $\langle 1, 2 \rangle$ and the velocity vector at time $t > 0$ is given by $\left\langle \dfrac{1}{t^2}, 4t \right\rangle$.

   a) Find the acceleration vector at time $t = 2$.

   b) Find the position of the particle at time $t = 3$.

   c) For what time $t > 0$ does the tangent line to the path of the particle at $\langle x(t), y(t) \rangle$ have slope 32?

3.   During the time period from $t = 0$ to $t = 6$ a particle moves along a path given by $x = t^2 - t$ and $y = t^3 - 2t^2$.

   a) Find the position of the particle at time $t = 2$.

   b) Find $\dfrac{dy}{dx}$ in terms of t.

   c) Find an equation of the tangent line to C at $t = 2$

   d) Find $\dfrac{d^2 y}{dx^2}$ at time $t = 1$.

4.   A particle has position $\langle x(t), y(t) \rangle$ at time t. The position of the particle at time $t = 1$ is $\langle \ln 2, 0 \rangle$ and the velocity vector at time $t > 0$ is given by $\left\langle \dfrac{1}{t+1}, 2t \right\rangle$.

   a) Find the acceleration vector at time $t = 2$.

   b) Find the position of the particle at time $t = 3$.

   c) Find the average rate of change of y with respect to x from $t = 0$ to $t = 4$.

   d) Find the instantaneous rate of change of y with respect to x at $t = 1$

5.   A curve C is defined by the parametric equations $x = \cos t + t \sin t$ and $y = \sin t - t \cos t$.

   a) What are the coordinates of the point on the curve C when $t = \dfrac{\pi}{2}$?

   b) Find $\dfrac{dy}{dx}$ in terms of t.

   c) Calculate the length of the curve over the interval $0 \le t \le \pi$

6.   A curve C is defined by the parametric equations $x = \frac{1}{8}t^3 - \frac{3}{2}t$ and $y = \frac{3}{8}t^2$.

   a) Find $\frac{dy}{dx}$ in terms of t.

   b) For what values of t are the tangents to the curve vertical?

   c) For what values of t are the tangents to the curve horizontal?

7.   The position of a particle at any time $t \ge 0$ is given by $x(t) = e^t + e^{-t}$ and $y(t) = e^t - e^{-t}$.

   a) Find the position of the particle at time $t = \ln 2$.

   b) Find $\frac{dy}{dx}$ at time $t = \ln 2$.

   c) Find an equation of the tangent line to the path of the particle at time $t = \ln 2$.

   d) Approximate the total distance traveled by the particle from $t = 0$ to $t = 1$.

8.   A particle moves in the plane so that its position at time $t \ge 0$ is determined by the vector-valued function

$$R(t) = \frac{1}{\sqrt{1+t^2}} \cdot i \oplus \frac{t}{\sqrt{1+t^2}} \cdot j.$$

   a) Find the velocity vector at time $t = 1$.

   b) At what time t is the speed of the particle a maximum?

9.   Find a rule for $R(t)$, given that $R'(t) = 2\cos t \cdot i \oplus t \sin t^2 \cdot j$ and $R(0) = i \oplus j$.

10.   Given the polar equation $r \sin \theta = r^2 \cos^2 \theta$, find the corresponding equation in rectangular coordinates and sketch its graph.

11.   Given the polar equation $r = 2\sin\theta + 2\cos\theta$, find the corresponding equation in rectangular coordinates and sketch its graph.

12.   A particle moves in the plane so that its position at time t is given by the parametric equations $x(t) = \sqrt{t-2}$ and $y(t) = \sqrt{6-t}$

   a) Find the magnitude of the velocity of the particle at $t = 3$.

   b) Find the x- and y -components of the acceleration of the particle at $t = 3$.

   c) Find a single equation in x and y for the path of the particle. Sketch the path.

13.   Given the polar curve $r = 1 + \cos\theta$.

   a) Sketch the polar curve.

   b) Find the area of the region enclosed by the polar curve.

   c) Find the length of the polar curve.

14.   Let R be the region inside the cardioid $r = 1 - \cos\theta$ and outside the circle $r = \sin\theta$.

   a) Sketch the two polar curves and shade region R.

   b) Find the area of region R.

15.   Find the length of the arc from $\theta = 0$ to $\theta = \dfrac{\pi}{4}$ for the polar curve $r = \sec\theta$

16 .   Given the polar curve $r = 7 + 3\cos(2\theta)$

   a) Sketch the polar curve.

   b) Find the area of the region enclosed by the polar curve.

   c) Find the length of the arc from $\theta = 0$ to $\theta = 2\pi$ for the polar curve.

17.   Find   a) $\lim\limits_{x \to 0} \dfrac{\ln(x+1)}{\tan x - 2x}$ b) $\lim\limits_{x \to \infty} \dfrac{x^2 - x}{2x^2 - 1}$

18.   Find the values of a and b such that $\lim\limits_{x \to 0} \dfrac{\sin 3x + ax + bx^3}{x^3} = 0$.

19.   Let f be the function defined by $f(x) = \dfrac{1}{x(\ln x)^2}$ for $x \ge e$ and let R be the region
   between the graph of f and the x-axis. Determine whether R has finite area. Justify your
   answer.

20.   Let f be the function defined by $f(x) = \dfrac{1}{1 + x^2}$ for $x \ge 1$ and let R be the region between
   the graph of f and the x-axis.

   a) Determine whether R has finite area. Justify your answer.

   b) Determine whether the solid generated by revolving region R about the y-axis has finite
       volume. Justify your answer.

21.   Let f be the function defined by $f(x) = \dfrac{1}{\sqrt{x}}$ for $0 < x \le 1$ and let R be the region between
   the graph of f and the x-axis.

   a) Determine whether R has finite area. Justify your answer.

   b) Determine whether the solid generated by revolving region R about the x-axis has finite
       volume. Justify your answer.

22.   Let f be the function defined by $f(x) = \dfrac{4}{x^2 - 2x - 3}$ for $4 < x \le 6$ and let R be the region
   between the graph of f and the x-axis.

   a) Find the area of R.

   b) Find the average value of f over the interval [4, 6].

23.   Consider the logistic differential equation $\dfrac{dy}{dt} = 2y(3 - y)$ with initial conditions $y(0) = \dfrac{1}{2}$.

a) Solve the differential equation.

b) At what value of y is $\dfrac{dy}{dt}$ a maximum?

24.   The number of flu-infected individuals in a population at a time is a solution of the logistic differential equation $\dfrac{dy}{dt} = 0.6y - 0.0002y^2$, where y is the number of infected individuals in the population and t is time in days.

a) Given that 10 people were infected at the initial time $t = 0$, find a solution for the differential equation.

b) How many days will it take for half of the population to be infected?

25.   Assume that the rate of growth of a population of fruit flies is proportional to the size of the population at each instant of time. If 100 fruit flies are present initially and 200 are present after 5 days, how many will be present after 10 days?

26.   Find the third-degree Taylor polynomial about $x = 0$ for the function f defined by $f(x) = \ln(1 - x)$.

27.   Let f be the function defined by $f(x) = \dfrac{1}{1 - 3x}$.

a) Find the first four terms of the Taylor series for f about $x = 0$.

b) Find the interval of convergence for the series in part a).

c) Find the first four terms of the series $g(x) = \dfrac{x}{1 + 3x}$.

28.   The Taylor series about $x = 1$ for a certain function f converges for all x in the interval of convergence. The nth derivative of f at $x = 1$ is given by $f^{(n)}(1) = \dfrac{n!}{n \cdot 2^n}$ and $f(1) = 0$.

a) Write the third degree Taylor polynomial for f about $x = 1$.

b) Find the radius of convergence for the Taylor series for f about $x = 1$.

29.   Let f be the function defined by $f(x) = \dfrac{1}{(1 + x)^2}$

a) Find the first four terms of the Taylor series for f about $x = 0$.

b) Find the interval of convergence for the series in part a)

c) Evaluate $f\left(\dfrac{1}{4}\right)$.

d) How many terms of the series are needed to approximate $f\left(\dfrac{1}{4}\right)$ with an error not exceeding 0.05? Justify your answer.

30.   Find the average value of the function f defined by $f(x) = \ln x$ on the interval $[1, e]$.

31.   Explain why each of the following statements is false. Giving a counterexample or an appropriate graph is satisfactory.

a) If the function f is continuous, then f is differentiable.

b) If the function f is increasing on an interval I, then its second derivative f'' is positive valued on I.

c) If $f''(a) = 0$, then $(a, f(a))$ is a point of inflection.

d) For all real numbers a, $\displaystyle\int_{-a}^{a} \sqrt{1 + x^2}\, dx = 0$.

e) If $F(x) = \displaystyle\int_{0}^{x} f(t)\, dt$ then $F'(x) = f(x)$.

32.   Find the equations of the lines through the point $(2, 5)$ that are tangent to the curve $y = 1 + 2x - x^2$.

33.   Find the minimum value of the function F defined for all $x, 0 \le x \le 1$,

by $F(x) = \displaystyle\int_{0}^{x} t\, e^t\, dt$. Justify your answer.

34.   Consider the family of polar curves defined by $r = 1 + \cos(k\theta)$ where k is a positive integer.

a) Find the area of the region enclosed by polar curve if $k = 3$.

b) Show that the area of the region enclosed by the polar curve does not depend on the value of k.

35.   Let f be a polynomial function with $f(0) = 2$ and derivative given by

$$f'(x) = (x^2 - 4)(x^2 - 3x + 2)(x^2 - 2x + 1).$$

a) Find the slope of the graph of f at the point where $x = 0$.

b) Write an equation for the line tangent to the graph of f at $x = 0$ and use it to approximate $f(0.2)$.

c) Determine the intervals on which f is increasing.

d) Determine the x-coordinates of the inflection points on the graph of f.

36.   Let f be the function defined by $f(x) = \begin{cases} x^2, & x < 2 \\ -(x - 3)^2 + k, & x \ge 2 \end{cases}$

(a) For what value of k will f be continuous at $x = 2$? Justify your answer.

(b) For the value of k found in part (a), determine whether f is differentiable at $x = 2$. Show your reasoning.

(c) Using the value of k found in part (a), find the average value of f on the interval $[0, 2]$.

37.   If the function f is defined by $f(x) = x^2 e^{-x^2}$ then its second derivative is
$f''(x) = 2e^{-x^2}(1 - 5x^2 + 2x^4)$.

a) Show that $f'(x) = 2xe^{-x^2}(1 - x^2)$.

b) Write an equation for the line tangent to the graph of f at $x = 1$.

c) Find all values of x at which f has a relative minimum. Justify your answer.

d) Give the x-coordinate of any one of the points of inflection.

38.   Let f be defined by the following graph and the table below contains some values of the continuous function g.

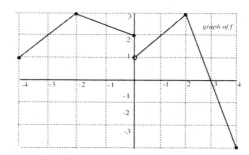

graph of f

x	−4	−3	−2	−1	0	1	2	3	4
$g(x)$	10	9	5	−1	0	2	6	0	−3

a) Find $f'(1)$.

b) Approximate $g'(2)$. Show your work.

c) If the function h is defined by $h(x) = g[f(x)]$, evaluate: i) $h(2)$ and ii) $h'(1)$

d) Approximate $\displaystyle\int_{-2}^{0} f(x)\, dx$.

39.   A particle moves along the x-axis (units in feet) in such a way that its velocity v at time $0 \le t \le 6$ is graphed in the figure. Its initial position at $t = 0$ sec is $x(0)=2$.

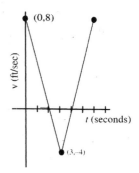

a) What is the particle's displacement between $t = 0$ and $t = 6$?

b) What is the total distance traveled by the particle between $t = 0$ and $t = 6$?

c) Give the position of the particle at $t = 3\sec$.

d) What is the acceleration of the particle at

i) $t = 1\sec$ and   ii) $t = 4\sec$.

40.   A 13 foot ladder is leaning against a wall so that the foot of the ladder is 1 foot from the wall. A gust of wind causes the ladder to begin sliding down the wall. The motion of the top of the ladder as it slides down the wall is described by

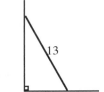

$y = -16t^2 + .05t + \sqrt{168}$, where t is measured in seconds.

a) When does the top of the ladder reach the ground?

b) Determine the velocity of the end of the ladder that is resting on the ground when it is 5 ft from the wall.

41.   The function f is defined on the closed interval [–3, 5]. The graph of f , shown in the figure, consists of three line segments. Let G be the function defined by

$G(x) = \int_{-3}^{x} f(t)\, dt.$

graph of f

a) Find $G(-3)$ and $G'(-3)$

b) On which interval or intervals is the graph of G concave down?

c) Find the input x at which G has its maximum value on the closed interval [–3, 5]. Justify your answer.

42.   Consider the following table of values for the differentiable function f.

x	1.0	1.2	1.4	1.6	1.8
$f(x)$	5.0	3.5	2.6	2.0	1.5

a) Estimate $f'(1.4)$.

b) Give an equation for the tangent line to the graph of f at $x = 1.4$.

c) What is the sign of $f''(1.4)$? Explain your answer.

d) Using the data in the table, find a midpoint approximation with 2 equal subdivisions for

$$\int_{1.0}^{1.8} f(x)\, dx.$$

43.   A graph of $y^3 + y^2 - 5y - x^2 = -4$ is shown at the right.

a) Find $\dfrac{dy}{dx}$ in terms of x and y.

b) Write an equation for the line tangent to the curve at the point (2, 0).

c) Find the x-coordinates of the points at which the graph of the equation has vertical tangent lines or horizontal tangent lines. Justify your answer.

Answers to Selected Problems

1.1 Exercises (pp. 5-8)

1. a) input: days of year, {1, 2, 3, ..., 365}
       output: hours of daylight, [5.5, 18.5]
       at 60° latitude

    b) input: time in minutes, [0, 60]
       output: degrees F, [32, 80]

    c) input: time in seconds, [0, 10]
       output: height in feet, [0, 200]

    d) input: time in hours, [0, 24]
       output: depth of water in feet, [3, 7]

3. a)   b)

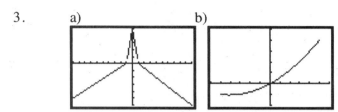

5. a) reals   b) [–5, –3] ∪ [–1, 2]   c) $f(1) \approx 1.8$, $f(4) = 0$

    d) (–5, –3) ∪ (–3, 1) ∪ (4, 5)   e) increasing

7.   9.   11.   13.

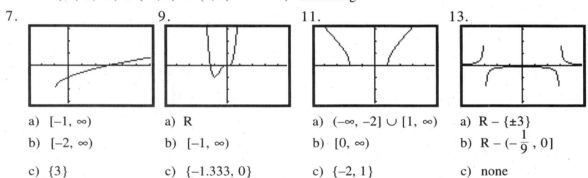

    7. a) [–1, ∞)   9. a) R   11. a) (–∞, –2] ∪ [1, ∞)   13. a) R – {±3}

       b) [–2, ∞)   b) [–1, ∞)   b) [0, ∞)   b) $R - (-\frac{1}{9}, 0]$

       c) {3}   c) {–1.333, 0} c) {–2, 1}   c) none

15.   17. a) $V = x(8.5 - 2x)(11 - 2x)$; $0 < x < 4.25$

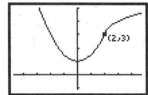

                                     b) x = 1.585 inches

19. a) $N(p) = (p - 60)\left(1000 + \frac{300 - p}{10} \cdot 50 \right)$ b) $2.80 gives max profit

21. a) h = 3/2 – 2e   b) $S = 6e - 6e^2$, $0 < e < .75$

23. a) max = –2.264 at x = 1   b) max = 1.619 at x = 1   25. $A = \frac{\sqrt{3}}{4} s^2$ 27. yes

1.2 Exercises (pp. 12-14)

1. $f(x) = x^3 + 2$ 3. $f(x) = \frac{1}{x + 1}$ 5. $f(x) = (x + 3)^2 - 9$

13. $y = 3|x| + 4$ 15. $y = \frac{1}{x - 1} + 3$ 17. $y = \sqrt{x + 2} + 3$

19. $y = 2\sqrt{x} - 3$ 21. b) reflects graph of f about the y-axis

23. g(x)   a) & b)   c)

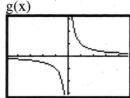   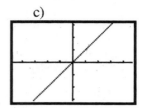

25. a) shifts g down 1; $-2 \le x \le 4$; $-3 \le y \le 2$

b) shifts g right 1; $-1 \le x < 5$; $-2 \le y \le 3$

c) reflects g about x-axis; $-2 \le x \le 4$; $-3 \le y \le 2$

d) shifts g left 2; $-4 \le x \le 2$; $-2 \le y \le 3$

e) reflects g about x-axis then shifts up 2; $-2 \le x \le 4$; $-1 \le y \le 4$

f) stretched g vertially by 2; $-2 \le x \le 4$; $-4 \le y \le 6$

g) stretches g vertically by 1/2; $-2 \le x \le 4$; $-1 \le y \le 1.5$

h) shifts g left 1 then down 2; $-3 \le x \le 3$; $-4 \le y \le 1$

1.3   Exercises (pp. 22-24)

1. $m = -1.5$; $(0, 6)$; $(4, 0)$ 3. $4x - 3y = -12$ 5. yes; $y = 2.1(x - 4.2) + 7.82$

7.   a) $m(-1) = -1$; $m(1) = 1$; $m(2) = 2$ b) $m(x) = x$ c) $(x \ne 0)$

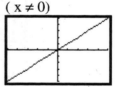

9. $F = \frac{9}{5}C + 32$ 11. a) ii   b) $y = 0.5(x - 1.01) + 1.505$ c) $y = 2.325$

13. a)   b) $d(1.3) \approx 4.7$; $d(1.8) \approx 8.95$

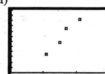   c) $y = 7.13x - 4.39$

d) $d(1.3) = 4.88$; $d(1.8) = 8.45$

15. a) $S = 2000 + 1.75x$; $F = 1500 + 2x$ b)

c) when flying less than 2,000 miles

17. b) $p(7) = 98.6$

1.4   Exercises (pp. 30-33)

1.   a) y_1, y_3, y_6 are increasing; y_2, y_4, y_5 3.   a) shifts graph 1 to right
are decreasing
                                                              b) reflects graph about y-axis
b) greatest y_6; least y_3
                                                              c) shifts graph 2 to left
c) y_5
                                                              d) reflects graph about both x and y-axes
d) yes; all when $x = 0$
                                                              e) shift graph right 1 and down 1
e) when $0 < b < 1$, f is decreasing;
     when $b > 0$, f is increasing.   f) stretches graph horizontally by a factor of 2

5. a) $y = -(3^x + 3)$ b) $y = 3^{-x}$ 7. a) $y = 3(2^x)$ b) $y = 2(5^{-x})$

    c) $y = -(3^{-x})$ d) $y = 3^{x-2} + 3$ c) $y = 1.5(3^x)$ d) $y = 2(0.9)^x$

9. $y = 120(1.15^x)$ 11. a) $y = 14000 - 1680x$; $x = 5$, $y = \$5600$ b) $7388.25

13. a) $A = 7000(1.0575^t)$ b) $t = 3$ years, $A = \$8278.26$ c) $A = \$14{,}000$, $t = 12.4$ years

15. linear; $y = 0.4x + 2.14$ 17. linear; $y = 1.70 - 1.175x$

19. $r = 6.5\%$ 21. a) $C = 830t + 15780$ b) $C = 15720(1.045)^t$ c) 23,250 d) Prairie U

1.5 Exercises (pp. 37-38)

1. $f(x) = 13(1.2214)^x$, increasing   3. $p(t) = 0.5(0.5117)^t$, decreasing

5. $h(x) = 1705(0.135)^x$, decreasing   7. $A(x) = A_0(e^{0.693x})$

9. $P(x) = 16.5(e^{0.1823x})$ 11. $k(x) = 275(e^{-1.0986x})$

13. a) $A = 100(1.07)^t$ b) $A = 100(1.0175)^{4t}$ c) $A = 100\left(1 + \dfrac{0.07}{12}\right)^{12t}$

    d) $A = 100\left(1 + \dfrac{0.07}{365}\right)^{365t}$ e) $A = 100(e^{.07t})$

15. a) $P = 200e^{.11t}$ exceeds 10,000 in 35.6 days   17. $n = 414$ 19. $263.34

1.6 Exercises (pp. 42-44)

1. $y = f^{-1}(x)$

x	0	2	10	26	50
y	–1	1	3	5	7

3. $y = f^{-1}(x)$

x	0	1/2	$\sqrt{2}/2$	$\sqrt{3}/2$	1
y	0	$\pi/6$	$\pi/4$	$\pi/3$	$\pi/2$

5. domain $f = [-3, 2]$; domain $f^{-1} = [-3, 3]$ 7. domain $f = [-3, 3]$; domain $f^{-1} = (-2, 0] \cup [1, 2]$

9. a) and d) are one-to-one   11. $K^{-1}(x) = \sqrt{x} + 1$

13. no, either $-3 \le x \le -1$ or $-1 \le x \le 4$ 15. no, either $-3 \le x \le 0$ or $0 \le x \le 3$

17. $F^{-1}(x) = x^3$; domain $F^{-1} =$ domain $F = R$ 19. $H^{-1}(x) = x^2 + 1$; domain H^{-1}: $x \ge 0$; domain H: $x \ge 1$

21. $G^{-1}(x) = \sqrt{x + 2}$; domain G^{-1}: $x \ge -2$; domain G: $x \ge 0$

23. $J^{-1}(x) = \dfrac{1}{x}$; domain J^{-1}: $x \ne 0$; domain J: $x \ne 0$ 25. $g^{-1}(x) = \ln(x)$

27. $g(x) = \begin{cases} \dfrac{x + 1}{2}, & x < -1 \\ \sqrt{x + 1}, & x > -1 \end{cases}$ 31. a) $f^{-1}(x) = \sqrt[3]{1 - x}$ b) 5

1.7 Exercises (pp.49-50)

1. a) 3   b) 4   3. a) $y\log x - \log z$

    c) –4   d) 7   b) $2\log x + 3\log y - \log z$

                  c) $4\log x + \log y + 0.5\log z$

                   d) $\log z - 0.5\log x - 0.5\log y$

5. $\dfrac{\ln 7}{\ln 5}$ 7. $\dfrac{\ln 9.41}{2\ln 3}$ 9. $\ln 14756 - \ln 5320$

11. 1   13. 57.4 days   15. 8.8 years   17. a) $V = 4300(0.75)^t$ b) 13 years

19. $y = \dfrac{\ln x}{\ln 2}$ 21. $y = \dfrac{2\ln x}{\ln 0.5}$ 23. $y = 2^x,\; y = \log_2 x$ 25. $y = e^x,\; y = \ln x$

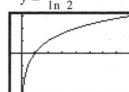 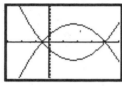 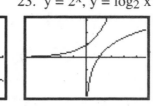 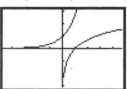

29. a) $(4, 13]$ b) No solution   c) $x = \dfrac{33}{7} \approx 4.714$ d) $x = 16, \dfrac{1}{2}$

1.8   Exercises (pp. 58-61)

1. a) 1   b) 0   c) 14 d) 4   e) $\{0, 3\}$

3. a) even   b) odd   c) odd   d) even   e) neither   f) neither

5. $p(x) = (x + 1)(x - 2)(x - 3)(x^2 + 1)$

7. a) 1, 4   b) 0.517, −1.472   c) −1, 2.414

9. a)   b) the same   c) reflections about x–axis

11. $f(x) = \dfrac{(x + 2)(x - 3)}{(x + 1)(x - 2)}$ 13. f is the slant asymptote for g

15. $y = -x^2 + 3x + 6$ 17.   a) good on [0,1], poor on [1, 2], [0, 1.04]

19. a) 2   b) f   c) g   d) $\{4, 2\}$

1.9 Exercises (pp. 66-68)

1. a) 0   b) −1   c) 3   d) −1   e) −1   f) −3, −2, −1, 0, 5

3. a) 1   b) −1   c) undefined   d) $\{-1, 0\}$ e) $\{-2, 0, 2, 4\}$

5. a) $f(g(x)) = (5x + 1)^3$; Dom $f \circ g = \mathbb{R}$ b) $h(f(x)) = 2^{x^3}$; Dom $h \circ f = \mathbb{R}$

   c) $h(g(x)) = 2^{5x+1}$; Dom $h \circ g = \mathbb{R}$ d) $g(h(x)) = 5(2^x) + 1$; Dom $g \circ h = \mathbb{R}$

   e) $g(f(x)) = 5x^3 + 1$; Dom $g \circ f = \mathbb{R}$ f) $f(g(h(x))) = [5(2^x) + 1]^3$; Dom $f \circ g \circ h = \mathbb{R}$

7.

	a	b	c	d	e	f
inner	$7x - 3$	$x^2 + 4$	$3x - 5$	$\ln x$	e^x	$\ln x$
outer	x^3	$\ln x$	2^x	e^x	$7x+13$	$x^2 + 1$

11. a) graph of f shifted up 1   13. a) $f^{-1}(x) = \dfrac{x + 2}{5}$

   b) graph of f shifted right 2 then reflected in the x-axis   b) $f^{-1}(x) = \dfrac{\ln x + 1}{2}$

   c) graph of f stretched vertically by 2 then shifted down 3

   d) graph of f reflected about y-axis then about x-axis   c) $f^{-1}(x) = \sqrt[3]{e^x - 4}$

                                                                d) $f^{-1}(x) = 1 + \dfrac{1}{x}$

15. a) $[f \circ g](x) = (x - 2)^2 - 4$ b) $[g \circ f](x) = x^2 - 6$ c) $[h \circ g(x) = |x - 2|$

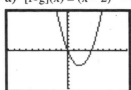 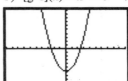 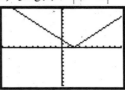

d) $[h \circ f \circ g](x) = |(x - 2)^2 - 4|$ e) $[h \circ g \circ f](x) = |x^2 - 6|$

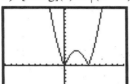 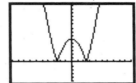

1.10 Exercises (pp. 73-75)

1. a)   b) amplitude changes by a factor of |A|

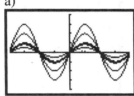

3. a) graph shifts horizontally to left   b) graph shifts horizontally to right

5. Not the same; $2\cos x + 3 \neq 2(\cos x + 3)$

7. a) 7   b) 3   9. a) $\{0, \pi, 2\pi\}$ b) $x = 0$ or $x = \frac{\pi}{2} \pm 2n\pi,\ n \in N$

11. a) $\{-\frac{7\pi}{4}, -\frac{3\pi}{4}, \frac{\pi}{4}, \frac{5\pi}{4}\}$ b) $[-\frac{7\pi}{4}, -\frac{3\pi}{4}] \cup [\frac{\pi}{4}, \frac{5\pi}{4}]$ c) $\{-2\pi, -\frac{3\pi}{2}, 0, \frac{\pi}{2}, 2\pi\}$

   d) $[-2\pi, -\frac{3\pi}{2}] \cup [-\pi, -\frac{\pi}{2}] \cup [0, \frac{\pi}{2}] \cup [\pi, \frac{3\pi}{2}]$ e) $\{0, \pm 1.895\}$ f) 1.030

13. Max = June 21; Min = Dec 21   15. $\sin(\frac{\pi}{4}) = \frac{\sqrt{2}}{2}$; $\cos(\frac{\pi}{4}) = \frac{\sqrt{2}}{2}$

17. a) $-3\sin[3(x - \frac{\pi}{3})]$ b) $-\cos[\frac{1}{2}(x + \frac{\pi}{4})]$

Chapter 1 Supplementary Problems (pp. 76-82)

1. a) T   b) T   c) F   d) T   e) F

3. a) $y = 2^{x-1}$ b) $y = (x - 2)^2 - 2$ c) $y = -\sin x$ d) $y = \frac{x}{x - 1}$

5. a) (1, 0), (0, 1/8)   c)

   b) $y = 0$; $x = 2, x = -2$

7. a) R   b) 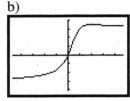   9. a) exponential   c)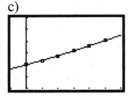

c) $y = 3$, $y = -3$

b) $y = 8(1.046)^x$

d) $(-3, 3.464]$

d) 2.6 hours

11.   a) exponential; $y = 3.12(0.705)^x$ b) linear; $y = 2.46x + 2.71$

13.   a) $f(x) = x^2$; $g(x) = \sin x$ b) $f(x) = \sin x$; $g(x) = x^2$

c) $f(x) = x^3$; $g(x) = x + 1$ d) $f(x) = e^x$; $g(x) = \cos x$

15.   a) $-2 \le x < \infty$; $0 \ge y > -\infty$ b) $R - \{2\}$; $R - \{1\}$

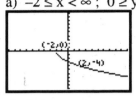   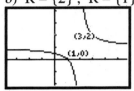

c) $0 < x < \infty$; $-\infty < y < \infty$ d) $-\infty < x < \infty$; $-2 \le y \le 2$

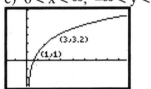   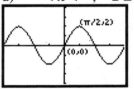

17.   a) i) –2 ii) 24 iii) undefined iv) 2.5   b) domain $f + g = \{-3, -2, -1, 1, 2, 3, 4\}$

domain $f/g = \{-3, -2, -1, 2, 3, 4\}$

19.   a) $f^{-1}(x) = \dfrac{5x}{3 - x}$ b) $\{0, -2\}$

21.   a) $(-0.246, -1.216)$, $(0.960, 4.020)$, $(4.161, -2.361)$

b) $\{-5, -\pi, 0, \pi, 5\}$ c) $[0, 4.758]$

23.   a) 1.253, 1.386 b) $-\infty < x < \infty$, $[\ln 2, \ln 4]$ c) neither   d) yes, 2π

25.   a) i) 6π ii) 2   iii) $[0, 6\pi]$, $[-3, 3]$

b) i) 2   ii) 1/2   iii) $[0, 6]$, $[-2, 2]$

c) i) π ii) 3   iii) $[3\pi/4, 19\pi/4]$, $[-4, 4]$

27. $-\infty < x \le -2$ or $12 \le x < \infty$; $0 \le y < \infty$; $\{-2, 12\}$

29.   a) 15   b) 7/8   c) 5   d) 1/17   e) 6

31.   a) R   b) 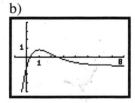   c) $y = -1$ d) $(-\infty, .839]$

2.1   Exercises   (pp. 87-89)

1.
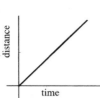

3.
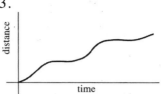

5. a) 50 mph   b) 40 mph   c) speeding up   d) 50 mph; 0 mph   e) t ≈ .4 hr, 1 hr , 1.75 hr

7. a) B   b) C   c) 23.3   d) 30 mph; 40 mph

9. a) 80 ft/sec   b) 64 ft/sec   c) 3 sec

11. a)
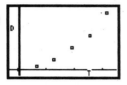

b) $\dfrac{\text{rise}}{\text{run}} = \dfrac{\text{distance}}{\text{time}}$

c) i) 42.83 ft/sec   ii) 69.13 ft/sec

iii) 55.98 ft/sec

13. a) avg speed = 5/3 m/sec; avg velocity = 0 m/sec   b) 1.61 m/sec

2.2   Exercises   (pp. 97-99)

1. a) –1   b) y = –x + 3   c)
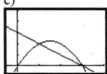

3. a) -4   b) -4   c) y = -4(x – 4) + 7   d) f(4.5) ≈ 5

5. $m_{sec} = 5 + 2h$; 5.002; 4.998; 5   7. $m_{sec} = \dfrac{2^{2+h} - 4}{h}$; 2.7735; 2.7716; 2.77

9. $m_{sec} = \dfrac{\sin(1+h)^2 - \sin 1}{h}$; 1.0795; 1.0817; 1.08

11. a) Derivative exists everywhere except at (3, 0). b) Derivative exists everywhere except at (1, f(1)).

13. 101.11; 110.0; 105.56   15. a) 1.5   b) 0.375   c) –1

17. Zooming in reveals graph is locally linear except at (0, 0).

19. Derivative does not exist at x = 0, 1, 2, 2.5, 4, 5.

2.3   Exercises   (pp. 105-108)

1.
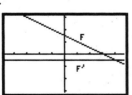

3.
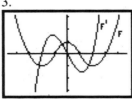

5.
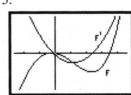

7.
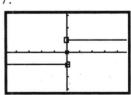

9. a)

x	-3	-2	-1	0	1
f '(x)	4	2	0	-2	-4

b) linear

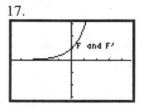

c) $f'(x) = -2x - 2$; $f'(-1.5) \approx 1$

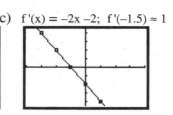

11. a) $f'(1.7) \approx 7.5$ b) $y = 7.5(x - 1.7) + 23$ c) 23.75

13.

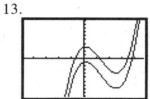

Graphs differ by only
a vertical shift.
$f'(x) = g'(x)$ for all x.

15.

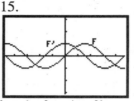

17.

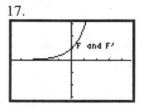

19. $\dfrac{f(x+h)-f(x)}{h} = \dfrac{[3(x + h) + 1] - [3x + 1]}{h} = \dfrac{3x + 3h + 1 -3x -1}{h} = \dfrac{3h}{h} = 3.$

Then $\dfrac{dy}{dx} = \lim_{h\to 0} \dfrac{f(x + h) - f(x)}{h} = \lim_{h\to 0} 3 = 3.$

21. $\dfrac{f(x+h)-f(x)}{h} = \dfrac{[(x+h)^2 - (x + h)] - [x^2 - x]}{h} = \dfrac{x^2 + 2xh + h^2 - x - h - x^2 + x}{h} = \dfrac{2xh + h^2 - h}{h} = 2x + h - 1.$

Then $\dfrac{dy}{dx} = \lim_{h\to 0} \dfrac{f(x+h)-f(x)}{h} = \lim_{h\to 0} (2x + h - 1) = 2x - 1.$

23. a) The graph of g is f-graph shifted down 3 units.   b) They are the same. c) g'(2) =12

25. a) The graph of g is f-graph stretched vertically by a factor of 3.

b) The graph of g' is f'-graph stretched vertically by a factor of 3.   c) $g'(1) = 9$

27. a) 2, 1, 0.67, 0.5, 0.4   b)

c) $f'(x) = \dfrac{1}{x}$

29. b) Use the average of left and right secant slopes (where possible) to approximate f '(t).

t	1985	1986	1987	1988	1989	1990	1991	1992
f '(t)	-1.52	-0.83	0.35	1.14	0.42	-1.35	-2.025	-1.95

c) graph of f   graph of f '

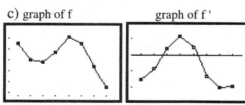

2.4   Exercises   (pp. 112-113)

1. $y = 2x$

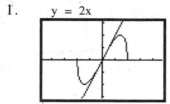

3. $y = 2(x - e) + e$

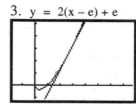

5. $y = 1.08(x - 1) + \sin 1$

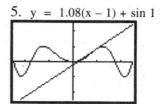

7.   a) 0.85; 0.55   b) $y - 0.19 = 0.85(x - 0.1)$

9.   a) f '(0) is undefined; f '(−1) = 1; f '(1) = −2.   13. a)

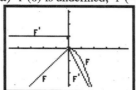

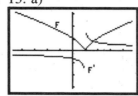

b) Graph of f is not locally
    linear at (1, 0).

c) 0; same type of error as
    in abs(x) at (0, 0). This
    function graph is
    symmetric across the
    line x = 1.

2.5   Exercises   (pp. 116-120)

1. $f(x) = x^4 - 2x^2$ f'

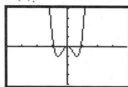

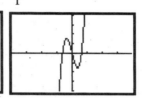

critical num	pos to neg	neg to pos	local min	local max
−1		X	X	
0	X			X
1		X	X	

3. $f(x) = \sin x - \cos x$ f'

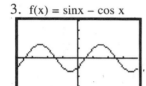

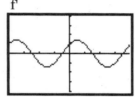

crit num	pos to neg	neg to pos	local min	local max
-3.927	X			X
-0.785		X	X	
2.356	X			X
5.498		X	X	

5. $f(x) = \frac{2}{3} x^3 - x^2$ f '

critical num	pos to neg	neg to pos	local min	local max
0	X			X
1		X	X	

7. a) If c is a critical number and f'(x) changes sign from positive to negative (as x goes from being less
        than c to being greater than c), then f has a local maximum at (c, f(c)). If c is a critical number and
        f'(x) changes sign from negative to positive, then f has a local minimum at (c, f(c)).
    b) No. If f'(x) does not change sign, then it is possible for f to fail to have a local
        maximum or minimum at (c, f(c)). For example: $f(x) = x^3$ at x = 0.

9. a) y = e^x   b) y = e^{-x}   11. a) −1, 2   e)

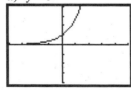

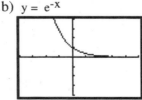

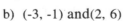

b) (-3, -1) and(2, 6)

c) (-3, 1) and (4.5, 6)

d) f '(0.5) ≈ 2.5

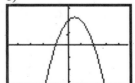

13.   a) T   b) F   c) F   d) F

15.   a) increasing -4 < x < 2.2 and 3.6 < x < 4;   d)
          decreasing 2.2 < x < 3.6

      b) −3, 2.2, 3.6

      c) max (2.2, g(2.2)); min (3.6, g(3.6))

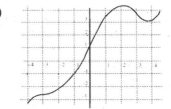

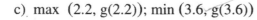

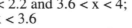

17. three; $0, \pi, 2\pi$

19. a) $(-\infty, -1)$ and $(2, \infty)$ d)

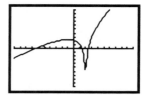

b) Local max at $x = -1$

c) $(-\infty, 2)$ and $(2, \infty)$

21. a) $0 < x < \frac{\pi}{2}$; $\pi < x < \frac{3\pi}{2}$, etc   d)

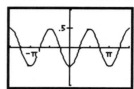

b) max $\pm\frac{\pi}{2}, \frac{3\pi}{2}$, etc;   min $\pm\pi, 2\pi$, etc

c) $\frac{\pi}{4} < x < \frac{3\pi}{4}$, etc

23. a) $(-4, -1)$ and $(1, 4)$, because f appears to be increasing on these intervals.
     b) Between -2 and 0, because f changes from increasing to decreasing there, also between
         0 and 2 , because f changes from decreasing to increasing there.

25. a) $(-2, -1.1)$ and $(1.5, 2)$, where $f'(x) > 0$.   c)

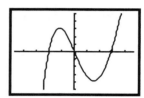

b) Between 1 and 1.5, where $f'(x)$ changes from
    negative to positive.

2.6   Exercises   (pp. 125-129)

1. a) $-3 < x < 1$, $1 < x < 5$, $x = 1$ b) $f''(2) < f'(2) = f'(0) < f(2)$

3. a) incr $0 < x < 0.505$ c)

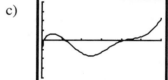

and $2.45 < x < 6$

b) up $1.442 < x < 3.326$

and $4.543 < x < 6$

5. a) $2 < x < 4$ b) $(2, g(2))$ c) increasing; $g'(2) > 0$

7. a) i) incr $(2, 4]$ b) i) incr $-4.71 < x < -1.57$, $0 < x < 1.57$, $4.71 < x \le 2\pi$
       ii) up $[-4,1)$ and $(1.667, 4]$ ii) up $-2\pi \le x < -3.43$, $-0.86 < x < 0.86$, $3.43 < x \le 2\pi$
       iii) $x = 1$ and $x = 1.667$ iii) ± 3.43, ± 0.86
    c) i) incr $(-1.62, -1)$ and $(.62, 2]$,
       ii) up $[-2, -1.33)$ and $(0, 2]$
       iii) $x = -1.33$ and $x = 0$

9.   11.

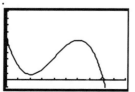

$x < 0$:   f is increasing and concave up

$x = 0$:   point $(0, 3)$ is on curve; function is
             increasing; concavity changes
$0 < x < 2$: f is increasing and concave down
$x = 2$:   graph has a horizontal tangent
$x > 2$:   f is decreasing and concave down

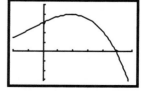

13. a) N is increasing but the rate of increase is decreasing   b) N is decreasing more and more
         rapidly   c) N is increasing more and more rapidly.

15. a) 5 mph; west   b) – 17.5 mph   c) t = 2.5

    d) i)   The car is 12 miles west of Worcester after 2 hours.
        ii) After two hours the car is traveling east at 30 mph.
        iii) After 2.5 hours the car's velocity is neither increasing nor decreasing.

17. The graph of f is the a-graph; the graph of f ' is the b-graph; the graph of f" is the c-graph

19. a) V '(t) is the rate the volume is changing. Since water is flowing into the tank at a
    constant rate, V '(t) is a positive constant. H '(t) is the rate at which the depth of water is
    changing. Since the water is flowing into the tank, the depth is always increasing. Thus
    H '(t) > 0.

    b) Since V '(t) is constant, we know that V "(t) = 0.

    c) i.   When the tank is one-quarter full, the depth is increasing, but more slowly than just
              before that moment, Hence H "(t) < 0 at that moment.
        ii.   When the tank is half full, the depth is increasing, but that increase is changing form
              slowing down to speeding up. Hence H "(t) = 0 then.
        iii. When the tank is three-quarters full, the depth is increasing, but since the radius of the
              cross section is shrinking, the depth will be increasing at ever-increasing rates. Hence,
              H "(t) > 0 at that moment.

2.7 Exercises (pp. 137-139)

1. 0.5   3. 2.718   5. 0   7. 0   9. 2.718

11. a) 5.545; $f(x) = 2^x$; $a = 3$ b) -1.98; $f(x) = 2\sin x$; $a = 3$

    c) 0.5; $f(x) = \sqrt{x}$; $a = 1$ d) 2; $f(x) = \tan x$; $a = \frac{\pi}{4}$

    e) 4; $f(x) = x^2$; $a = 2$ f) 1; $f(x) = \ln x$; $a = 1$

13. a) –1   b) 1   c) –1   d) –1   e) 1   f) does not exist

15. 0   17. 1   19. 6   21. 0.368   23. k/2   25. 1   27. 1

29. a) 0   b) 0   c) 0

2.8 Exercises (pp. 145-148)

1. a) –3, –1, 1   b) [–4,–3], (–3, –1), (–1,1), (1, 4]   3.

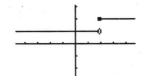

5. –1   7. 2   9. 0   11. b = 2

13. b = 0, 2   15. $\lim\limits_{x \to 3} \frac{x^2 - 9}{5} \neq g(3)$

17.

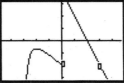

21. a) 1

    b) –2 < x < 0 and 1 < x < 4

    c) $\lim\limits_{x \to 0} g(x) = g(0) = 2$

    d) $\frac{2}{3}$

    f) min value = 0; max value = 2

19.

e)

Chapter 2 Supplementary Problems   (pp. 149-156)

1. a) 6 m   b) –3 m/sec   c) –2 m/sec   d) t = 2.5 sec
   e) $0 \le t < 2.5$ sec   f) 12.5 meters

3. a) 45 m   5.

   b) 2.82 sec

   c) 3.03 sec

   d) –14.7 m/sec

   e) –26.19 m/sec

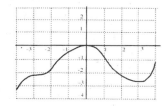

7. a) T   b) T   c) T   d) T   e) F   f) F

9.   f and f' are graphed in a) and c).

11. a) y = -0.063x + 12.45   b) 1.94, 4.47, 8.03   c) 1.94 < x < 4.47; 8.03 < x < 10

13. a) f '(3) = -0.25   b) 0 < x < 2

15. a) 0.33 million people per year   b) 0.36 million people per year

17.   a) –4 < x < –3 and –3 < x < 0 and 3 < x < 4 d)

    b) x = –3, 0, 3

    c) x = 0

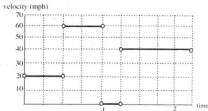

19. a) neg -1 < x < 4;   pos -4 < x < -1; zero –1   b) f"(2) < f '(2) < f '(0) < f(2)

21.   a) 1   b) –1   c) –1   d) 1   e) 0

23.   a)   b)   avg rate = $35\frac{5}{9}$ mph

                                             c)

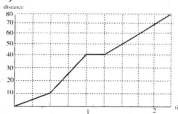

25.   i)   ii)   b)

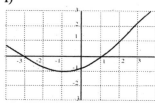

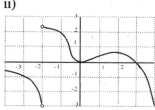

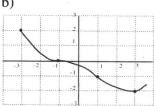

27. b = –1, m = 1   29. a) 7   b) $-\frac{2}{3}$ c) 1   d) 4

31.   a) f(0) = 1 and f(1) = –1. Sign change implies f(c) = 0 for some c in (0, 1).

       b) g(1) ≈ –0.37 and g(2) ≈ 0.56 . Sign change implies g(c) = 0 for some c in (1, 2).

33. a) 1, 3   b) –1.6   c) –2   d) –3, 1, 2, 3   e) –4 < x < –3; 0 < x < 1; 1 < x < 2   f) –1

3.1 Exercises (pp. 163-165)

1. a)

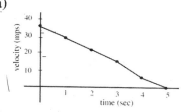

b)

lower estimate = 30·1 + 23·1 + 15·1 + 6·1 + 0·1
= 74 m

upper estimate = 36·1 + 30·1 + 23·1 + 15·1 +
6·1 = 110 m

c)

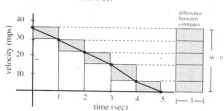

3.   a) 200 miles   b) $\frac{850}{2} = 425$ mph

5.   lower estimate = 0·1 + 10·1 + 20·1 + 25·1 30·1 + 32·1 = 117 meters
      upper estimate = 10·1 + 20·1 + 25·1 30·1 + 32·1 + 35·1 = 152 meters

      total distance $\approx \frac{117 + 152}{2} = 134.5$ meters

7.

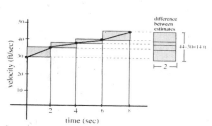

a) Minimum distance = 30x2 = 60 ft
   Maximum distance = 36x2 = 72 ft
b) lower estimate = 30·2 + 36·2 + 38·2 + 40·2 = 288ft
   upper estimate = 36·2 + 38·2 + 40·2 +44·2 = 316 ft
c) guess $\approx \frac{288 + 316}{2} = 302$ ft; Max diff = 14 ft
d) difference = 14 ft

9.   a)   lower estimate = 0.1 + 1·1 + 3·1 + 5·1 + 6·1 + 7·1 + 7·5·1 + 8·1 + 8.5·1 + 8.75·1 = 54.75 gals
          upper estimate = 1.1 + 3·1 + 5·1 + 6·1 + 7·1 + 7·5·1 + 8·1 + 8.5·1 + 8.75·1 + 9·1 = 63.75 gals
     b) $\frac{54.75 + 63.75}{2} = 59.25$ gals;   error < 9/2 = 4.5 gals

3.2 Exercises (pp 169-170)

1.   f(x) = x+1, [0, 4]; Area $\approx$ 12 units^2

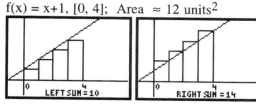

3. f(x) = x^2 – 8x + 20, [0, 4]; Area $\approx$ 38 units^2

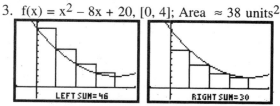

5.   f(x) = ln x, [1, 5]; Area ≈ 3.983 units^2

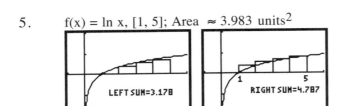

LEFT SUM=3.178   RIGHT SUM=4.787

7. f(x) = $\sqrt{4 - x^2}$, [0, 2]

n	Left sum	Right sum
4	3.4957	2.4957
8	3.3398	2.8398
16	3.2483	2.9983

area approximation = 3.123
area region = 3.142

9. f(x) = sin x, [0, π/2]

n	Left sum	Right sum
4	0.79077	1.1835
8	0.89861	1.0950
16	0.95011	1.0483

area approximation = 0.9992
area region = 1

11.   Left sum = 23; Right sum = 17; Area = 20

3.3 Exercises (pp 176-178)

1. a) 8   b) 1   c) $\frac{\pi}{2}$

3.

	i)	ii)	iii)
a	6	12	7.5
b	4	6	7.5
c	6	2	20

5.

n	Left	Right
4	−0.950	−1.283
8	−1.020	−1.187
16	−1.058	−1.141

guess −1.100

7.

n	Left	Right
4	−2.203	4.547
8	−0.832	2.543
16	−0.067	1.620

guess 0.776

9.   guess 0.5, TI-83: 0.524   11. n = 60; 2.333   13. n = 35; 1

15.   a) 34   b) 24   c) 38   d) 26   17.   left sum = 1·(.22) + 1·(.30) + 1·(.38) = 0.90

19.   right sum = .5[(−.96) + (−.67) + (.13) + (1.67)] = 0.085

3.4 Exercises (pp. 184-186)

1. a) L(50) = 1.0076, R(50) = 0.9926   b)

13. a) L(50) = 2.7010, R(50) = 2.7354   b) e ≈ 2.71828

5. a) $\int_0^4 2^t \, dt$ b) 21.64 hundred bacteria   7. 6   9. $e^4 - 1$

11.   a)   2, 5, 7.5   b)

time (minutes)

13. a) $\int_0^3 500(t-6)\,dt$ b) –6750   c) $8750   15. 2.932   17. 6.151

19. a)
x	0	1	2	3	4
F(x)	0	1	2	11/4	3

b)

Chapter 3 Supplementary Problems (pp. 187-192)

1. a) $\frac{3}{5}$ b) 0   c) 4   d) 5.6

3. a) 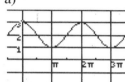   b)   RSum = 5π

$$\int_0^{3\pi} (2 + \cos x)\,dx = 6\pi$$

LSum = 7π

5. a) i) 4   b) i) 6   c) i) 0
    ii) 2   ii) 10   ii) –2
    iii) $\frac{25}{2}$ iii) 10   iii) $-\frac{3}{2}$

7. 15.5 words   9. a) Maxima 5 sec, Lexus 6 sec   b) Maxima 220 ft ; Lexus ≈ 353 ft

11. a) t = 4 b) 135 miles c) t = 4   d) t = 0   e) no

13. a) 2   b) does not exist   c) 1.5   d) 5.3

15. a)   b)   c)   d)
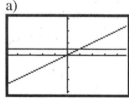 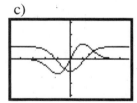

17. a) –.376   b) 1.12

19. a) $f(x) = \cos x; a = \frac{\pi}{2}; f'(\frac{\pi}{2}) = -1$ b) $f(x) = x^{3/2}; a = 4; f'(4) = 3$

21. a) $(-\infty, -3) \cup (1, \infty)$ d)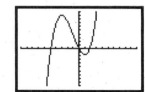
     b) –3
     c) $(-\infty, -1)$

23. a) yes   b) 0   c) yes   25. i, ii, iii

4.1   Exercises   (pp. 200-202)

1. a) $m = 0.5$ b) $m = 3$ c) $m = 1.5$ 3. $\frac{dy}{dx} = -4$ 5. $f'(x) = 3x^2 + 6x$ 7. $\frac{dy}{dx} = 6x - 8$

9. $g'(t) = 6t^2 - 8t$ 11. $\frac{dy}{dx} = 5x^4 - x^2 - 2$ 13. $f'(x) = -3x^{-4} + 2x^{-3} + x^{-2}$

15. $f'(x) = -2x^{-5/3} + \frac{3}{4}x^{-1/4}$ 17. $\frac{dy}{dx} = -\frac{4}{3}x^{-1/3} - \frac{4}{5}x^{-9/5}$ 19. a) $(0, 2)$, $\left(\pm\sqrt{\frac{3}{2}}, -\frac{1}{4}\right)$

b) $(1, 2), (-1, -2)$ c) $(1, 4), (3, 0)$ d) $\left(\frac{1}{4}, -\frac{1}{4}\right)$ 21. $t = 2$, $t = 7$

23. a) increasing; $f'(4) = 2.25$ b) decreasing; $f'(1) = -5$ 25. $(1, 0)$

27. a) $-4 < x < 0$ and $3 < x < 4$ b) $x = 0$, $x = 3$ c) decreasing d) increasing

29. $\left(\frac{1}{4}, 1\right)$ 31. $v(8) = -328$; $a(8) = \frac{-167}{3}$ 33. $k = -4$

4.2   Exercises   (pp. 206-207)

1. $3e^x$ 3. $2e^x + 2x$ 5. $6^t \ln 6$ 7. $2^x \ln 2 + 2x$ 9. $e^x + ex^{e-1}$ 11. $(\ln 2)e^x$

13. $9x^8 + 9^x\ln 9$ 15. $20x^3 - 2^{x+1} \ln 2$ 17. $x - y + 1 = 0$ 19. a) $e^{x+2} = e^2 \cdot e^x$; multiply

each ordinate of $y = e^x$ by e^2 b) $f'(x) = e^2 \cdot e^x$ 21. a) $2e - 4$ b) $t \approx 0.910$ and 3.733

c) $(.204, .910)$ and $(2.833, 3.733)$ 23. a) 12.5 people/year   b) $y = 111.12(1.0975)^x$,

where x is the number of years since 1990.   c) when $x = 2$, $\frac{dy}{dx} = 12.45$ people/year

25. $y' = -2e^x$; $y' - y = (-2e^x) - (1 - 2e^x) = -1$.

4.3   Exercises   (pp. 215-217)

1. a) $5x^4 - 2x$ b) $2x + 6$ c) $24x - 5$ d) $4x^3$ 3. $x \cos x + \sin x$ 5. $(x^2 + 1)\cos x + 2x \sin x$

7. $e^x(x + 1)$ 9. $-4 + 2 \sec^2 x$ 11. $\dfrac{2}{(x + 1)^2}$ 13. $\dfrac{-5}{2(t - 1)^2}$ 15. $\dfrac{(x + 1)\cos x - \sin x}{(x + 1)^2}$

17. $\dfrac{\sin x - x \cos x}{\sin^2 x}$ 19. $\dfrac{-2 \cos x}{(1 + \sin x)^2}$ 21. $\dfrac{-x^2 + 2x + 1}{(1 - x)^2}$ 23. $\dfrac{2^x(\ln 2 - \ln 3)}{3^x} = \left(\frac{2}{3}\right)^x \ln\left(\frac{2}{3}\right)$

25. $g'(x) = f(x)f'(x) + f(x)f'(x) = 2f(x) f'(x)$ 27. a) $x < 2$ b) $x > 3$ 29. a) $2(\cos^2 x - \sin^2 x)$

$= 2 \cos (2x)$ b) $-2 \sin (2x)$ 31. $y = x$ 33. a) -3 b) $f(1) = f(0) + \displaystyle\int_0^1 f'(x)\, dx$ and, using area

to approximate $\displaystyle\int_0^1 f'(x)\, dx$, we get $f(1) \approx 5 + 1.6 = 6.6$.   c) $g'(1) = \dfrac{f'(1) - f(1)}{e} \approx \dfrac{1.2 - 6.6}{e} \approx -2$,

so g decreases at $x = 1$ d) $g''(0) = \dfrac{f''(0) - 2 f'(0) + f(0)}{1} \approx 0 - 2\cdot2 + 5 = 1$

35. $(-3, 50)$, $\left(\frac{5}{3}, \frac{-22}{27}\right)$ 37. a) -1.9 b) -6 c) 0.2 d) -0.025 e) 6

39. a) $-\csc^2(x)$ b) $\sec(x)\tan(x)$ c) $-\csc(x) \cot(x)$

4.4   Exercises   (pp. 223-225)

1. a) $10(5x - 3)$ b) $2e^x(e^x + 1)$ c) $2(2x + x^2)(2 + 2x)$ d) $1 + \dfrac{1}{\sqrt{x}}$ 3. $6x(x^2 - 3)^2$

5. $\dfrac{1}{\sqrt{2x}}$ 7. $\dfrac{x}{\sqrt{x^2 + 1}}$ 9. $2e^{2x}$ 11. $-\sin(\sin x)\cdot\cos x$ 13. $3t^2\cos(t^3 - 1)$

15. $12x^2(x^3 - 4)^3$ 17. $2x \cos(x^2)$ 19. $\dfrac{\sec^2\sqrt{x}}{2\sqrt{x}}$ 21. $e^{2x}(1 + 2x)$ 23. $2\left(x - \dfrac{1}{x^3}\right)$

25. $\dfrac{1}{x} - \dfrac{1}{x+1} = \dfrac{1}{x(x+1)}$ 27. $\dfrac{1}{2x}$ 29. $\dfrac{2x}{x^2 + 1}$ 31. $\dfrac{1}{x \ln x}$ 33. $x(1 + 2\ln x)$

35. a) $\dfrac{1}{5}$ b) $h'(-2) \approx (-1)(.5) = -.5$; negative   c) $h'(-1) = f'[g(-1)]\, g'(-1) \approx (-1)(-.5) = .5$

     d) $-4, -3, -1.5, 0, 1, 2.5, 5$ 37. $k'(0) \approx g'(1)(-1) \approx (-1.5)(-1) = 1.5$; increasing

39. $12x + y - 32 = 0$ 41. $x - 2y + 2 = 0$ 43. a) 23 b) $\dfrac{7}{9}$ c) -20 d) 60

45. $g'(-1) = -2$ and $k'(-1) = 2$; $h(0) = 1$ and $h'(0) = 0$; $f(1) = -1, g'(1) = 2$, and $k'(1) = -2$

4.5   Exercises   (pp. 229-230)

1. $-\dfrac{x}{y}$ 3. $\dfrac{2x - y}{x}$ 5. $\dfrac{1 - y}{x + 2y - 1}$ 7. $\dfrac{2x}{2 - \cos y}$ 9. $-\sqrt{\dfrac{y}{x}}$ 11. $x - 4y - 12 = 0$

13. $y = x$ 15. $x + 2y - 10 = 0$ 17. $x - ey - e = 0$ 19. a) $x = 0, y' = 0$; $x = 4, y' = \dfrac{4}{0}$:

undefined, vertical tangent.   b) -1 21. a) $\dfrac{2x}{x^2 + y^2 - 2y}$ b) $y = 2x - 2$

23. a) $\dfrac{e^{5y}}{3 - 5xe^{5y}}$ b) $y = \dfrac{1}{3}x$ 25. $-\dfrac{25}{64}$ 27. Use implicit differentiation on the equation $y^5 = x^2$

4.6   Exercises   (pp. 235-237)

1. 12π 3. 6 5. $\dfrac{6\sqrt{10}}{5}$ 7. $\dfrac{3}{4\pi}$ in/sec   9. 3 ft/sec   11. $\left(\dfrac{1}{4}, \dfrac{1}{2}\right)$ 13. a) $v(h) = 2h^2\sqrt{3}$

b) $\dfrac{\sqrt{3}}{3}$ ft/sec   15. a) $s = \dfrac{2}{3}d$ b) $\dfrac{10}{3}$ ft/sec   17. $7\sqrt{2}$ in/min   19. a) $.902$ sec

b) 18.827 ft/sec   21. $\dfrac{dv}{dt} = 4\pi r^2 \dfrac{dr}{dt} \Rightarrow 9\pi \approx 4\pi(3^2)\dfrac{dr}{dt} \Rightarrow \dfrac{dr}{dt} \approx 0.25$ 23. 0.28 ft/min

4.7   Exercises   (pp. 243-244)

1. $L(x) = 9 + 6(x - 3)$; $L(4) = 15$, $f(4) = 16$;   error $= 1$

3. $L(x) = -x + \pi$; $L(\pi + 1) = -1$, $f(\pi + 1) \approx -.84$;   error ≈ 0.16

5. $L(x) = 1 - x$; $L(1) = 0$, $f(1) = \dfrac{1}{2}$;   error $= \dfrac{1}{2}$

7. $L(x) = 2 - 4\left(x - \dfrac{1}{2}\right)$; $L\left(\dfrac{3}{2}\right) = -2$, $f\left(\dfrac{3}{2}\right) = \dfrac{2}{3}$;   error $= \dfrac{8}{3} \approx 2.67$

9. a) $f(x) = \sqrt{x}$, $a = 100$; $L(x) = 10 + \dfrac{1}{20}(x - 100)$, $L(103) = 10.15$; error $= L(103) - f(103) \approx .0011$

b) $f(x) = \sqrt[3]{x}$, $a = 27$; $L(x) = 3 + \dfrac{1}{27}(x - 27)$, $L(29) \approx 3.074$;   error $\approx .0018$

11. $x \approx 0.682$ 13. $x \approx -1.167$ 15. When $x_0 = 1.9$, zero ≈ 0.947. When $x_0 = 2$ or $x_0 = -0.9$,

Newton 's Method fails to converge.   19. $x \approx 1.293$

Chapter 4   Supplementary   Problems   (pp. 245-252)

1. a) Left-sum $= .385$;   Right-sum $= .351$ b) $\displaystyle\int_0^1 \dfrac{1 - x}{e^x}\, dx = \dfrac{x}{e^x}\Big]_0^1 = \dfrac{1}{e} \approx 0.3679$

3. $\left(\dfrac{1}{4}, 1\right)$ 5. $\dfrac{1}{6}$ 7. a) $f'(2) \approx 6.773$ b) $y \approx 6.773(x - 2) + 4$ c) $L(2.2) \approx 5.355$

d) $f(2.2) - L(2.2) \approx 0.312$ 9. b) $0 < x < 2$ c) $.59 < x < 3.41$

11. a) 2 zeros   b) $x_0 = 1.4 \Rightarrow x_1 \approx 1.3654$ c) horizontal tangent at $x = 1$

13. a) $y - \dfrac{1}{e^6} = -\dfrac{3}{e^6}(x-2)$; $L(2.2) = .00099$ 15. a) dom $f = R$ b) $f'(x) = \dfrac{-\cos x}{2\sqrt{1-\sin x}}$

c) $\sin x \neq 1$ if $x \neq \dfrac{\pi}{2} + 2n\pi$ d) $y = 1 - \dfrac{1}{2}x$ 17. a) $x > \sqrt{3}$ or $x < -\sqrt{3}$ b) $\pm\sqrt{6}$

19. a) $f'(x) = \dfrac{-1}{(1+x^2)}$; $f'(0) = -1$; $L(x) = 1 - x$ b) $y = \dfrac{1}{2}x - \dfrac{3}{2}$

21. a) increase $(0, 1)$; decrease $(-\infty, 0)$ and $(1, \infty)$ b) conc up $(-\infty, -\frac{1}{2})$; conc down $(-\frac{1}{2}, 0)$ and $(0, \infty)$

25. $(\frac{7}{4}, 2)$ 27. a) $\dfrac{-4+\sqrt{10}}{3} \approx -.279$, $\dfrac{-4-\sqrt{10}}{3} \approx -2.387$ 29. a) $(2, 4.5$ b) $(1, 2)$ and $(6.25, 8)$

5.1 Exercises (pp. 260–263)

1. $f'(x) = x^2 - 2x - 3$; $CN = -1, 3$

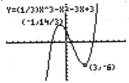

3. $f'(x) = 4(x^3 - 3x^2)$; $CN = 0, 3$

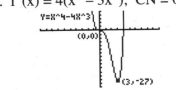

5. $f'(x) = x^{-1/3}(\frac{5}{3}x - \frac{10}{3})$; $CN = 0, 2$

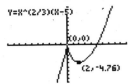

7. $f'(x) = \dfrac{4}{3}x^{-2/3}(x+1)$; $CN = 0, -1$

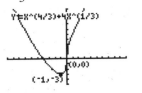

9. $f'(x) = e^x(x+1)$; $CN = -1$

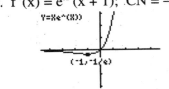

11. $f'(x) = (2x - x^2)e^{-x}$; $CN = 0, 2$

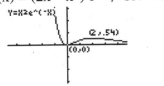

13. local max $(-.5, -.8)$
local min $(0, -1.0)$

15. local max $(1.2, 1.6)$ and $(-5.1, 7.9)$
local min $(-1.2, -1.6)$ and $(5.1, -7.9)$

17. local max $(2.8, 0.07)$ and $(-4.3, .02)$
local min $(-.3, -6.2)$
global max $(1, 2)$, local min $(3, 0)$

19. a) global max $(1, 2)$ and $(4, 2)$
global min $(-2.1, -1.5)$; local min $(3, 0)$

b)   global max $(3, 2.5)$
global min $(-4, -2)$

c) global max at $(k, 1)$ where
$-4 \leq k \leq 2$ or $k = 4$
global min $(1, -2)$

d)   local max $(-2, 2)$, global max $(4, 3)$
global min $(-4, 0)$ and $(0, 0)$
local min $(-4, 0)$ and $(0, 0)$

21. global max $= 11$
global min $= 2$

23. global max $= .125$
global min $= -.25$

25. global max = 1.299
     global min = −1.299

27. global max = 14
     global min = −.472

29. global max = .61
     global min = −27.05

31. global max = 1.25
     global min = −1

33. a) increasing on $(4, \infty)$

     b) local min at x = 4

     c)

35. a) increasing on $(-\infty, 2)$ and $(4, \infty)$

     b) local max at x = 2 ; local min at x = 1 and 4

     c)

37. a) x + y − 1 = 0 b) 1,3 c) (1, 3) d) local max at x = 1

39.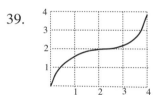

41. one local min at x = 0

43. a = −1.5, b = 2.5

45. 5

5.2 Exercises (pp. 268-270)

1. i) a, c, e, h   ii) d, g, i   iii) c, e

3. a) (−2, 1)

    b) −2, 1

    c) graph at right.

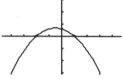

5. $f'(x) = 3x^2 - 6x$; $f''(x) = 6x - 6$;
    conc down on $(-\infty, 1)$, up on $(1, \infty)$;
    ip: x = 1

7. $f'(x) = 4x^3 - 12x$; $f''(x) = 12x^2 - 12$; conc up on $(-\infty, -1)$ and $(1, \infty)$; down on $(-1, 1)$; ip: x = ± 1

9. $g'(x) = \dfrac{3x - 2}{2\sqrt{x - 1}}$; $g''(x) = \dfrac{3x - 4}{4(x-1)^{3/2}}$; conc up on $(\frac{4}{3}, \infty)$, down on $(1, \frac{4}{3})$; ip: $x = \frac{4}{3}$

11. $g'(x) = (2x - x^2)e^{-x}$; $g''(x) = (2 - 4x + x^2)e^{-x}$; conc up on $(-\infty, 2 - \sqrt{2})$ and $(2 + \sqrt{2}, \infty)$,

     down on $(2 - \sqrt{2}), 2 + \sqrt{2})$; ip: $x = 2 - \sqrt{2}, 2 + \sqrt{2}$ 13. $f'(x) = 2x + \dfrac{1}{x}$; $f''(x) = \dfrac{2x^2 - 1}{x^2}$;

conc down on $(0, \sqrt{2}/2)$; conc up on $(\sqrt{2}/2, \infty)$; ip: $x = \sqrt{2}/2$ 15. loc min at x = 2; loc max at $x = \frac{2}{3}$

17. loc min at $x = \dfrac{7\pi}{4}$; loc max at $x = \dfrac{3\pi}{4}$ 19. loc min at x = 0   21. loc min at x = 0; loc max at x = 4

23. a) v(1) = −3 ft/sec

     b) a(1) = 0 ft/sec^2

     c) v(1) = −3 ft/sec

25. a) v(1) = 1 ft/sec

     b) a(1) = 0 ft/sec^2

     c) 1 ft/sec at t =1; −1 ft/sec at t = − 1

27. a) v(1) = π ft/sec

     b) a(1) = π^2 ft/sec^2

     c) $v(\frac{1}{4}) = -\pi\sqrt{2}$ ft/sec, $v(\frac{5}{4}) = \pi\sqrt{2}$, ...

29. a) Since f(−1) = −1 and f(1) = $\frac{1}{8}$, f is <u>not</u> even.   b) $y = \frac{1}{8}$ c) local maximum at $(1, \frac{1}{8})$ d) 3

31. a) $f' = \dfrac{1 - 3x}{e^{3x}}$; $f''(x) = \dfrac{9x - 6}{e^{3x}}$ b) increasing on $(-\infty, \frac{1}{3})$ c) concave up on $(\frac{2}{3}, \infty)$

5.3 Exercises (p. 275)

1. $\frac{3}{5}$ 3. 2   5. No limit   7. No limit   9. $\frac{3}{4}$

11. Horizontal: y = 0   13. Horizontal: y = 0   15. Horizontal: y = 1

Vertical: x = 2, x = –2   Vertical: x = 1, x = –1   Vertical: x = –3, x = 1

17. Horizontal: y = 0   19. Horizontal: y = 0

Vertical: None   Vertical: x = 2, x = –2

5.4 Exercises (pp. 281–282)

1. 128 in^3 3. a) $A(h) = \frac{h}{2}(16 - h)$ b) h = 8   5. 2.128   7. 108 in^2 9. 4 x 3

11. a) $A(x) = 2x \cos x,\ 0 \le x \le \frac{\pi}{2}$ b) 1.12 units^2 13. $\frac{16}{\sqrt{3}} \approx 9.238$ units^2

15. a) $A(1) = \frac{2}{e^2}$ b) $A(x) = 2x\, e^{-2x^2}$, $\frac{1}{\sqrt{e}} \approx .6065$ 17. 2.33

5.5 Exercises (p. 285)

1. 1.5 and 5.5   3. yes; x = 3   5. yes: $x = \frac{3\pi}{4}$ 7. yes; x = 1   9. yes; x = 6.25

11. f is not continuous at x = 1; but, $\frac{f(2) - f(-2)}{2 - (-2)} = 1$ and f '(x) = 1 for any input $-2 < x < 1$.

5.6 Exercises (pp. 291-292)

1. 4x + C   3. $x^3 + x^2 + x + C$ 5. $e^x + \frac{2}{3}x^3 + C$ 7. $\frac{x^4}{4} + \frac{1}{x} + C$ 9. tan x + x + C

11. $\frac{x^4}{4} - \frac{2}{3}x^3 + 1$; $\frac{x^4}{4} - \frac{2}{3}x^3 + 2$; $\frac{x^4}{4} - \frac{2}{3}x^3 + 3$ 13. $g(x) = x^2 - 3x$ 15. a) $s(t) = -16t^2 + 96t + 144$

b) t = 3   c) $t = 3 + 3\sqrt{2}$ sec; –135.8 ft/sec   17. $\frac{-88}{9}$ ft/sec^2 19. yes; $s(t) = -4t^2 + 66t$ and since v = 0

at $t = \frac{33}{4}$, we have s $(\frac{33}{4}) = 272.25$ feet.   21. a effects the local maximum and minimum points of f

23. a effects the shape of the graph and the horizontal stretch of the graph.   25. $y = 1 - x^2$

27. a) The critical numbers for each family are x = ln k c) There is on local minimum point which moves downward and away from the y-axis

Chapter 5 Supplementary Problems (pp. 293–298)

1. a) $\frac{1}{2}$ b) y – 1 = 3(x + 2)   3. a) 1, 4.5 b) 4.5 c) (–1, 1) d) 6.9   5. a) p = –2, q = 5

b) p = –6 c) $p^2 < 3q$ 7. a) –1 ft/sec b) increasing at 1.471 ft/sec   9. $(\frac{1}{k}, \frac{1}{ke})$

11. a) y = x + 4 b) x = 3 c) (–4, –3), (–1, 2)   13. a) $\pm \sqrt{c/d}$ b) local min at $x = \sqrt{c/d}$; local max at $x = -\sqrt{c/d}$ 15. a) –1, 1, 3   b) (–1, 1)   c) x = –1   d) (–3, 0) and (2, 3)

17. $5\sqrt{5} \approx 11.18$ ft   19. a = –3, b = 6   21. $330

23. a) h(–2) = 1, h(1) = –2   b) h'(–1) = 0, h'(0) = –1   c) x = –2.5, –2, –1, 0.5, 1.4, 3

25. g'(0.1) ≈ 3.84   27. a) $v(t) = 6t^2 - 6t - 12$ b) t = 2   c) $s(t) = 2t^3 - 3t^2 - 12t$

29. t = 3.792 and t = 5.084   31. 206.37 cm^3

32. a) y = 3x+5   b) x = 2   c) x = 0, 4   d) absolute max at f(2) = 8

6.1 Exercises (pp. 304-306)

1. a) Left-sum = 66.765; Right-sum = 71.265   b) $\int_1^4 (3x^2 + 2)\,dx = \left[x^3 + 2x\right]_1^4 = (64 + 8) - (3) = 69$

3. a) Left-sum = 1.026; Right-sum = .974   b) $\int_{\pi/2}^{\pi} \sin x\,dx = [-\cos x]_{\pi/2}^{\pi} = -[\cos \pi - \cos \frac{\pi}{2}] = 1$

5. $\int_0^{\pi/2} \cos t\,dt = [\sin t]_0^{\pi/2} = \sin \frac{\pi}{2} - \sin 0 = 1$

7. a) $\int_0^5 (20 + 4t)\,dt$ b) $\int_0^5 (20 + 4t)\,dt = \left[20t + 2t^2\right]_0^5 = (100 + 50) - 0 = 150$ gallons

9. a) runner 1   b) runner 2;   Distance run = area under velocity curve.

11. a) $\int_{-1}^3 [\,f(x) + g(x)]\,dx = \int_{-1}^3 f(x)\,dx + \int_{-1}^3 g(x)\,dx = 7 + 4 = 11$ b) $\int_{-1}^3 -2\,g(x)\,dx = -2\int_{-1}^3 g(x)\,dx$

$-2(4) = -8$ c) $\int_{-1}^3 [\,2f(x) - 3g(x)]\,dx = 2\int_{-1}^3 f(x)\,dx - 3\int_{-1}^3 g(x)\,dx) = 2(7) - 3(4) = 2$

d) $\int_{-1}^3 [\,f(x) + 1]\,dx = \int_{-1}^3 f(x)\,dx + \int_{-1}^3 1\,dx = 7 + [x]_{-1}^3 = 7 + [3 - (-1)] = 11$

13. a) If $0 \le x \le 1$, then $3x \ge x$ and $3x + 1 \ge x + 1$, so $\int_0^1 (3x + 1)\,dx \ge \int_0^1 (x + 1)\,dx$

b) If $1 \le x \le 2$, then $x^2 \ge x$ and $x^2 + 1 \ge x + 1$, so $\int_1^2 (x^2 + 1)\,dx \ge \int_1^2 (x + 1)\,dx$

15. a) $\int_{-2}^0 f(x)\,dx = -5$ b) $\int_{-2}^2 f(x)\,dx = 0$ c) $\int_0^2 [2f(x)+3]\,dx = 16$ d) $\int_{-2}^2 |f(x)|\,dx = 10$

17. a) $\int_0^3 r(t)\,dt$, where $r(t) = W'(t)$ is the given rate at which water is leaking out.

b) midpoint: $(37)(1) + (20)(1) + (10)(1) = 67$ gallons

6.2 Exercises (pp. 312-314)

1. a)

x	−1	0	1	2	3	4
F(x)	0	1.5	4	6.25	7	6.5

b)

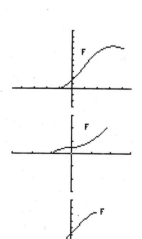

3. a) $F(-1) = 0$, $F(0) = .5$, $F(2) = 2.5$ b)

5. a) $F(-2) = 0$, $F(0) = \pi$, $F(2) = 2\pi$ b)

7. a)

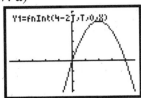

b)

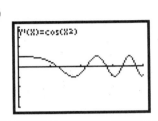

c) $Y_1(0) = 0$, $Y_1(2) = 4$, $Y_1(4) = 0$

9. a)

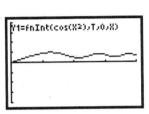

b)

	(0, 1.25)	(1.25, 2.17)	(2.17, 2.80)	(2.80, 3.32)	(3.32, 3.76)	(3.76, 4)
A	increasing	decreasing	increasing	decreasing	increasing	decreasing
f	positive	negative	positive	negative	positive	negative

11. b) Each graph is a vertical shift of the others; they are part of the same family of curves.
   c) The derivatives are all the same.

6.3 Exercises (pp. 318-319)

1. a) $\frac{dy}{dx} = x^2 - \sqrt{x}$ **b)** $\frac{dy}{dx} = \frac{1}{x^2 + 4}$ **c)** $\frac{dy}{dx} = \sqrt{x^2 - 1}$ **d)** $\frac{dy}{dx} = -\sqrt{3x + 1}$

3. a) $0, \frac{\pi}{2}, \pi, \frac{3\pi}{2}, 2\pi$ **b)** concave up: $(0, \frac{\pi}{4})$, $(\frac{3\pi}{4}, \frac{5\pi}{4})$, $(\frac{7\pi}{4}, 2\pi)$; concave down: $(\frac{\pi}{4}, \frac{3\pi}{4})$, $(\frac{5\pi}{4}, \frac{7\pi}{4})$

5. global maximum at $x = 1$, $F(1) \approx 0.693$ **7.** $v(2) = \frac{2}{3}$, $a(2) = \frac{1}{9}$ **9.** $(-1 - \sqrt{2}, -1 + \sqrt{2})$ **11. B**

13. D   15. D

6.4 Exercises (p. 324)

1. $\frac{1}{12}$ units^2 **3.** $\frac{32}{3}$ units^2 **5.** $\frac{131}{4}$ units^2 **7.** 1.385 units^2 **9.** .690 units^2 **11.** .947 unit^2

13. 7.36 units^2 **15.** $\frac{37}{6}$ units^2 **17.** .828 units^2 **19.** $k = 3$ or -3 **21.** 8 units^2

Chapter 6 Supplementary Problems   (pp. 325–328)

1. .382 units^2 **3.** $y - 4 = -\frac{3}{4}(x - 4)$ and $y + 4 = \frac{3}{4}(x - 4)$

5. a) no critical numbers in $(0, \pi)$, but $f'(\pi) = 0$ **b)** $f'(x) > 0$, f is monotone increasing on $[0, \pi]$; minimum at $x = 0$; $f(0) = 0$

7. $\frac{1}{3}$ units^2 **9.** $a = 0$ **11.** 1455 gallons   **13. a)** $f'(-1) \approx -2$ **b)** 5.5   **c)** 2.3

15. $v = 0$ if $32t = 10$, so $t = \frac{5}{16}$ and $s(\frac{5}{16}) \approx 7.563$ feet; hits ground when $s(t) = 0$, $t = 1$ sec.

17. a) $v(t) = -6t^2 + 18t + 24$ **b)** $x(t) = -2t^3 + 9t^2 + 24t - 10$ **c)** $x(4) = 102$ meters   **19. a)** $(-2, 1)$ and $(3.25, 4)$ **b)** $x = -2$ or $x = 3.25$; but global min. is at $x = -2$ **c)** $(-2, -.5)$ and $(2, 4)$

21. a) T   **b)** T   **c)** T   **d)** F   **e)** T   **23. a)** $(0, 0)$, $(0.964, 0)$, $(1.684, 0)$
b) $(0, 0.398)$, $(1.351, 3)$ **c)** min = $-.098$, max = 1.366

7.1 Exercises (pp. 332-333)

1. $\frac{3}{2} x^2 + 5x + C$ 3. $\frac{x^2}{2} - 2\sqrt{x} + C$ 5. $\frac{x^2}{2} + \frac{1}{x} + C$ 7. $2\sqrt{x} + C$ 9. $x - \ln|x| + C$

11. $e^x + 5x + C$ 13. $x + C$ 15. $-3\cos x - 5\sin x + C$ 17. $\tan x + C$ 19. $-\frac{15}{2}$

21. $\ln 2 + \frac{3}{2}$ 23. $2e - 2$ 25. 1   27. 2   29. $\frac{1}{e} \approx .368$

31. a) $v(t) = \sin t + \cos t - 1$ b) $x(t) = \sin t - \cos t - t + 1$ c) $t \approx 2.412$ 33. $\frac{16 - 4\sqrt{2}}{3}$

35. $\int \cos^2 x \ dx = \int [\frac{1}{2} + \frac{1}{2}\cos(2x)] \ dx = \frac{1}{2} x + \frac{1}{2}\int \cos(2x) \ dx = \frac{1}{2} x + \frac{1}{4}\sin(2x) + C$

7.2 Exercises (pp. 336-337)

1. $\frac{2}{3}(x^2 + 1)^{3/2} + C$ 3. $\frac{(2x + 3)^4}{8} + C$ 5. $\frac{1}{2}\tan(2x) + C$ 7. $\frac{2}{3}(x + 2)^{3/2} + C$

9. $\frac{1}{2} e^{2x-1} + C$ 11. $e^{\sin x} + C$ 13. $\frac{1}{3}\tan^3 x + C$ 15. $2\sin\sqrt{x} + C$ 17. $\frac{1}{8}(x^2 - 1)^4 + C$

19. $-\frac{2}{3}(\cos x)^{3/2} + C$ 21. $\ln|\sin x + 1| + C$ 23. 4   25. $\frac{1}{4}$ 27. $\ln 3$

29. $\ln(\frac{e + 2}{2})$ 31. $\ln\sqrt{5}$ 33. 1   35. $-\ln|\cos x| + C$ 37. $-\frac{1}{3}\ln|\cos(3x)| + C$

7.3 Exercises (pp 342-343)

1. $\frac{1}{2}(\ln x)^2 + C$ 3. $\frac{1}{5}\ln(5x^2 + 7) + C$ 5. $-e^{1/x} + C$ 7. $\frac{1}{3}\tan(x^3) + C$

9. $\frac{1}{4}[\frac{1}{5 - x^4}] + C$ 11. $\frac{2}{5}(x + 2)^{5/2} - \frac{4}{3}(x + 2)^{3/2} + C$ 13. $\frac{1}{6}(2x + 3)^{3/2} - \frac{3}{2}\sqrt{2x + 3} + C$

15. $\frac{4}{3}(x + 3)^{3/2} + 14\sqrt{x + 3} + C$ 17. $\frac{26}{3}$ 19. $\frac{1}{2}$ 21. $\frac{16}{15}$

23. $2\ln\frac{3}{2}$ 25. $\frac{2 - \sqrt{2}}{3}$ 27. $\frac{1}{8}(\ln x)^8 + C$ 29. $\sec(\ln x) + C$

31. $\frac{2}{45}(3x - 6)^{5/2} + \frac{4}{9}(3x - 6)^{3/2} + C$ 33. $\frac{4}{15}\ln(3 + 5e^{3x}) + C$

7.4 Exercises (p. 346)

1. $xe^x - e^x + C$ 3. $x\ln(2x) - x + C$ 5. $-x\cos x + \sin x + C$ 7. $\frac{1}{2} x(\ln x - 1) + C$

9. $x\tan x + \ln|\cos x| + C$ 11. $\frac{(\ln x)^3}{3} + C$ 13. $\frac{2}{3} x(1 + x)^{3/2} - \frac{4}{15}(1 + x)^{5/2} + C$

15. $\frac{1}{4} - \frac{3}{4e^2} \approx 0.148$ 17. $\frac{\pi}{4}$ 19. $14\ln 2 - 3 \approx 6.704$ 21. $e^2 + 1 \approx 8.389$ units^2

7.5 Exercises (pp. 354-355)

a) $\frac{\pi}{2}$ b) $\frac{\pi}{4}$ c) $\frac{\pi}{2}$ d) $-\frac{\pi}{6}$ e) $\frac{\pi}{4}$ f) $-\frac{\pi}{3}$ g) $\frac{2}{3}\pi$ h) $\frac{\sqrt{3}}{2}$ i) $-\frac{\pi}{4}$ j) 1.018

k) 1.551   l) 2.426   3. $\frac{2}{\sqrt{1 - 4x^2}}$ 5. $\frac{1}{x^2 + 2x + 2}$ 7. $\frac{1}{(1 + x^2)\arctan x}$

9. $\frac{e^{\arcsin x}}{\sqrt{1 - x^2}}$ 11. $\frac{1}{(1 + 4x^2)\sqrt{\arctan(2x)}}$ 13. $\frac{\cos x}{1 + \sin^2 x}$ 15. $e^x(\arcsin x + \frac{1}{\sqrt{1 - x^2}})$

21. $\frac{\pi}{4}$ 23. $\frac{\pi}{12}$ 25. .099   27. $\frac{\pi}{4} \approx .785$ 29. $\frac{-\sin y(1 + y^2)}{x\cos y(1 + y^2) - 1}$ 31. π

7.6 Exercises (pp. 363-366)

1. $M_{10} = 0.788$; $T_{10} = 0.776$; fnInt = 0.785

3. $M_{10} = 1.553$; $T_{10} = 1.566$; fnInt = 1.557

5. 1.111

7. 0.659

9. 0.506 units^2

11. $\int_1^2 \frac{1}{x} dx \approx 0.694$ $\int_1^8 \frac{1}{x} dx \approx 2.138$; $\int_1^{16} \frac{1}{x} dx \approx \int_1^2 \frac{1}{x} dx + \int_1^8 \frac{1}{x} dx = 2.832$ $\int_1^{16} \frac{1}{x} dx \approx 3.013$ (n = 8)

13. 14.4 cu ft

15. 1.601

17. a) 25   b) 24.5

19. a) right: 1.562; midpoint: 1.726; trapezoid: 1.735; left: 1.908   b) 1.726, 1.735

21. c) n = 10

23. n = 7

25. n = 10

27. n = 10

Chapter 7   Supplementary Problems   (pp. 367-372)

1. $\frac{-x^{-4}}{4} + C$

3) $\frac{-2}{t^5} + C$

5. $\frac{(3x + 2)^5}{15} + C$

7. $4\sqrt{x} - 2\sqrt{x^3} + C$

9. $\frac{(x^3 + 1)^3}{9} + C$

11. $\frac{x^3 \ln x}{3} - \frac{x^3}{9} + C$

13. $\frac{1}{6}(2 + x^4)^{3/2} + C$

15. $\frac{2}{5}(1 - x)^{5/2} - \frac{2}{3}(1 - x)^{3/2} + C$

17. $2 \arcsin(x) + C$

19. $2 \arcsin(3x) + C$

21. 5

23. $3 \ln 3 - 2$

25. $\frac{16}{9}$

27. $\frac{5}{16}$

29. 2

31. $\frac{2}{3}$

33. 81 in

35. 2 units^2

37. a) L = 9.410, R = 11.158, T = 10.284, and M = 10.366   b) Between 10.284 and 10.366.

39. $\frac{22}{3}$

41. a) $T_4 = 9$; $M_4 = 8$

43. $\frac{d\theta}{dt} = 0.333$ rad/sec

45. a) $f'(x) = 2x^{-1/3} - 2$ b) {0, 1}

c) max = 5; min = –4

47. $\frac{8}{3}$ units^2

49. a) Increasing $(-\infty, \infty)$ b) Concave up $(0, \infty)$ 51. 47.5

ft

53. $x_1 = .667$; $x_2 = .652$

55. a) 27   b) $a = \frac{5}{4}$, $b = \frac{5}{4}$

57. a) $f^n(x) = e^x(x + n)$ b) $f^n(x) = \frac{(-1)^n \cdot n!}{x^{n+1}}$

59. 1 rad/sec; $\lim_{x \to \infty} \frac{d\theta}{dt} = 0$

61. a) f(1)=4; f(7)=5+2π   b) x = 3   c) x = 2,5   d) absolute max at f(7)=5+2π

8.1 Exercises (pp. 376-377)

1. net distance 3; total distance 5

3. net distance $10\frac{2}{3}$; total distance $11\frac{1}{3}$

5. net distance –.362; total distance .782

7. $v(t) = 3t^2 - 18t + 24$; net distance 4; total distance 12

9. $v(t) = 1 - \sin(2\pi t)$; net distance 1; total distance 1

11. 2515 feet

13. a) 1 km/min   b) t = 7, 11, 19, 23, ...   c) 12.82 km   d) 12.82/9 ≈ 1.42 km/min

15. a) 3 km

b) t = 4; car moves to the right when $0 < t < 4$; moves to the left when $4 < t < 6$; is stopped when $6 \le t \le 7$; moves to the right when $7 < t \le 8$.

c)

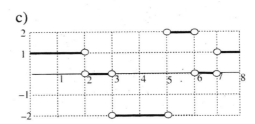

8.2 Exercises (pp. 383-384)

1. 216 in^3 3. 132 ft^3 5. $\frac{\pi}{3}\text{ units}^3$ 7. $8\pi\text{ units}^3$ 9. $\frac{2\pi}{35}\text{ units}^3$ 11. $\frac{16384\pi}{9}\text{ units}^3$

13. $\frac{3\pi}{4}\text{ units}^3$ 15. $\frac{32\pi}{15}\text{ units}^3$ 17. a) $\frac{256\pi}{5}\text{ units}^3$ b) $\frac{1088\pi}{15}\text{ units}^3$ c) $\frac{512\pi}{15}\text{ units}$

19. 516 cm^3 21. a) $T_5 = 157\text{ units}^3$ b) $y = -0.074x^3 + 0.528x^2 - 0.112x + 0.667$

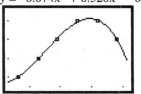

c) $\text{fnInt}(\pi\ Y_1{}^2, X, 1, 6) =$
  160.3 units^3

8.3 Exercises (p. 388)

1. $\frac{128\pi}{5}\text{ units}^3$ 3. $2\pi e^2 + 2\pi\text{ units}^3$ 5. $\frac{28\pi}{3}\text{ units}^3$ 7. $\frac{64\pi}{5}\text{ units}^3$ 9. $\frac{2\pi}{5}\text{ units}^3$

11. $\pi - \frac{\pi}{\sqrt{e}}\text{ units}^3$ 15. a) $T_8 = 1526.814\text{ units}^3$ b) $y = 0.012x^4 - 0.280x^3 + 1.862x^2$
  $- 3.179x + 4.667$

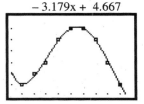

13. $\frac{\pi^2}{6}\text{ units}^3$ c) $\text{fnInt}(2\pi\ XY_1, X, 1, 9) =$
  1537.298 units^3

8.4 Exercises (pp. 391-392)

1. $\frac{1}{3}$ 3. $\frac{-17}{4}$ 5. $\frac{2}{\pi}$ 7. $\frac{45}{28}$ 9. 1.266 11. a) 6.9 b) 2.3

13. 53 mph 15. 0.854 17. a) $\frac{1}{n+1}$ b) $\lim\limits_{n\to\infty}\frac{1}{n+1} = 0$

8.5 Exercises (pp. 398-401)

1. a) $a(t_1)\cdot\Delta t + a(t_2)\cdot\Delta t + \dots + a(t_n)\cdot\Delta t$ where $0 < t_1 < t_2 < \dots < t_{n-1} < t_n = 8$ and $\Delta t = \frac{8}{n}$.

   b) $\int_0^8 a(t)\ dt = 0\text{ ft/sec}$

3. a)

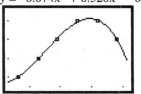

$A'(t) = 100e^{-0.2t}$

b) $\Delta A \approx A'(t_i)\cdot\Delta t = 100e^{-0.2t_i}\cdot\Delta t$

c) $A \approx A'(t_1)\cdot\Delta t + A'(t_2)\cdot\Delta t + \dots + A'(t_n)\cdot\Delta t$
  where $0 < t_1 < t_2 < \dots < t_{n-1} < t_n = b$ and $\Delta t = \frac{b}{n}$.

$A = \int_0^b 100e^{-0.2t}\ dt = 500 - 500e^{-0.2b}$

d) $\int_0^5 100e^{-0.2t}\ dt \approx 316.06\text{ tons}$

5. $\int_0^2 t\sqrt{4 - t^2}\,dt = \frac{8}{3}\text{ gallons}$

7. a) $d(x_1)\pi x_1 \Delta x + d(x_2)\pi x_2 \Delta x + \ldots + d(x_n)\pi x_n \Delta x$ where $0 < x_1 < x_2 < \ldots < x_{n-1} < x_n = 5$ and $\Delta x = \frac{5}{n}$.

   b) $\int_0^5 d(x)\pi x \, dx = \int_0^5 20000 \, e^{-.13x} \pi x \, dx \approx 515{,}387$ people.

9. a) Assuming uniform linear density determined by the x-coordinate at a strip's left-hand edge, we have

   $[f(0) \cdot 0] \cdot 1 + [f(1) \cdot 1] \cdot 1 + [f(2) \cdot 2] \cdot 1 + [f(3) \cdot 3] \cdot 1 \quad = \frac{0}{1+0} \cdot 1 + \frac{1}{1+1} \cdot 1 + \frac{2}{1+2} \cdot 1 + \frac{3}{1+3} \cdot 1 = \frac{23}{12}$

   b) $M = \int_0^4 \frac{x}{1+x} \, dx$ c) $(4 - \ln 5)$ grams

11. a) $W(3) \cdot 05 = 10.55$ walkers   b) $W(x) \cdot \Delta x$ c) $W(x_1) \cdot \Delta x + W(x_2) \cdot \Delta x + \cdots + W(x_n) \cdot \Delta x$ where $0 = x_0 < x_1 < x_2 < \ldots < x_n = 6$

   d) $\int_0^6 (0.17x^3 - 3.1x^2 + 14.8x) \, dx \approx 98$ walkers

13. $\ln\sqrt{5} \approx 0.805$ foot-pounds   15. 44 foot-pounds   17. $\int_0^{1/2} 36x \, dx = 4.5$ foot-pounds

Chapter 8 Supplementary Problems   (pp. 402-408)

1. a) $\frac{4}{3}$ units^2 b) $\frac{64\pi}{15}$ units^3 c) $\frac{8\pi}{3}$ units^3 d) $\pi \int_0^2 [(2x+2)^2 - (x^2 + 2)^2] dx$

3. a) $\frac{13}{6}$ units^2 b) $\frac{158\pi}{15}$ units^3 c) $\frac{3\pi}{2}$ units^3 d) $\pi \int_0^1 [(6 - x^2)^2 - (3x + 2)^2] \, dx$

5. a) $\int_0^1 [(x + 1) - 2x^3] \, dx$ b) $\pi \int_0^1 [(x + 1)^2 - (2x^3)^2] \, dx$ c) $2\pi \int_0^1 x[(x + 1) - 2x^3] \, dx$

7. b) 1 units^2 c) $\pi [\frac{e^2 + 1}{2}] \approx 13.177$ 9. 127.235 units^3 11. $\frac{16}{15}$ units^3

13. a) $x(t) = t^3 - 3t^2 + 8$ b) $t = 1 + \sqrt{3}$ c) 22   15. a) $\int_0^2 \sqrt{1 + e^{-x}} \, dx$

   b) $T_4 = 2.391$ c) $\pi(3 - e^{-2})$ units^3 17. a) –0.0375 ft per ft   b) 27.772 ft^3

19. Subdividing the interval [0, 80] into 8 equal strips, each strip has width $\Delta x = \frac{80}{8} = 10$ meters and length 100 meters. Using the left-endpoint of each strip as the distance from the source of the spill, the corresponding Riemann sum is

   $$\frac{5000 \, x_1}{1 + x_1^2} \Delta x + \frac{5000 \, x_2}{1 + x_2^2} \Delta x + \ldots + \frac{5000 \, x_8}{1 + x_8^2} \Delta x$$

   where $x_1 = 0$, $x_2 = 10$, $x_3 = 20$, $x_4 = 30$, $x_5 = 40$, $x_6 = 50$, $x_7 = 60$, $x_8 = 70$ and $\Delta x = 10$

   b) $\int_0^{80} \frac{5000x}{1 + x^2} \, dx$ c) $2500 \ln(1 + 80^2) \approx 21910.524$ kilogram

21. 1.213

23.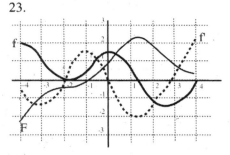

25. Assuming $\frac{dA}{dt} \approx 0.6\pi$, we have $\frac{dr}{dt} = .173$ miles per day.

27. a)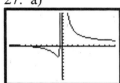

b) vertical: x = 0; horizontal: y = 1

c) range of f = $[-\frac{37}{12}, \infty)$

d) $x = -\frac{9}{7}$

29. a) $\frac{\pi}{4}$ feet   b) $\sqrt{3}$ minutes   c) no   31. 3 meters x 6 meters x 2 meters ; $270

33. The cost will be minimized if the landfall of the pipeline is approximately 2.683 miles down the shoreline.

9.1   Exercises (pp. 411-412)

1. a) (ii)   b) (iii)   c) (i)   3. $y = Ce^{-x^2}$, $y' = -2Cxe^{-x^2}$ and $-2Cxe^{-x^2} + 2x(Ce^{-x^2}) = 0$

5. $y = \frac{x^3}{3} - 4x + 5$ 9. a) $y = -2e^x$ b) $y = \frac{5}{e}e^x$ c) $y = 8e^3e^x$ 11. C = -1   13. b) $y = 5e^{2 \sin x}$

9.2   Exercises (pp. 416-418)

1. a) 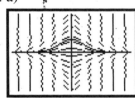 b) 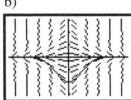   3. a)    b)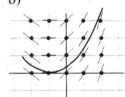

5. a) 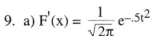   b) 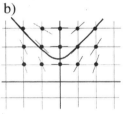   7. a) $y' = 0.25y(4 - y)$

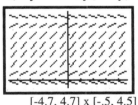

[-4.7, 4.7] x [-.5, 4.5]

9. a) $F'(x) = \frac{1}{\sqrt{2\pi}} e^{-.5t^2}$ b) 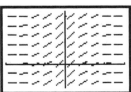   c)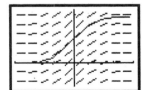

11. a) (iv)   b) (i)   c) (ii)   d) (iii)

9.3 Exercises   (pp. 423-424)

1. a) $y(2) \approx 0.7595$ b) $\ln 2 \approx 0.6931$ c) bigger   3. a) $y' = \sec^2 x = \tan^2 x + 1 = y^2 + 1$

5. 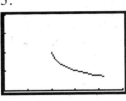   7.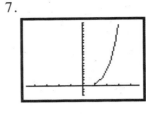

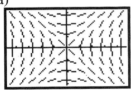   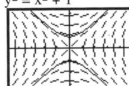   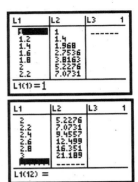

9. $y(2) \approx 6$ 11. a) $y(2) \approx 2.5$ b) $y(x) = \ln \sqrt{x^2 + 1} + 2$; $y(2) \approx 2.805$;

13. a) $y(0.8) \approx 1.28$ b) $y(0.8) \approx 1.36$ d) $y(0.8) \approx 1.47$

9.4   Exercises   (p. 427)

1. $2\sqrt{y} = x^2 + C$ 3. $\sin y = x^2 + C$ 5. $\tan y = -\cot x + C$ 7. $\ln |y| = x \ln x - x + C$

9. $y = 3e^{x^2/4}$ 11. $y = e^{\sin x}$ 13. $y^2 = x^2 - 2x + 25$ 15. $y = x + 1$

17. a)   b) $y^2 = x^2 + 1$ c) $y^2 = x^2 + C$

19. a) $\dfrac{dT}{dt} = -.05(T - 70)$ b) $T = 70 + 20e^{-0.05t}$ c) 2 PM

Chapter 9   Supplementary Problems   (pp. 429–434)

1. $y = \sqrt{3 + 2e^x}$ 3. $y = 5e^{x^4}$ 5. b) $y = \dfrac{1}{1+\cos x}$ 7. a)

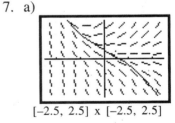

$[-2.5, 2.5] \times [-2.5, 2.5]$

9. a) $y(0.8) \approx -.0736$ c) $y(0.8) \approx -.226$

11. a) $y(12) = 5196$ b) $t = 10.095$ hours

13. a) $y(t) = \sqrt{6t + 1}$ b) $t = \dfrac{1}{2}$ hours   15. $p(t) = (2t + 50)^2$

17. a) Right Secant Slope $= \dfrac{102 - 111.8}{6} \approx -1.633$;   b) $y = 157.034(0.980)^t$ c) $y'(16) = -2.263$

Left Secant Slope $= \dfrac{111.8 - 121.3}{4} \approx -2.375$

Rate of change $_{(t = 16)} \approx \dfrac{-1.663 + -2.375}{2} = -2.004$

19. a) $1, -1$ b) $(-1, 1)$ c) $-\sqrt{3}, 0, \sqrt{3}$ 21. a) $F'(x) = 2x\sqrt{x^4 + 8}$ b) $\dfrac{22}{3}$

23. (0.218, 1.14)   25. a) -2.1875 gals/min   b) $t = 0$

29. $9.50 when attendance is 8000   31. a) (0, 3) and (5, 7)   b) $T_9 = 26$

33. $f(2.2) \approx 4.6$ 35. a) $f'(1.4) \approx -3.75$ b) $y = -3.75(x - 1.4) + 2.6$
                                   c) $f''(1.4) > 0$ because f' is increasing. d) $M_2 = 2.2$

37. $k = \sqrt{2e^2 + 2}$

10.1   Exercises   (pp. 440–441)

1. 3   3. $\dfrac{1}{24}$ 5. 1   7. $-\pi$ 9. $\ln (2/3)$ 11. 1   13. 1

15. 2   17. $\dfrac{1}{40}$ 19. $-\dfrac{7}{5}$ 21. 0   23. ∞ 25. 0   27. e^2

29. e^{-2} 31. $-\dfrac{1}{4}$ 33. does not exist

10.2   Exercises   (p. 447)

1. $\dfrac{1}{2}$ 3. 1   5. 1   7. $\dfrac{1}{2}$ 9. ∞ 11. $2\sqrt{2}$ 13. 1

15. 0   17. 4π units ^{2} 19. 4π units ^{2} 21. π units^3

10.3   Exercises   (pp. 450-451)

1. c) $\ln\left|\dfrac{1+x}{1-x}\right| + C$ 3. $\ln\dfrac{|x-1|}{|x|} + C$ 5. $\ln\dfrac{|x-4|^3}{(x-3)^2} + C$ 7. $\dfrac{1}{6}\ln\left|\dfrac{x-3}{x+3}\right| + C$

9. $x - \dfrac{4}{3}\ln|x+2| + \dfrac{1}{3}\ln|x-1| + C$ 11. $\ln\dfrac{|2x-1|^3}{|x+1|^4} + C$ 13. $\ln\left(\dfrac{3}{2}\right)$ 15. $\dfrac{1}{2}\ln\left(\dfrac{8}{3}\right) + \dfrac{5}{2}$

17. $\ln\left(\dfrac{9}{2}\right) \approx 1.504$ feet   19. $\dfrac{1}{2}\ln 3 \approx 0.549$ units^2

10.4   Exercises   (pp. 456-458)

1. a) 600   b) 0   3. No zeros   5. a) $(-\infty, 69)$ b) $(69, \infty)$

7. 69 hours   9. a) $y' = .03y\left(1 - \dfrac{y}{150}\right)$ b) $k = 0.03;\ L = 150$ c) $y = \dfrac{150}{1 + 6.5e^{-0.03t}}$

11. $y = \dfrac{6000}{1 + 5e^{-0.0015t}}$ 13. $y = \dfrac{150}{1 + 4e^{-0.225t}}$

15. a) $y' = .01y\left(1 - \dfrac{y}{1000}\right)$ b) $y = \dfrac{1000}{1 + 19e^{-0.01t}}$ d) $P(10) \approx 54.97;\ P(100) \approx 125.161$

17. a) $y' = .07y\left(1 - \dfrac{y}{1000}\right)$ b) $y = \dfrac{1000}{1 + 24e^{-.07t}}$ c) $P(25) \approx 193;\ P(100) \approx 979$

19. a) $y = \dfrac{300}{1 + 29e^{-0.06t}}$ b) 56 days   21. a) 511   b) 2.976 days

Chapter 10 Supplementary Problems   (pp. 429–464)

1. a) $\dfrac{2}{3}$ b) 0   3. $\dfrac{1}{2}$ 7. 2π units^3

9. a) $2\ln3 \approx 2.197$ units^2 b) $4\pi\ln3 \approx 13.806$ units^3

11. a) $\ln\frac{9}{2}$ units^2 b) $2\pi(\ln 2 + 2) \approx 16.922$ units^3

13. a) $y' = 0.04y(1 - \frac{y}{800})$ b) $y = \dfrac{800}{1 + 19e^{-.04t}}$ c) $P(25) \approx 100.129$; $P(100) \approx 593.473$

15. i) c   ii) d   iii) b   iv) a   17. a) $y = \dfrac{1200}{1 + 11e^{-0.26t}}$ b) 1131

19. a) 14 m/sec   b) $8 + 2h$ m/sec   c) 8 m/sec   21. $(\frac{7}{4}, 2)$ 23. 13.86 years

25. 8   27. $k = 4$ 29. a) $4k^{3/2}$ b) $6k + \frac{2}{3}$ 31. 3 times   33. $m = -4$

35. a) LHS=8.25; RHS=6.25   b) 7.25   37. b) $y = 2e^{x^2/4}$ c) $y(2) = 2e$ 39. π units^3

11.1 Exercises (pp. 472-473)

1. $p_2(x) = 1 + \frac{x}{2} - \frac{x^2}{8}$ 3. $p_3(x) = 1 + 2x + 2x^2 + \frac{4}{3}x^3$ 5. $p_2(x) = 1 + x + x^2$

7. $p_3(x) = 3 - 4x + 4x^2 + \frac{2}{3}x^3$; $f(0.2) \approx 2.365$ 9. $p_3(x) = 1 + 2x + 4x^2 + 8x^3$; $f(0.2) \approx 1.624$

11. $p_3(x) = 2 + \dfrac{(x-8)}{12} - \dfrac{(x-1)^2}{288}$ 13. $p_3(x) = 2 + \dfrac{(x-4)}{4} - \dfrac{(x-4)^2}{64} + \dfrac{(x-4)^3}{512}$

15. $\sin(.4) \approx p_7(.4) = 0.389418$; $\sin_{cal}(.4) = 0.389418$

     $\sin(\frac{\pi}{12}) \approx p_7(\frac{\pi}{12}) = 0.258819$; $\sin_{cal}(\frac{\pi}{12}) = 0.258819$

17. a) $p_4(x) = 1 - x^2 + \frac{1}{3}x^4$ b) $p_4(x) = 1 - (x - \pi)^2 + \dfrac{(x-\pi)^4}{3}$ 19. $[-1.665, 1.665]$

21. a) $p_5(x) = 1 + x - \dfrac{x^2}{2} + \dfrac{x^3}{3} - \dfrac{x^4}{4} + \dfrac{x^5}{5}$ b) $f(0.25) \approx p_5(0.25) = 1.223177$ c) $f(0.25) = 1.22314$

11.2 Exercises (pp. 483-485)

1. $S_1 = 1$, $S_2 = 0.5$, $S_3 = 0.750$ $S_4 = 0.625$, $S_5 = 0.688$; $S = \lim_{n \to \infty}\{S_n\} = \frac{2}{3}$

3. $S_1 = 2$, $S_2 = 2.5$, $S_3 = 2.625$ $S_4 = 2.656$, $S_5 = 2.672$; $S = \lim_{n \to \infty}\{S_n\} = \frac{8}{3}$

5. $S_1 = .3$, $S_2 = .33$, $S_3 = .333$ $S_4 = .3333$, $S_5 = 3.3333$; $S = \lim_{n \to \infty}\{S_n\} = \frac{1}{3}$

7. $\lim_{n \to \infty} a_n = 1$; a) convergent b) divergent   9. $\lim_{n \to \infty} a_n = \frac{3}{2}$; a) convergent b) divergent

11. a) $a_1 = \frac{1}{2}$, $a_2 = \frac{1}{6}$, $a_3 = \frac{1}{12}$, $a_4 = \frac{1}{20}$ b) $a_1 = \frac{1}{2}$, $a_2 = -\frac{1}{6}$, $a_3 = -\frac{1}{12}$, $a_4 = -\frac{1}{20}$

13. a) $r = \frac{3}{2} > 1$ b) $\lim_{n \to \infty} a_n = 2 > 0$ c) $\lim_{n \to \infty} a_n \neq 0$ 15. Sum $= \frac{1}{2}$ 17. diverges

19. Sum $= 0$ 21. $r = 2 > 1$;diverges   23. Sum $= -\frac{9}{4}$ 27. $-\frac{1}{2} < x < \frac{1}{2}$

29. $-\frac{1}{2} < x < \frac{1}{2}$ 31. $a = \frac{1}{4}$ and $r = \frac{1}{4}$ so sum $= \dfrac{1/4}{1 - 1/4} = \dfrac{1}{3}$ 33. \$3 billion

11.3 Exercises (pp. 493-494)

1. a) $p_3(x) = 2x - 2x^2 + \frac{8}{3}x^3$ b) $x - x^2 + \frac{1}{2}x^3$ 3. a) $1 + \frac{x}{2} - \frac{x^2}{8} + \frac{x^3}{16}$ b) $-1 < x < 1$

5. a) $x - \frac{x^2}{2} + \frac{x^3}{3} - \frac{x^4}{4}$ b) $-1 < x < 1$ 9. $R = 1$ 11. $R = \infty$ 13. $R = \infty$

15. $R = 0$ 17. $R = 1$ 19. $R = 4$

21. $\sum_{n=0}^{\infty}(-1)^n(n+1)x^n$, $R = 1$ 23. $\sum_{n=0}^{\infty}\frac{(-1)^n x^{2n}}{4^{n+1}}$, $R = 2$

25. a) $p_3(x) = 1 + 3(x-1) + \frac{9}{2}(x-1)^2 + \frac{9}{2}(x-1)^3$ b) $\sum_{n=0}^{\infty}\frac{3^n(x-1)^n}{n!}$, $R = \infty$

27. a) $p_3(x) = 1 - 4(x-6) + 16\sqrt{2}(x-6)^2 - 64\sqrt{3}(x-6)^3$ b) $R = \frac{1}{4}$

11.4 Exercises (pp. 501-502)

1. $2\sum_{n=0}^{\infty}(-1)^n x^n$ $R = 1$ 3. $\sum_{n=0}^{\infty}x^{2n}$ $R = 1$ 5. $\sum_{n=0}^{\infty}(-1)^n x^{n+2}$ $R = 1$ 7. $\sum_{n=2}^{\infty}\frac{x^n}{n}$ $R = 1$

9. $\sum_{n=0}^{\infty}\frac{(2x)^n}{n!}$ $R = \infty$ 11. $\sum_{n=1}^{\infty}(-1)^{n+1}\frac{x^n}{n!}$ $R = \infty$ 13. $\sum_{n=0}^{\infty}\frac{x^{2n}}{(n+1)!}$ $R = \infty$

15. $\sum_{n=0}^{\infty}(-1)^n\frac{(x-\pi/2)^{2n}}{2n!}$ $R = \infty$ 17. $\sum_{n=0}^{\infty}\frac{(-1)^n e^2 (x+2)^n}{n!}$ $R = \infty$ 19. $2\sum_{n=0}^{\infty}x^{2n}$ $R = 1$

21. $\sqrt{2} + \frac{\sqrt{2}(x-1)}{4} - \frac{\sqrt{2}(x-1)^2}{32}$ 23. $1 + 2x + 3x^2 + 4x^3 + \cdots + (n+1)x^n + \cdots$ $R = 1$

25. $x + \frac{x^3}{3!} + \frac{x^5}{5!} + \cdots = \sum_{n=0}^{\infty}\frac{x^{2n+1}}{(2n+1)!}$ 27. $\lim_{x\to 0}\frac{e^x - 1}{x} = \lim_{x\to 0}\left(1 + \frac{x}{2!} + \frac{x^2}{3!} + \frac{x^3}{4!} + \cdots\right) = 1$

29. $\frac{d}{dx}\left[\sin(x^2)\right] = \frac{d}{dx}\left(x^2 - \frac{x^6}{6} + \frac{x^{10}}{120} - \cdots\right) = 2x - x^5 + \frac{x^9}{12} - \cdots = (1 - \frac{x^4}{2!} + \frac{x^8}{4!} - \cdots)2x = \cos(x^2)2x$

31. a) $g(x) = \frac{1}{1-3x} = 1 + 3x + (3x)^2 + (9x)^3 + \cdots = \sum_{n=0}^{\infty}3^n x^n$ b) $(-\frac{1}{3}, \frac{1}{3})$

c) $\frac{3/2}{1-3x} - \frac{1/2}{1-x} = \left(\frac{3}{2} + \frac{9}{2}x + \frac{27}{2}x^2 - \frac{81}{2}x^3\right) - \left(\frac{1}{2} + \frac{1}{2}x + \frac{1}{2}x^2 + \frac{81}{2}x^3\right) = 1 + 4x + 13x^2 + 40x^3 + \cdots = \sum_{n=0}^{\infty}\frac{3^{n+1}-1}{2}x^n$

11.5 Exercises (pp. 513-514)

1. Convergent   3. Divergent   5. Divergent   7. a) ($p = 3 > 1$) Convergent

   b) ($p = \frac{3}{2} > 1$) convergent   c) ($p = \frac{2}{3} < 1$) Divergent   d) ($p = \frac{4}{3} > 1$) Convergent

   e) ($p = \frac{1}{4} < 1$) Divergent   f) ($p = \frac{5}{3} > 1$) convergent   9. Divergent

11. Conditionally convergent   13. Divergent

15. $a_5 = \frac{1}{120}$ b) $n = 5$ 17. $a_5 = \frac{1}{25}$ b) $n = 14$ 19. $a_5 = 0.00032$ b) $n = 3$

21. $[-1, 1]$ 23. $[-1, 5)$ 25. $(-\infty, \infty)$ 27. $(-4, 2)$ 29. $[-3, -1)$

31. $\displaystyle\sum_{n=1}^{\infty}(-1)^{n+1}\frac{x^n}{n}$, $(-1, 1]$ 33. $\displaystyle\sum_{n=0}^{\infty}\frac{(-3)^n x^n}{\sqrt{n+1}}$, $(-\frac{1}{3}, \frac{1}{3}]$ 35. a) $[-1, 1)$ b) $f(0)=0, f'(0)=\frac{\ln 2}{2}$

11.6 Exercises (pp. 519-521)

1. a) $6+8x+15x^2+8x^3$ b) $f(0.2)\approx 8.264$.   c) max error $= 0.005$

3. lerrorl $<\dfrac{(0.01)^2}{8}=0.0000125$ 5. lerrorl $<\dfrac{e^{0.1}(0.1)^3}{6}=0.000184$ 7. lerrorl ≤ 0.00344

9. a) $p_2(x)=2+\dfrac{1}{4}(x-4)-\dfrac{1}{64}(x-4)^2$ b) $f(4.2)\approx p_2(4.2)=2.04938$

   c) lerrorl $<\dfrac{3(0.2)^3}{6\cdot 256}=0.000015625$ 11. a) $(-2,2)$ b) $\dfrac{13}{16}$ c) $\dfrac{1}{64}$ 13. 0.905

15. a) $1+2(x-1)+\dfrac{1}{4}(x-1)^2$ b) $f(1.5)\approx p_2(1.5)=\dfrac{33}{16}$ c) lerrorl $\le\dfrac{3}{32}$ 17. $[-0.282, 0.275]$

Chapter 11 Supplementary Problems (pp. 522–526)

1. a) $1-4x^2+8x^4-\dfrac{32}{3}x^6$; $\displaystyle\sum_{n=0}^{\infty}(-1)^n\frac{(2x)^{2n}}{n!}$ b) $(-\infty,\infty)$

   c) $\dfrac{d}{dx}(1-4x^2+8x^4-\dfrac{32}{3}x^6+\cdots+(-1)^n\dfrac{(2x)^{2n}}{n!}+\cdots)$

   $=\ \ 0-8x+32x^3-64x^5+\cdots+(-1)^n\dfrac{2n(2x)^{2n-1}}{n!}+\cdots$

   $=\ -8x(1-4x^2+8x^4-\dfrac{32}{3}x^6+\cdots)=-8xe^{-4x^2}$

3. a) $f(t)=2+2t^2+2t^4+2t^6+\cdots+2t^{2n}+\cdots$

   b) $G(x)=2x+\dfrac{2}{3}x^3+\dfrac{2}{5}x^5+\cdots+\dfrac{2}{2n+1}x^{2n+1}+\cdots$ c) $[-1, 1)$

5. a) $\dfrac{1}{2}$ b) 1 c) $\dfrac{1}{3}$ 7. no; $f'(0)\ne 1$

9. a) $\dfrac{1}{3}(x-2)+\dfrac{1}{18}(x-2)^2+\dfrac{1}{81}(x-2)^3$ b) $R=3$

13. a) $x^2-\dfrac{x^6}{3!}+\dfrac{x^{10}}{5!}-\cdots=\displaystyle\sum_{n=0}^{\infty}(-1)^n\frac{x^{4n+2}}{(2n+1)!}$ b) 0.3102814

15. a) $1-\dfrac{1}{2}x-\dfrac{1}{8}x^2$ b) $f(0.2)\approx 0..491$; max error $= 0.5$

17. $[-4, 4)$ 19. $[-1, 1]$ 21. b) 0.1157   23. 0.0000010125

25. a) $-1+2x-2x^2$ b) $f(0.5)\approx p_2(0.5)=-0.5$ c) $|\text{error}|<\dfrac{12(0.5)^3}{6}=0.25$

27. a) $1+\dfrac{3}{2}(x+3)+\dfrac{3}{8}(x+3)^2-\dfrac{1}{16}(x+3)^3$ b) $8+3x+\dfrac{3}{16}x^2$

c) $|\text{error}| \le \dfrac{Mx^3}{3!}$ where $M = \max|f'''(x)| = \dfrac{3}{24}$. So $|\text{error}| \le \dfrac{3x^3}{64 \cdot 6} = \dfrac{x^3}{128}$

29. a) 2   b) –2   c) $\ln a - \ln b$ 31. a) $\dfrac{LHS + RHS}{2} = 508.25\,\text{ft}$ b) $\dfrac{6600}{13} = 507.7\,\text{ft}$

33. $k = 2$ 35. a) –3.75   b) $y - 2.6 = -3.75(x - 1.4)$ c) $\dfrac{\Delta y}{\Delta x}$ increasing, $\therefore\ f''(1.4) > 0$
   d) 2.2

12.1 Exercises (pp. 535-537)

1.
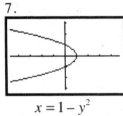

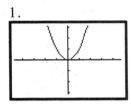
$x = t$
$y = t^2$

3.
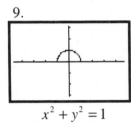

$x = e^t$
$y = e^{2t}$

5.
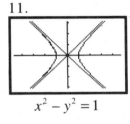

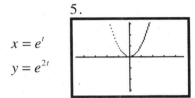
$x = \sqrt{t - 1}$
$y = t - 1$

7.
$x = 1 - y^2$

9.
$x^2 + y^2 = 1$

11.
$x^2 - y^2 = 1$

13.
$x^2 + y^2 = 1$

17.   y=-.25x+1   19. $y - 3 = 8(x - 2)$ 21. $y - 1 = -\dfrac{2\sqrt{3}}{3}\left(x - \dfrac{3\sqrt{3}}{2}\right)$

23. $\dfrac{dy}{dx} = 2t^2 - t;\ \ \dfrac{d^2y}{dx^2} = 4t^2 - t$ 25. $\dfrac{dy}{dx} = t + t\ln t;\ \ \dfrac{d^2y}{dx^2} = 2t + t\ln t$

27. a) (0, 8)   b)

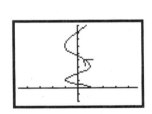

c) $\dfrac{dy}{dx} = \dfrac{t}{\pi \cos \pi t}$ d) $y = \dfrac{4}{\pi}x + 8$

29. a) (0, 1)   b)

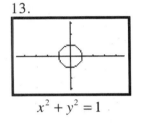

c) $\dfrac{dy}{dx} = -\cot \pi t$ d) $y = 1$ e) 3 times

31.   a) $\dfrac{dy}{dx} = \dfrac{-2}{\sqrt{t + 1}}$ b) $y = \dfrac{5}{4} - x$ c) $y = 1 - x^2$ for $x > 0$

12.2 Exercises   (p. 541)

1. $\left(\frac{1}{27}\right)\left(13^{3/2} - 8\right) \approx 1.440$ 3. $\frac{10}{3}$ 5. $\sqrt{2}\left(1 - e^{-\pi}\right) \approx 1.353$ 7. $\ln(\sqrt{2} + 1) \approx 0.881$

9. a) $\frac{dy}{dt} = -\tan t$ b) $y = 2\sqrt{2} - x$ c) 2π

12.3 Exercises   (p 547)

1. a) $\langle 0, 2 \rangle$ b) $\langle -1, 3 \rangle$ c) $\langle 0, 3 \rangle$ d) $\langle 0, -1 \rangle$ e) $\langle -2, 6 \rangle$ 3. 1

7. a) $\langle 6, 4 \rangle$ b) $\sqrt{13}$ c) $\sqrt{5}$ 9. $x - y = 2$ 11. $y = 2 - \frac{2}{x+1}$

12.4 Exercises   (p 551)

1. $\langle 1, 0 \rangle$ 3. $R'(t) = i \oplus 4t \cdot j$; $R''(t) = 4 \cdot j$; speed $= \sqrt{17}$;

5. $R'(t) = 2 \cdot i \oplus -2e^{-t} \cdot j$; $R''(t) = 2e^{-t} \cdot j$; speed $= 2\sqrt{2}$

7. $R'(t) = 2t \cdot i \oplus e^{t} \cdot j$; $R''(t) = 2 \cdot i \oplus e^{t} \cdot j$; speed $= \sqrt{4 + e^{2}}$

9. $R'(t) = \sqrt{1+t} \cdot i \oplus (1 + \ln t) \cdot j$; $R''(t) = \frac{1}{2\sqrt{1+t}} \cdot i \oplus \frac{1}{t} \cdot j$; speed $= \sqrt{3}$

13. $(3, -4)$ 15. a) $R(t) = \ln(t + 1) \cdot i \oplus (t^2 - 1) \cdot j$ b) 4.335

17. a) $\langle -2, 8 \rangle$ b) $\left\langle -8, \frac{53}{3} \right\rangle$ c) $t = 3$ d) 20.415

12.5 Exercises   (p 560)

3. a) $(4, \frac{\pi}{3})$ b) $(2, -\frac{\pi}{6})$ c) $(4\sqrt{2}, \frac{5\pi}{4})$ d) $(2, \frac{\pi}{6})$ e) $(3\sqrt{2}, \frac{3\pi}{4})$ f) $(4, \frac{2\pi}{3})$

5. a) $x^2 + (y - 1)^2 = 1$ b) $(x - 1)^2 + (y - \frac{3}{2})^2 = \frac{13}{4}$ c) $y^2 = \frac{x^4}{4 - x^2}$

d) $\left(x^2 + y^2\right)^{3/2} = x^2 - y^2$ e) $y^2 = 4x + 4$ f) $x^2 - y^2 = 4$

7.

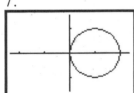

$r = 2\cos\theta$

9.

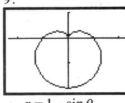

$r = 1 - \sin\theta$

11.

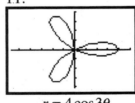

$r = 4\cos 3\theta$

13.

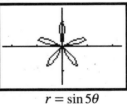

$r = \sin 5\theta$

15.

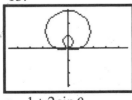

$r = 1 + 2\sin\theta$

17.

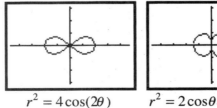

$r^2 = 4\cos(2\theta)$

19.

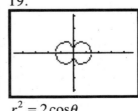

$r^2 = 2\cos\theta$

21.

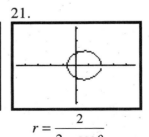

$r = \frac{2}{2 - \cos\theta}$

23.

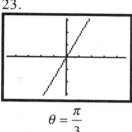

$$\theta = \frac{\pi}{3}$$

25. a) π b) π

12.6 Exercises (p. 565)

1. $\frac{\pi}{4}$ 3. $\frac{3}{2}\pi$ 5. $\frac{\pi}{4}$ 7. 2   9. 18π 11. $\frac{\pi^3}{6}$ 13. $\frac{\pi}{2} \approx 1.57$

15. $\frac{1}{2}(\pi-2)\approx0.57$ 17. $\frac{\sqrt{5}}{2}(e^\pi - 1) \approx 24.75$ 19. 16   21. 8|a|

Chapter 12 Supplementary Problems (pp. 522–526)

1. a) $\frac{dy}{dx} = \frac{4}{(t-4)^2}$ b) $y = x - 3$ c) $y = 1 - \frac{4}{x}$

3. a) $\langle 2,0 \rangle$ b) $\frac{dy}{dx} = \frac{3t^2 - 4t}{2t - 1}$ c) $y = \frac{4}{3}x - \frac{8}{3}$ d) 4

5. a) $\left(\frac{\pi}{2}, 1\right)$ b) $\frac{dy}{dx} = \tan t$ c) $\frac{\pi^2}{2} \approx 4.935$

7. a) $\left\langle \frac{5}{2}, \frac{3}{2} \right\rangle$ b) $\frac{5}{3}$ c) $y = \frac{5}{3}x - \frac{8}{3}$ d) 2.634

9. $R(t) = (2\sin t + 1) \cdot i \oplus \left(-\frac{1}{2}\cos t^2 + \frac{3}{2} \right) \cdot j$ 11. $(x-1)^2 + (y-1)^2 = 2$

13. b) $\frac{3\pi}{2} \approx 4.71$ c) 8   15. 1   17. a) –1   b) $\frac{1}{2}$

19. 1   21. a) 2   b) volume is not finite   23. a) $y = \frac{3}{1 + 5e^{-6t}}$ b) $y = \frac{3}{2}$

25. 400   27. a) $1 + 3x + 9x^2 + 27x^3$ b) $\left(-\frac{1}{3}, \frac{1}{3} \right)$ c) $x - 3x^2 + 9x^3 - 27x^4$

29. a) $1 - 2x + 3x^2 - 4x^3$ b) $-1 < x < 1$ c) $f\left(\frac{1}{4}\right) \approx 0.64$ d) Four terms

33. 0   35. a) –8   b) $y = 2 - 8x$; $f(0.2) \approx 0.4$ c) $(-\infty, -2)$ and $(1, 2)$ and $(2, \infty)$
    d) $\{-1.446,\ 1.613,\ 2\}$

37. b) $y = \frac{1}{e}$ c) Relative minimum at $x = 0$; $f''(0) = 2 > 0$ d) –1.510

39. a) 14 ft   b) 20 ft   c) $x(3) = 8$ d) i) –4 ft / sec^2 ii) 4 ft / sec^2

INDEX